Animal Evolution

Interrelationships of the
Living Phyla

CLAUS NIELSEN

Zoologisk Museum,
University of Copenhagen

Oxford   New York   Tokyo

OXFORD UNIVERSITY PRESS

1995

Oxford University Press, Walton Street, Oxford OX2 6DP
Oxford   New York
Athens   Auckland   Bangkok   Bombay
Calcutta   Cape Town   Dar es Salaam   Delhi
Florence   Hong Kong   Istanbul   Karachi
Kuala Lumpur   Madras   Madrid   Melbourne
Mexico City   Nairobi   Paris   Singapore
Taipei   Tokyo   Toronto
and associated companies in
Berlin   Ibadan

Oxford is a trade mark by Oxford University Press

Published in the United States
by Oxford University Press Inc., New York

A catalogue record for this book is available from the British Library

Library of Congress Cataloging in Publication Data
Nielsen, Claus.
Animal evolution: interrelationships of the living phyla/Claus Nielsen.
Includes bibliographical references and indexes.
1. Phylogeny.   2. Evolution (Biology).   I. Title.
QH367.5.N53 1995   591.3'8–dc20   94-37478
ISBN 0 19 854868 0 (Hbk)
ISBN 0 19 854867 2 (Pbk)

Typeset by EXPO Holdings, Malaysia

Printed and bound by
Information Press Ltd,
Oxford, England

PREFACE

No naturalist can avoid being fascinated by the diversity of the animal kingdom and by the sometimes quite bizarre specializations which have made it possible for the innumerable species to inhabit almost all conceivable ecological niches.

However, comparative anatomy, embryology and especially molecular biology demonstrate a striking unity among organisms, and show that the sometimes quite bewildering diversity is the result of variations over a series of basic themes, some of which are even common to all living beings.

To me, this unity of the animal kingdom is just as fascinating as the diversity, and in this book I will try to demonstrate the unity by tracing the evolution of all the 31 living phyla from their unicellular ancestor.

All modern books on systematic zoology emphasize phylogeny, but space limitations usually preclude thorough discussions of the characteristics used to construct the various phylogenetic trees. I will try to document and discuss all the characters which have been considered in constructing the phylogeny – both those which corroborate my ideas and those which appear to detract from their probability.

In the study of many phyla I have come across several important areas where the available information is incomplete or uncertain and yet other areas which have not been studied at all; on the basis of this I have for each phylum given a list of some interesting subjects for future research, and I hope that these lists will serve as incentives to further investigations.

It should be stressed that this book is not meant as an alternative to the several recent textbooks of systematic zoology, but as a supplement, one which I hope will inspire not only discussions between colleagues but also seminars on phylogeny – of the whole animal kingdom or of selected groups – as an integrated part of the teaching of systematic zoology.

The ideas put forward in this book have developed over a number of years, and during that period I have benefitted greatly from interactions with many colleagues. Some have been good listeners when I have felt the need to talk about my latest discovery, some have discussed new or alternative ideas, names or concepts with me, some have provided eagerly sought pieces of literature or given me access to their unpublished results, and some have sent me photos for publication; to all these friends I extend my warmest thanks; no names are mentioned, because such a list will inevitably be incomplete. A number of colleagues have read one to several chapters (the late Robert D. Barnes (Gettysburg) and Andrew Campbell (London) have read them all) and given very valuable and constructive comments which I have often but not always followed; I want to mention them all, not to make them in any way responsible, but to thank for the help and support which is necessary during an undertaking such as this: Quentin Bone (Plymouth), Kristian Fauchald (Washington, DC), Gary Freemann (Austin), Jens T. Høeg (Copenhagen), Åse Jespersen (Copenhagen), Niels Peder Kristensen (Copenhagen), Margit Jensen (Copenhagen), Reinhardt Møbjerg Kristensen (Copenhagen), Barry S.C. Leadbeater (Birmingham), Jørgen Lützen (Copenhagen), George O. Mackie (Victoria), Mary E. Petersen (Copenhagen), Mary E. Rice (Fort Pierce), Edward E. Ruppert (Clemson), Amelie H. Scheltema (Woods Hole), George L. Shinn (Kirksville), Volker Storch (Heidelberg), Ole S. Tendal (Copenhagen), and Russell L. Zimmer (Los Angeles).

The Danish Natural Science Research Council and the Carlsberg Foundation are thanked heartily for their continued support covering travel expenses, instrumentation and laboratory assistance; the Carlsberg Foundation has given a special grant to cover the expenses of the illustrations for this book.

Financial support from 'Højesteretssagfører C.L. Davids Legat for Slægt og Venner' is gratefully acknowledged.

Mrs Birgitte Rubæk and Mrs Beth Beyerholm are thanked for their excellent collaboration on the artwork.

My warmest thanks go to Kai and Hanne (Olsen & Olsen, Fredensborg) for undertaking the typesetting of the book and for fine work during the preparatory stages.

Dr Mary E. Petersen (Copenhagen) is thanked for her meticulous reading of the first set of proofs.

Finally, my thanks go to Oxford University Press for a positive and constructive collaboration.

Copenhagen   C.N.
July 1994

CONTENTS

1	Introduction	1
2	The trochaea theory	9
3	Kingdom ANIMALIA (= METAZOA)	18
4	Prelude: Phylum CHOANOFLAGELLATA	32
5	Phylum PORIFERA	35
6	Phylum PLACOZOA	42
7	EUMETAZOA (= GASTRAEOZOA)	45
8	Phylum CNIDARIA	53
9	BILATERIA	61
10	PROTOSTOMIA (= GASTRONEURALIA)	75
11	SPIRALIA	89
12	TELOBLASTICA	99
13	Phylum SIPUNCULA	102
14	ARTICULATA	109
15	Phylum MOLLUSCA	110
16	EUARTICULATA	124
17	Phylum ANNELIDA	126
18	PANARTHROPODA	149
19	Phylum ONYCHOPHORA	152
20	Phylum ARTHROPODA	159
21	Phylum TARDIGRADA	176
22	BRYOZOA	182
23	Phylum ENTOPROCTA	185
24	Phylum ECTOPROCTA	192
25	PARENCHYMIA	211

26 Phylum PLATYHELMINTHES   215

27 Phylum NEMERTINI   225

28 ASCHELMINTHES   234

29 Phylum ROTIFERA   238

30 Phylum ACANTHOCEPHALA   248

31 Phylum CHAETOGNATHA   254

32 CYCLONEURALIA   261

33 Phylum GASTROTRICHA   264

34 INTROVERTA   272

35 Phylum NEMATODA   275

36 Phylum NEMATOMORPHA   287

37 CEPHALORHYNCHA   291

38 Phylum PRIAPULA   294

39 Phylum KINORHYNCHA   299

40 Phylum LORICIFERA   303

41 PROTORNAEOZOA   307

42 Phylum CTENOPHORA   310

43 DEUTEROSTOMIA (= NOTONEURALIA)   318

44 Phylum PHORONIDA   337

45 Phylum BRACHIOPODA   346

46 NEORENALIA   356

47 Phylum PTEROBRANCHIA   359

48 Phylum ECHINODERMATA   366

49 CYRTOTRETA   382

50 Phylum ENTEROPNEUSTA   387

51 CHORDATA   396

52 Phylum UROCHORDATA   407

53 Phylum CEPHALOCHORDATA   422

54	Phylum VERTEBRATA (= CRANIATA)	431
55	Five enigmatic taxa	436
56	Concluding remarks	438
	Postscript	441
	Glossary	442
	Systematic index	445
	Subject index	454

1

Introduction

Modern understanding of biological diversity goes back to Darwin (1859), who created a revolution in biological thought by regarding the origin of species as the result of 'descent with modification'. As a consequence of this idea he also stated that the 'natural system' (i.e. the classification) of the organisms must be strictly genealogical ('like a pedigree'), and that the 'propinquity of descent' is the cause for the degree of similarity between organisms. The term homology had already been in use for some time, and Owen (1848) had used it in a practical attempt to create a common anatomical nomenclature for the vertebrates, but it was Darwin's ideas about evolution which gave the word its present meaning and importance: features are homologous in two or more species when they are derived from one feature in the species' most recent common ancestor.

Shortly after, Haeckel (1866) drew the first phylogenetic tree ('Stammbaum'; Fig. 1.1) based on Darwin's ideas, and coined the words phylogeny and ontogeny. His tree was labelled 'monophyletic' and his definition of a phylum as embracing an ancestor and all its living and extinct descendants agrees completely with the cladistic use of the word monophyletic. Haeckel leaned toward the opinion that the 19 phyla in his tree had evolved separately from unorganized organic substances, but a common ancestry was also considered a possibility; this should not detract from the general validity of his definition of the term monophyletic, which is now used on all systematic levels.

The conceptual base for phylogenetic work is thus more than a century old, and it could perhaps be expected that such studies had reached a level where only details of genealogy remained to be cleared up, but this is far from being the case. There are several reasons for this.

Darwin's comprehensive theory of evolution was actually five interwoven theories (Mayr 1982), and his theories about speciation and selection mechanisms were soon attacked from several sides. So although the idea of evolution and speciation became accepted rather easily, the attacks on his explanation of speciation focused on one side of the theory which has turned out to need a good deal of modification. Some of Darwin's followers carried their arguments to extremes, which undoubtedly detracted from the credibility of the whole field. Finally, the growing interest in experimental biology turned the spotlight away from phylogeny.

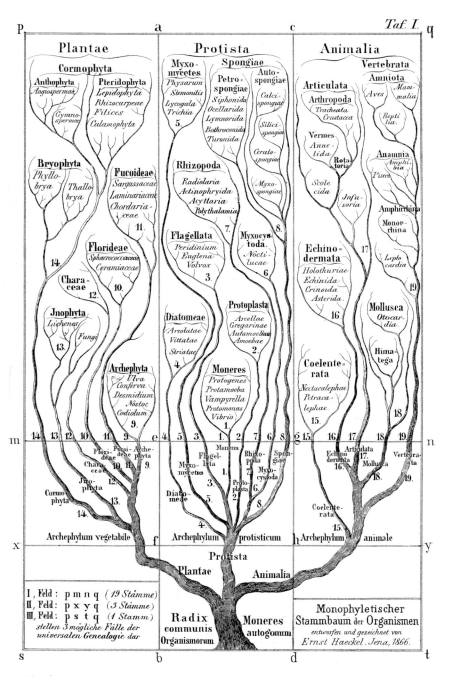

Fig. 1.1. The first phylogenetic tree of the animal kingdom, drawn by Haeckel (1866); note the word 'Monophyletischer' in the legend.

However, since the 1950s, a revival of the phylogenetic interests has taken place and the field is again producing a steady flow of interesting results. Whole new areas of information have been added, the most important being ultrastructure and bio-chemistry/molecular biology. Also, there has been important progress in methodo-logy; phylogenetic reasoning has been sharpened by the methods proposed by Hen-nig (1950), and the homology criteria formulated by Remane (1952) have eased the identification of homologous structures.

A number of new phylogenies for all the living beings as well as for narrower categories, such as phyla, classes and orders, have been proposed during recent decades, and there appears to be considerable agreement about main points of the earliest evolutionary history of living beings. But there is little agreement about the phylogeny of the animal kingdom – compare for example the phylogenetic trees in Salvini-Plawen (1982), Barnes (1987), Pearse *et al.* (1987), Margulis & Schwartz (1988), Brusca & Brusca (1990; Fig. 1.2), Willmer (1991), and Schram (1991; Fig. 1.3). One important reason for the disagreements between earlier trees is that differ-ent methods were used: some were built on the more traditional method which grouped animals on the base of more general similarity, whereas especially the newest trees are based on the cladistic method, which recognizes only monophyletic

Fig. 1.2. Animal phylogeny according to Brusca & Brusca (1990; redrawn); cladogram generated by the computer program PAUP. Note that the Aschelminthes are lacking.

3

Fig. 1.3. Animal phylogeny according to Schram (1991; redrawn); cladogram generated by the computer program PHYSIS (WAGNER.S).

groups. Another reason is that some of the earlier trees put special emphasis either on adult or larval characters instead of considering all characters of all ontogenetic stages (as already pointed out by Darwin). It should be evident that all characters are of importance, the only question being at which level they contain phylogenetic information. Another weakness has been that the presence of 'advanced' characters, such as coelom and metanephridia, has been used to characterize higher taxa without discussing the possibility that these characters could have evolved a number of times. Most of the discrepancies between the phylogenetic trees obtained by Brusca & Brusca (1990) and Schram (1991), who used cladistic programs such as PAUP and PHYSIS (WAGNER.S) on large character matrices, and the phylogeny which has been the result of my considerations using a more traditional method (see below; Fig. 1.4) are probably due to this. In my opinion, coeloms have evolved several times (Chapter 9), and dealing with this character in only two states, absent and present, is bound to lead to unreliable results. The possibility of convergency is discussed in the following whenever it has seemed necessary.

Strong interest in the results obtained by the new molecular methods, especially the sequencing of DNA/RNA (which appears to be the only appropriate method at the phylum level; Hillis & Moritz 1990), has resulted in several phylogenetic trees based solely on these data. Several zoologists believe that the sequencing techniques will eventually provide an 'unequivocal' phylogenetic tree of the whole animal kingdom. The earliest results of the DNA sequencing were based on a few species only, and trees presented for example by Field *et al.* (1988), showing the relationships of some major groups, demonstrated that different trees can be obtained with different selections of the analysed material. Further discussion of the same material (Field *et al.* 1989, Lake 1990) showed that different statistical methods lead to different shapes of trees (or different probabilities). In a recent overview of the applications of molecular systematics, Hillis & Moritz (1990, p. 502) state 'The rapid development and power of these techniques has produced an euphoria in evolutionary biology; because so many new problems can be addressed, it is a commonly held misconception that all evolutionary problems are solvable with molecular data. This is clearly not the case.' It seems obvious that the molecular characters should be used together with the morphological data and with just as much caution and tact. This book aims at constructing a phylogeny based on morphological evidence alone, and this tree can then be used as a reference for the results of the molecular methods.

It should also be remembered that the molecular data can give only 'naked' trees without morphological characters, and as stated by Raff *et al.* (1989, p. 258) 'The use of rRNA sequences to infer distant phylogenetic relationships will not displace morphology and embryology from the study of the evolutionary history of animal life: *after all, it is the history of morphological change that we wish to explain*' (my italics).

In this book cladistic principles have been used to determine phylogeny. They can perhaps best be explained by the following illustration: phylogeny is seen as a tree, which is one species in cross section everywhere except at the branching points (almost exclusively bifurcations; anastomoses are almost totally absent because hybridization appears to be a very rare method of forming new species among the animals). The living species form the tips of the branches reaching a horizontal plane representing the present time. Genera, families, orders and so on are the (monophyletic) branches which fall off after pruning at different levels. The levels of the various cuts which define the categories are chosen by the specialists on rather subjective grounds, and orders in one phylum need not be cut at the same level as those of other phyla. The phyla do not represent equal levels either, and the reason why I have chosen to work with that category is that the phylum is the highest category (= deepest cut) where each member is generally accepted as monophyletic (the arguments for regarding the 31 phyla as monophyletic are given for each phylum). The resulting phylogeny of the 31 phyla is presented in Fig. 1.4.

Cladistic classification has one great advantage over the evolutionary classification, namely that it aims only at establishing the sequence of branching points of the phylogenetic tree, and there is only one tree which reflects the genealogy of the organisms, whereas the evolutionary approach must in addition evaluate the degrees of deviation of the various branches. There is no objective measure of the deviations, and the evolutionary systematists are therefore likely to get lost in futile discussions about degrees of similarity/deviation (see also Eldredge & Cracraft 1980 and Ridley 1986).

5

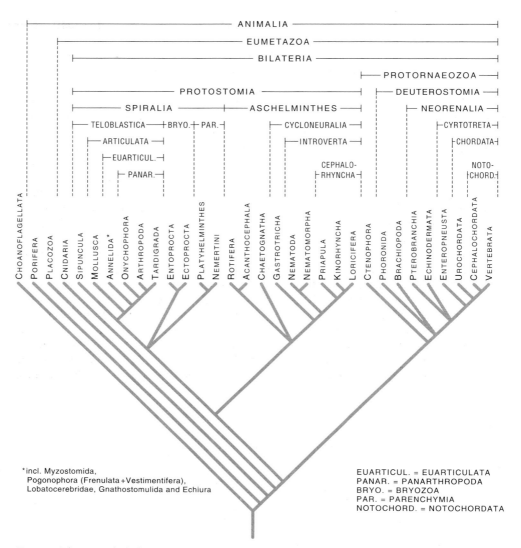

Fig. 1.4. The animal phylogeny proposed in this book. The tree is a summary of Figs 3.6, 9.1, 11.4, 12.2, 18.1, 28.1, 32.2, 43.8, and 49.2, where the apomorphies can be found. In the text, each taxon is described and discussed in a separate chapter. A number of new supraphyletic taxa are introduced here, for example Teloblastica, Cycloneuralia, Introverta, Neorenalia, and Notochordata, and some of the better-known groups have been given slightly modified contents; the definition of each taxon can be found in the respective chapter.

My aim has thus been to give a reasoned representation of the early part of the animal evolution, viz. the part of the phylogenetic tree which is left when all living phyla have been cut off leaving only their ancestral species. In constructing the tree, as many characters as possible have been taken into account. It has been necessary to make a number of generalizations about the phyla, and these generalizations reflect similar considerations about the phylogeny of the individual phyla.

The method used in this book has been iterative, but a number of steps can be outlined. The first step is the identification of monophyletic phyla; if a phylum (for example Hemichordata) is found to be polyphyletic it is broken up into monophyletic phyla. The next step is the identification of ancestral characters of the phylum, and this has in some cases, such as Mollusca and Urochordata, made it necessary to make cladistic analyses of single phyla. The phyla can now be compared and sister groups identified, and increasingly more comprehensive, monophyletic groups can be defined. This process finally results in a phylogenetic tree comprising all (or almost all) phyla, where the distribution of the various characters can be analysed. This analysis may result in changed opinion about some characters, sometimes necessitating remodelling of parts of the tree followed by a new analysis. A 'final' tree makes it possible to trace all characters to their origin, and therefore to define the point where a character is apomorphic. The various characters can then be discussed and their reliability (weight) in phylogenetic analyses evaluated. These evaluations sometimes gain information from information about the intraphyletic stability of the characters; an example is the fate of the blastopore which has turned out to be highly variable both within annelids and molluscs, and this character can therefore be expected to show variation also between phyla within a more comprehensive group, such as the Protostomia.

The analysis of the resulting phylogeny has finally focused on the necessary functional continuity between the ancestors and their descendants. With the wording of Frazzetta (1975, p. 20) the analysis should ascertain if the proposed evolution has proceeded like 'the gradual improvement of a machine *while it is running*' (my italics).

References

Barnes, R.D. 1987. Invertebrate Zoology, 5. ed. – Saunders, Philadelphia.

Brusca, R.C. & G.J. Brusca 1990. Invertebrates. – Sinauer Associates, Sunderland.

Darwin, C. 1859. On the Origin of Species by Means of Natural Selection. – John Murray, London.

Eldredge, N. & J. Cracraft 1980. Phylogenetic Patterns and the Evolutionary Process. – Columbia Univ. Press, New York.

Field, K.G., G.J. Olsen, D.J. Lane, S.J. Giovannoni, M.T. Ghiselin, E.C. Raff, N.R. Pace & R.A. Raff 1988. Molecular phylogeny of the animal kingdom. – Science **239**: 748-753.

Field, K.G., G.J. Olsen, S.J. Giovannoni, E.C. Raff & R.A. Raff 1989. Phylogeny and molecular data: response. – Science **243**: 550-551.

Frazzetta, T.H. 1975. Complex Adaptations in Evolving Populations. – Sinauer Associates, Sunderland, MA.

Haeckel, E. 1866. Generelle Morphologie der Organismen. 2 vols. – Georg Reimer, Berlin.

Hennig, W. 1950. Grundzüge einer Theorie der phylogenetischen Systematik. – Deutscher Zentralverlag, Berlin.

Hillis, D.M. & C. Moritz (1990). An overview of applications of molecular systematics. – *In* D.M. Hillis & C. Moritz (eds): Molecular Systematics. Sinauer Associates, Sunderland, MA.

Lake, J.A. 1990. Origin of the Metazoa. – Proc. natl Acad. Sci. USA **87**: 763-766.

Margulis, L. & K.V. Schwartz 1988. Five Kingdoms. An Illustrated Guide to the Phyla of Life on Earth, 2nd ed. – W.H. Freeman, New York.

Mayr, E. 1982. The Growth of Biological Thought. Diversity, Evolution, and Inheritance. – Harvard Univ. Press, Cambridge, MA.

Owen, R. 1848. On the Archetype and Homologies of the Vertebrate Skeleton. – Richard & John E. Taylor, London.

Pearse, V., J. Pearse, M. Buchsbaum & R. Buchsbaum 1987. Living Invertebrates. – Blackwell, Palo Alto, CA.

Raff, R.A., K.G. Field, G.J. Olsen, S.J. Giovannoni, D.J. Lane, M.T. Ghiselin, N.R. Pace & E.C. Raff 1989. Metazoan phylogeny based on analysis of 18S ribosomal RNA. – *In* B. Fernholm, K. Bremer & H. Jörnvall (eds): The Hierarchy of Life. Molecules and Morphology in Phylogenetic Analysis, pp 247-260. Excerpta Medica/Elsevier, Amsterdam.

Remane, A. 1952. Die Grundlagen des natürlichen Systems, der vergleichenden Anatomie und der Phylogenetik. – Akademische Verlagsgesellschaft, Leipzig.

Ridley, M. 1986. Evolution and Classification. – Longman, London.

Salvini-Plawen, L.v. 1982. A paedomorphic origin of the oligomerous animals ? – Zool. Scr. 11: 77-81.

Schram, F.R. 1991. Cladistic analysis of metazoan phyla and the placement of fossil problematica. – *In* A.M. Simonetta & S. Conway Morris (eds): The Early Evolution of Metazoa and the Significance of Problematic Taxa, pp 35-46. Cambridge Univ. Press, Cambridge.

Willmer, P. 1991. Invertebrate Relationships. Patterns in Animal Evolution. – Cambridge Univ. Press, Cambridge.

2

The trochaea theory

The present book grew out of my phylogenetic considerations around the trochaea theory, and since reference will often be made to parts of the theory, I have found it useful to give a summary of the earlier publications. The theory makes a number of predictions about characters in the various phyla, and the concluding chapter (Chapter 56) will reevaluate its usefulness.

The forerunner of the trochaea theory (Nielsen 1979) was inspired by an appreciation that adult rotifers and several types of spiralian larvae have ciliary feeding structures of similar structure and function (Strathmann, Jahn & Fonseca 1972) and that this could be of phylogenetic significance. It stressed the difference between the downstream-collecting ciliary systems of the protostomian (gastroneuralian) larvae and the upstream-collecting systems of the deuterostomian (notoneuralian) larvae, but a phylogenetic explanation was not proposed.

The trochaea theory was developed in full by Nielsen & Nørrevang (1985) and the corresponding phylogeny and the details of its background in the structure and function of the ciliary bands were discussed by Nielsen (1985, 1987).

The 'backbone' of the trochaea theory shows the early evolution of the animal phyla as the evolution of a series of holoplanktonic ancestors (Fig. 2.1). Three of these ancestors then gave rise to pelago-benthic ancestors, with sessile or creeping adults (Fig. 2.2).

The early part of the phylogeny corresponds to the gastraea theory of Haeckel (1874). The ancestor of the Metazoa was a spherical, planktonic colony of choanoflagellates, and the first metazoan, blastaea, came into existence when the cells of a colony came into close contact and formed occluding junctions between the cells. This new organization had the advantage of an inner cavity, the blastocoel, which was isolated more or less completely from the environment so that organic molecules could easily be exchanged between the cells without loss to the outside. This structure made the establishment of a division of labour between the cells possible. The blastaea was without polarity and rolled through the water much like the green alga *Volvox* (the similarity being entirely superficial). The first division of labour between the cells probably evolved when the organism established a preferred direction of swimming, with special sensory cells at the anterior pole. This evolutionary step established a polarity in the organism with a longitudinal axis parallel to the swim-

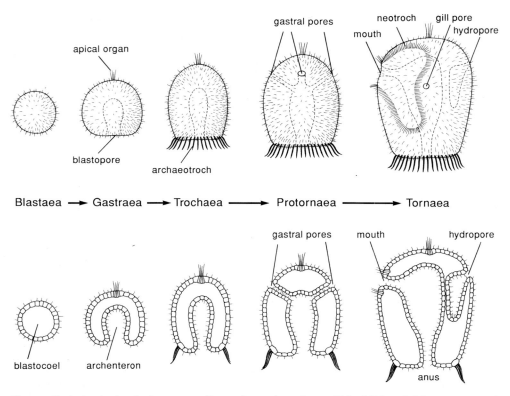

Fig. 2.1. Evolution in the plankton, according to the trochaea theory. (After Nielsen & Nørrevang 1985.)

ming direction and an anterior end, the apical pole, with a primitive sense organ, the apical organ. The leeward side at the posterior pole offered the possibility of retaining larger food particles and an invagination, the archenteron, with absorbing and digesting cells enhanced the food uptake. The epithelium of the outer side of this organism kept its locomotory function and is called ectoderm, while the inner, digestive epithelium is called endoderm. All cells had one cilium, as in the choanoflagellates. This organism is called the gastraea, and its main axis was the apical-blastoporal axis.

The cnidarians are organized like a gastraea, and the sessile stage, the polyp, is a gastraea which has settled with the apical pole and developed tentacles around the blastopore (mouth; Fig. 2.2); this is clearly seen in the ontogeny of many recent cnidarians.

The holoplanktonic gastraea was swimming with separate cilia of the ectoderm, and the power of the separate cilia must have set some upper limits in size; the formation of compound cilia (see for example Fig. 43.5) in a ring around the posterior part of the body gave increased powers of swimming and made larger body size possible. It is believed that this ring, which is called the archaeotroch, was also capable of capturing food particles as a downstream-collecting system (Fig. 2.3) and transferring them to the area of separate cilia immediately around the blastopore and the archen-

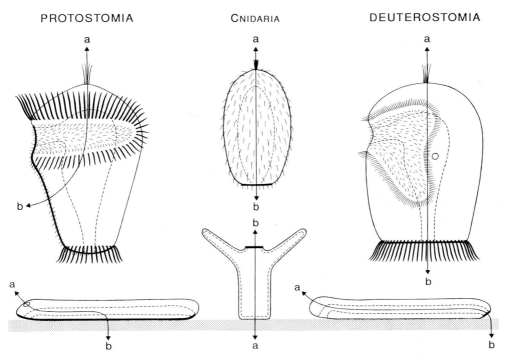

PROTOSTOMIA   CNIDARIA   DEUTEROSTOMIA

Fig. 2.2. Ancestral life cycles of protostomes (gastroneuralians), cnidarians and deuterostomes (notoneuralians) according to the trochaea theory. a-b, the apical-blastoporal axis. The blastoporal region is indicated by a heavy line. (Modified from Nielsen & Nørrevang 1985; a planula (gastrula) larva has been drawn instead of a medusa to emphasize that the medusa is not regarded as a larval or juvenile stage.)

teron. It is further believed that this organism had an apical sensory organ, a ring nerve along the cells of the archaeotroch and a few nerves connecting the apical organ with the ring nerve, so that information from the apical organ at the frontal pole of the animal could be used to control the movements of the archaeotroch and hence the swimming. This holopelagic ancestor is called trochaea.

If adults of this organism went down to the bottom of the sea and stopped the swimming movements of the archaeotroch they could crawl on the single cilia of the zone immediately surrounding the blastopore, and these cilia could also collect particles from the sediment and transport them to the gut. At first, the movements on the bottom were probably at random, but a preferred direction of crawling could easily become established giving the adult a new antero-posterior main axis perpendicular to the apical-blastoporal axis of the larva. With the new front end, an elongation of the blastopore and a one-way traffic through the archenteron could be established, and the function of the gut could become more efficient if the lateral blastopore lips were pressed together. The archaeotroch had thus lost its function in the adult and was soon lost, and the advanced trochaea had acquired a pelago-benthic life cycle (Fig. 2.4, left).

downstream-collecting system   upstream-collecting system

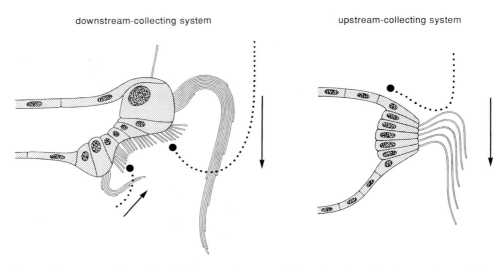

Fig. 2.3. Structure and function of the two main types of ciliary bands used in locomotion and filter feeding. The arrows indicate the direction of the effective stroke of the cilia, and thus of the water currents, and the dotted lines indicate the paths of particles which are being captured. (After Nielsen 1987.)

The modifications needed for reaching the ancestor called the gastroneuron (Fig. 2.4, right) are small: the blastopore lips fused permanently in the adult creating a tube-shaped gut, and the circumblastoporal nerve ring developed into nerve rings around mouth and anus and a pair of ventral nerves along the fused blastopore lips. The nervous centre connected with the apical organ was retained in the adult and became the brain. The blastopore closure became expressed already in the pelagic larva, where the part of the archaeotroch and the adoral ciliary zone surrounding the mouth became expanded into a pair of lateral loops, so that the well-known system of a preoral prototroch and a postoral metatroch of compound cilia (still both functioning as downstream-collecting systems; Fig. 2.4) became established; the compound cilia around the anus, the telotroch, lost the particle-collecting function, and the compound cilia along the fused blastopore lips were lost altogether. The area of separate cilia surrounding the mouth was pulled out between prototroch and metatroch (to form the adoral ciliary zone) and its continuation through the ventral break in the metatroch along the ventral mid-line formed the gastrotroch. Gastroneuron thus had a pelago-benthic life cycle with a planktotrophic trochophora larva and a creeping benthic adult which was feeding on sediment particles (Fig. 2.4).

Gastroneuron is believed to be the ancestor of all the protostomians (gastroneuralians), a group which has been established since the turn of the century and which in the present book consists of the two groups Spiralia (comprising several phyla with trochophore larvae) and Aschelminthes (comprising the rotifers, some of which have a ciliary feeding system which is identical to that of a trochophore).

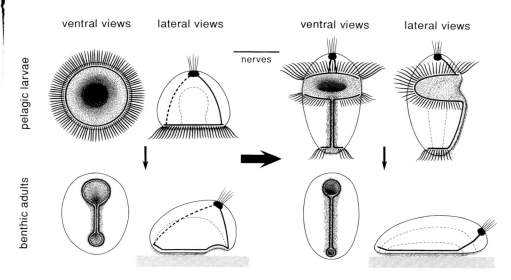

Fig. 2.4. Evolution of the protostomian ancestor (gastroneuron) from an advanced pelago-benthic trochaea. Small arrows, ontogeny; large arrows, phylogeny. (After Nielsen & Nørrevang 1985.)

In the plankton, a further specialization of the trochaea took place along a line which led towards the deuterostomes (notoneuralians). As a first step, four new openings from the upper part of the archenteron developed, and these openings could function as outlets for indigestible particles, as the pores of some recent comb-jellies (Main 1928). The immediate adaptive advantage of these new structures is difficult to imagine, but it must be remembered that secondary openings from the gut are formed in a number of other invertebrates (Nielsen & Nørrevang 1985; Fig. 48.3), so the process of forming such secondary openings is realized also today. This ancestor is called the protornaea.

The next step of the evolution in the plankton consisted in an organization of cilia around one of the new openings, which became specialized as a mouth, into the shape of an upstream-collecting band, the neotroch. The gut became a tube with food particles passing only in one direction, from the new mouth to the blastopore, which became the anus. The archaeotroch lost its particle-collecting function, and the canal from the gut to the exterior opposite to the mouth became pinched off from the gut and later developed into the coelomic system; the remaining two, lateral openings became gill pores. This holopelagic organism, which is called the tornaea, has almost the same structure as the tornaria larva of some recent enteropneusts.

Adults of tornaea-like organisms may have colonized the sea-bottom, started to creep on cilia between mouth and anus, and begun to collect particles from the sediment. The function of the archaeotroch became superfluous in the adults and it was lost; the new pelago-benthic organism, notoneuron, had a planktotrophic larva like the tornaea and a creeping, benthic adult (Fig. 2.2).

From the above description it should be clear that the area called ventral in the protostomes, namely the area of the laterally compressed blastopore, is not the same as the area called ventral in the deuterostomes, namely an area between the blastopore and the apical organ.

According to the trochaea theory, the two large taxa Protostomia (Gastroneuralia) and Deuterostomia (Notoneuralia) show characteristic distinguishing characters both in the fate of the blastopore, the shape and structure of the larval ciliary bands and the shape of the central nervous systems.

The protostomes should have: 1) a blastopore which becomes divided into the adult mouth and anus through lateral fusion of the blastopore lips (Fig. 2.4), 2) larvae of the trochophore type, with prototroch and metatroch of compound cilia functioning as downstream-collecting systems around the mouth (Fig. 2.5), and 3) a main nervous system consisting of a brain derived from the apical organ and a pair of ventral main nerves (Fig. 2.6). The deuterostomes should have: 1) a blastopore which becomes the anus in the adult, whereas the adult mouth forms anew, 2) larvae of the tornaria type, with a neotroch of single cilia functioning as an upstream-collecting system around the mouth (Fig. 2.5), and 3) an adult main nervous system which is a dorsal longitudinal structure formed behind the larval apical organ (Fig. 2.6).

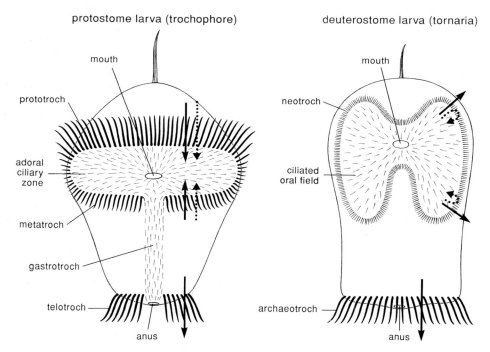

Fig. 2.5. The ancestral larval types of protostomes (trochophora larva) and deuterostomes (tornaria larva) according to the trochaea theory. (After Nielsen 1987.)

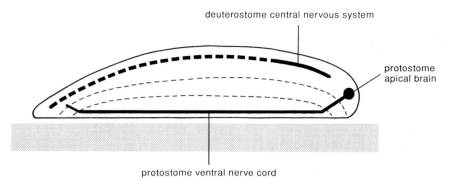

deuterostome central nervous system

protostome
apical brain

protostome ventral nerve cord

Fig. 2.6. Ancestral central nervous systems of adult protostomes (gastroneuralians) and deutero-stomes (notoneuralians). (After Nielsen 1985.)

Not all representatives of all phyla agree with this scheme in all characters, and the agreements/disagreements will be documented and discussed in the chapters about the individual phyla.

One of the objections to the trochaea theory has been that small eggs and pelagic, planktotrophic early development are only observed in animals of a comparatively large size which enables them to produce a high number of eggs. Olive (1985) has pointed out that very small organisms rarely, or perhaps never, have planktotrophic larvae and that this detracts from the probability of phylogenetic theories which regard pelagic, ciliated larva-like organisms as ancestors of major metazoan groups. It is well documented that modern marine invertebrates with planktotrophic development have high numbers of eggs and that very small organisms have brood protection. The explanation given for this is usually that the small planktotrophic larvae stay in the plankton, where there is a heavy predation, for a rather long time; only species which produce large batches are likely to get larvae through to metamorphosis (Thorson 1950).

However, it must be remembered that the planktotrophic ancestral forms, be they holoplanktonic forms or larvae of pelago-benthic forms, which are important in the various phylogenetic theories must have lived about 600 million years ago, at a time when there were no (or at least very few) multicellular predators in the plankton; the heavy predation which creates many constraints on the life-styles in the plankton today did not exist, so small animals producing small numbers of small eggs giving rise to planktotrophic larvae may indeed have existed.

A specific attack on the trochaea theory and on the phylogeny derived from the theory is presented by Ivanova-Kazas & Ivanov (1988). This is not a place for a detailed reply, and some of their arguments are dealt with in the chapters on the respective phyla or supraphyletic groups. However, their main idea about the origin of the ciliary bands of invertebrate larvae (also expressed by Ivanova-Kazas 1985a-c, 1986a-c and Salvini-Plawen 1980) should be addressed here.

Ivanova-Kazas, Ivanov and Salvini-Plawen are of the opinion that the various types of planktotrophic as well as lecithotrophic larvae with ciliary bands are derived

15

independently from lecithotrophic larvae with a uniform ciliation (atrochal larvae); such atrochal larvae are known in almost all phyla. To me it appears very improbable that the highly characteristic ciliary bands of the trochophore-type larvae of polychaetes, molluscs and entoprocts, as well as those of some of the adult rotifers should have developed convergently. These bands show detailed similarities in structure and function as well as in their position relative to other larval organs (Nielsen 1987; Chapter 10). Even the idea that both a gut and the complicated ciliary structures should evolve without having any adaptive value until fully formed appears as a straight example of orthogenesis, an evolutionary principle which is now generally rejected in the biological sciences (see for example Mayr 1982).

References

Haeckel, E. 1874. Die Gastraea-Theorie, die phylogenetische Classification des Thierreichs und die Homologie der Keimblätter. – Jena. Z. Naturw. 8: 1-55, pl. 1.

Ivanova-Kazas, O.M. 1985a. The origin and phylogenetic significance of the trochophoran larvae. 1. The larvae of coelomate worms and molluscs. – Zool. Zh. 64: 485-497 (In Russian, English summary).*

Ivanova-Kazas, O.M. 1985b. The origin and phylogenetic significance of the trochophoran larvae. 2. Evolutionary significance of the larvae of coelomate worms and molluscs. – Zool. Zh. 64: 650-660 (In Russian, English summary).*

Ivanova-Kazas, O.M. 1985c. Origin and phylogenetic significance of the trochophoran larvae. 3. Larvae of flatworms and nemerteans. – Zool. Zh. 64: 1765-1776 (In Russian, English summary).*

Ivanova-Kazas, O.M. 1986a. Origin and phylogenetic significance of the trochophoran larvae. 4. Kamptozoa larvae. General ideas. – Zool. Zh. 65: 165-174 (In Russian, English summary).*

Ivanova-Kazas, O.M. 1986b. Analysis of larval development in Tentaculata. 1. Larvae of Phoronida and Brachiopoda. – Zool. Zh. 65: 757-770 (In Russian, English summary).*

Ivanova-Kazas, O.M. 1986c. Analysis of larval development in Tentaculata. 2. Bryozoa larvae. – Zool. Zh. 65: 1445-1456 (In Russian, English summary).*

Ivanova-Kazas, O.M. 1987. The origin, evolution and phylogenetic significance of ciliated larvae. – Zool. Zh. 66: 325-338 (In Russian, English summary).*

Ivanova-Kazas, O.M. & A.V. Ivanov 1988. The trochaea theory and phylogenetic significance of ciliate larvae. – Soviet J. mar. Biol. 13: 67-80.

Main, R.J. 1928. Observations of the feeding mechanism of a ctenophore, Mnemiopsis leidyi. – Biol. Bull. Woods Hole 55: 69-78.

Mayr, E. 1982. The Growth of Biological Thought. Diversity, Evolution and Inheritance. – Harvard Univ. Press, Cambridge, Mass.

Nielsen, C. 1979. Larval ciliary bands and metazoan phylogeny. – Fortschr. zool. Syst. Evolutionsforsch. 1: 178-184.

Nielsen, C. 1985. Animal phylogeny in the light of the trochaea theory. – Biol. J. Linn. Soc. 25: 243-299.

Nielsen, C. 1987. Structure and function of metazoan ciliary bands and their phylogenetic significance. – Acta zool. (Stockh.) 68: 205-262.

Nielsen, C. & A. Nørrevang 1985. The trochaea theory: an example of life cycle phylogeny. – In S. Conway Morris, J.D. George, R. Gibson & H.M. Platt (eds): The Origin and Relationships of Lower Invertebrate Groups, pp 28-41. Oxford Univ. Press, Oxford.

Olive, P.J.W. 1985. Covariability of reproductive traits in marine invertebrates: implications for the phylogeny of the lower invertebrates. – In S. Conway Morris, J.D. George, R. Gibson & H.M. Platt (eds): The Origins and Relationships of Lower Invertebrate Groups, pp 42-59. Oxford Univ. Press, Oxford.

Salvini-Plawen, L.v. 1980. Was ist eine Trochophora? Eine Analyse der Larventypen mariner Protostomier. – Zool. Jb., Anat. 103: 389-423.

Strathmann, R.R., T.L. Jahn & J.R. Fonseca 1972. Suspension feeding by marine invertebrate larvae: clearance of particles by ciliated bands of a rotifer, pluteus, and trochophore. – Biol. Bull.
Woods Hole 142: 505-519.

Thorson, G. 1950. Reproductive and larval ecology of marine bottom invertebrates. – Biol. Rev. 25:
1-45.

* English translations available from: Library, Canadian Museum of Nature, P.O. Box 3443, Stn. D,
Ottawa, Ontario, Canada K1P 6P4.

Kingdom ANIMALIA (= METAZOA)

In the first edition of 'Systema Naturæ' Linnaeus (1735) defined the Kingdom Animalia as natural objects which grow, live and sense, in contrast to the plants, which grow and live but do not sense, and the minerals, which grow but neither live nor sense. This definition of the animal kingdom, which goes back to the classical period, was retained almost unchanged in the 10th edition of 'Systema Naturæ' (Linnaeus 1758), which forms the baseline for the zoological nomenclature. His arrangement of the species in classes, families and genera reflects the similarity of the organisms, but of course without a causal explanation. His division of the organisms into animals and plants was almost unchallenged for more than a century.

The first classification of living beings based on Darwin's (1859) evolutionary thoughts was presented by Haeckel (1866; Fig. 1.1). He gave a remarkably modern definition of the kingdom Animalia, which was separated from the new kingdom Protista by the possession of tissues and organs. This definition excluded the sponges from the animals, but he later (Haeckel 1874) included the sponges in the group Metazoa.

The word 'animal' is still used in the wide, Linnean sense, but the kingdom Animalia is now usually restricted to comprise only the multicellular animals, i.e. the Metazoa.

The separation of the prokaryotes from the eukaryotes is now generally accepted, and the schemes defining five or more kingdoms (see for example Whittaker 1969 and Margulis 1981) are widely accepted. Most of these schemes are attractive because they try to define (mainly) monophyletic groups, such as Plantae (green plants), Animalia and Fungi, but leave a large paraphyletic group of mostly unicellular eukaryotic organisms called Protista/Protoctista.

My earlier approach to a definition of the animal kingdom (Nielsen 1985) was a little different from that of Whittaker and Margulis in that I tried to follow the animal line as far down as possible instead of cutting off at the grade of multicellularity. This made me include the choanoflagellates in the kingdom Animalia, but I will here return to the more conventional definition, and only discuss the choanoflagellates as the most probable sister group of the Animalia.

There are a few textbooks (for example Pearse *et al.* 1987) which favour the polyphyletic origin of the metazoan phyla from protist ancestors, but most modern authors believe that multicellular animals have evolved only once (Gruner 1980, Möhn 1984, Barnes 1985, 1987, Siewing 1985, Margulis & Schwartz 1988, Kozloff 1990, Brusca & Brusca 1990); most of these authors favour the traditional idea that the ancestor of the metazoans resembled a colonial choanoflagellate. The first attempts at cladistic analyses of the interrelationships of the animal phyla have appeared recently (Brusca & Brusca 1990; Fig. 1.2; Schram 1991; Fig. 1.3), and most of the synapomorphies used in constructing these trees will be discussed in the following chapters.

The phylogenetic tree resulting from my own analysis of the animal phyla is shown in Fig. 1.4. The synapomorphies of the various parts of the tree can be found on the phylogenetic diagrams in Figs 3.6, 9.1, 11.4, 12.2, 18.1, 28.1, 32.2, 43.8, and 49.2.

The most important synapomorphy of the metazoans is their multicellularity, as opposed to the coloniality shown by many choanoflagellates. In colonies the cells may have different shapes and functions, but they all feed because there is no contact between the cells enabling transport of nutrients between cells. In multicellular organisms the cells are engaged in a division of labour and may be specialized to serve different functions, such as sensation, contraction or secretion; this is only possible because nutrients can be transported from feeding to non-feeding cells. The multicellular level of organization is reflected in a series of advanced characters both at the morphological and biochemical levels. A number of characters unrelated to multicellularity can also be identified, but some of these characters have not been investigated in the choanoflagellates and may thus prove to be shared characters of both groups (see Table 3.1).

There are three sets of characters which can be used to define the Metazoa: 1) Characters directly connected with the function of a multicellular organism, such as certain types of cell junctions and complex organic molecules which mediate cell communication. 2) Specific structures or molecules inside the cells or at their surface. 3) Structures connected with the sexual life cycle, i.e., the origin and structure of the haploid gametes, the fertilization and the development from the zygote to the adult organism.

Cell junctions are specialized areas of contact between cell membranes. Three main functional types can be recognized: 1) Occluding junctions which more or less completely seal off the intercellular spaces from the environment so that the composition of the fluid in the spaces can be controlled; these junctions must also have an adhesive function. The septate junctions, which occur in all invertebrates and the tight junctions of vertebrates, tunicates and some arthropods belong to this type. 2) Desmosomes, comprising spot or belt desmosomes, which strengthen the cell to cell adhesion, and hemidesmosomes, which strengthen cell to substrate adhesion (to the extracellular matrix between cell layers or to an external skeleton). 3) Gap junctions which serve the chemical and electrical communication between the cells; this type of junction occurs only in hydrozoans and bilaterians (Mackie, Anderson & Singla 1984) and will be discussed in the chapter on Eumetazoa (Chapter 7).

Occluding junctions are known from all metazoan phyla (Mackie 1984).

The septate junctions have the shape of a series of parallel septa bridging the intercellular spaces. In epithelia, the septa are oriented parallel to the outer surface

Table 3.1. The distribution of important characters in the 'lower' animal phyla and the Bilateria. The more detailed information is given in the chapters on the respective groups. Septate junctions are only found between choanocytes and between scleroblasts in the Porifera and the presence of this type of cell junction in the Placozoa is somewhat doubtful. The two uppermost lines comprise complexes which are generally regarded as characteristic of the Metazoa, but this is uncertain as long as sexual reproduction has not been observed in the choanoflagellates. Advanced character states in boldface.

	Choano-flagellata	Pori-fera	Placo-zoa	Cnid-aria	Bila-teria
Meiosis, diploidy	?	+	?	+	+
Spermatozoa of the animal type	?	+	?	+	+
Synapses with acetylcholine/cholinesterase	−	−	−	−	+
Hemidesmosomes	−	−	−	+	+
Basement membrane	−	−	−	+	+
Gap junctions	−	−	−	+	+
Sensory cells, nerve cells + synapses	−	−	−	+	+
Synapses with RFamide	−	−	−	+	+
Cross-striated ciliary rootlets	−	−	+	+	+
Belt desmosomes	−	−	+	+	+
Septate junctions	−	(+)	'+'	+	+
Collagen	−	+	?	+	+
Ciliary necklace with 3(-4) strings	−	+	?	+	+
pp60$^{v\text{-}src}$ kinase activity	?	+	−	+	+
Choanocytes or choanosyncytia with contractile microvilli	+	+	−	−	−
Crested mitochondria	+	+	+	+	+
Cilia with perpendicular accessory centriole	+	+	+	+	+

of the cell and are therefore seen as lines between cell membranes in transverse sections of epithelia. Such junctions have been found in a wide variety of tissues and in all metazoan epithelia (Green 1984). In sponges they occur only occasionally, and they have only been observed between the cells secreting the spicules (Ledger 1975; Fig. 3.1) where their function probably is to isolate the fluid in the extracellular space where the spicules are formed, and between choanocytes (Green & Bergquist 1979) where their function appears to be structural; they have not been observed in freshwater sponges (personal communication, Dr N. Weissenfels, Univ. Bonn). The hexactinellid sponges are organized with large syncytia, but septate junctions have been observed between the collar bodies and the trabecular tissue (Mackie & Singla 1983). The outer epithelial covering of sponges is quite loosely organized and has no cell junctions; it can apparently not control the composition of the intercellular fluid, and this is clearly very important for the whole organization of the sponge. *Trichoplax* has structures resembling septate junctions just below the belt desmosomes (Ruthmann, Behrendt & Wahl 1986; Fig. 3.1). In cnidarians (Fig. 3.1) and bilaterians, septate junctions are a well-known feature of most types of tissues. The junctions of the invertebrate phyla can be classified in a number of subtypes (Green & Bergquist 1982), but the phylogenetic interpretation is still uncertain.

Tunicates and vertebrates have tight junctions, where the cell membranes are in close contact, instead of septate junctions, but those of the tunicates are of a somewhat different structure (Green 1984); tight junctions are also found in certain organs of arthropods (Huebner & Caveney 1987).

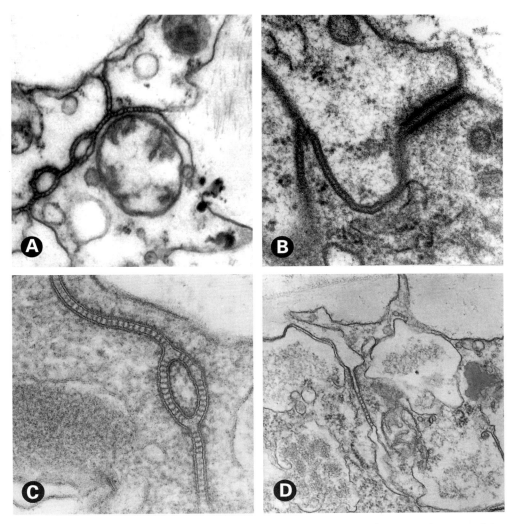

Fig. 3.1. Septate junctions in 'primitive' metazoans. – A, septate junctions between sclerocytes of the calcareous sponge *Sycon ciliatum* (from Ledger 1975). – B, belt-desmosomes and septate-like junctions of *Trichoplax adhaerens* (TEM courtesy of Dr A. Ruthmann, Ruhr-Univ. Bochum, FRG; see Ruthmann, Behrent & Wahl 1986). – C, septate junctions between ectodermal cells of *Hydra* (TEM courtesy of Dr R.L. Wood, Univ. Southern California, Los Angeles, CA, USA; see Wood 1985). – D, septate junctions between blastomeres in a late blastula of the starfish *Asterina pectinifera* (TEM courtesy of Dr M. Dan-Sohkawa, Osaka City Univ., Japan).

In the eumetazoans, septate junctions develop just after the embryos have passed from the morula to the blastula stage (see for example van den Biggelaar & Guerrier 1983 and Dan-Sohkawa & Fujisawa 1980; Fig. 3.1), i.e. just at the developmental stage corresponding to the evolutionary stage where the cell junctions are believed to have evolved (Chapter 2).

Belt desmosomes usually occur distal to (i.e., on the outer side of) the occluding junctions and have been described from placozoans (Ruthmann, Behrendt & Wahl 1986; Fig. 3.1) and all eumetazoans (Green 1984). Spot desmosomes are known to occur in sponges (Pavans de Ceccatty 1985) and in all eumetazoans (Green 1984). Hemidesmosomes are apparently restricted to eumetazoans (Chapter 7).

Another very important function in multicellular organisms is the communication between the cells, for example the activation of contractile cells by cells which are sensory or nervous. The conduction may be through chemical synapses or by direct electrical coupling; a primitive communication system built on stretch-sensitive channels has been proposed for the slow, non-electrical conduction in demosponges (Mackie 1990). Conduction through neurotransmitters is known from all eumetazoans, where the substances are located in synapses. RFamide has been found in the synapses of almost all eumetazoans, whereas acetylcholine appears to be restricted to the bilaterians (Chapter 9). Acetylcholinesterase, catecholamines and serotonin are found in sponges, but they are apparently not engaged in cell-communication (Mackie 1990).

The terms cilia and flagella are currently used for organelles of essentially identical structures (characterized by the presence of an axoneme consisting of $9 \times 2 + 2$ microtubules) occurring in most eukaryotes, and the term flagella is used also for the much simpler structures (without microtubuli) found in bacteria. Botanists have preferred the word flagella for the structures found in algae (Moestrup 1982), whereas zoologists generally have used the word flagella when only one or a few appendages are found per cell and cilia when many occur. Margulis (1980) introduced the word undulipodia for the structures found in the eukaryotes to eliminate the ambiguity, but instead of accepting a new name for a well-known structure I have chosen to extend the meaning of the word cilia to all the structures which contain one axoneme and to restrict the term flagella to the simpler structures found in the bacteria (Nielsen 1987).

The undulating cilia of choanoflagellates, choanocytes and spermatozoa transport water away from the cell body or propel the cell through the water with the cell body in front, whereas most protists swim in the opposite direction. The simple structure with only an axoneme and no hairs or other extracellular specializations (except the vane observed in choanoflagellates and many sponges, see Chapters 4 and 5) is also characteristic of choanoflagellates and metazoans, whereas for example most of the unicellular algae and the swarmers of multicellular algae have extra rods or other structures along the axoneme or intricate extracellular ornamentations (Moestrup 1982). The effective-stroke cilia characteristic of most metazoan epithelia show well defined patterns of movement with various types of metachronal waves (Nielsen 1987). Such patterns are rare in the protists with the ciliates as a conspicuous exception.

The basal structures of the cilia of some choanoflagellates, metazoan spermato-zoa and almost all monociliate metazoan cells show specific similarities with an ac-cessory centriole situated perpendicular to the basal body of the cilium (Fig. 3.2). None of the other unicellular organisms show a similar structure. Star-shaped ar-rangements of microtubuli, possibly with an anchoring and strengthening function, surround the basal bodies in choanoflagellates (Hibberd 1975) while similar patterns in spermatozoa of many metazoans are formed by microfilaments (Franzén 1987). Placozoans and bilaterians have cross-striated rootlets, but these structures have not been observed in choanoflagellates and sponges (Fig. 3.2). These structures are highly important for the understanding of relationships between the groups.

Freeze-fracture studies of the cell membrane of cilia have also revealed charac-ters which appear to be of phylogenetic importance. Bardele (1983, in Nielsen 1991) has shown that the intermembranous particles in metazoan cilia show 3(-4) rings of particles at the base (4-5 in vertebrates) while the cilia of choanoflagellates and a number of other protozoans show very different patterns.

The general morphology of the mitochondria has also been used in characteriz-ing various higher taxonomic units; choanoflagellates and metazoans have flat cris-tae (with exceptions in certain organs in various groups), while mitochondria with tubular cristae are characteristic of almost all heterotrophic protists; mitochondria with flat cristae occur also in chlorophytes, rhodophytes and non-flagellate fungi (Taylor 1978), but none of these groups are believed to be closely related to the Animalia.

Glycoproteins incorporating hydroxyproline and hydroxylysine are found in metazoans, higher plants, higher algae, and certain fungi, but only the metazoans have the highly characteristic form called collagen, while higher plants and green algae have extensins (Towe 1981). The cross-banded fibrils described from dinofla-gellates and foraminiferans are not collagens (Garrone 1978). This underlines the differences between the metazoans and the green plants and makes attempts at deriv-ing the metazoans from *Volvox*-like ancestors rather improbable (Kazmierczak 1981). Collagen occurs both between the cells in the shape of basement membranes

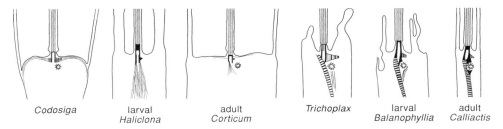

| Codosiga | larval Haliclona | adult Corticum | Trichoplax | larval Balanophyllia | adult Calliactis |

Fig. 3.2. Diagrams of ciliary basal complexes showing the accessory centrioles and the various types of ciliary rootlets. Choanoflagellata: *Codosiga botrytis* (based on Hibberd 1975); Porifera: larva of *Haliclona* sp. (based on Nielsen 1987) and choanocyte of *Corticum candelabrum* (based on Boury-Esnault *et al.* 1984); Placozoa: *Trichoplax adhaerens* (redrawn from Ruthmann, Behrent & Wahl 1986); Cnidaria: ectodermal cell of a larva of *Balanophyllia regia* (based on Lyons 1974) and pha-ryngeal cell of an adult *Calliactis parasitica* (based on Holley 1984).

or more elaborate connective tissues, such as the organic matrix of vertebrate skeletons, and as filaments in the cuticle of the outer side of epithelia. It occurs in the mesohyl of sponges (both cellularian sponges (Green & Bergquist 1979) and hexactinellids (Mackie & Singla 1983)) and in all eumetazoans. It appears to be a synapomorphy of the metazoans, but has to my knowledge not been searched for in the choanoflagellates.

Also some of the structural glycoproteins found in the intercellular matrix of connective tissues are very similar from sponges to vertebrates (Junqua, Fayolle & Robert 1975).

The occurrence of pp60$^{v\text{-}src}$ kinase activity has been investigated in a variety of eukaryotic organisms: green algae, such as *Euglena*, *Chlorogonium* and *Volvox*, ciliates, red algae, higher plants, and in a number of metazoans. The activity was only detected in metazoans and was highest in the central nervous systems; the activity was difficult to detect in whole organism preparations and this is probably the reason why a positive reaction was not obtained in rotifers, sea-anemones and *Trichoplax*; however, a high activity was found in whole preparations of sponges. This investigation indicates that the pp60$^{v\text{-}src}$ kinase activity is a characteristic of the metazoans (Schartl & Barnekow 1982).

Another feature which is characteristic of metazoans is that all cells except eggs and sperm are diploid. There are so few exceptions to this that it appears beyond doubt that the early metazoans were diploid with meiosis directly preceding the formation of eggs and sperm. Sexual reproduction has never been observed in choanoflagellates and their chromosome numbers have not been investigated.

The spermatozoa of metazoans show many specializations in the various phyla, but there is one type which is considered primitive and which has been found in representatives of the major part of the animal phyla (Fig. 3.3). This primitive type of sperm consists of an ovoid head with an apical acrosome (possibly absent in cnidarians) and a body of nuclear material, a mid-piece with usually four spheres of mitochondria surrounding the basal part of a long cilium with a perpendicular accessory centriole, and a tail which is the long undulating cilium (Bacetti & Afzelius 1976).

The metazoan eggs develop from one of the four cells of a meiosis, while the other three cells become the polar bodies and degenerate. Only in some parasitic wasps do the polar bodies divide further and form a tissue (trophamnion) which surrounds the embryo, but they do not contribute cells to the embryos (Martin 1914).

When the zygotes divide, the blastomeres form cleavage patterns and later on embryos/larvae of types characteristic of larger systematic groups. Representatives of many phyla go through a blastula stage, and this has been considered one of the important apomorphies of the Animalia (Margulis 1990). The larvae may be planktotrophic or lecithotrophic, and the planktotrophic larvae usually have feeding structures which are different from those of the adults.

The multicellularity and the several complicated apomorphies shared by sponges, placozoans and eumetazoans (for example septate junctions, belt desmosomes, collagen, and ciliary necklace with three rings of particles, see Table 3.1) strongly indicate

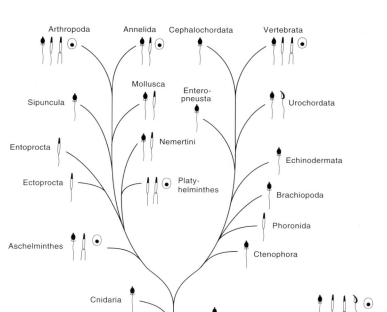

Fig. 3.3. The occurrence of different types of spermatozoa in the metazoan phyla. 1, primitive type; 2, modified; 3, biflagellate; 4, ascidian type; 5, aflagellate. (Based on Franzén 1987.)

that the Metazoa is a monophyletic group, and this opinion is as mentioned above shared by most modern authors.

The choanoflagellates share a number of important characters with some of the metazoans and they have therefore already been mentioned a number of times as the sister group of the Metazoa. The most important synapomorphy of the two groups is the collar complex of choanoflagellates and of the choanocytes/choanosyncytia in sponges (Fig. 3.4). This complex consists of a circle of retractile microvilli, which contain actin, surrounding an undulating cilium. The extended collar is a tubular or steep funnel-shaped structure with narrow slits; there is no fixed structural component keeping the microvilli in position, but mucous or fibrillar material forms narrow bridges between them. A long, undulating cilium is situated in the centre of the collar; it may have a pair of narrow, longitudinal ridges formed by the cell membrane, but characteristic of many species is a very thin, extracellular structure called the vane, which gives the whole structure the shape of a long band. The vane consists of two layers of very fine parallel fibres, which are oriented at angles to the cilium in the choanoflagellates (Chapter 4), while its structure appears less well defined in the sponges (Chapter 5), but this difference may be the result of different methods of study. The undulating movements of cilium and vane propel water out of the funnel

25

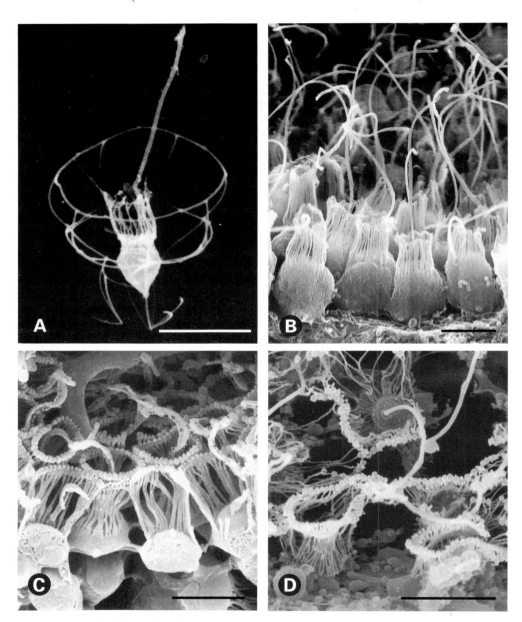

Fig. 3.4. SEM of a choanoflagellate and collar chambers of three types of sponges. – A, the solitary choanoflagellate *Pleurasiga minima* Throndsen (Monterey Bay, April 1992; SEM courtesy of Dr Kurt Buck, Monterey Bay Aquarium Research Institute, Pacific Grove, CA, USA). – B, the calcarean *Scypha* sp. (Friday Harbor Laboratories, WA, USA, July 1988). – C, the demosponge *Callyspongia diffusa* (SEM courtesy of Dr I.S. Johnston, Bethel College, MI, USA; see Johnston & Hildemann 1982). – D, the hexactinellid *Euplectella jovis* Schmidt (Tartar Bank (south of Cat Island), The Bahamas, depth 610 m, October 1990; Johnson-Sea-Link II, dive 2826). – Scale bars: 5 μm.

away from the cell body and new water flows between the microvilli into the funnel. Particles in the water are retained at the outer side of the mechanical filter formed by the collar and become ingested by the cell body.

Cells with a ring of shorter or longer microvilli surrounding one or more cilia are known from most metazoan phyla. Certain types, often with long microvilli of various specializations, function in excretion (cyrtocytes of the protonephridia; see for example Kümmel 1962 and Brandenburg 1966), while the types with shorter microvilli have various functions (Cantell, Franzén & Sensenbaugh 1982). One type of such cells lines coelomic cavities, for example in the tube feet of echinoderms (Fig. 3.5) where they generate circulation in the coelomic fluid probably aiding gas exchange. Mucus cells of a similar type have been reported from cnidarian larvae, and the nematocytes containing the nematocysts have a modified cilium surrounded by microvilli (Fig. 3.5). Absorbing functions have been postulated for other cells of this type, but this has never been proved. Many of these cells are known to be sensory (see for example Rieger 1976), and some ganglion cells have a small complex with a cilium surrounded by inverted microvilli (Westfall 1988; Fig. 3.5). It is important to note that these cells have functions different from those of the choanocytes and consequently different structures; all the just-mentioned choanocyte-like cells have microvilli which are not retractile and which, at least in most cases, have various supporting intracellular structures. Contractile microvilli containing actin, like those of the collared units, occur on intestinal cells of many animals (Remane, Storch & Welsch 1989), but they form a thick 'brush border' and are not engaged in particle capture. Many cell types can, thus, form various types of microvillar structures, and it is highly questionable to propose a homology between all the cell types which have the microvilli arranged in a circle. I would prefer to restrict the term choanocyte to cover only the structures found in choanoflagellates and sponges. Also Cantell, Franzén & Sensenbaugh (1982) suggested a restriction of the term choanocyte to cover only the feeding structures found in choanoflagellates and sponges, and they proposed to use the collective term 'collar cells' for all cells with a ring of microvilli around one or more cilia; this change in terminology would remove the unclarity of many comparative discussions and is highly recommended.

The several synapomorphies of the metazoan phyla (see Table 3.1) strongly indicate that the Animalia is a monophyletic group, and the specific characters shared with the choanoflagellates make it natural to consider the two groups as sister groups. The distribution of the characters discussed above further support the phylogenetic interrelationships indicated in Fig. 3.6.

References

Bacetti, B. & B.A. Afzelius 1976. The Biology of the Sperm Cell. (Monographs in Developmental Biology, vol. 10) – S. Karger, Basel.

Bardele, C.F. 1983. Comparative freeze-fracture study of the ciliary membrane of protists and invertebrates in relation to phylogeny. – J. submicrosc. Cytol. **15**: 263-267.

Barnes, R.D. 1985. Current perspectives on the origins and relationships of lower invertebrates. – *In* S. Conway Morris, J.D. George, R. Gibson & H.M. Platts (eds): The Origins and Relationships of Lower Invertebrates, pp 360-367. Oxford Univ. Press, Oxford.

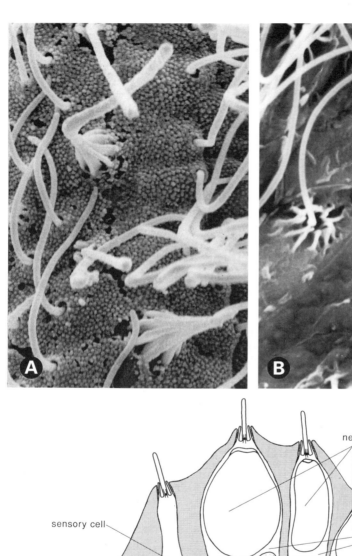

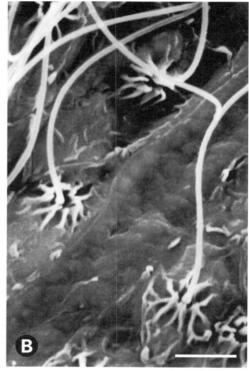

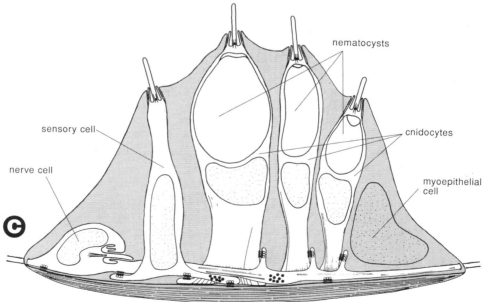

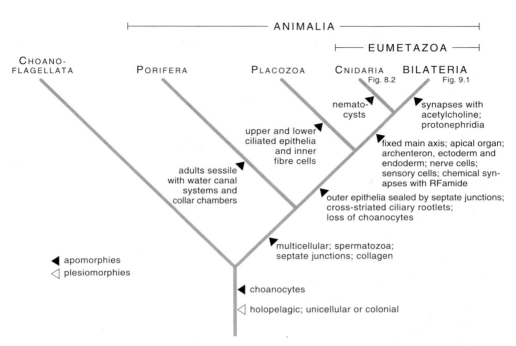

Fig. 3.6. Phylogeny of the basal animal groups (see also Table 3.1 and Fig. 41.2).

Barnes, R.D. 1987. Invertebrate Zoology, 5th ed. – Saunders, Philadelphia.

Boury-Esnault, N., L. de Vos, C. Donadey & J. Vacelet 1984. Comparative study of the choa-nosome of Porifera. I. The Homosclerophora. – J. Morph. 180: 3-17.

Brandenburg, J. 1966. Die Reusenformen der Cyrtocyten. Eine Beschreibung von fünf weiteren Reusengeisselzellen und eine vergleichende Betrachtung. – Zool. Beitr. 12: 345-417.

Brusca, R.C. & G.J. Brusca 1990. Invertebrates. – Sinauer Associates, Sunderland, MA.

Cantell, C.-E., Å. Franzén & T. Sensenbaugh 1982. Ultrastructure of multiciliated collar cells in the pilidium larva of *Lineus bilineatus* (Nemertini). – Zoomorphology 101: 1-15.

Dan-Sohkawa, M. & H. Fujisawa 1980. Cell dynamics of the blastulation process in the starfish, *Asterina pectinifera*. – Dev. Biol. 77: 328-339.

Darwin, C. 1859. On the Origin of Species by Means of Natural Selection. – John Murray, London.

Franzén Å. 1987. Spermatogenesis. – In A.C. Giese, J.S. Pearse & V.B. Pearse (eds): Reproduction of Marine Invertebrates, vol. 9, pp 1-47. Blackwell/Boxwood, Pacific Grove, CA.

Garrone, R. 1978. Phylogenesis of Connective Tissue. (Frontiers in Matrix Biology, vol. 5.) – S. Kar-ger, Basel.

Fig. 3.5. Collar cells which are not choanocytes. – A, sensory laterofrontal cells on a tentacle of the actinotrocha larva of a phoronid (plankton, off Phuket Marine Biological Center, Thailand, March 1982). – B, cells lining the coelom in a tube foot of the sea-urchin *Strongylocentrotus* (SEM courtesy of Drs M.A. Cahill and E. Florey, Univ. Konstanz). – C, diagram of an epitheliomuscular cell of a *Hydra* tentacle with cnidocytes, a sensory cell and a ganglion cell which all have collars (modified from Westfall 1988, Holstein & Hausmann 1988 and Westfall & Kinnamon 1984). – Scale bars A, B: 5 μm.

Green, C.R. 1984. Intercellular junctions. – *In* J. Bereiter-Hahn, A.G. Matoltsy & S.K. Richards (eds): Biology of the Integument, vol. 1. Invertebrates, pp 5-16. Springer, Berlin.

Green, C.R. & P.R. Bergquist 1979. Cell membrane specialisations in the Porifera. – Colloques int. Cent. natn. Res. scient. **291**: 153-158.

Green, C.R. & P.R. Bergquist 1982. Phylogenetic relationships within the Invertebrata in relation to the structure of septate junctions and the development of 'occluding' junctional types. – J. Cell Sci. **53**: 279-305.

Gruner, H.-E. 1980. Einführung. – *In* H.-E. Gruner (ed.): A. Kaestner's Lehrbuch der speziellen Zoologie (4. ed.), 1. Band, 2. Teil, pp 15-156. Gustav Fischer, Stuttgart.

Haeckel, E. 1866. Generelle Morphologie der Organismen. 2 vols. – Georg Reimer, Berlin.

Haeckel, A. 1874. Die Gastraea-Theorie, die phylogenetische Classification des Thierreichs und die Homologie der Keimblätter. – Jena. Z. Naturw. **8**: 1-55, pl. 1.

Hibberd, D.J. 1975. Observations on the ultrastructure of the choanoflagellate *Codosiga botrytis* (Ehr.) Saville-Kent with special reference to the flagellar apparatus. – J. Cell Sci. **17**: 191-219.

Holley, M.C. 1982. The control of anthozoan cilia by the basal apparatus. – Tissue Cell **14**: 607-620.

Holstein, T. & K. Hausmann 1988. The cnidocil apparatus of hydrozoans: a progenitor of higher metazoan mechanoreceptors. – *In* D.A. Hessinger & H.M. Lenhoff (eds): The Biology of Nematocysts, pp 53-73. Academic Press, San Diego.

Huebner, E. & S. Caveney 1987. Invertebrate cell junctions. – *In* A.H. Greenberg (ed.): Invertebrate Models. Cell Receptors and Cell Communications, pp 190-219. Karger, Basel.

Johnston, I.S. & W.H. Hildemann 1982. Cellular organization in the marine demosponge *Callyspongia diffusa*. – Mar. Biol. (Berl.) **67**: 1-7.

Junqua, S., J. Fayolle & L. Robert 1975. Structural glycoproteins from sponge intercellular matrix. – Comp. Biochem. Physiol. 50B: 305-309.

Kazmierczak, J. 1981. The biology and evolutionary significance of Devonian volvocaceans and their Precambrian relatives. – Acta palaeont. Polon. **26**: 299-337, pls 26-31.

Kozloff, E.N. 1990. Invertebrates. – Saunders College Publishing, Philadelphia.

Kümmel, G. 1962. Zwei neue Formen von Cyrtocyten. Vergleich der bischer bekannten Cyrtocyten und Erörterung des Begriffes 'Zelltyp'. – Z. Zellforsch. **57**: 172-201.

Ledger, P.W. 1975. Septate junctions in the calcareous sponge *Sycon ciliatum*. – Tissue Cell **7**: 13-18.

Linnaeus, C. 1735. Systema Naturæ sive Regna Tria Naturæ systematice proposita per Classæ, Ordines, Genera, & Species. – Theod. Haack, Lugdunum Batavorum.

Linnaeus, C. 1758. Systema Naturæ, 10th edition. 10 vols. – Laurentius Salvius, Stockholm.

Lyons, K.M. 1973. Collar cells in planula and adult tentacle ectoderm of the solitary coral *Balanophyllia regia* (Anthozoa Eupsammiidae). – Z. Zellforsch. **145**: 57-74.

Mackie, G.O. 1984. Introduction to the diploblastic level. – *In* J. Bereiter-Hahn, A.G. Matoltsy & S.K. Richards (eds): Biology of the Integument, vol. 1. Invertebrates, pp 43-46. Springer, Berlin.

Mackie, G.O. 1990. The elementary nervous system revisited. – Am. Zool. **30**: 907-920.

Mackie, G.O., P.A.V. Anderson & C.L. Singla 1984. Apparent absence of gap junctions in two classes of Cnidaria. – Biol. Bull. Woods Hole **167**: 120-123.

Mackie, G.O. & C.L. Singla 1983. Studies on hexactinellid sponges. I. Histology of *Rhabdocalyptus dawsoni* (Lambe, 1873). – Phil. Trans. R. Soc. B **301**: 365-400, 12 pls.

Margulis, L. 1980. Undulipodia, flagella and cilia. – BioSystems **12**: 105-108.

Margulis, L. 1981. Symbiosis in Cell Evolution. – W.H. Freeman, San Francisco.

Margulis, L. 1990. Kingdom Animalia: the zoological malaise from a microbiological perspective. – Am. Zool. **30**: 861-875.

Margulis, L. & K.V. Schwartz 1988. Five Kingdoms. An Illustrated Guide to the Phyla of Life on Earth. 2nd ed. – W.H. Freeman, New York.

Martin, F. 1914. Zur Entwicklungsgeschichte des polyembryonalen Chalcidiers *Ageniaspis* (*Encyrtus*) *fuscicollis* Dalm. – Z. wiss. Zool. **110**: 414-479, pls 15-16.

Moestrup, Ø. 1982. Flagellar structure in algae: a review, with new observations particularly on the Chrysophyceae, Phaeophyceae (Fucophyceae), Euglenophyceae, and *Reckertia*. – Phycologia **21**: 427-528.

Möhn, E. 1984. System und Phylogenie der Lebewesen, vol. 1. – E. Schweitzerbart'sche Verlags-buchhandlung, Stuttgart.

Nielsen, C. 1985. Animal phylogeny in the light of the trochaea theory. – Biol. J. Linn. Soc. **25**: 243-299.

Nielsen, C. 1987. Structure and function of metazoan ciliary bands and their phylogenetic significance. – Acta zool. (Stockh.) **68**: 205-262.

Nielsen, C. 1991. The origin of the Metazoa. – *In* E.C. Dudley (ed.): The Unity of Evolutionary Biology, pp 445-446. Dioscorides Press, Portland, OR.

Pavans de Ceccatty, M. 1985. Les éponges et la pluricellularité des métazoaires. – Année biol. **24**: 275-288.

Pearse, V., J. Pearse, M. Buchsbaum & R. Buchsbaum 1987. Living Invertebrates. – Blackwell Scientific Publ., Palo Alto, CA.

Remane, A., V. Storch & U. Welsch 1989. Kurzes Lehrbuch der Zoologie, 6. ed. – Gustav Fischer, Stuttgart.

Rieger, R.M. 1976. Monociliated epidermal cells in Gastrotricha: significance for concepts of early metazoan evolution. – Z. zool. Syst. Evolutionsforsch. **14**: 198-226.

Ruthmann, A., G. Behrendt & R. Wahl 1986. The ventral epithelium of *Trichoplax adhaerens* (Placozoa): Cytoskeletal structures, cell contacts and endocytosis. – Zoomorphology **106**: 115-122.

Schartl, M. & A. Barnekow 1982. The expression in eukaryotes of a tyrosine kinase which is reactive with pp60^{v-src} antibodies. – Differentiation **23**: 109-114.

Schram, F.R. 1991. Cladistic analysis of metazoan phyla and the placement of fossil problematica. – *In* A.M. Simonetta & S. Conway Morris (eds): The Early Evolution of Metazoa and the Significance of Problematic Taxa, pp 35-46. Cambridge Univ. Press, Cambridge.

Siewing, R. 1985. Subregnum: Metazoa. – *In* R. Siewing (ed.): H. Wurmbach's Lehrbuch der Zoologie (3. ed.), vol. 2, pp 112-125. Gustav Fischer, Stuttgart.

Taylor, F.J.R. 1978. Problems in the development of an explicit hypothetical phylogeny of the lower eukaryotes. – BioSystems **10**: 67-89.

Towe, K.M. 1981. Biochemical keys to the emergence of complex life. – *In* J. Billingham (ed.): Life in the Universe, pp 297-306. MIT Press, Cambridge, MA.

van den Biggelaar, J.A.M. & P. Guerrier 1983. Origin of spatial organization. – *In* K.M. Wilbur (ed.): The Mollusca, vol. 3, pp 179-213. Academic Press, New York.

Westfall, J.A. 1988. Presumed neuronematocyte synapses and possible pathways controlling discharge of a battery of nematocysts in *Hydra*. – *In* D.A. Hessinger & H.M. Lenhoff (eds): The Biology of Nematocysts, pp 41-51. Academic Press, San Diego.

Westfall, J.A. & J.C. Kinnamon 1984. Perioral synaptic connections and their possible role in the feeding behavior of *Hydra*. – Tissue Cell **16**: 355-365.

Whittaker, R.H. 1969. New concepts of kingdoms of organisms. – Science **163**: 150-160.

Wood, R.L. 1985. The use of *Hydra* for studies of cellular ultrastructure and cell junctions. – Archs Sci., Genève **38**: 371-383.

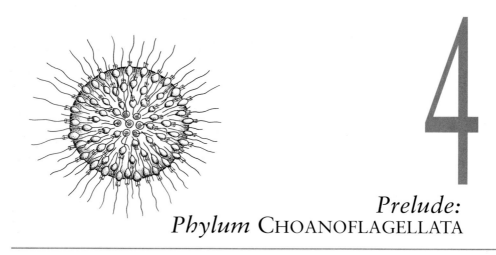

Prelude:
Phylum Choanoflagellata

The choanoflagellates are a small phylum, only about 140 species have been described, of unicellular, solitary or colony-forming 'flagellates' without chloroplasts (the two observations of choanoflagellates with chloroplasts are now considered misinterpretations; Hibberd 1986). The cells of some of the colonial species are united by cytoplasmic bridges, which have not been studied in detail but which appear to be remains of incomplete cell divisions so that the colonies are actually plasmodia; cell junctions have not been reported. The colonies are platelike or spherical; the more well-known spherical type, such as *Sphaeroeca* (see the chapter vignette) has the collar complexes on the outer side of the sphere, but *Diaphanoeca* (Fig. 4.1) has the collar complexes facing an internal cavity, and the colonies resemble free-swimming collar chambers of a sponge.

Choanoflagellates occur in most aquatic habitats and are either pelagic or sessile. The apparently most primitive species are naked or ensheathed in a gelatinous envelope, but a large group of marine and brackish-water species have an elaborate lorica consisting of siliceous costae united into an elegant bell-shaped meshwork (Fig. 3.4).

The ovoid cell body has a circle of 15-50 microvilli or tentacles forming a funnel surrounding a long undulating cilium (Fig. 3.4). The microvilli are retractile and contain actin, and the whole structure is held in shape by a mucous or fibrillar meshwork (Leadbeater 1983b). The single cilium (usually called flagellum) is usually much longer than the collar and has an extracellular vane consisting of two layers of parallel microfibrils forming angles of 65-70° with the long axis of the cilium (observed only in a few species such as *Codosiga botrytis*; Petersen & Hansen 1954 and Hibberd 1975). The cilium forms sinusoidal waves travelling towards the tip of the cilium pumping water between the microvilli; bacteria and other food items are retained on the outside of the collar and become engulfed by pseudopodia formed at the base of the collar (Leadbeater 1977, 1983b). In pelagic species, the ciliary activity is locomotory too. The basal body of the ciliary axoneme is surrounded by an intricate system of radiating microtubules and there is an accessory centriole, which is oriented at right

Chapter vignette: *Sphaeroeca volvox.* (Redrawn from Leadbeater 1983a.)

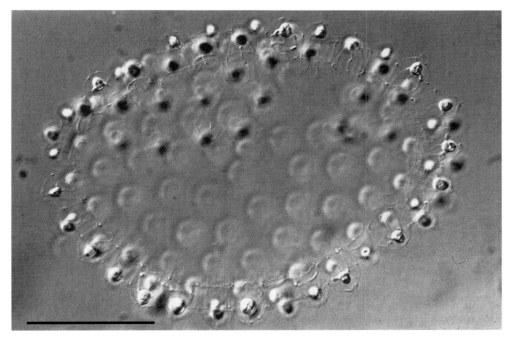

Fig. 4.1. A colony of *Diaphanoeca spherica* Thomsen; the colony is a hollow sphere with the collared units on the inside – as in the collared chambers of the sponges (plankton, Isefjord, Denmark, February 1985; interference contrast photo by Dr Helge Thomsen, Univ. Copenhagen). – Scale bar: 50 μm.

angles to the basal body in some species (Fig. 3.2); a striated rootlet as that found in the eumetazoans has not been observed (Hibberd 1975). The ciliary necklace shows a pattern quite distinct from that of the metazoans (Bardele, in Nielsen 1991).

The mitochondria have flattened cristae like those of most metazoans and chlorophytes but unlike those of chromophytes and most protozoans, and since there is no positive evidence supporting a closer relationship with the green plants, Leadbeater & Manton (1974, p. 274) concluded that the choanoflagellates can 'be treated as animals, related to sponges in any way that zoologists may decide'.

Reproduction is by binary fission; sexual processes have not been observed but since a complicated life cycle with different cell and colony types has only recently been discovered in one species (Leadbeater 1983a) it cannot be excluded that sexual reproduction occurs. It is not known whether choanoflagellates are haploid or diploid.

It is generally stated that collagen does not occur in unicellular organisms (Garrone 1978), but I am not aware of any specific investigations of choanoflagellates.

The list of synapomorphies between choanoflagellates and metazoans discussed in Chapter 3 (see also Table 3.1) may of course turn out to be incomplete if it turns out that the choanoflagellates have sexual reproduction. However, already the present knowledge of the detailed cytological similarities between choanoflagellates and various types of metazoan cells speaks strongly in favour of regarding Choanoflagellata and Animalia (Metazoa) as sister groups. Also the isolated position of the choanoflagellates among the protists (Corliss 1987) can be taken to support this interpretation.

Interesting subjects for future research

1. Life cycles: does sexual reproduction occur?
2. Chromosome numbers: haploid/diploid.
3. Does collagen occur?

References

Corliss, J.O. 1987. Protistan phylogeny and eukaryogenesis. – Int. Rev. Cytol. **100**: 319-370.
Garrone, R. 1978. Phylogenesis of Connective Tissue (Frontiers of Matrix Biology, vol. 5.) – S. Karger, Basel.
Hibberd, D.J. 1975. Observations on the ultrastructure of the choanoflagellate *Codosiga botrytis* (Ehr.) Saville-Kent with special reference to the flagellar apparatus. – J. Cell Sci. **17**: 191-219.
Hibberd, D.J. 1986. Ultrastructure of the chrysophyceae – phylogenetic implications and taxonomy. – *In* J. Christiansen & R.A. Andersen (eds): Chrysophytes: Aspects and Problems, pp 23-36. Cambridge Univ. Press, Cambridge.
Leadbeater, B.S.C. 1977. Observations on the life-history and ultrastructure of the marine choanoflagellate *Choanoeca perplexa* Ellis. – J. mar. biol. Ass. U.K. **57**: 285-301, 5 pls.
Leadbeater, B.S.C. 1983a. Life-history and ultrastructure of a new marine species of *Proterospongia* (Choanoflagellida). – J. mar. Biol. Ass. U.K. **63**: 135-160.
Leadbeater, B.S.C. 1983b. Distribution and chemistry of microfilaments in choanoflagellates, with special reference to the collar and other tentacle systems. – Protistologia **19**: 157-166.
Leadbeater, B.S.C. & I. Manton 1974. Preliminary observations on the chemistry and biology of the lorica in a collared flagellate (*Stephanoeca diplocostata* Ellis). – J. mar. biol. Ass. U.K. **54**: 269-276, 2 pls.
Nielsen, C. 1991. The origin of the Metazoa. – *In* E.C. Dudley (ed.): The Unity of Evolutionary Biology, pp 445-446. Dioscorides Press, Portland, OR.
Petersen, J.B. & J.B. Hansen 1954. Electron microscope observations on *Codonosiga botrytis* (Ehr.) James-Clark. – Bot. Tidsskr. **51**: 281-291.

Phylum PORIFERA

The sponges are a rather small, highly characteristic phylum of aquatic metazoans; about 5000 living species are recognized. The fossil record goes back to the Upper Vendian about 600 m.y. ago, but only some types of sponges fossilize easily, and the phylum may be considerably older.

The general morphology of the adult sponge with a system of branched canals with choanocyte chambers transporting water through the body, which has neither mouth nor anus, gives the phylum a very isolated position among the animals, and there is still discussion about their inclusion in the Metazoa.

Until recently, the presence of choanocytes was considered the unique character of all sponges, but the new knowledge on the ultrastructure of the hexactinellids (Mackie & Singla 1983) has made it necessary to modify the defining character to presence of collared units rising from single cells or large syncytia. The Porifera are now usually divided into two subphyla: Symplasma (consisting of the class Hexactinellida) and Cellularia (comprising the living classes Calcarea and Demospongiae; see also Van Soest 1991) and it has been necessary to distinguish between these two groups in the descriptions of most characters. The following description of hexactinellids is based on the papers on *Rhabdocalyptus* by Mackie, Lawn & Pavans de Ceccatty (1983), Mackie & Singla (1983), and Reiswig & Mackie (1983) and on *Farrea* by Reiswig & Mehl (1991) and Mehl & Reiswig (1991) unless otherwise stated.

The cells of adult cellularian sponges belong to differentiated cell types, such as choanocytes, myocytes, sclerocytes, porocytes, or spongocytes (Bergquist 1978), but contrary to the non-reversible differentiation of almost all cell types in the eumetazoans (Chapter 8) some cell types in sponges are able to de-differentiate and some types may even re-differentiate into other types; this is well established for choanocytes which de-differentiate and become oocytes or spermatocytes. Other cell types become amoeboid if the sponge is dissociated, and the cells move around and rearrange into a new sponge; however, such cells may re-differentiate into their original type.

Chapter vignette: *Euplectella aspergillum*. (Redrawn from Schulze 1887).

The body of a cellularian sponge is surrounded by the pinacoderm, a layer of epithelial cells which are not connected by cell junctions (Mackie 1984). Pores in the pinacoderm may lead directly to extensive canals lined by choanocytes (the *Ascon*-type of the Calcarea) or to pinacoderm-lined canals which open into better-defined choanocyte chambers. Excurrent canals are likewise lined by pinacoderm. The hexactinellid sponges consist mainly of syncytia, and a pinacoderm cannot be recognized.

The collared units of choanocytes/choanosyncytia have a funnel- or tube-shaped collar consisting of about 20-40 long microvilli (Fig. 3.4), the whole structure being stabilized by a fine meshwork of mucus or extracellular fibrillae; the microvilli contain a core of microfilaments (probably of actin; Bergquist 1978). The collar surrounds a long cilium, which may have an accessory centriole at the base but lacks striated ciliary rootlets (Brill 1973, Simpson 1984; Fig. 3.2). An extracellular structure in the shape of a pair of lateral wings forming a vane on the basal portion of the cilium has now been reported from all major groups (Brill 1973, Simpson 1984, Mehl & Reiswig 1991). The undulating movements of the cilium + vane propels water away from the cell body, thus creating a current between the microvilli into the funnel. The general orientation of the collared units in the collared chambers and their water currents create a flow of water through the sponge. Particles are captured both by pinacocytes in the incurrent canals and by the sieves formed by the collars. Particles captured by the collars become ingested by pseudopodia formed from an area around the funnel and may be passed on to mesenchyme for digestion (Langenbruch 1985), but both choanocytes, archaeocytes and pinacocytes are capable of endocytozing and digesting particles (Willenz & Van de Vyver 1984). Choanocytes are sometimes connected by septate junctions (Green & Bergquist 1979), but septate junctions have not been observed for example in freshwater sponges (pers. comm., Dr N. Weissenfels, Univ. Bonn).

In hexactinellids the collared units rise from the fine reticular choanosyncytium and protrude through openings in an overlying, more robust meshwork called the trabecular tissue, which is believed to be continuous through the whole sponge (it is absent in *Dactylocalyx*; Reiswig 1991). The small cytoplasmic masses which bear a collared unit (collar bodies) lack nuclei and are isolated from the common syncytium by peculiar intracellular plugs (perforate septal partitions), which resemble some structures described from the placozoan *Trichoplax*. It appears that the collared units together with their basal cytoplasmic masses degenerate periodically and become replaced through budding from the choanosyncytium. Septate junctions occur between the collar bodies and the adjacent trabecular tissue (Mackie & Singla 1983).

Pinacoderm cells and choanocytes together surround the internal tissue, mesohyl, which consists of several cell types imbedded in a matrix with collagen fibrils; a basal lamina is lacking (Garrone 1978, Pedersen 1991). The hexactinellids have a mesolamella between syncytial layers, and this lamella consists mainly of collagen. The skeleton of most sponges is secreted by cells of the mesohyl. The spicules are secreted in small lumina surrounded by sclerocytes, which are in some cases connected by septate junctions (Ledger 1975; Fig. 3.1). In the cellularians some cell types contain parallel actin filaments and can contract (Pavans de Ceccatty 1981); the hexactinellids apparently lack contractile elements.

Special sensory cells, nerve cells conducting electrical impulses and gap junctions have not been observed in any type of sponge (Bergquist 1978, Mackie 1984, 1990), but the hexactinellids conduct electrical impulses along the syncytia, and these impulses arrest the activity of the cilia. Acetylcholin and cholinesterase have been reported to occur in demosponges, with the cholinesterase restricted to the myocytes (Bergquist 1978), but there is no evidence that they should be involved in cell communication, and similar molecules are found also in plants (Mackie 1990). It appears that the myocytes form a network of contractile cells which can conduct stimuli and thereby coordinate for example rhythmic activity (Reiswig 1971). The nature of the conduction is not known, but Mackie (1990) suggested that direct mechanical stimulation of stretch-sensitive channels may produce ion fluxes which activate neighbouring cells. This mechanism is obviously not an early stage in the evolution of the nervous system known from the eumetazoans.

In the cellularians, eggs develop from archaeocytes or from choanocytes which lose their cilium and collar and move away from the choanocyte chambers; the spermatozoa develop from choanocytes (or whole choanocyte chambers become transformed into spermatogonia) (Bergquist 1978, Diaz & Connes 1980, Paulus 1989). Several species have spermatozoa with the elements of a typical metazoan spermatozoon (Reiswig 1983). The sperm is shed into the exhalant channels and expelled from the sponge. When spermatozoa enter another sponge they become trapped by choanocytes, which then shed the collar and cilium and transport the single sperm head to an egg. In the calcareans, the future main axis of the larva is perpendicular to the choanocyte layer with the anterior pole in contact with the choanocytes; the polar bodies are given off at the equator (Tuzet 1970).

Development to the mature larval state takes place inside the maternal sponge in almost all species, but a few demosponges shed the eggs in gelatinous masses (Levi 1956); it is not known if the eggs are fertilized before or after spawning in these species.

In some of the calcareans (Franzén 1988; Fig. 5.1), cleavage leads to the formation of a coeloblastula having unciliated cells at the pole facing the maternal choanocytes and monociliated cells at the opposite half, with the cilia at the interior side of the sphere. An opening then forms between the unciliated cells and the blastula turns inside out. Other species show a direct development of an embryo with ciliated cells at the outer side. The embryo now becomes compact, either by elongation of the ciliated cells and immigration of some of the unciliated cells, or through immigration of ciliated cells. The larvae are released from the maternal sponge and swim with the ciliated pole in front. The cilia beat with the normal effective stroke and an accessory centriole has been observed at the base of each cilium; there is a ciliary rootlet in the shape of a tight bundle of non-striated fibrils but also a system of radiating fibrils (Borojevic 1969, Gallissian 1983). At settling, the ciliated cells invaginate and the larvae settle with the anterior pole against the substratum. The large cavity thus formed cannot be compared to an archenteron (Lemche & Tendal 1977).

In the demosponges, the cleavage leads more directly to the formation of a compact or hollow embryo with monociliate cells covering the whole surface except the posterior pole. The larvae swim for a short period and their cilia show the effective-

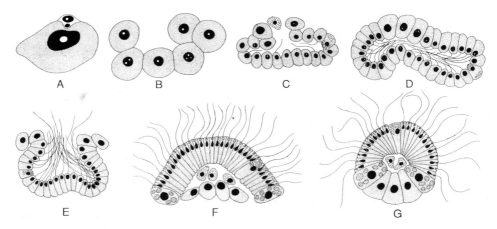

Fig. 5.1. Embryology of the calcarean sponge *Scypha ciliata*; transverse sections; the upper sides are facing the collar chamber. – A, fertilization with the sperm cell still inside the carrier cell (above). – B, 8-cell stage. – C, young blastula. – D, blastula with cilia on the inside. – E, inversion of the blastula which brings the cilia to the outer side and in contact with the cavity of the collar chamber through an opening in the choanocyte layer. – F, amphiblastula; the ciliated half of the larva bulges into the collar chamber. – G, free-swimming larva. (Modified from Franzen 1988.)

stroke beating pattern characteristic of all planktonic larvae, with conspicuous metachronal waves (Green & Bergquist 1979; Fig. 5.2). At least some demosponge larvae lack the accessory centriole and the cilia are anchored directly to a bundle of slightly diverging fibrils (Nielsen 1987; Fig. 3.2). The larvae have a clear polarity, but the anterior pole does not show a sensory organ. Internally, the larvae are more or less filled with cells, some of which have already started to secrete spicules, and small choanocyte chambers may already be present. The inner cells have migrated from the outer cell layer, and there is never any sign of a gastrulation by invagination. The larvae settle with the anterior pole, and a re-arrangement of the cell layers takes place; it has been assumed that the ciliated cells become the choanocytes, but new investigations have shown that this is not the case at least in some demosponges, where the choanocytes differentiate from archaeocytes (Misevic & Burger 1982, Bergquist & Glasgow 1986).

Sexual reproduction and larval development of the hexactinellids are poorly known, but Okada (1928) described spermatozoa with a rounded head and a cilium and cleavage stages with blastomeres as in the cellular sponges. Later stages had spicules and cells arranged around what he believed would become choanocyte chambers. It is not known if the larvae are ciliated, and the formation of the syncytia has not been studied.

Asexual reproduction through budding, fission and the special resting stages called gemmulae are important in the life cycles of many demosponges.

Reiswig & Mackie (1983) treated the hexactinellids and the cellular sponges as two subphyla, Symplasma and Cellularia, but expressed uncertainty about the monophyly of the Porifera, and Bergquist (1985) regarded the hexactinellids as a separate phylum. There are obvious differences between the organization of the two groups,

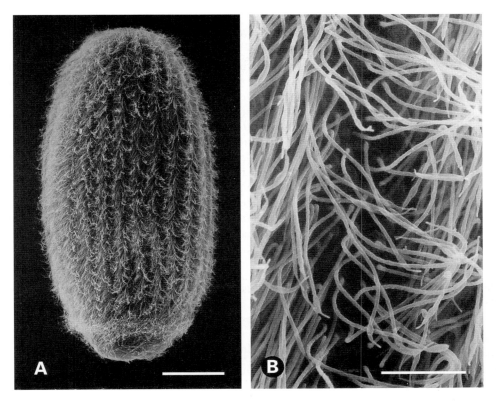

Fig. 5.2. Larva of the demosponge *Haliclona* sp. – A, a whole larva. – B, detail of the ciliated surface. (From Nielsen 1987.) – Scale bar A: 50 μm; scale bar B: 5 μm.

and it can indeed be speculated that the two groups evolved from different choanoflagellate-like ancestors. However, if the two groups evolved multicellularity independently, septate junctions and collagen should have evolved convergently in the two groups.

Several questions (some of which are listed below) must be answered before a well founded choice between the possible phylogenies can be made, but I have chosen to follow the conservative view which regards the two groups as subphyla of the Porifera, and this phylum in turn as the sister group of Placozoa + Eumetazoa.

The choanocytes/choanosyncytia which characterize the sponges so clearly among the metazoans are obviously a plesiomorphic character, because the choanoflagellates have the same essential structure as a choanocyte. However, several apomorphies characterize the sponges (although knowledge about these characters is in several cases lacking for the hexactinellids): The multicellularity combined with the sporadic occurrence of septate junctions, and the lack of basal lamina can be regarded as intermediate grades in a transformation series from the colonial grade of the choanoflagellates to the fully integrated multicellular grade of the eumetazoans. The architecture of the adult sponges with the branched water canals instead of a gut with mouth and anus is an unquestionable autapomorphy. The transport of the spermatozoon to the egg by a choanocyte and the equatorial position of the polar bodies may be further autapomorphies.

According to the gastraea/trochaea theories, the sponges represent the evolutionary stage of the blastaea, but since all species are sessile it is to be expected that the ancestral body plan cannot be recognized in the adults. However, a blastula stage is commonly recognized during ontogeny, while the gastrulation-like processes which occur during the settling of several species do not lead to the formation of blastopore and archenteron, and there is no cell layer which can be identified as endoderm. An antero-posterior axis is clearly established in the larva and the metachronal pattern of its ciliary movements resembles that of many lecithotrophic larvae of spiralians and deuterostomes (Nielsen 1987). However, there is no sense organ at the anterior pole and the ciliated outer layer of cells should not be homologized with the ciliated ectoderm of the bilaterian larvae. Also the lack of a basal lamina is characteristic of the grade of the blastaea.

Interesting subjects for future research

1. Sexual reproduction of the hexactinellids: spermatozoa, fertilization, polar bodies, cleavage, larval structure, formation of syncytia, etc.
2. Ciliary basal structures in different types of larvae and adults.
3. Comparisons between formation of the skeletal elements in the two subphyla.
4. Cell communication in the two subphyla.

References

Bergquist, P.R. 1978. Sponges. – Hutchinson, London.
Bergquist, P.R. 1985. Poriferan relationships. – *In* S. Conway Morris, J.D. George, R. Gibson & H.M. Platt (eds): The Origins and Relationships of Lower Invertebrates, pp 14-27. Oxford Univ. Press, Oxford.
Bergquist, P.R. & K. Glasgow 1986. Developmental potential of ciliated cells of ceractinomorph sponge larvae. – Exp. Biol. **45**: 111-122.
Borojevic, R. 1969. Étude du développement et de la différenciation cellulaire d'éponges calcaires calcinéennes (genres *Clathrina* et *Ascandra*). – Annls Embryol. Morphogen. 2: 15-36.
Brill, B. 1973. Untersuchungen zur Ultrastruktur der Choanocyte von *Ephydatia fluviatilis* L. – Z. Zellforsch. **144**: 231-245.
Diaz, J.-P. & R. Connes 1980. Étude ultrastructurale de la spermatogenèse d'une Démosponge. – Biol. cell. **38**: 225-230.
Franzen, W. 1988. Oogenesis and larval development of *Scypha ciliata* (Porifera, Calcarea). – Zoomorphology **107**: 349-357.
Gallissian, M.-F. 1983. Étude ultrastructurale du développement embryonnaire chez *Grantia compressa* F (Porifera, Calcarea). – Archs Anat. micr. Morph. exp. **72**: 59-75.
Garrone, R. 1978. Phylogenesis of Connective Tissue. (Frontiers of Matrix Biology, vol 5.) – S. Karger, Basel.
Green, C.R. & P.R. Bergquist 1979. Cell membrane specializations in the Porifera. – Colloques int. Cent. natn. Res. scient. **291**: 153-158.
Langenbruch, P.-F. 1985. Die Aufnahme partikulärer Nahrung bei *Reniera* sp. (Porifera). – Helgoländer wiss. Meeresunters. **39**: 263-272.

Ledger, P.W. 1975. Septate junctions in the calcareous sponge *Sycon ciliatum.* – Tissue Cell **7**: 13-18.

Lemche, H. & O. Tendal 1977. An interpretation of the sex cells and the early development in sponges, with a note on the terms acrocoel and spongocoel. – Z. zool. Syst. Evolutionsforsch. **15**: 241-252.

Levi, C. 1956. Étude des *Halisarca* de Roscoff. Embryologie et systématique des Démosponges. – Archs Zool. exp. gén. **93**: 1-181.

Mackie, G.O. 1984. Introduction to the diploblastic level. – *In* J. Bereiter-Hahn, A.G. Matoltsy & K.S. Richards (eds): Biology of the Integument, vol. 1, pp 43-46. Springer Verlag, Berlin.

Mackie, G.O. 1990. The elementary nervous system revisited. – Am. Zool. **30**: 907-920.

Mackie, G.O., I.D. Lawn & M. Pavans de Ceccatty 1983. Studies on hexactinellid sponges. II. Excitability, conduction and coordination of responses in *Rhabdocalyptus dawsoni* (Lambe, 1873). – Phil. Trans. R. Soc. B **301**: 401-418.

Mackie, G.O. & C.L. Singla 1983. Studies on hexactinellid sponges. I. Histology of *Rhabdocalyptus dawsoni* Lambe, 1873). – Phil. Trans. R. Soc. B **301**: 365-400, 12 pls.

Mehl, D. & H.M. Reiswig 1991. The presence of flagellar vanes in choanomeres of Porifera and their possible phylogenetic implications. – Z. zool. Syst. Evolutionsforsch. **29**: 312-319.

Misevic, G.N. & M.M. Burger 1982. The molecular basis of species specific cell-cell recognition in marine sponges, and a study on organogenesis during metamorphosis. – Prog. clin. biol. Res. **85B**: 193-209.

Nielsen, C. 1987. Structure and function of metazoan ciliary bands and their phylogenetic significance. – Acta zool. (Stockh.) **68**: 205-262.

Okada, Y. 1928. On the development of a hexactinellid sponge, *Farrea sollasii.* – J. Fac. Sci. Tokyo Imp. Univ., Sect. **4**, 2: 1-27, pls 1-8.

Paulus, W. 1989. Ultrastructural investigations of spermatogenesis in *Spongilla lacustris* and *Ephydatia fluviatilis* (Porifera, Spongillidae). – Zoomorphology **109**: 123-130.

Pavans de Ceccatty, M. 1981. Demonstration of actin filaments in sponge cells. – Cell Biol. internatl Rep. **5**: 945-952.

Pedersen, K.J. 1991. Invited review: Structure and composition of basement membranes and other basal matrix systems in selected invertebrates. – Acta zool. (Stockh.) **72**: 181-201.

Reiswig, H.M. 1971. Particle feeding in natural populations of three marine demosponges. – Biol. Bull. Woods Hole **141**: 568-591.

Reiswig, H.M. 1983. Porifera. – *In* K.G. Adiyodi & R.G. Adiyodi (eds): Reproductive Biology of Invertebrates, vol. 2, pp 1-21. John Wiley, Chichester.

Reiswig, H.M. 1991. New perspectives on the hexactinellid genus *Dactylocalyx* Stutchbury. – *In* J. Reitner & H. Keupp (eds.): Fossil and Recent Sponges, pp 7-20. Springer-Verlag, Berlin.

Reiswig, H.M. & G.O. Mackie 1983. Studies on hexactinellid sponges III. The taxonomic status of Hexactinellida within the Porifera. – Phil. Trans. R. Soc. B **301**: 419-428.

Reiswig, H.M. & D. Mehl 1991. Tissue organization of *Farrea occa* (Porifera, Hexactinellida). – Zoomorphology **110**: 301-311.

Schulze, F.E. 1887. Report on the Hexactinellida. – Rep. scient. Results Voy. Challenger **53** (Zool. 23): 1-514, 105 pls.

Simpson, T.L. 1984. The Cell Biology of Sponges. – Springer-Verlag, New York.

Tuzet, O. 1970. La polarité de l'oeuf et la symmétrie de la larve des éponges calcaires. – *In* W.G. Fry (ed.): The Biology of the Porifera (Symp. zool. Soc. Lond. 25), pp 437-448. Academic Press, London.

Van Soest, R.W.M. 1991. Demosponge higher taxa classification re-examined. – *In* J. Reitner & H. Keupp (eds): Fossil and Recent Sponges, pp 54-71. Springer-Verlag, Berlin.

Willenz, P. & G. Van de Vyver 1984. Ultrastructural localization of lysosomal digestion in the freshwater sponge *Ephydatia fluviatilis.* – J. Ultrastr. Res. **87**: 13-22.

41

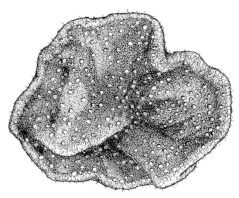

6

Phylum PLACOZOA

The phylum Placozoa comprises only one known species, *Trichoplax adhaerens*; an additional species recorded only once is generally regarded as a misinterpretation. The usually 1-2 mm large, rounded, flat organism creeps on algae and has been recorded from warm waters from many parts of the world.

Trichoplax consists of one layer of epithelial cells surrounding a rather narrow, flat space containing a meshwork of fibre cells (Grell & Ruthmann 1991); the fibre cells contain actin filaments and are therefore undoubtedly responsible for the sometimes quite rapid changes of shape (Thiemann & Ruthmann 1989). The epithelial cells are diploid (2n = 12) and the fibre cells tetraploid; the cells can be dispersed after various treatments and re-aggregate to apparently normal individuals with epithelial cells surrounding fibre cells (Ruthmann & Terwelp 1979).

The epithelium of the lower side (the side facing the substratum) consists of rather tall cells of two types: ciliated cells and glandular cells. The ciliated cells have one cilium, which rises from a pit with a ring of supporting rods; the basal complex comprises a long, cross-striated rootlet, short lateral rootlets, and a perpendicular accessory centriole (Ruthmann, Behrendt & Wahl 1986; Fig. 3.2). The glandular cells are filled with secretion droplets. The upper epithelium consists of flat monociliate cells and spectacular cells with refringent inclusions originating from degenerating cells. The epithelial cells are connected by belt desmosomes and by what looks like septate desmosomes (Ruthmann, Behrendt & Wahl 1986; Fig. 3.1).

The fibre cells are here and there connected by small disc-shaped, osmiophilic structures which may be temporary, but which in some cases appear to be intracellular plugs surrounded by the cell membrane (Grell & Benwitz 1974b).

The food consists of small algae or protozoans, but larger organisms may be ingested too. Small cells become digested extracellularly by the lower epithelium where indications of endocytosis are seen in the ciliated cells (Ruthmann, Behrendt & Wahl 1986). Other food items may be transported to the upper side by the cilia and become transported through the epithelium to the fibre cells, where digestion occurs (Wenderoth 1986).

Chapter vignette: *Trichoplax adhaerens*. (Drawing based on Rassat & Ruthmann 1979.)

Asexual reproduction is by fission or by formation of spherical swarmers from the upper side. The swarmers are surrounded by cells resembling the upper epithelium and have a central cavity lined by cells resembling the lower cells; fibre cells occur between the two layers of ciliated cells. The spheres open at one side and stretch out so that the normal upper and lower epithelia become established (Thiemann & Ruthmann 1988).

Sexual reproduction is incompletely known. In cultures, one or a few large eggs have been observed to develop in each animal (Fig. 6.1); they differentiated from a cell of the lower epithelium and became surrounded by a layer of fibre cells which functioned as nurse cells. Meiosis and spermatozoa were not observed. Some of the egg cells formed a fertilization membrane and started dividing, but the embryos soon degenerated (Grell 1972, Grell & Benwitz 1974a).

The placozoans are here regarded as sister group of the eumetazoans (Fig. 3.6) because they have epithelial cells connected by belt desmosomes and structures which resemble septate junctions, whereas cell junctions between cells of an outer 'ectoderm' are lacking in the sponges (Mackie 1984), and the cilia have cross-striated ciliary rootlets like those of all the eumetazoans.

Interesting subjects for future research

1. Sexual reproduction (spermatozoa, cleavage, cell differentiation, larva).
2. Cell contacts and cell communication.

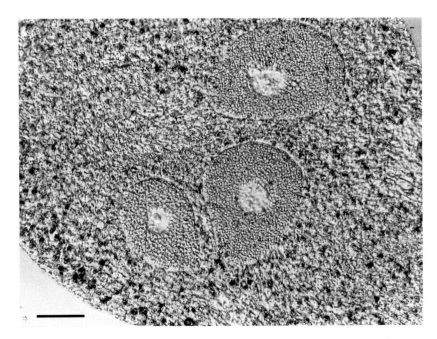

Fig. 6.1. Trichoplax adhaerens. Detail of a specimen with oocytes. (Photomicrograph courtesy of Dr K.G. Grell, Univ. Tübingen, FRG; see also Grell 1972.) – Scale bar: 50 µm.

3. Ciliary necklace.
4. Is collagen present?

References

Grell, K.G. 1972. Eibildung und Furchung von *Trichoplax adhaerens* F.E. Schultze (Placozoa). – Z. Morph. Tiere **73**: 297-314.

Grell, K.G. & G. Benwitz 1974a. Elektronenmikroskopische Untersuchungen über das Wachstum der Eizelle und die Bildung der 'Befruchtungsmembran' von *Trichoplax adhaerens* F.E. Schulze (Placozoa). – Z. Morph. Tiere **79**: 295-310.

Grell, K.G. & G. Benwitz 1974b. Spezifische Verbindungsstrukturen der Faserzellen von *Trichoplax adhaerens* F.E. Schulze. – Z. Naturforsch. **29**c: 790-790a.

Grell, K.G. & A. Ruthmann 1991. Placozoa. – *In* F.W. Harrison (ed.): Microscopic Anatomy of Invertebrates, vol. 2, pp 13-27. Wiley-Liss, New York.

Mackie, G.O. 1984. Introduction to the diploblastic level. – *In* J. Bereiter-Hahn, A.G. Matoltsy & K.S. Richards (eds): Biology of the Integument, vol. 1, pp 43-46. Springer, Berlin.

Rassat, J. & A. Ruthmann 1979. *Trichoplax adhaerens* F.E. Schulze (Placozoa) in the scanning microscope. – Zoomorphology **93**: 59-72.

Ruthmann, A., G. Behrendt & R. Wahl 1986. The ventral epithelium of *Trichoplax adhaerens* (Placozoa): Cytoskeletal structures, cell contacts and endocytosis. – Zoomorphology **106**: 115-122.

Ruthmann, A. & U. Terwelp 1979. Disaggregation and reaggregation of cells of the primitive metazoan *Trichoplax adhaerens*. – Differentiation **13**: 185-198.

Thiemann, M. & A. Ruthmann 1988. *Trichoplax adhaerens* F.E. Schulze (Placozoa): the formation of swarmers. – Z. Naturforsch. **43**c: 955-957.

Thiemann, M. & A. Ruthmann 1989. Microfilaments and microtubules in isolated fiber cells of *Trichoplax adhaerens* (Placozoa). – Zoomorphology **109**: 89-96.

Wenderoth, H. 1986. Transepithelial cytophagy by *Trichoplax adhaerens* F.E. Schulze (Placozoa) feeding on yeast. – Z. Naturforsch. **41**c: 343-347.

EUMETAZOA (= GASTRAEOZOA)

The eumetazoans constitute a very well defined group which is considered mono-phyletic by almost all authors. Haeckel (1866) pointed out that, in contrast to the sponges, the eumetazoans (his Animalia) are organized with tissues which form or-gans, but it may be a semantic question whether for example the choanocyte cham-bers should be called organs.

Much more significant are the facts that all eumetazoans consist of characteristic cell layers which rest on basement membrane and that the high integration of the cells and tissues has led to the development of a number of special cell types, for example sensory cells and nerve cells, and a number of cytological specializations not known in the 'lower' animals. Another significant apomorphy is the establishment of a primary, antero-posterior axis through the anterior apical sense organ and the posterior blastopore (see below).

As in the sponges, the ontogeny of the eumetazoans leads from the zygote through various embryological and larval stages before the adult form is reached. In species with free spawning the cleavage often results in the formation of a sphere of cells, each with one or several cilia on the outer surface. The sphere may consist of only one layer of cells surrounding a cavity (a coeloblastula with the blastocoel) or compact (a sterroblastula). Septate junctions form between the cells of the late blas-tula in several groups (Chapter 3; Fig. 3.1), and a basement membrane has also been observed in coeloblastulae (Wessel, Marchase & McClay 1984).

The subsequent ontogenetic stage is a gastrula often formed through an invagina-tion of one side of a coeloblastula (Fig. 7.1). This process leads to a differentiation of two areas of the ciliated epithelium, viz. an outer epithelium, ectoderm, which retains the locomotive function, and an inner epithelium, endoderm, which surrounds the archenteron and becomes the gut epithelium. The two cell layers, often called the pri-mary germ layers, are only connected around the edge of the blastopore.

The ciliated gastrulae swim with the pole opposite the blastopore in front, and a concentration of sensory cells with long cilia, the apical organ (Fig. 7.2), is found in almost all types; this organ can be recognized in almost all free larvae, and it is an im-portant marker for the orientation of the ontogenetic stages. Primitive nervous cells

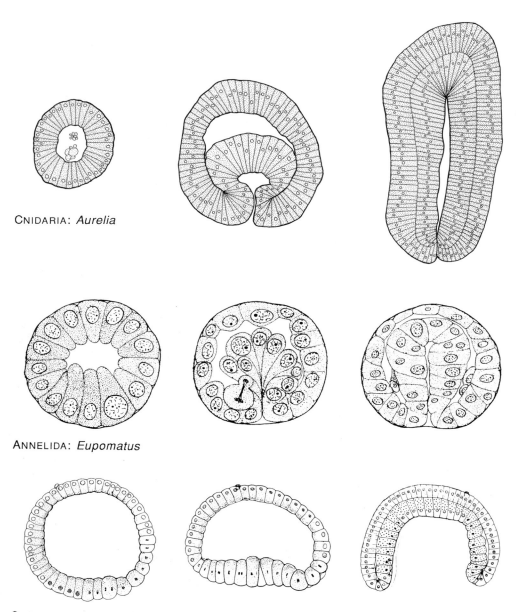

CNIDARIA: *Aurelia*

ANNELIDA: *Eupomatus*

CEPHALOCHORDATA: *Branchiostoma*

Fig. 7.1. Gastrulation in major metazoan groups. – The cnidarian *Aurelia aurita* (redrawn from Hein 1900). – A protostome: the annelid *Eupomatus uncinatus* (from Shearer 1911). – A deuterostome: the cephalochordate *Branchiostoma lanceolatum* (modified from Conklin 1932).

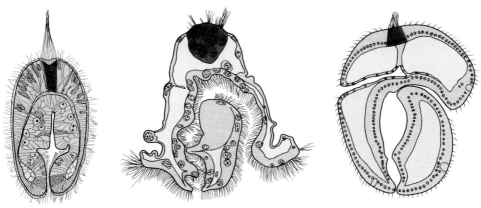

CNIDARIA: *Aiptasia* ENTOPROCTA: *Loxosoma* ENTEROPNEUSTA: *Balanoglossus*

Fig. 7.2. Longitudinal sections of larvae showing the apical organs (dark shading). – Planula larva (gastrula) of the cnidarian *Aiptasia mutabilis* (modified from Wiedersten 1968). – Trochophora larva of the entoproct *Loxosoma pectinaricola* (after Nielsen 1971). – Tornaria larva of the enteropneust *Balanoglossus clavigerus* (redrawn from Stiasny 1914).

in connection with the apical organ can be recognized in cnidarian larvae of this stage (Chia & Crawford 1977) and in slightly older stages of echinoderm larvae (Burke 1983). The blastocoel often becomes completely obliterated so that the basement membrane of ectoderm and endoderm become closely apposed.

Adult cnidarians are organized as gastrulae, while the remaining eumetazoans, the Bilateria, go through a gastrula stage but differentiate further by developing a third cell layer, the mesoderm, surrounded by the basement membrane of the two primary cell layers.

Archenteron and blastopore directly become gut and mouth in the cnidarians (in which the mouth simultaneously functions as anus). In most bilaterians the archenteron develops into the mid-gut and the fate of the blastopore varies between the major groups.

The early ontogeny just described is considered characteristic of all eumetazoans, but in fact there is an enormous variation both in the processes leading from one stage to the next and between comparable stages of different species.

It has been mentioned already that the blastulae may be hollow or solid, and the processes leading to a two-layered stage with or without an archenteron show many variations. The inner cell-layer may be formed through invagination, delamination or ingression, or through combinations of these (Fig. 7.3). It is important to notice, however, that this variation can be observed within many phyla, and the variation can even be observed within each of the major cnidarian classes (Tardent 1978). This shows that presence or absence of one developmental type cannot be used in phylogenetic argumentation. The hollow blastula and the invagination leading to a gastrula with an archenteron are observed within many phyla and are presumed to be primitive.

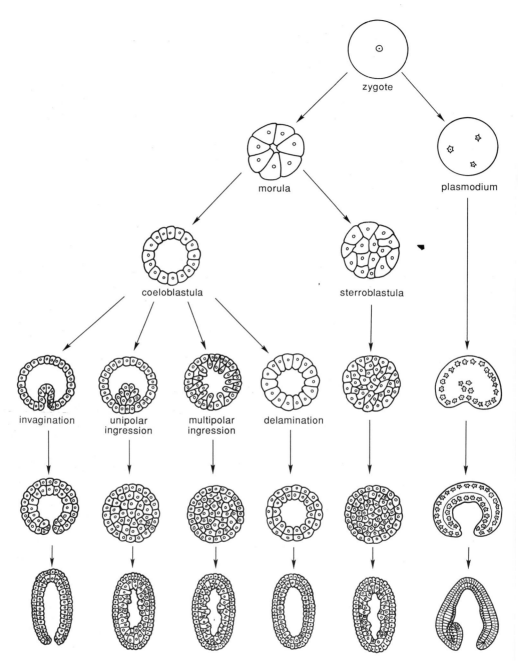

Fig. 7.3. Gastrulation types in the Cnidaria; diagrams of the main types of gastrulation following the holoblastic cleavage are shown, together with one of the several types of plasmodial development. (Modified from Tardent 1978.)

48

The higher integration of the cells and tissues of the eumetazoans is expressed in a number of characteristic cell types and in ultrastructural and biochemical characteristics which are regarded as synapomorphies (see Table 3.1).

Special sensory cells of many different types occur in epithelia of all eumetazoans. Photoreceptor cells were once believed to be of high phylogenetic importance, with a 'ciliary' type characteristic of the cnidarians and deuterostomes and a 'rhabdomeric' type characteristic of the protostomes, but both types have now been found in several phyla (Coomans 1981).

The nervous systems control and coordinate the various activities of the eumetazoans and make quick adaptive responses to external events possible. Action potentials pass rapidly along the axons and from one cell to another via gap junctions (see below) and with a small delay through synapses. A synapse (Fig. 7.4) is a specialized area of a nerve cell where neurotransmitters are secreted from small vesicles into the narrow space separating it from a neighbouring cell; the neurotransmitter excites the neighbouring cell and is immediately broken down or re-absorbed so that the excitation stops. Various RFamides function as neurotransmitters in almost all eumetazoans, whereas the acetylcholine/cholinesterase system is found only in the synapses of the bilaterians (Grimmelikhuijzen et al. 1987, Eldefrawi & Eldefrawi 1988).

In the cnidarians, the nervous system is partially organized as a nerve net, in which some of the synapses are symmetric, i.e. they can transmit both ways (Jha & Mackie 1967). This is often considered a primitive character, but could just as well be a specialization (Chapter 8).

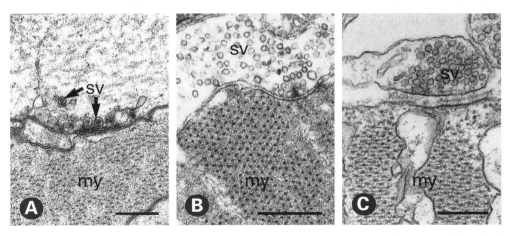

Fig. 7.4. Neuromuscular synapses in cnidarians, protostomes and deuterostomes. – A, synapse between a nerve cell and a myoepithelial cell of the scyphozoan *Haliclystus auricula* (TEM courtesy of Dr J. Westfall, Kansas State Univ., Manhattan, KA, USA; see also Westfall 1973). – B, synapse between a nerve cell and a limb muscle of the shore crab *Pachygrapsus crassipes* (collected near Los Angeles, CA, USA; TEM courtesy of Dr A.L. Atwood, Univ. Toronto, Canada). – C, synapse between a nerve cell and a sonic swimbladder muscle of the fish *Porichthys notatus* (TEM courtesy of Dr A.H. Bass, Cornell Univ., NY, USA; see also Lindholm & Bass 1993). – my, cross-sectioned myofibrils – sv, synaptic vesicles in nerve cell. – Scale bars: 0.5 μm.

Gap junctions are specialized areas of contact between cell membranes which permit diffusion of small molecules (up to about 1000 dalton) and conduction of action potentials without a synapse between neighbouring cells. Each junction is supposed to be formed by six ca. 27 000-dalton polypeptide molecules arranged in a ring in each cell membrane, together forming an intercellular channel which can be opened and closed (Hertzberg 1985). There seems to be some variation in size between structures found in the different phyla, but variation due to different methods of study may account for some of this variation (Huebner & Caveney 1987). Gap junctions are formed as early as at the 2-cell stage in some molluscs (van den Biggelaar *et al.* 1981), and are believed to be important in regulating developmental processes (Caveney 1985).

The gap junctions have been considered a synapomorphy of all eumetazoans, but it now appears that they do not occur in anthozoans and scyphozoans. This is nicely

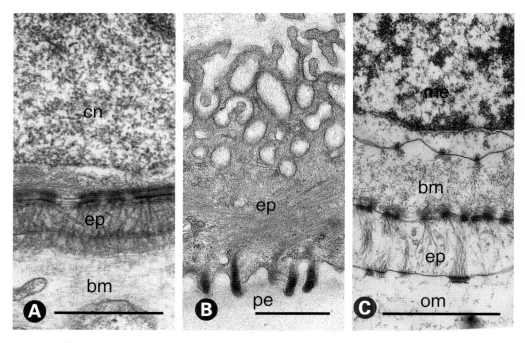

Fig. 7.5. Hemidesmosomes in cnidarians and a bilaterian. – A, hemidesmosomes and tonofibrils in the connection between a cnidocyte, an epithelial cell and the basement membrane in a hydra (TEM courtesy of Dr R.D. Campbell, Univ. California, Irvine, CA, USA; see Campbell 1987). – B, hemidesmosomes between epithelial cells and the perisarc in the hydrozoan polyp *Cordylophora caspia* (TEM courtesy of Dr B.A. Marcum, California State Univ., Chico, CA, USA; see Marcum & Diehl 1978). – C, hemidesmosomes and tonofibrils in the connections between a mesodermal cell – the basement membrane – an epithelial cell – and the organic matrix of the calcified skeleton in the ectoproct *Crisia eburnea* (see Nielsen & Pedersen 1979). – bm, basement membrane – cn, cnidocyte – ep, epithelial cell – me, mesodermal cell – om, organic matrix of calcified wall – pe, perisarc. – Scale bars: 1 µm.

correlated with differences in electrophysiology between the hydrozoans and the other cnidarians (Mackie, Anderson & Singla 1984). It is still being debated whether the hydrozoans are the most 'primitive' cnidarians, which will imply that the gap junctions have been lost in the 'higher' cnidarians, or whether the gap junctions have evolved independently in hydrozoans and bilaterians (this is discussed further in Chapter 8).

The basement membrane is a thin layer of organic material comprising laminin, fibronectin and various types of collagen (Pedersen 1991). It is found in almost all eumetazoans, the only exceptions being some platyhelminths (Nemertodermatoida and Acoela, see Ehlers 1985 and Chapter 26).

Firm connections between ectodermal cells and intercellular matrix on the basal side of the cell and the organic matrix or other extracellular coverings on the apical side may be secured by hemidesmosomes (often connected through intercellular tonofibrils). These structures have been observed in cnidarians and most other eumetazoan phyla and appear to be an apomorphy of the eumetazoans (Fig. 7.5).

Choanocytes with the characteristic ring of contractile microvilli are not found in any eumetazoan, but representatives of most phyla show various types of monociliate cells with the cilium surrounded by a circle of microvilli. None of these cells are involved in particle collection and, as argued above (Chapter 3), they should preferably be called collar cells rather than choanocytes.

The gastrula with a locomotory ectoderm and a digestive endoderm separated by basement membrane, an archenteron in which larger particles can be digested, and an apical sensory organ with nerves originating from the sensory cells is an ontogenetical stage not present in the 'lower' animals, and it must be considered a highly important set of synapomorphies of the Eumetazoans; the presence of hemidesmosomes in the same groups further corroborates this interpretation.

Both larval and adult cnidarians and larval stages of most 'higher' metazoans are thus of the same structure as the hypothetical ancestor gastraea, and the gastraea theory, which is at the same time the early part of the trochaea theory, appears very well founded.

References

Burke, R.D. 1983. Development of the larval nervous system of the sand dollar, *Dendraster excentricus*. – Cell Tissue Res. **229**: 145-154.

Campbell, R.D. 1987. Organization of the nematocyst battery in the tentacle of hydra: arrangement of the complex anchoring junctions between nematocytes, epithelial cells, and basement membrane. – Cell Tissue Res. **249**: 647-655.

Caveney, S. 1985. The role of gap junctions in development. – Ann. Rev. Physiol. **47**: 319-335.

Conklin, E.G. 1932. The embryology of *Amphioxus*. – J. Morph. **54**: 69-151.

Chia, F.-S. & B. Crawford 1977. Comparative structural studies of planulae and primary polyps of identical age of the sea pen, *Ptilosarcus gurneyi*. – J. Morph. **151**: 131-158.

Coomans, A. 1981. Phylogenetic implications of the photoreceptor structure. – Atti Conv. Lincei **49**: 162-174.

Ehlers, U. 1985. Das phylogenetische System der Plathelminthes. – Gustav Fischer, Stuttgart.

Eldefrawi, A.T. & M.E. Eldefrawi 1988. Acetylcholine. – In G.G. Lunt & R.W. Olsen (eds): Comparative Invertebrate Neurochemistry, pp 1-41. Croom Helm, London.

Grimmelikhuijzen, C.J.P., D. Graff, A. Groeger & I.D. McFarlane 1987. Neuropeptides in Invertebrates. – *In* M.A. Ali (ed.): Invertebrate Nervous Systems (NATO ASI, Ser. A 141), pp 105-132. Plenum Press, New York.

Haeckel, E. 1866. Generelle Morphologie der Organismen. 2 vols. – Georg Reimer, Berlin.

Hein, W. 1900. Untersuchungen über die Entwicklung von *Aurelia aurita*. – Z. wiss. Zool. **67**: 401-438, pls 24-25.

Hertzberg, E.L. 1985. Antibody probes in the study of gap junctional communication. – Ann. Rev. Physiol. **47**: 305-318.

Huebner, E. & S. Caveney 1987. Invertebrate cell junctions. – *In* A.H. Greenberg (ed.): Invertebrate Models: Cell Receptors and Cell Communication, pp 190-219. Karger, Basel.

Jha, R.K. & G.O. Mackie 1967. The recognition, distribution and ultrastructure of hydrozoan nerve elements. – J. Morph. **123**: 43-62.

Lindholm, M.M. & A.H. Bass 1993. Early events in myofibrillogenesis and innervation of skeletal sound-generating muscle in a teleost fish. – J. Morph. **216**: 225-239.

Mackie, G.O., P.A.V. Anderson & C.L. Singla 1984. Apparent absence of gap junctions in two classes of Cnidaria. – Biol. Bull. Woods Hole **167**: 120-123.

Marcum, B.A. & F.A. Diehl 1978. Anchoring filaments (desmocytes) in the hydrozoan polyp *Cordylophora*. – Tissue Cell **10**: 113-124.

Nielsen, C. & K.J. Pedersen 1979. Cystid structure and protrusion of the polypide in *Crisia* (Bryozoa, Cyclostomata). – Acta zool. (Stockh.) **60**: 65-88.

Pedersen, K.J. 1991. Invited review: Structure and composition of basement membranes and other basal matrix systems in selected invertebrates. – Acta zool. (Stockh.) **72**: 181-201.

Shearer, C. 1911. On the development and structure of the trochophore of *Hydroides uncinatus* (*Eupomatus*). – Q. Jl microsc. Sci., N.S. **56**: 543-590, pls 21-23.

Stiasny, G. 1914. Studien über die Entwicklung des *Balanoglossus clavigerus* Delle Chiaje. I. Die Entwicklung der Tornaria. – Z. wiss. Zool. **110**: 36-75, pls 4-6.

Tardent, P. 1978. Coelenterata, Cnidaria. – *In* F. Seidel (ed.): Morphogenese der Tiere, Deskriptive Morphogenese, 1. Lieferung, pp 69-415. VEB Gustav Fischer, Jena.

van den Biggelaar, J.A.M., A.W.C. Dorresteijn, S.W. de Laat & J.G. Bluemink 1981. The role of topographical factors in cell interaction and determination of cell lines in molluscan development. – *In* G.H. Schweiger (ed.): International Cell Biology 1980-1981, pp 526-538. Springer, Berlin.

Wessel, G.M., R.B. Marchase & D.R. McClay 1984. Ontogeny of the basement membrane in the sea urchin embryo. – Dev. Biol. **103**: 235-245.

Westfall, J.A. 1973. Ultrastructural evidence for neuromuscular systems in coelenterates. – Am. Zool. **13**: 237-246.

Widersten, B. 1968. On the morphology and development in some cnidarian larvae. – Zool. Bidr. Upps. **37**: 139-182.

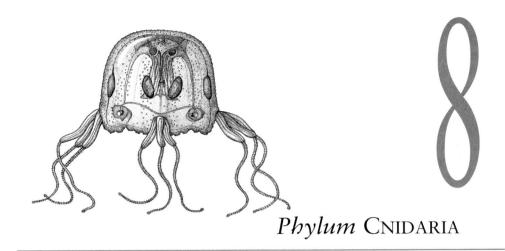

Phylum CNIDARIA

Cnidarians are a well defined phylum comprising about 8000 living species, which occur in aquatic, mainly marine habitats. The oldest fossils date back to the late Precambrian (Conway Morris 1985).

The most conspicuous apomorphy of the phylum is the presence of nematocysts (cnidae), which are highly complicated structures formed inside special cells called cnidocytes (= nematoblasts) (Werner 1984, Mariscal 1984). The cnidocytes differentiate from interstitial cells. The nematocyst is formed as a small cup-shaped structure inside the cell; it increases in size and becomes pear-shaped; a long hollow thread forms from the narrow end of the capsule and this thread, which in most types has rows of spines at the base, finally invaginates and becomes coiled up inside the capsule. The nematocyst is now fully formed and when the cell has reached the position where it is to function, the final differentiation of the cell with the cnidocil takes place; when the nematocyst explodes the tube everts again (Fig. 8.1). A few other organisms contain nematocysts, but in these cases the nematocysts originate from cnidarian prey organisms, which may be identified by studying the 'stolen nematocysts' (cleptocnidia). Some protozoan groups have structures which superficially resemble nematocysts, but these structures are clearly not homologous with those of the cnidarians (Robson 1985). The colloblasts of the ctenophores are cells of a completely different structure (Chapter 42).

Another characteristic which sets the cnidarians apart from all other eumetazoans is their primitive (plesiomorphic) main structure which in principle is that of a gastrula, i.e. only an outer cell layer, ectoderm (epidermis), and an inner cell layer, endoderm (gastrodermis), the latter surrounding the digestive cavity (archenteron); the blastopore functions both as mouth and anus. The two cell layers are separated by a basement membrane, which in several types is elaborated to form a gelatinous mesogloea. The symmetry is generally radial, but bilaterality is seen in certain hydroids (for example *Branchiocerianthus* and many siphonophores) and the anthozoans are clearly bilateral; a head with a brain is not developed (probably because of

---

Chapter vignette: The cubomedusa *Tripedalia cystophora*. (Redrawn after Werner 1973.)

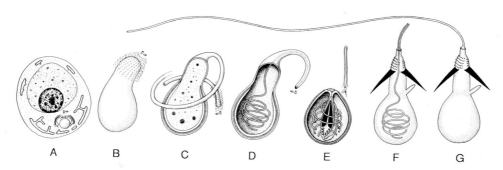

Fig. 8.1. Formation (A-E) and discharge (F-G) of a nematocyst. (Redrawn from Holstein 1981 and Tardent & Holstein 1982.)

the sessile or pelagic life styles), but the gastrula/planula larvae have a concentration of sensory cells at the anterior (apical) pole indicating a nervous centre (Chia & Bickell 1978). The main axis of the cnidarians is the primary, apical-blastoporal axis.

There are two main types of body structure: the sessile polyp, usually with a rather thin mesogloea and with the mouth surrounded by tentacles, and the pelagic medusa with a thick gelatinous mesogloea. Many cubozoans, scyphozoans and hydrozoans have life cycles comprising both polyp and medusa stages, while other species of these groups have only polyps or medusae; the anthozoans have only polyps. Most cnidarians have asexual reproduction with budding, which may occur both from polyps and medusae. The buds may become released, or the budding leads to the formation of colonies of species-specific shapes. Some hydrozoans, for example the siphonophores, consist of several types of units which resemble polyps and medusae and which serve different functions.

Ectodermal and endodermal cells form sealed epithelia with septate junctions and belt desmosomes (Hündgen 1984). Gap junctions (electrical synapses, which are well known in all 'higher' eumetazoans) are only described from hydrozoans where both interneuronal, neuromuscular and intermuscular gap junctions have been observed (Mackie, Anderson & Singla 1984, Mackie, Nielsen & Singla 1989). The gap junctions of *Hydra* are made up of a protein which can be recognized by antibodies to rat liver gap-junction proteins and these proteins must thus be very similar (Fraser *et al.* 1987).

Many of the cells carry one cilium with an accessory centriole; some of these cilia are locomotory while others are sensory. Scattered examples of cells with more than one cilium are known: the endodermal cells of certain hydropolyps and anthozoans have two or several cilia per cell (Bouillon 1968, Lentz 1966). The ectoderm of labial tentacles of some anthozoans comprises cells with several cilia each with an accessory centriole (Tiffon & Hugon 1977). The tentacles of the hydromedusa *Aglantha* have a pair of lateral ciliary bands which are formed from multiciliate cells (Mackie, Nielsen & Singla 1989); these ciliary bands propel water past the tentacles but do not collect particles like the ciliary bands of many bilaterian larvae. Cilia generally beat as single units, but compound cilia in the shape of wide, oblique membranelles occur in the anthozoan larva called zoanthina (Nielsen 1984).

The nervous system is intraepithelial, forming nerve nets which may be concentrated in nerve rings in both polyps and medusae and in ganglia in scyphomedusae. The apical nervous concentration in the larva disappears at metamorphosis. Special sensory cells of various types occur at characteristic positions, for example in ocelli and statocysts, in both polyps and medusae. Observations on endoderm-free embryos indicate that the nervous cells originate from the endoderm and the sensory cells from the ectoderm (Thomas, Freeman & Martin 1987). Interneuronal and neuromuscular chemical synapses of the unidirectional type (the usual type in the 'higher' metazoans) are known from all cnidarian classes (Westfall 1987). Bi-directional (symmetrical) interneuronal synapses have also been described from all cnidarian classes (Jha & Mackie 1967, Anderson 1985, Passano 1982, Westfall 1987). Acetylcholine, the neurotransmitter occurring in all bilateral metazoans, has never been observed in cnidarians, which appear to rely on transmitters such as FMRFamide only (Grimmelikhuijzen et al. 1987).

Epithelio-muscular cells occur in all groups and is the only type of muscle cell in several groups (Werner 1984); these cells occur both in the ectoderm and the endoderm. Special muscle cells (myocytes) without an epithelial portion occur in scyphozoan and cubozoan polyps; some cubozoan genera (for example *Carybdea*) have only myocytes, while *Tripedalia* has both types of contractile cells with all types of intermediary stages (Chapman 1978).

The anthomedusa *Podocoryne* has cross-striated myocytes in the subumbrellar epithelium. These cells can be induced to transdifferentiate into smooth muscle cells and to divide and give rise to nerve cells and cnidocytes; small regenerates may resemble manubria, but the further development was not followed (Schmid 1988). The significance of this apparent totipotentiality of one type of apparently fully differentiated cells is not known.

The epithelial cells are sometimes anchored both to the mesogloea and to an external cuticle, perisarc, by special cells with tonofilaments and hemidesmosomes like those of the other eumetazoans (Van-Praët 1977, Marcum & Diehl 1978; Fig. 7.5).

Ectoderm and endoderm are rather closely apposed in most scyphopolyps and hydropolyps, but in all other forms a more or less thick, gelatinous to almost cartilaginous, hyaline layer, the mesogloea, separates the two cell layers. The mesogloea consists mainly of water which is stabilized by a web of collagenous fibres; it is generally without cells in hydrozoans and in cubomedusae, but cells of various types enter the mesogloea from both ectoderm and endoderm in scyphopolyps, cubo- and scyphomedusae and especially in the anthozoans (Werner 1984). The cells of the anthozoan mesogloea enter the gelatinous matrix in the form of tubes or solid cell strings and differentiate into isolated, star-shaped cells, scleroblasts, myocytes and a number of other cell types (Tixier-Derivault 1987). It is clear from all recent descriptions that the mesogloea with its varying content of cells moving in from both ectoderm and endoderm cannot be compared to the mesoderm of the bilaterians. The 'real' mesoderm forms epithelia and other tissues in which the cells are connected by cell junctions and which are isolated from ectoderm and endoderm by basement membranes.

Eggs and sperm differentiate from endodermal cells in anthozoans, scyphozoans and cubozoans and from ectodermal cells in hydrozoans; they are usually shed freely in the water. The spermatozoa are of the primitive metazoan type (Fig. 3.3). In hydrozoans the main (apical-blastoporal) axis of the larva becomes determined already during oogenesis with the presumptive apical pole situated at the side of the egg which is in contact with the endoderm or mesogloea (Freeman 1990). Fertilization and polar body formation take place at the oral (blastoporal) pole where cleavage is also initiated (this end of the embryo is therefore called the animal pole). It appears that a similar orientation is found in anthozoans and scyphozoans (Tardent 1978) but the details are not so well documented. The position of the polar bodies at the oral pole also appears to be a cnidarian characteristic; almost all bilaterians have the polar bodies situated at the apical pole.

Cleavage is usually total, but superficial cleavage occurs in some hydrozoans. In species with total cleavage, the first cleavage cuts towards the apical pole, and a very unusual stage with the two incipient blastomeres connected with a handle-like apical bridge is reached just before the separation of the blastomeres (Tardent 1978). There is a considerable variation in the subsequent development (Tardent 1978; Fig. 7.3). Several species have the presumably primitive coeloblastula which invaginates to form a gastrula. The larvae are ciliated, often with an apical tuft of longer cilia (Fig. 7.2). Some of these larvae are planktotrophic, but the feeding structures never include ciliary bands engaged in filter feeding. Development through a feeding gastrula larva occurs in several anthozoans and scyphozoans. In other species, the coeloblastula develops into a ciliated, lecithotrophic planula larva, which consists of ectoderm and endoderm; the endoderm may be compact or surround a gut; a mouth is usually lacking in the early stages but develops before the larva settles. Many authors regard this larval type as characteristic of the cnidarians. Development leading to the planula shows much variation with a coeloblastula or a sterroblastula stage and with formation of endoderm through various types of immigration or delamination, in some cases combined with a small invagination, followed by rearrangement of cells (see Fig. 7.3). The mosaic-like occurrence of the different developmental types in the classes makes it impossible to identify the ancestral (plesiomorphic) type of development directly.

The larva usually settles with the apical pole and becomes a polyp, but the larvae of a few hydrozoans and scyphozoans develop directly into medusae, and a polyp stage is lacking. All primary polyps are capable of one or more types of asexual reproduction, for example lateral budding, often leading to the formation of colonies, frustule formation from a basal plate, or transverse fission giving rise to medusae.

The medusae are formed through different processes in the three classes. In the cubozoans the polyp goes through a metamorphosis and becomes the medusa; the tentacles of the polyp lose the nematocysts and become sense organs and new tentacles develop (Werner 1973, 1984). In the hydrozoans, the polyps typically form lateral medusa buds, which may detach as medusae or remain attached as variously reduced medusoid reproductive individuals. In the scyphozoans the medusae are formed through transverse fission (strobilation) of the polyp.

The few cubozoan species studied so far and all the anthozoans have the life cycles mentioned as characteristic of the classes, while much variation is encountered in the two other classes. The hydrozoans comprise many examples of reduction of either polyp or medusa (Petersen 1990). Some orders comprise only holopelagic species (for example the Trachylina), and holopelagic species are known in some of the other orders too. Reduction of the medusae can be followed in many transformational series from the eumedusoid stage, where the medusa is relatively little reduced, through cryptomedusoid and styloid stages to the type represented by *Hydra*, which has testes and ovaries situated in the ectoderm of the polyp. Also the scyphozoans show some variation with a few holopelagic genera, such as *Pelagia*, and the stauromedusae, which lack the medusa stage.

One of the main questions about the phylogeny of the cnidarians is whether the polyp or the medusa represents the ancestral type. There are several hydrozoans and scyphozoans which lack a polyp stage completely, and some authors have interpreted such life cycles as ancestral and accordingly regarded the polyp stage as a larval/juvenile specialization; these authors regard the hydrozoans or the scyphozoans as the most 'primitive' cnidarians. Other authors regard the polyp as the ancestral cnidarian type and interpret the medusa as a specialized sexual stage. Werner (1973) came to the conclusion that the ancestral cnidarian was a tetraradial polyp from which one line should lead to the bilateral anthozoans and another line to the radial medusozoans with medusae as the reproductive stage.

However, if the gastraea theory is accepted, there can be no doubt that the ancestral cnidarian was a holopelagic, advanced gastrula, probably with nematocysts on small tentacles. This organism must have been one of the very first metazoan carnivores. If this gastrula type attached by the apical pole, we have an organism resembling Werner's hypothetical ancestor of the cnidarians. The interpretation of the Anthozoa as the sister group of the Medusozoa (see Fig. 8.2) would be strongly supported if other distinguishing characters than just the life cycle could be identified. One such character is apparently found in the nematocysts, which lack an operculum in the anthozoans, while an operculum is present in all the three medusozoan classes (Werner 1984). Differences in the cnidocytes may exist too, but the information is incomplete. The anthozoan cnidocil is a normal cilium, while those of hydrozoans and probably scyphozoans are specialized, non-motile cilia with an elevated number of central tubuli; the cubozoans have not been studied in detail (Holstein & Hausman 1988). The anthozoans exhibit several apomorphies, including the bilateral symmetry with an ectodermal pharynx.

The evolution of the medusozoans could have gone through the following steps (Werner 1973, Fig. 8.2). 1a) Development of a tetraradial theca with four triangular peridermal flaps closing the upper opening by retraction of the corresponding interradial septal muscles. 1b) Evolution of strobilation with the medusa functioning as the sexual phase. The extinct group Conulata had the theca with the opercular flaps, but it is not known if it had strobilation, so it cannot be ascertained which of these steps came first. 2) Loss of the opercular flaps, but retention of the septal muscles; lateral budding; this is the life cycle type represented by the scyphozoans, which have polydisc strobilation. 3) Only monodisc strobilation – in extreme cases the whole polyp becomes transformed into a medusa; this is the life cycle type of the cubozoans. 4) Metagenesis – an obligatory change between the sexually reproducing medusae and the asexual polyps: the hydrozoans.

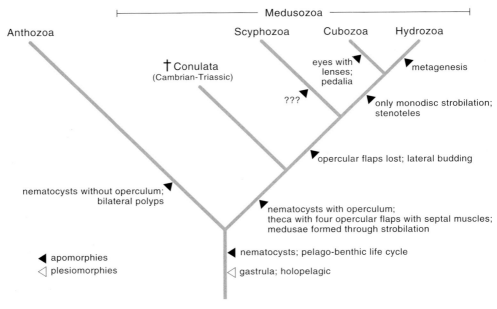

Fig. 8.2. Phylogeny of the Cnidaria.

Werner (1973) expressly described the cnidarians as a dead-end line of animal evolution. The occurrence of multiciliate cells and of compound cilia in certain cnidarians must thus be interpreted as isolated specializations, and this is also indicated by their sporadic occurrences. The Hydrozoa clearly comprise some very specialized types with highly developed nervous systems with gap junctions rendering complex behaviour possible (Mackie 1984). It appears improbable that gap junctions were present in the gastraea and lost in anthozoans, scyphozoans and cubozoans, so their presence in hydrozoans and all bilaterians must be regarded as a remarkable example of convergent molecular evolution.

The cnidarians are at the gastraea stage with only ectoderm and endoderm and have the nematocysts as the unquestionable apomorphy; they form the logical sister group of the bilaterians, which all have mesoderm and which in many phyla show larval types feeding by way of complicated ciliary bands. The lecithotrophic planula larvae can easily be interpreted as a specialization from the planktotrophic larval type; similar transitions from planktotrophic to lecithotrophic development with concurrent delay of development of the gut and various feeding structures are well known in almost all invertebrate groups.

Interesting subjects for future research

1. Comparisons between gap junctions of hydrozoans and bilaterians.
2. Ultrastructure of cubozoan and scyphozoan cnidocytes.

58

References

Anderson, A.V. 1985. Physiology of a bidirectional, excitatory, chemical synapse. – J. Neurophysiol. **53**: 821-835.

Bouillon, J. 1968. Introduction to the coelenterates. – *In* M. Florkin & B.T. Scheer (eds): Chemical Zoology, vol. 2, pp 81-147. Academic Press, New York.

Chapman, D.M. 1978. Microanatomy of the cubopolyp, *Tripedalia cystophora* (Class Cubozoa). – Helgoländer wiss. Meeresunters. **31**: 128-168.

Chia, F.-S. & L.R. Bickell 1978. Mechanisms of larval attachment and the induction of settlement and metamorphosis in coelenterates: a review. – *In* F.-S. Chia & M.E. Rice (eds): Settlement and Metamorphosis of Marine Invertebrate Larvae, pp 1-12. Elsevier, New York.

Conway Morris, S. 1985. Non-skeletized lower invertebrate fossils: a review. – *In* S. Conway Morris, J.D. George, R. Gibson & H.M. Platts (eds): The Origins and Relationships of Lower Invertebrates, pp 343-359. Oxford Univ. Press, Oxford.

Fraser, S.E., C.R. Green, H.R. Bode & N.B. Gilula 1987. Selective disruption of gap junctional communication interferes with a patterning process in *Hydra*. – Science **237**: 49-55.

Freeman, G. 1990. The establishment and role of polarity during embryogenesis in hydrozoans. – *In* D.L. Stocum & T.L. Karr (eds): The Cellular and Molecular Biology of Pattern Formation, pp 3-30. Oxford Univ. Press, Oxford.

Grimmelikhuijzen, C.J.P., D. Graff, A. Groeger & I.D. McFarlane 1987. Neuropeptides in invertebrates. – *In* M.A. Ali (ed.): Invertebrate Nervous Systems (NATO ASI, Ser. A 141), pp 105-132. Plenum Press, New York.

Holstein, T. 1981. The morphogenesis of nematocytes in *Hydra* and *Forskålia*: an ultrastructural study. – J. Ultrastruct. Res. **75**: 276-290.

Holstein, T. & K. Hausman 1988. The cnidocil apparatus of hydrozoans: a progenitor of higher metazoan mechanoreceptors. – *In* D.A. Hessinger & H.M. Lenhoff (eds): The Biology of Nematocysts, pp 53-73. Academic Press, San Diego.

Hündgen, M. 1978. The biology of colonial hydroids. I. The morphology of the polyp of *Eirene viridula* (Thecata: Campanulinidae). – Mar. Biol. (Berl.) **45**: 79-92.

Hündgen, M. 1984. Cnidaria: Cell types. – *In* J. Bereiter-Hahn, A.G. Matoltsy & K.S. Richards (eds): Biology of the Integument, vol. 1, pp 47-56. Springer, Berlin.

Jha, R.K. & G.O. Mackie 1967. The recognition, distribution and ultrastructure of hydrozoan nerve elements. – J. Morph. **123**: 43-62.

Lentz, T.L. 1966. The Cell Biology of *Hydra*. – North Holland, Amsterdam.

Mackie, G.O. 1984. Fast pathways and escape behavior in Cnidaria. – *In* R.C. Eaton (ed.): Neural Mechanisms and Startle Behavior, pp 15-42. Plenum Press, New York.

Mackie, G.O., P.A.V. Anderson & C.L. Singla 1984. Apparent absence of gap junctions in two classes of Cnidaria. – Biol. Bull. Woods Hole **167**: 120-123.

Mackie, G.O., C. Nielsen & C.L. Singla 1989. The tentacle cilia of *Aglantha digitale* (Hydrozoa: Trachylina) and their control. – Acta zool. (Stockh.) **70**: 133.141.

Marcum, B.A. & F.A. Diehl 1978. Anchoring cells (desmocytes) in the hydrozoan polyp *Cordylophora*. – Tissue Cell **10**: 113-124.

Mariscal, R.N. 1984. Cnidaria: Cnidae. – *In* J. Bereiter-Hahn, A.G. Matoltsy & K.S. Richards (eds): Biology of the Integument, vol. 1, pp 57-68. Springer, Berlin.

Nielsen, C. 1984. Notes on a *Zoanthina*-larva (Cnidaria) from Phuket, Thailand. – Vidensk. Meddr dansk naturh. Foren. **145**: 53-60.

Passano, L.M. 1982. Scyphozoa and Cubozoa. – *In* G.A.B. Shelton (ed.): Electrical Conduction and Behaviour in 'Simple' Invertebrates, pp 149-202. Oxford Univ. Press, Oxford.

Petersen, K.W. 1990. Evolution and taxonomy in capitate hydroids and medusae (Cnidaria: Hydrozoa). – Zool. J. Linn. Soc. **100**: 101-231.

Robson, E.A. 1985. Speculations on coelenterates. – *In* S. Conway Morris, J.D. George, R. Gibson & H.M. Platt (eds): The Origins and Relationships of Lower Invertebrates, pp 60-77. Oxford Univ. Press, Oxford.

Schmid, V. 1988. The potential for transdifferentiation and regeneration of isolated striated muscle of medusae in vitro. – Cell Differ. **22**: 173-182.

Tardent, P. 1978. Coelenterata, Cnidaria. – *In* F. Seidel (ed.): Morphogenese der Tiere, Deskriptive Morphogenese, 1. Lieferung, pp 69-391. VEB Gustav Fischer, Jena.

Tardent, P. & T. Holstein 1982. Morphology and morphodynamics of the stenothele nematocyst of *Hydra attenuata* Pall. (Hydrozoa, Cnidaria). – Cell Tissue Res. **224**: 269-290.

Thomas, M.B., G. Freeman & V.J. Martin 1987. The embryonic origin of neurosensory cells and the role of nerve cells in metamorphosis in *Phialidium gregarium* (Cnidaria, Hydrozoa). – Int. J. Invert. Reprod. Dev. **11**: 265-287.

Tiffon, Y. & J.S. Hugon 1977. Ultrastructure de l'ectoderme des tentacules labiaux et marginaux de *Pachycerianthus fimbriatus* McMurrich. – J. exp. mar. Biol. Ecol. **29**: 151-159.

Tixier-Derivault, A. 1987. Sous-classe des Octocoralliaires. – Traité de Zoologie, vol. 3(3): 3-185. Masson, Paris.

Van-Praët, M. 1977. Étude histocytologique d'*Hoplangia durotrix* Gosse (Anthozoa, Scleractinaria). – Annls Sci. nat., Zool., **12**, Sér. 19: 279-299.

Werner, B. 1973. New investigations on systematics and evolution of the class Scyphozoa and the phylum Cnidaria. – Publ. Seto mar. biol. Lab. **20**: 35-61.

Werner, B. 1984. Stamm Cnidaria, Nesseltiere. – *In* H.-E. Gruner (ed.): A. Kaestner's Lehrbuch der speziellen Zoologie (4th ed.), vol. 2, pp 11-305. Gustav Fischer, Stuttgart.

Westfall, J.W. 1987. Ultrastructure of invertebrate synapses. – *In* M.A. Ali (ed.): Invertebrate Nervous Systems (NATO ASI, Ser. A 141), pp 3-28. Plenum Press, New York.

9

BILATERIA

Since last century it has been customary to contrast two main groups within the Eumetazoa. One group comprises mainly radial animals organized in principle as a gastraea, i.e. consisting of only ectoderm and endoderm and with a sac-shaped gut without an anus. This group has generally been called Coelenterata and has comprised the two phyla Cnidaria and Ctenophora. I believe that the Ctenophora have mesoderm and that they are the sister group of the Notoneuralia and not of the Cnidaria (Chapter 42; see Fig. 9.1); with the Ctenophora removed, the Coelenterata become synonymous with the Cnidaria, and I have chosen to use the name Cnidaria to avoid misunderstandings. The other main group has been given a series of different names reflecting the different authors' interpretation of the 'principal' distinguishing character (synapomorphy): Triploblastica, Bilateria, Coelomata, Trochaeozoa. This disagreement reflects the fact that several 'new' characters have evolved at the root of this large metazoan group.

However, it is uncertain whether the apomorphies stressed by the previous authors are homologous throughout the bilaterians or whether they have evolved convergently in two or more of the bilaterian groups.

Most authors divide the triploblastic animals into two groups, usually called Protostomia and Deuterostomia. The distinction between these two main types is based on several characters and goes back a century; also this pair of groups has been given different names by different authors to underline the differences each author considered most important. The name pairs introduced by some important authors are:

Hatschek (1888):	Zygoneura	Ambulacralia + Chordonia
Grobben (1908):	Protostomia	Deuterostomia
Hatschek (1911):	Ecterocoelia	Enterocoelia
Cuénot (1940):	Hyponeuralia	Epineuralia
Ulrich (1951):	Gastroneuralia	Notoneuralia

Hatschek's first system was based on the trochophora theory, which emphasized the unity of the Zygoneura; Grobben's division was based on the differences in the fate of the blastopore (see below); Hatschek's second scheme was based on differences in the formation of the mesoderm and coelom; and Cuénot's and Ulrich's names refer to the

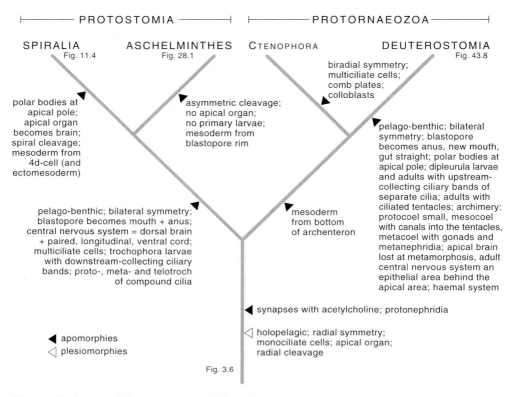

PROTOSTOMIA   PROTORNAEOZOA

SPIRALIA   ASCHELMINTHES   CTENOPHORA   DEUTEROSTOMIA
Fig. 11.4   Fig. 28.1   Fig. 43.8

biradial symmetry;
multiciliate cells;
comb plates;
colloblasts

polar bodies at   asymmetric cleavage;
apical pole;   no apical organ;
apical organ   no primary larvae;   pelago-benthic; bilateral
becomes brain;   mesoderm from   symmetry; blastopore
spiral cleavage;   blastopore rim   becomes anus, new mouth,
mesoderm from   gut straight; polar bodies at
4d-cell (and   apical pole; dipleurula larvae
ectomesoderm)   and adults with upstream-
                                                           collecting ciliary bands of
                                                           separate cilia; adults with
pelago-benthic; bilateral symmetry;   mesoderm   ciliated tentacles; archimery:
blastopore becomes mouth + anus;   from bottom   protocoel small, mesocoel
central nervous system = dorsal brain   of archenteron with canals into the tentacles,
+ paired, longitudinal, ventral cord;   metacoel with gonads and
multiciliate cells; trochophora larvae   metanephridia; apical brain
with downstream-collecting ciliary   lost at metamorphosis, adult
bands; proto-, meta- and telotroch   central nervous system an
of compound cilia   epithelial area behind the
                                                           apical area; haemal system

synapses with acetylcholine; protonephridia

apomorphies   ◁ holopelagic; radial symmetry;
◁ plesiomorphies   monociliate cells; apical organ;
                            radial cleavage

Fig. 3.6

Fig. 9.1. Phylogeny of the main groups of the Bilateria.

position of the main parts of the nervous systems. There are no nomenclatural rules for the highest systematic categories in zoology, and although I regard the names which emphasize the position of the nervous systems as those which best characterize the two groups I have chosen to use Protostomia and Deuterostomia in the following discussion because they have become so well established during the last decades.

The two groups have usually been regarded as evolved together from coelenterates, but Hatschek (1911) regarded the differences in the formation of mesoderm and coelom as so great that he pictured the two groups as being derived from different coelenterate types.

Another theory interprets the bilateral animals as consisting of the two sister groups Plathelminthomorpha and Eubilateria (see for example Ax 1987, 1989). The distinguishing character between the two groups should be that only the Eubilateria have acquired a one-way gut, while the Plathelminthomorpha are, so to speak, still at the gastraea-stage. This theory implies that spiral cleavage has evolved twice (in the Plathelminthomorpha and in the Annelida-Mollusca) or that the ancestor of all the bilateral animals had spiral cleavage. Both these alternatives appear very unlikely, and the theory will not be discussed further here (see also Chapter 11).

Theories which derive the deuterostomes from various positions within the protostomes will be discussed in the chapter on Deuterostomia (Chapter 43).

The most important characteristics of the triploblastic animals are discussed below in an attempt to evaluate information about the monophyly/polyphyly of the group.

Tube-shaped gut with mouth and anus. Almost all bilaterian phyla have members with a 'normal' gut with mouth and anus. Some parasitic groups, such as the nematomorphs, have non-functioning rudiments of a gut, while other parasites, such as the acanthocephalans and the cestodes, lack a gut completely; gutless parasitic groups occur within a number of other phyla and the lack of a gut can in all these examples with high certainty be interpreted as a specialization.

Free-living animals without an anus can be more difficult to evaluate. The articulate brachiopods all lack an anus, but as all other brachiopods have a rather normal gut, it appears most probable that the articulates are the specialized group also in this respect (Chapter 45). Most of the gnathostomulids have no anus, but a small connection between the posterior end of the gut and the dorsal epithelium may function as a temporary anus in species of *Haplognathia* and *Gnathostomula* (this is discussed in more detail in Chapter 17). Almost all free-living platyhelminths lack an anus, but one or a number of temporary or permanent openings from the intestine may function as anal pores in a few species (Ehlers 1985); this is generally seen as secondary adaptation, but the interpretation is always strongly coupled with a phylogenetic theory.

The ctenophores have a sac-shaped gut which in some species shows four small canals leading towards the apical area; two of these open on the surface and may function as anal openings (Chapter 42).

The next question is, of course, whether the tube-shaped guts of the bilaterian phyla are homologous. Their position, their general structure/function and their phylogenetic origin as specialization from the archenteron of a gastraea appear similar, but ontogenetically there appear to be two fundamentally different ways of deriving a tube-shaped gut from the sac-shaped gut of a gastraea: 1) The lateral lips of the blastopore fuse so that the blastopore becomes divided into mouth and anus (as observed in many protostomes), and 2) the blastopore becomes the anus while the mouth is formed as a new opening from the bottom of the archenteron (as observed in many deuterostomes). This interpretation is not generally accepted, but it does at least demonstrate that the alimentary canals of the two main groups of bilateral animals, viz. Protostomia and Deuterostomia, cannot *a priori* be regarded as homologous. The fate of the blastopore and the development of the adult gut in the two groups are discussed in Chapters 10 and 43.

Bilaterality. All bilaterians except the ctenophores are bilateral and it is thus almost always easy to distinguish an anterior-posterior main axis, a right and a left side, and a dorsal and a ventral side; a head with a nervous concentration, cerebral ganglion or brain, is usually a conspicuous structure at the anterior end. However, as shown in the above section about the alimentary canal, the mouth of protostomes and deuterostomes may not be homologous and the orientation of the two groups may therefore be incommensurable (see also Fig. 2.2). Since bilaterality may thus have evolved independently in the two groups it is a highly questionable synapomorphy of the bilaterians.

The ctenophores are biradial, i.e. they possess two planes of symmetry which are perpendicular and run through the apical-oral axis. This symmetry can be related neither to that of the protostomes nor to that of the deuterostomes, and there are no indications that bilaterality has ever been present. This phylum does therefore not fit into the 'Bilateria', but if my interpretation of the germ-layers is accepted, they do belong to the 'Triploblastica'. They could be treated as the sister group of the 'Bilateria', but this appears to be contradicted by their mesoderm formation (Chapters 41, 42).

Mesoderm. The tissue situated between the basement membrane of ectoderm and endoderm is called mesoderm (or the secondary germ layer). It may be organized as epithelia surrounding various types of cavities, as various types of muscles and other mesodermal organs, or it may form compact, apolar, mesenchymatous tissue. The epithelia show a pronounced polarity with a basement membrane at the basal surface and cell junctions, Golgi apparatus and in some cases one or more cilia at the apical pole. The apolar tissues lack a basement membrane and uniform orientation of the cells and have only narrow cell contacts with rather extensive intercellular matrix (Rieger 1986). The mesodermal tissues are easily distinguished from the rather isolated cells in the mesogloea of the cnidarians (Chapter 8).

The mesoderm originates through ingression or invagination from the ectoderm or the endoderm or from the blastopore region where the two primary germ layers are in contact. Three (or four) main types of mesoderm formation can be recognized: 1) ingression of ectodermal cells, 2) ingression of one or a few cells at the blastopore rim, 3) compact or hollow evaginations from the endoderm; the neural crest cells of the vertebrates may be regarded as a fourth type (see Chapter 54).

Mesoderm formed through the first-mentioned process is called ectomesoderm and has been reported from many of the spiralians, where it usually originates from the 2nd and 3rd micromere quartet (see Table 11.1). Ectomesoderm is usually not reported from aschelminths, but since the cleavage of the nematodes has now been described in great detail, it is possible that some of the cells which form muscles could be characterized as ectomesoderm (Chapter 35).

There are only few reports of ectomesoderm in the deuterostomes (except for the material from the neural crest) (Salvini-Plawen & Splechtna 1979), and most of the reports of normal ectodermal origin of parts of coeloms are connected with coelomoducts, which are discussed below. Organs which are normally formed by mesoderm and endoderm may come from the ectoderm under regeneration or budding, but these special cases are discussed under the respective phyla.

Cilia have now been reported from many different types of mesodermal tissues, such as muscle cells and coelomic epithelia, and in many phyla, such as annelids, sipunculans, and echinoderms, so it must be concluded that a single cilium is just a 'normal' organelle also of mesodermal cells. It is not an indication that the tissue is ectomesodermal.

Mesoderm formed from the blastoporal lips or from one cell located at the posterior side of the blastopore (the 4d-cell in the spiral cleavage) is found in most protostomes and has not been documented in deuterostomes. This type of mesoderm mixes freely with ectomesoderm so that the origin of various mesodermal structures from these two sources cannot be distinguished. The ectomesoderm may take over completely so that the formation of 4d-mesoderm becomes suppressed (Chapter 11).

Mesoderm formed from various parts of the archenteron is characteristic of deuterostomes, but the chaetognaths (which are here classified as aschelminths) have mesoderm originating from the archenteron (Chapter 31).

It is obvious that the cells which give rise to mesoderm have different origin in different phyla, and there is nothing which indicates that the mesoderm of all phyla is homologous.

Coelom. A coelom can simply be defined as a cavity surrounded by mesoderm. Very often, the mesoderm lining the cavity has the morphology of an epithelium with basement membrane and apical cell junctions, a peritoneum, but this is not always the case. The discussions about the origin and homologies of the mesodermal cavities in the various phyla have been very extensive, but here it may suffice to say that since the mesoderm surrounding the coelomic cavities appears to be of different origin in protostomes and deuterostomes the cavities cannot be homologous.

Only in cases where the coeloms have an identical origin and an identical, characteristic structure, as for example the endodermally derived proto-, meso- and metacoel of the deuterostomes, can one speak of well-founded homology.

The origin of the coelomic compartments has been related to the locomotory habits of the organisms, with the coeloms functioning as hydrostatic skeletons (Clark 1964). This interpretation appears very well founded and has been accepted by most authors. Fluid-filled compartments which function as hydrostatic skeletons may be primary body cavities (blastocoels; as for example in nematodes, rotifers and many pelagic larvae), intracellular vacuoles (such as in the pharynx of gymnolaemate bryozoans) or coeloms. There is nothing to indicate that the various coeloms are homologous, and this opinion is now shared by most authors (Clark 1964, Ruppert 1991).

The function of the coelomic cavities as hydrostatic organs does not necessitate the presence of coelomoducts, and the association between gonads and coeloms (see below) is probably secondary. Coelomoducts may have originated in connection with this association between gonads and coelom, so that the original function of coelomoducts is that of gonoducts.

Many textbooks follow Hyman's (1951) division of the bilaterians into acoelomates, pseudocoelomates and coelomates. However, it should be stressed that nothing is known about the morphology of the mesoderm in the ancestral bilaterians; it may have been compact, as in the parenchymians (Chapter 25), or it may, perhaps more probably, have had the shape of scattered contractile cells between ectoderm and endoderm; such an organization is seen in many larval forms today, and it can best be classified as pseudocoelomate, although the adults are in many cases coelomate. There is nothing to indicate that the acoelomate condition is ancestral.

65

Gonads. The origin of the germ cells has been studied in most phyla, and it appears that it is almost always mesodermal. The origin is not established in ctenophores, but it is believed to be from the endodermal wall of the meridional canals (Chapter 42). Chaetognaths and nematodes have clearly recognizable germ cells already at the early cleavage stages; the positions of the blastomeres in relation to the germ layers are somewhat uncertain in the aschelminthes, but the position of the gonads is clearly mesodermal in both groups (Chapters 31, 35).

Spawning of the mesodermal gametes may have been through rupture of the external body wall in the earliest triploblastic organisms, but this method is rare among extant organisms (amphioxus is one example; see Chapter 53), and the isolated examples must be interpreted as apomorphies. Gonoducts of acoelomate organisms are typically formed by fusion of an ectodermal invagination with an extension of the gonadal wall (for example in nemertines; Chapter 27). In coelomates, the gametes are shed via the coelom and the gonoducts are therefore coelomoducts. These structures are usually formed through fusion between an ectodermal invagination and an extension from the coelom (see below). Although there is thus an intimate connection between gonoducts and coelomoducts it does not imply that the secondary body cavities have evolved as gonocoels (see below).

Circulatory systems. Special fluid transport systems are absent in aschelminths and in the larvae of many of the spiralian and deuterostome phyla where diffusion and circulation of the fluid in the primary body cavity appear to be sufficient for the transport of gases and metabolites. In adults of larger compact or coelomate organisms these methods are supplemented by special transport systems, which fall into two main types: coelomic and haemal (Ruppert & Carle 1983).

Coelomic circulatory systems are lined by coelomic epithelia and the circulation is caused by cilia of the epithelia, which have now been observed in many phyla (see Ruppert & Carle 1983), or by muscles. The coelomic cavities of some of the polychaetes are short and cannot transport substances along the body, whereas other articulates have extensive coelomic cavities formed by fusion of segmental coelomic sacs so that the coelomic fluid, which contains respiratory pigments in many forms, can circulate through most of the animal. In the leeches, the coelomic cavities have been transformed into a system of narrow canals which are continuous throughout the body and which function as a circulatory system. Coelomic specializations such as the tentacle coelom with its one or two tubular, contractile 'compensation sacs' of the sipunculans and the water-vascular system of the echinoderms have respiratory functions too.

Haemal systems, often called blood vascular systems, are cavities between the basal laminae of two mesothelia, for example blood vessels in the dorsal and ventral mesenteria in annelids, or between a mesothelium and the endoderm (Rähr 1981, Ruppert & Carle 1983). This position could indicate that the blood spaces are remnants of the blastocoel, but the blood vessels arise *de novo* between cell layers in most cases (Ruppert & Carle 1983). The only major deviations from this structure appear in vertebrates, which have blood vessels with endothelial walls, in cephalopods, which have endothelium in some vessels, and in nemertines, which have a very special system of blood vessels (Chapter 27).

The blood vessels may be well defined in the whole organism, as for example in annelids, or there may be smaller or larger blood sinuses or lacunae in addition to a heart and a few larger vessels, such as in molluscs. These two types are sometimes called closed and open haemal systems, respectively, but the phylogenetic value of this distinction is dubious.

The arthropods show a very peculiar organization of coelom and haemal system. During ontogeny, parts of the coelomic sacs fuse with the haemal system so that a large mixocoel is formed.

Contractile blood vessels are found in many animals, and their muscular walls are in all cases derivatives of the surrounding mesoderm, usually coelomic walls. The well-defined hearts are small, usually paired coelomic pouches which surround a blood vessel and each have a muscular inner wall and a thin outer wall, separated by a coelomic space, the pericardial cavity, which facilitates the movements of the heart.

Unique circulatory systems of various types are found in a few phyla. Many platyhelminths have lacunae between the mesenchymatic cells and some digenean trematodes have a well defined system of a right and a left channel lined by mesodermal syncytia and surrounded by muscle cells (Strong & Bogitsch 1973). The channels have branches to the gut, the reproductive organs and the suckers and are supposed to transport nutrients between the organs (Willey 1930). Ruppert & Carle (1983) included this system in their category of coelomic circulatory systems, but there is nothing to indicate that it is homologous with other coelomic organ systems.

Carle & Ruppert (1983) interpreted the funiculus of ectoproct bryozoans as a haemal system, because it usually consists of one or more hollow strands of mesoderm where the lumen is sometimes lined by a basement membrane. It is possible that this structure transports nutrients from the gut to the testes, which are usually located on the funiculus (and to the developing statoblasts of phylactolaemates), but neither position, structure or function of the funiculus bears resemblance to blood vessels of other metazoans, so I cannot accept the idea of a homology (Chapter 24).

The rhynchocoel and the lateral channels of nemertines are discussed in Chapter 27.

The occurrence of the two main types of circulatory systems are of considerable phylogenetic importance. Coelomic circulatory systems are found for example in sipunculans, which lack a blood vascular system, while haemal systems are found in most other bilateral animals which lack a coelomic system. The echinoderms are a remarkable exception in that they have both a haemal system and a coelomic circulatory system – the water-vascular system clearly transports for example oxygen from the podia.

The morphology of the haemal system may also be used in phylogenetic considerations in cases where the blood vessels are parts of more complicated organs, such as the axial organ of a large group of deuterostomes, the Neorenalia (Chapter 46).

On the other hand it must be emphasized that there is no indication of the existence of a haemal system in the common ancestor of protostomes and deuterostomes, and the haemal systems of for example annelids and echinoderms can therefore not be homologous even though they have the same morphological position, viz. between basal laminae of the various cell layers, and function in the two groups. Only by regarding the haemal systems as modified blastocoels can a homology be postulated (Ruppert & Carle 1983), but this will in my opinion stretch the homology concept so far that it becomes rather useless.

Nephridia/coelomoducts/gonoducts. Half a century ago Goodrich (1946) summarized the available knowledge about nephridia/coelomoducts/gonoducts and concluded that there are two main types of nephridia: protonephridia with a closed inner part, and metanephridia with a ciliated funnel opening into a coelom. He further stated that the coelomoducts appear to have originated as gonoducts, and that these canals may secondarily have become engaged in excretion and eventually have lost the primary function as gonoducts. These different types of organs may fuse in some species so that compound organs (nephromixia) are formed.

The last fifty years have added new dimensions to our understanding of both structure and function of these organs, especially through the use of the electron microscopes, but most of Goodrich's general conclusions still appear to hold.

With the present knowledge of structure and function of the various nephridia (excluding excretory glands) the two main types may be defined as follows:

1. Protonephridia are ectodermally derived canals with inner (terminal) cells with narrow, slit-like gaps, with or without an extracellular filtration membrane, which function as ultrafilters in the formation of the primary urine; the filtration is primarily from the blastocoel or from interstices between the mesodermal cells, but in coelomic organisms the terminal cells with the filters may penetrate the coelomic lining so that the primary urine can be filtered from the coelomic fluid. The primary urine is probably always modified during the passage through the canal cells, and the duct opens to the exterior through a nephridiopore cell. The cyrtopodocytes of amphioxus are not believed to be homologous with the protonephridia (Chapter 53).
2. Metanephridia are coelomoducts which transport and modify coelomic fluid through an open canal which is often of mixed ectodermal and mesodermal origin (for example in leeches, see Weisblat & Shankland 1985). The coelomic fluid can be regarded as the primary urine, and it is in most organisms filtered from the blood through ultrafilters formed by cells of blood vessels specialized as podocytes. There is, however, one important exception to this, namely the sipunculans, which do not have a haemal system at all (Chapter 13). The coelomic compartment may be large and have retained the primary functions (as hydrostatic skeleton and as gonocoel) or very restricted so that the blood vessel(s), podocytes, coelomic cavity, and sometimes also the modifying nephridial canal are united into a complex organ, as for example the axial complex of enteropneusts and echinoderms, the antennal gland of the crustaceans, and the vertebrate nephron.

These definitions (which agree with Goodrich's in the main parts) avoid the confusion which is related with more functionally based definitions, like that of Ruppert & Smith (1988) which makes for example the same excretory organ of the echinoderms protonephridia in the larvae and metanephridia in the adults and which also has to interpret the podocytes at the compensation sacs of the tentacle coelom of sipunculans as ultrafiltration areas, which appears less likely (Chapter 13).

Protonephridia are found in almost all bilaterian phyla, but the morphology of the organs is highly variable (reviews in Wilson & Webster 1974, Hay-Schmidt 1987, Bartolomaeus & Ax 1992). Some gastrotrichs have very simple protonephridia consisting of a monociliate terminal cell, a duct cell and a nephridiopore cell. Other bilaterians have several terminal cells, which may be mono- or multiciliate, and several branched or coiled duct with many cilia. However, both ontogeny and ultrastructure of many types of protonephridia indicate that the bilaterian ancestor had one pair of ectodermal protonephridia consisting of a monociliate terminal cell with a number of microvilli (possibly eight), a duct cell and a nephridiopore cell (Bartolomaeus & Ax 1992). The rapidly growing literature about ultrastructure of various invertebrate protonephridia still reveals new and unexpected details, and at present it appears impossible to use protonephridia in phylogenetic considerations.

Metanephridia, especially the more complex types, appear to be important phylogenetic markers, but it is clear that since the coeloms of for example protostomes and deuterostomes are not homologous, the metanephridia of these groups cannot be homologous either.

Nervous systems/neurotransmitters. Bilaterians have a brain, which is connected with the apical organ in the larvae; this larval brain is incorporated into the adult brain in the protostomes (although the aschelminthes generally lack larval stages with apical organs). The central nervous systems of adult protostomes and deuterostomes are bilateral, but the various components are apparently not homologous. Most bilaterians appear to have acetylcholine as one of their neurotransmitters, whereas the cnidarians rely on other substances (Chapters 3, 8). Acetylcholine has been detected in ctenophores (Anctil 1985), and this supports the interpretation of ctenophores as bilaterians (Chapter 42).

Monociliate/multiciliate cells. It is now generally accepted that the metazoan ancestor was monociliated (see for example Barnes 1985, Nielsen 1987). It seems certain that the multiciliate condition has evolved more than once outside the animal kingdom (polymastigine flagellates, ciliates, sperm of ferns) and it is therefore possible that it has evolved more than once also within the animals (Fig. 9.2). It is more questionable whether reversals from the multiciliate to the monociliate condition have taken place. None of these questions can be answered independent of a phylogenetic theory, but there are several independent pieces of information which throw light on the second question.

Almost all protostome phyla which have locomotory ciliary bands consisting exclusively of multiciliate cells have other types of ciliated cells which may be either monociliate or multiciliate: Sensory cells of molluscs and annelids may carry from one to 20 cilia (see for example Bubel 1984 and Welsch, Storch & Richards 1984) and protonephridial cells may have from one to many cilia (Hay-Schmidt 1987); also developing stages of mesodermal cells of organisms with multiciliate epithelial cells may have cells with one small cilium (*Magelona*, see Turbeville 1986). Most of these monociliate cells have the accessory centriole and rootlets characteristic of monociliate locomotory cells.

69

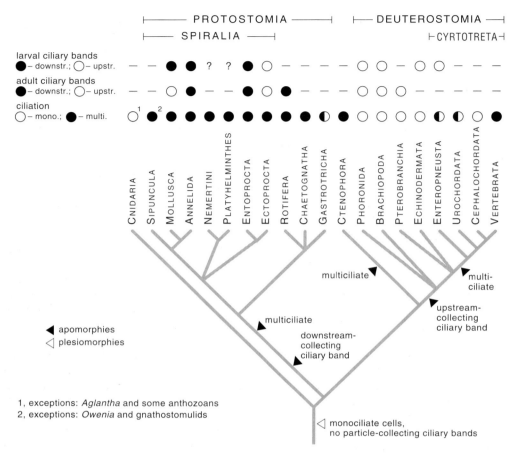

Fig. 9.2. Occurrence of mono- and multiciliate cells and particle-collecting ciliary bands in the eumetazoan phyla; phyla without ciliated epithelia are omitted.

Among deuterostomes phoronids, brachiopods, echinoderms, and cephalochordates have only monociliate cells whereas both mono- and multiciliate cells are found in enteropneusts, urochordates and vertebrates. The tornaria larva of the enteropneusts has a posterior, perianal ring of compound cilia from multiciliate cells (Fig. 43.5), while the ciliary band surrounding the mouth (neotroch) consists of single cilia on monociliate cells (Fig. 43.3), and also the adults have both mono- and multiciliate cells (Chapter 50). In urochordates mono- and multiciliate cells are found in well defined zones in the endostyle (Chapter 52). The monociliate cells of all these groups have the normal accessory centriole. The epithelia of vertebrates consist of multiciliate cells, but monociliate sensory cells occur for example in the lateral line system of fish and in the inner ear of mammals (Jørgensen 1989).

Furthermore, most of the animal phyla have sperm with a single cilium with an accessory centriole (Chapter 3).

Thus, all the phyla which have multiciliate cells in their epithelia also possess monociliate cells fulfilling other functions. It is furthermore clear that the centriolar apparatus, which forms the basal body of the monociliate cells, is characteristic of all animal cells. The accessory centriole at the ciliary base is absent in most multiciliate cells (exceptions among anthozoans, gastrotrichs and enteropneusts; Chapters 8, 33, 50), but the importance for the 'original' set of centrioles for the formation of the multiple basal bodies in multiciliate cells is not known (Dirksen 1991).

Ciliogenesis in multiciliate metazoan cells has been studied in a number of vertebrates, but only in very few invertebrates (Dirksen 1991). Some cells in mammalian organs have a small, primary or abortive cilium and such cilia are also formed in lung cells of foetal rats; these cells later become multiciliate (Sorokin 1968). The primary cilia have an accessory centriole, but many of them lack the central microtubules and beat irregularly or not at all; they may disappear when the cells are fully differentiated. In platyhelminths there is some variation in the development of the cilia on the multiciliate cells, but the first part of the cilium to be formed is apparently always a centriole which in most cases is formed in close connection with another centriole (Tyler 1984). The protonephridial cells of the nemertean *Lineus* go through a monociliate stage where the cilium has a rootlet and an accessory centriole; later stages show a multiplication of the centrioles and subsequent formation of several cilia with rootlets but without accessory centrioles (Bartolomaeus 1985). The multiple basal bodies of vertebrate multiciliate cells apparently develop independently of the original pair of centrioles (Dirksen 1991).

It appears that the multiciliate cells usually retain a pair of centrioles after the last cleavage and that some of these cells go through a monociliate stage, and this makes it easy to accept a complete reversal from the multiciliate condition to the monociliate condition. Such an evolutionary step can be regarded as an abbreviation, and it may well have happened in several smaller or larger groups of metazoans. The presence of epithelia of monociliate cells can therefore not *a priori* be regarded as a plesiomorphy. A strict adherence to the interpretation of all monociliate cells as plesiomorphic would for example mean that the polychaete *Owenia* would be the sister group of all other annelids, including *Myriochele* which is usually placed in the same family as *Owenia*, and this appears most unlikely (Chapter 17).

Cleavage/developmental patterns. Cnidarians have a cleavage pattern which is radial in most cases, and this is apparently the primitive cleavage pattern. The spiral cleavage is here regarded as the apomorphy of the supraphyletic group Spiralia (Chapter 11), and this cleavage type shows a whole series of advanced characteristics. The cleavage of the Aschelminthes (Chapter 28) shows more variation, comprising both apparently radial types, for example the chaetognaths (Chapter 31), and types with an asymmetric pattern, sometimes with a rhomboidal 4-cell stage. It appears that only the spiral pattern is sufficiently complex to be of any help in phylogenetic considerations, and even this pattern has obviously been lost within spiralian phyla, such as Mollusca (cephalopods; Chapter 15). Both Anderson (1973) and Dohle (1989) have demonstrated that obviously homologous organs can be formed from different blastomeres, for example in the spiral cell lineage. It is important to remember that very different embryologies can result in rather similar developmental stages and further to quite different adults, as demonstrated by the vertebrate classes (see for example Elinson 1987, figure 1). Also seemingly profound differences between cleavage patterns and development of most organ systems may result in adults which are so similar that they are classified in the same genus (for example the echinoderm genus *Heliocidaris*, see Raff 1987). This underlines once again that whole life cycles must be considered and that classifications must be based on characters which have been shown to be conservative or on patterns of characters rather than on single characters such as cleavage patterns or fates of the blastopore.

The widespread occurrence of a gastrula-stage and larval apical organs (Chapter 7) has been accepted as proof that the bilaterians (and the cnidarians) are descendants from a gastraea. The above discussions have shown that most of the characters which are usually regarded as synapomorphies of the Bilateria have evolved more than once and this could perhaps lead to the conclusion that for example the protostomes and the deuterostomes have evolved independently from the gastraea. However, there are a few characters which appear to be true synapomorphies for the group and which together characterize the Bilateria as the sister group of the Cnidaria:

1. The characteristic ectodermal protonephridial cells, which show a good deal of variation, but which nevertheless may be derived from one ancestral type, appear to be a synapomorphy of all the bilaterians.
2. The presence of acetylcholine/cholinesterase as one of the neurotransmitters in most bilaterian phyla is probably also a synapomorphy.

These few synapomorphies are perhaps not too convincing, but until further information becomes available it appears reasonable to regard the Bilateria as a monophyletic group and as the sister group of the Cnidaria. A more detailed discussion of the Protornaeozoa (= Ctenophora + Deuterostomia) is given in Chapter 41.

References

Anctil, M. 1985. Cholinergic and monoaminergic mechanisms associated with control of bioluminescence in the ctenophore *Mnemiopsis leidyi*. – J. exp. Biol. 119: 225-238.

Anderson, D.T. 1973. Embryology and Phylogeny in Annelids and Arthropods. – Pergamon Press, Oxford.

Ax, P. 1987. The Phylogenetic System. The Systematization of Organisms on the Basis of Their Phylogenesis. – John Wiley, Chichester.

Ax, P. 1989. Basic systematization of the Metazoa. – *In* B. Fernholm, K. Bremer & H. Jörnvall (eds): The Hierarchy of Life, pp 229-245. Excerpta Medica (Elsevier), Amsterdam.

Barnes, R.D. 1985. Current perspectives on the origin and relationships of lower invertebrates. – *In* S. Conway Morris, J.D. George, R. Gibson & H.M. Platt (eds): The Origin and Relationships of Lower Invertebrates, pp 360-367. Oxford Univ. Press, Oxford.

Bartolomaeus, T. 1985. Ultrastructure and development of the protonephridia of *Lineus viridis* (Nemertini). – Microfauna Mar. 2: 61-83.

Bartolomaeus, T. & P. Ax 1992. Protonephridia and metanephridia – their relation within the Bilateria. – Z. zool. Syst. Evolutionsforsch. 30: 21-45.

Bubel, A. 1984. Epidermal cells. – *In* J. Bereiter-Hahn, A.G. Matoltsy & K.S. Richards (eds): Biology of the Integument, vol. 1, pp 400-447. Springer, Berlin.

Carle, K.J. & E.E. Ruppert 1983. Comparative ultrastructure of the bryozoan funiculus: a blood vessel homologue. – Z. zool. Syst. Evolutionsforsch. 21: 181-193.

Clark, R.B. 1964. Dynamics in Metazoan Evolution. – Clarendon Press, Oxford.

Cuénot, L. 1940. Essai d'arbre généalogique du règne animal. – C.r. hebd. Séanc. Acad. Sci., Paris 210: 196-199.

Dirksen, E.R. 1991. Centriole and basal body formation during ciliogenesis revisited. – Biol. cell. 72: 31-38.

Dohle, W. 1989. Differences in cell pattern formation in early embryology and their bearing on evolutionary changes in morphology. – Geobios, Mém. spéc. 12: 145-155.

Ehlers, U. 1985. Das phylogenetische System der Plathelminthes. – Gustav Fischer, Stuttgart.

Elinson, R.P. 1987. Change in developmental patterns: embryos of amphibians with large eggs. – In R.A. Raff & E.C. Raff (eds): Development as an Evolutionary Process, pp 1-21. Alan R. Liss, New York.

Goodrich, E.S. 1946. The study of nephridia and genital ducts since 1895. – Q. Jl microsc. Sci., N.S. 86: 113-392.

Grobben, K. 1908. Die systematische Einteilung des Tierreichs. – Verh. zool.-bot. Ges. Wien 58: 491-511.

Hatschek, B. 1888. Lehrbuch der Zoologie, 1. Lieferung (pp 1-144). – Gustav Fischer, Jena.

Hatschek, B. 1911. Das neue zoologische System. – W. Engelmann, Leipzig.

Hay-Schmidt, A. 1987. The ultrastructure of the protonephridium of the actinotroch larva (Phoronida). – Acta. zool. (Stockh.) 68: 35-47.

Hyman, L.H. 1951. Platyhelminthes and Rhynchocoela. The Acoelomate Bilateria. The Invertebrates, vol. 2. – McGraw-Hill, New York.

Jørgensen, J.M. 1989. Evolution of octavolateralis sensory cells. – In S. Coombs, P. Görner & H. Münz (eds): The Mechanosensory Lateral Line, pp 115-145. Springer, New York.

Nielsen, C. 1987. Structure and function of metazoan ciliary bands and their phylogenetic significance. – Acta zool. (Stockh.) 68: 205-262.

Raff, R.A. 1987. Constraint, flexibility, and phylogenetic history in the evolution of direct development in sea urchins. – Dev. Biol. 119: 6-19.

Rähr, H. 1981. The ultrastructure of the blood vessels of Branchiostoma lanceolatum (Pallas) (Cephalochordata). – Zoomorphology 97: 53-74.

Rieger, R.M. 1986. Über den Ursprung der Bilateria: die Bedeutung der Ultrastrukturforschung für ein neues Verstehen der Metazoenevolution. – Verh. Dt. zool. Ges. 79: 31-50.

Ruppert, E.E. 1991. Introduction to the aschelminth phyla: a consideration of mesoderm, body cavities, and cuticle. – In F.W. Harrison (Ed.): Microscopic Anatomy of Invertebrates, vol. 4, pp 1-17. Wiley-Liss, New York.

Ruppert, E.E. & K.J. Carle 1983. Morphology of metazoan circulatory systems. – Zoomorphology 103: 193-208.

Ruppert, E.E. & P.R. Smith 1988. The functional organization of filtration nephridia. – Biol. Rev. 63: 231-258.

Salvini-Plawen, L.v. & H. Splechtna 1979. Zur Homologie der Keimblätter. – Z. syst. Zool. Evolutionsforsch. 17: 10-30.

Sorokin, S.P. 1968. Reconstructions of centriole formation and ciliogenesis in mammalian lungs. – J. Cell Sci. 3: 207-230.

Strong, P.A. & B.J. Bogitsch 1973. Ultrastructure of the lymph system of the trematode Megalodiscus temperatus. – Trans. Am. microsc. Soc. 92: 570-578.

Turbeville, J.M. 1986. An ultrastructural analysis of coelomogenesis in the hoplonemertine Prosorhochmus americanus and the polychaete Magelona sp. – J. Morph. 187: 51-60.

Tyler, S. 1984. Development of cilia in embryos of the turbellarian Macrostomum. – Hydrobiologia 84: 231-239.

Ulrich, W. 1951. Vorschläge zu einer Revision der Grosseinteilung des Tierreichs. – Zool. Anz., Suppl. 15: 244-271.

Weisblat, D.A. & M. Shankland 1985. Cell lineage and segmentation on the leech. – Phil. Trans. R. Soc. B 312: 39-56.

Welsch, U., V. Storch & K.S. Richards 1984. Annelida. Epidermal cells. – *In* J. Bereiter-Hahn, A.G. Matoltsy & K.S. Richards (eds): Biology of the Integument, vol. 1, pp 269-296. Springer, Berlin.

Willey, C.H. 1930. Studies on the lymph system of digenetic trematodes. – J. Morph. Physiol. 50: 1-37.

Wilson, R.A. & L.A. Webster 1974. Protonephridia. – Biol. Rev. 49: 127-160.

10

PROTOSTOMIA (= GASTRONEURALIA)

The Protostomia is a large group of bilateral animals characterized by a mosaic of features, but few species exhibit them all. As mentioned in Chapter 9, classification has often been based on one character only, but the following complex of characters should be considered as apomorphies of the protostomes: 1) the blastopore becomes divided by the fusing lateral lips, leaving only mouth and anus, 2) the nervous system includes the apical nervous centre (or a centre formed in the same area), lateral connectives around the oeophagus and a ventral longitudinal nerve trunk (single or paired) formed from the fused lateral blastopore lips, 3) the larvae are trochophore-types with downstream-collecting ciliary systems consisting of bands of compound cilia on multiciliate cells, and 4) the mesoderm is formed from the blastopore rim, often supplemented by ectomesoderm (groups with coeloms generally form the coelomic cavities through schizocoely). Only some polychaetes show both the life cycle and the larval and adult structures which are considered ancestral. However, Table 10.1 shows that a number of these protostome characters occur in every phylum (except the ectoproct bryozoans which are discussed in Chapter 24). The cleavage patterns are included in the table because the spiral cleavage, which as shown in Chapter 11 is very well defined, has been used as an additional character indicating that the Platyhelminthes and the Nemertini belong to the Protostomia.

The fate of the blastopore has, for almost a century, been used as the characterizing feature of the protostomes (see Chapter 9), usually with the understanding that the blastopore should become the mouth (as indicated by the name). There are, however, a number of species in which the blastopore becomes divided into mouth and anus by the fusion of the lateral blastopore lips, and this type of blastopore closure is indicated in the development of many protostomes, both species with embolic gastrulation and planktotrophic development and species with epibolic gastrulation (see Fig. 10.1).

The fate of the blastopore is still used by some authors as the major character, for example in discussions of the position of phoronids (Ivanova-Kazas 1986). However, it has been known for a long time that the blastoporal fate varies within well-defined phyla and classes: some polychaetes have a blastopore which becomes mouth and anus, some have a blastopore which becomes the adult mouth, and some have a blastopore which becomes the adult anus (Chapter 17). It should be clear that a char-

Table 10.1. Characteristics of protostomes. The aschelminths lack an apical organ, but their brain is situated in a position similar to that of the spiralians. The parenchymians (platyhelminths + nemertines) lack the ventral part of the nervous system.

	Trochophora larva	Blastopore becomes mouth+anus	Apical brain+ mid-ventral nerve cords	Spiral cleavage
Spiralia				
Sipuncula	+	−	+	+
Mollusca	+	−	(+)	+
Annelida	+	+	+	+
Onychophora	−	+	+	−
Tardigrada	−	?	+	?
Arthropoda	−	−	+	+
Entoprocta	+	−	−	+
Ectoprocta	−	−	−	−
Platyhelminthes	?	−	(+)	+
Nemertini	?	−	(+)	+
Aschelminthes				
Rotifera	(+)	−	(+)	−
Acanthocephala	−	−	(+)	−
Chaetognatha	−	−	+	−
Gastrotricha	−	?	+	−
Nematoda	−	+	+	−
Nematomorpha	−	?	+	−
Priapula	−	?	+	−
Kinorhyncha	−	?	+	−
Loricifera	−	?	+	?

acter showing such a degree of variation cannot be used alone when the position of a phylum is to be determined. On the other hand it should be remembered that a lateral blastopore closure dividing the primary mouth into mouth and anus has never been observed in any of the phyla assigned to the Deuterostomia.

The amount of yolk and the evolution of placental nourishment of the embryos have decisive influence both on the cleavage patterns and on the later embryological development. In many species a normal gastrulation can hardly be recognized and an archenteron is not formed at all. Many examples of epibolic gastrulation, superficial cleavage, discoidal cleavage, etc. will be mentioned in the discussions of the various phyla, and most of these cases can be interpreted as variations on the general theme of holoblastic cleavage followed by embolic gastrulation. In many groups whole series of intermediate stages between these apparently very different types are represented among the rather few species which have been studied so far.

The fate of the blastopore is intimately connected with the development of the ventral longitudinal part of the nervous system. The two longitudinal cords develop from the ectoderm along the lateral blastopore lips, and this can clearly be recognized not only in polychaetes with a 'typical' development with embolic gastrulation,

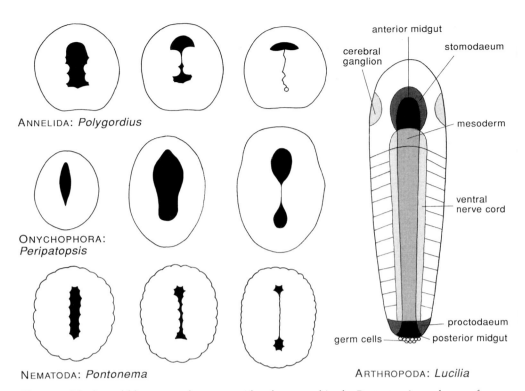

ANNELIDA: *Polygordius*

ONYCHOPHORA:
Peripatopsis

NEMATODA: *Pontonema*

ARTHROPODA: *Lucilia*

anterior midgut

cerebral
ganglion

stomodaeum

mesoderm

ventral
nerve cord

proctodaeum

germ cells

posterior midgut

Fig. 10.1. The lateral blastopore closure considered ancestral in the Protostomia; embryos of some annelids, onychophorans and nematodes with embolic gastrulation directly show the fusion of the lateral blastopore lips, leaving only mouth and anus; embryos with epibolic gastrulation or more derived types have fate maps which reflect the ancestral pattern with the ventral nerve cords developing from the ectoderm along the line between mouth and anus . – Annelida: *Polygordius* sp.; the white circle shows the position of the future anus (based on Woltereck 1904). – Onychophora: *Peripatopsis capensis* (redrawn from Manton 1949). – Nematoda: *Pontonema vulgare* (redrawn from Malakhov 1986). – Arthropoda: Fate map of the fly *Lucilia sericata* (based on Davis 1967 and Anderson 1973).

but also for example in the leeches, which have yolk-rich eggs and direct development (Fig. 17.7). In the nematodes, which have epibolic gastrulation enclosing a few endodermal cells, the lateral blastopore closure can nevertheless clearly be recognized (Fig. 10.1), and the development of the median nerve cord from the ectoderm of the blastopore lips has been documented in every detail (Fig. 35.3). The many intermediate stages between the development of the ventral nervous trunks from the blastopore lips, for example in the polychaete *Scoloplos* (Fig. 17.2) and the development from areas of the blastoderm corresponding to the blastopore lips, for example in leeches and insects (Fig. 10.1), indicate that these events are homologous. I therefore believe that the protostomes which have a (single or double) longitudinal ventral nerve trunk formed from parallel strips of ventral embryonic ectoderm are all derived from ancestors which had the lateral blastopore closure.

A larval apical organ which becomes incorporated in the adult brain has been described in most spiralian phyla (Chapter 11), but the exact position of the dorsal brain in the aschelminths is difficult to ascertain because the apical pole is more or less impossible to define in all the phyla in which the larval development has been studied (Chapter 28). However, it is clear that almost all the protostomes have a dorsal or circumoesophageal brain (see Fig. 10.2); the bryozoan phyla form the only exceptions, with the apical larval brain disappearing at metamorphosis in the entoprocts (Chapter 23), and the larval orientation becoming lost through the obligatory budding process in the ectoprocts (Chapter 24). Most deuterostome larvae have an apical organ, but the apical region is either cast off at metamorphosis or the organ disappears completely (Chapter 43).

The ventral, longitudinal nervous cord is perhaps the most stable character of the protostomes (and the name Gastroneuralia would therefore be preferable). As mentioned above, the cord differentiates from the lateral blastopore lips of the embryo or from comparable areas in species with modified embryology. This is well known from many studies of various spiralians, whereas the embryology of most of the aschelminth phyla is poorly studied or completely unknown. However, the detailed studies of the nematode *Caenorhabditis* (Chapter 35) have shown that the ventral nervous cells originate from the lateral blastoporal lips. The position of the ventral cord(s) is intraepithelial in the early stages, and it remains so in some phyla, whereas it becomes internalized in some types within other phyla (Fig. 17.2). A ventral cord in the primitive, intraepithelial/basiepithelial position, i.e. between the ventral epithelium and its basement membrane, is found in adults of annelids (both in several of the primitive 'polychaete' families and in some oligochaetes), arthropods, chaetognaths, gastrotrichs, nematodes, kinorhynchs, loriciferans, and priapulans. In some groups, for example nematomorphs, the cord is folded in and almost detached from the epithelium, but it is surrounded by the peritoneum and a basement membrane which is continuous with that of the ventral epithelium. A completely detached nerve cord surrounded by peritoneum and the basement membrane is found in sipunculans, molluscs, many annelids, onychophorans, tardigrades, and arthropods.

The sessile phyla Entoprocta and Ectoprocta lack a ventral cord. The entoprocts have a small ventral ganglion, which forms as an invagination from the ventral epithelium at metamorphosis or in a late stage of the budding process; its homology is uncertain. The ectoprocts form a similar ganglion during the budding process. These nervous systems are discussed in more detail in Chapters 23 and 24, but it should be mentioned here that ganglia formed by ectodermal invaginations are not an exclusive feature of deuterostomes (as sometimes believed, see below), but have been observed in such protostomes as pseudoscorpions and millipedes (Weygoldt 1964, Dohle 1964).

Platyhelminths and nemertines appear to lack the ventral component of the gastroneuralian nervous system (Chapters 25-27). Their main nervous system develops entirely from the apical organ, and there is no indication of ventral nerves originating from a blastopore closure. Their larvae have diminutive hypospheres, which can be interpreted as a reduction of the ventral part of the body.

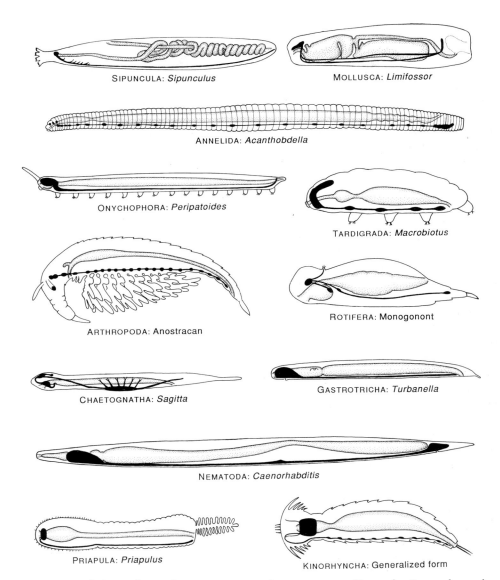

SIPUNCULA: *Sipunculus*

MOLLUSCA: *Limifossor*

ANNELIDA: *Acanthobdella*

ONYCHOPHORA: *Peripatoides*

TARDIGRADA: *Macrobiotus*

ARTHROPODA: Anostracan

ROTIFERA: Monogonont

CHAETOGNATHA: *Sagitta*

GASTROTRICHA: *Turbanella*

NEMATODA: *Caenorhabditis*

PRIAPULA: *Priapulus*

KINORHYNCHA: Generalized form

Fig. 10.2. Lateral views of central nervous systems of protostomes. – Sipuncula: *Sipunculus nudus* (based on Metalnikoff 1900). – Mollusca: *Limifossor talpoideus* (after an unpublished drawing by Drs A.H. Scheltema and M.P. Morse, based on Heath 1905). – Annelida: the hirudinean *Acanthobdella peledina* (based on Storch & Welsch 1991). – Onychophora: *Peripatoides novaezelandiae* (based on Snodgrass 1938). – Tardigrada: *Macrobiotus hufelandi* (modified from Cuénot 1949). – Arthropoda: generalized anostracan (redrawn from Storch & Welsch 1991). – Rotifera: generalized monogonont (redrawn from Hennig 1984). – Chaetognatha: *Sagitta crassa* (based on Goto & Yoshida 1986). – Gastrotricha: *Turbanella cornuta* (based on Teuchert 1977). – Nematoda: *Caenorhabditis elegans* (see Fig. 35.1). – Priapula: *Priapulus caudatus* (based on Apel 1885). – Kinorhyncha: generalized kinorhynch (redrawn from Hennig 1984).

It should be stressed that whereas not all protostomes (gastroneuralians) have the paired ventral nerve cords, none of the deuterostome phyla have ventral nervous concentrations of this type.

An alternative hypothesis for the origin and evolution of the nervous system of protostomes is the orthogon-theory (Reisinger 1925, 1972; Hanström 1928); it envisages a transformation series from the diffuse nervous net of the cnidarians via an orthogonal nervous system with a circumoral brain and typically eight longitudinal nerves to a nervous system with only a pair of mid-ventral longitudinal nerves. The problem with this theory is that a typical orthogon is only found in a few platyhelminths; it is definitely absent in nematodes (Chapter 35) (and other aschelminths) and its presence in annelids and molluscs is in the eye of the beholder. Platyhelminth nervous systems vary enormously (Chapter 26).

A number of authors (for example Reisinger 1972) have mentioned another difference between the nervous systems of protostomes and deuterostomes, namely that the main nerves should be subepidermal in the protostomes (at least 'primarily') and intraepithelial in the deuterostomes. This is clearly a misunderstanding: the ventral nerves are intraepithelial/basiepithelial in many protostome phyla (see above) and since this position is found in the larval stages of forms which in the adult stage have subepithelial nerves it must be concluded that the intraepithelial position is primitive also in the protostomes. Also some of the platyhelminths have intra-epidermal longitudinal nerves, and the Müller's larva of some polyclads has intraepithelial nerves (Chapter 26).

A further difference between the protostome and deuterostome nervous systems should be that a 'neurulation', i.e. an infolding of epithelium with intraepithelial nervous tissue to form a tube, should occur only in the deuterostomes (Reisinger 1972). This appears to hold true for the tube-shaped nervous centres, but invaginations which develop into one or a series of ganglia have been observed in both entoproct and ectoproct bryozoans and in some arthropods (see above).

The third feature considered characteristic of the Protostomia is the presence of trochophora larvae. The ancestral trochophore (Fig. 2.4) is believed to have been planktotrophic with a tube-shaped gut and the ciliary bands found for example in the planktotrophic larvae of some species of *Polygordius*: 1) a prototroch of compound cilia anterior to the mouth (a small dorsal break in this band is observed in early stages of several spiralians; see Chapter 11); 2) an adoral ciliary zone of separate cilia surrounding the mouth; 3) a metatroch of compound cilia behind the mouth, but with a break just behind the mouth; these three bands should be broken mid-dorsally; 4) a gastrotroch which is a mid-ventral band of separate cilia from the mouth to the anus; 5) a telotroch of compound cilia surrounding the anus except for a break at the ventral side. All these bands are formed from multiciliate cells. Prototroch, adoral ciliary zone and metatroch form the feeding organ of the larva, which is a downstream-collecting system in which the adoral ciliary zone transports the particles strained by the bands of compound cilia towards the mouth. The particles may be rejected at the mouth and transported along the gastrotroch to the anal region where they leave the water currents of the larva. The telotroch is locomotory. Other characteristics of trochophora larvae include the presence of an apical organ (see above) and probably also of a pair of protonephridia. Other features, such as the presence of mesoderm and its morphology, are not included in the definition.

The ciliary bands of invertebrate larvae have been discussed in detail elsewhere (Nielsen 1987) but some of the more general conclusions should be repeated here.

Planktotrophic larvae with the downstream-collecting ciliary complex consisting of prototroch, adoral ciliary zone and metatroch are known from many of the spiralian phyla: Annelida, Mollusca, Sipuncula, and Entoprocta (Figs 10.3, 10.4), whereas the larvae of Platyhelminthes and Nemertini are of a more dubious type (Chapters 26, 27). However, many types of planktonic larvae lack one or more of the ciliary bands described above, and it appears that only planktotrophic larvae have

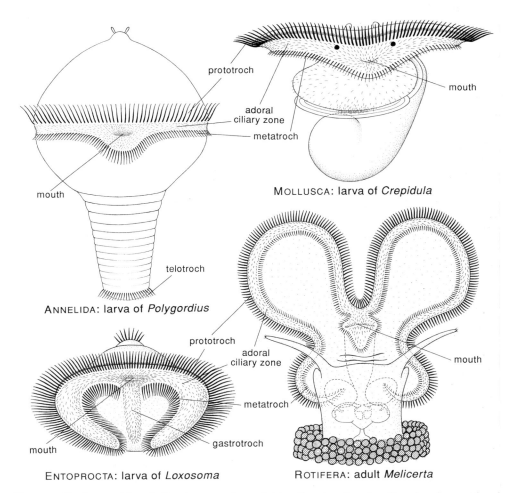

Fig. 10.3. Planktotrophic spiralian larvae and an adult rotifer, showing the ciliary bands considered ancestral in the Protostomia. (*Polygordius appendiculatus* based on Hatschek 1878; the other three drawings from Nielsen 1987.)

81

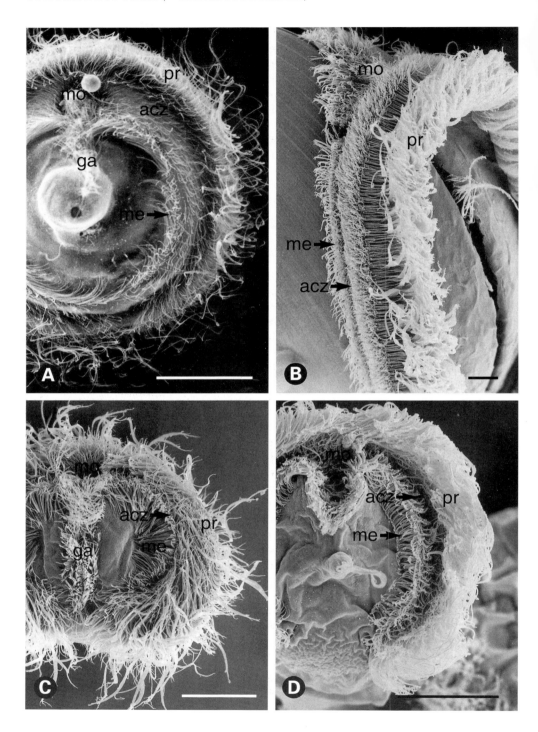

the metatroch. Among the phyla referred to the Protostomia only the ectoproct larvae have an upstream-collecting ciliary band, but this band is of a unique structure (and perhaps also function) which is discussed further in Chapter 24. The Aschelminthes lack primary larvae altogether, but some of the rotifers (Chapter 29) have a ciliary feeding organ which is of exactly the same structure and function as that of the *Polygordius* larva, and their ciliary bands are considered homologous with those of the spiralians (Figs 10.3, 10.4).

The original concept of the trochophora larva (Hatschek 1878, 1891) comprised also the actinotroch larva of *Phoronis*. The sharpened definition of structure and function of the ciliary bands given above (Nielsen 1985, 1987) excludes the actinotroch because it has an upstream-collecting ciliary system with the ciliary bands consisting of separate cilia on monociliate cells – features characteristic of deuterostome larvae (Chapter 43).

Salvini-Plawen (1980) restricted the term trochophora for the larvae of annelids and echiurans, and created new names for larvae of other phyla. The mollusc larvae, for example, were stated generally to lack a metatroch (as well as protonephridia), but a metatroch is present in almost all planktotrophic larvae of gastropods and bivalves (Nielsen 1987) (and protonephridia have recently been discovered in larvae of a polyplacophoran and a gastropod; Bartolomaeus 1989a). As defined above, the trochophora larva is a very useful concept in phylogenetic work, because it is easy to make references to the various, well-defined ciliary bands and other homologous structures.

Another term introduced by Salvini-Plawen (1972, 1980) is the pericalymma larva, which comprises trochophore-like larvae with most of the hyposphere covered by a usually ciliated expansion (often called serosa) from an anterior zone (these larvae are also called 'Hüllglocken', test-cell or serosa larvae). However, it is important to note that these expansions originate from different areas in different larvae (Fig. 10.5). In the larvae of molluscs such as *Yoldia* (Drew 1899) and *Neomenia* and in the sipunculan *Sipunculus* (Hatschek 1883), the expansion originates from the prototroch area (or from the episphere) (Type 1), whereas it originates from the the zone just behind the mouth in the annelids *Polygordius lacteus* and some unidentified phyllodocids (Dawydoff 1959), and in the mollusc *Lyrodus* (Type 2); in the oweniid polychaetes the serosa is formed from an area further away from the ciliary bands, viz. a zone between the two first segments with parapodia, so that the very long setae of the first segment are exposed (Type 3). It should be clear that the serosae of these three types of pericalymma larvae cannot be homologous. The fact that variations from the more usual planktotrophic trochophora larvae to pericalymma types can be observed within families makes it very unlikely that the evolution has gone from various lecithotrophic per-

Fig. 10.4. Oral area with prototroch, adoral ciliary zone, metatroch, and gastrotroch of larval and adult protostomes (SEM). – A, larva of the polychaete *Serpula vermicularis* (Friday Harbor Laboratories, WA, USA, July 1980). – B, larva of the bivalve *Barnea candida* (plankton, off Frederikshavn, Denmark, August 1984). – C, larva of the entoproct *Loxosoma pectinaricola* (Øresund, Denmark, October 1981). – D, adult of the rotifer *Conochilus unicornis* (Almind Lake, Denmark, May 1983). acz, adoral ciliary zone – ga, gastrotroch – me, metatroch – mo, mouth – pr, prototroch. (See also Nielsen 1987.) – Scale bars: 25 μm.

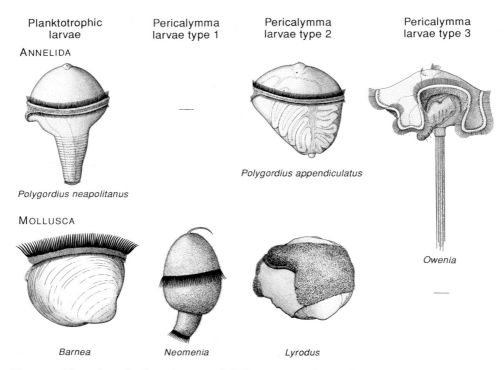

Planktotrophic larvae   Pericalymma larvae type 1   Pericalymma larvae type 2   Pericalymma larvae type 3

ANNELIDA

Polygordius appendiculatus

Polygordius neapolitanus

MOLLUSCA

Owenia

Barnea   *Neomenia*   *Lyrodus*

Fig. 10.5. Normal trochophore larvae and different types of pericalymma larvae of annelids and molluscs. – Annelida: planktotrophic trochophore of *Polygordius neapolitanus* (based on Woltereck 1902), pericalymma larva type 2: endolarva of *Polygordius appendiculatus* (based on Herrmann 1986), pericalymma larva type 3: endolarva of *Owenia fusiformis* (after Wilson 1932). – Mollusca: planktotrophic veliger larva of the bivalve *Barnea candida* (based on Nielsen 1987), pericalymma larva type 1: lecithotrophic larva of the solenogaster *Neomenia carinata* (based on Thompson 1960), pericalymma larva type 2: brooded larval stage of the bivalve *Lyrodus pedicellatus* (Long Beach, California, USA; after a scanning migrograph by Drs C.B. Calloway and R.D. Turner, Museum of Comparative Zoology, Harvard Univ., MA, USA). The presence of compound cilia in the larvae of *Neomenia* and *Polygordius neapolitanus* is inferred from the descriptions.

icalymma larvae to the structurally and functionally complicated planktotrophic trochophores which are very similar among the phyla. The pericalymma larvae must thus be interpreted as independent specializations within different phylogenetic lines.

None of the deuterostome larvae have downstream-collecting ciliary bands, and the upstream-collecting ciliary bands found in the larvae (and some adults) of several deuterostome phyla are in all cases made up of separate cilia on monociliate cells as opposed to the compound cilia on multiciliate cells observed in the protostomes.

The number of cilia per cell has been used in many discussions of the phylogeny of the protostomes. It is now generally accepted that the monociliate condition is ancestral in the metazoans (Chapter 9), and the question is then if the ancestral protostome, gastroneuron, was monociliate or multiciliate. Both mono- and multiciliate cells have been found in sensory and excretory organs in representatives of many phyla (for example Annelida, see Storch & Schlötzer-Schrehardt 1988 and Bartolo-maeus 1989b, respectively) and are apparently not of importance for this discussion.

I have earlier (Nielsen 1985) proposed that the hypothetical ancestor of all bilateral animals, trochaea, had a circumblastoporal ring of compound cilia forming a downstream-collecting band. The fact that the ring of compound cilia around the anus of the actinotroch is formed from monociliate cells has made me propose an alternative theory (Nielsen 1987) which I now find more probable: that the multiciliarity developed independently in Protostomia, in the Ctenophora and in the Cyrtotreta (Enteropneusta + Chordata) (Fig. 9.2). The ciliary bands of protostomes will be discussed here, while those of ctenophores and deuterostomes will be discussed later (Chapters 42, 43).

The occurrence of the two types of ciliated cells in the protostome phyla (Fig. 9.2) shows that the monociliate condition is restricted to three groups: the polychaete *Owenia*, the group Gnathostomulida and a number of gastrotrichs. The question is therefore whether the monociliate condition in each of these scattered groups is primary or secondary, i.e. whether the monociliate condition represents the plesiomorphic state of the Metazoa or whether it represents a reversal from the multiciliate condition (see also Chapter 9).

Owenia appears to be the most easy case. The highly characteristic family Oweniidae also comprises *Myriochele* which has multiciliate cells (Chapter 17), and if the monociliate condition of *Owenia* is plesiomorphic, this genus (together with the gnathostomulids) should be the sister group of *Myriochele* and all other annelids, or the multiciliate condition should have evolved many times within the Annelida – both choices appear highly improbable, so the monociliate condition in *Owenia* must be regarded as a specialization.

The Gnathostomulida apparently all have monociliate epidermal cells (Chapter 17), and Ax (1987, 1989) has argued strongly in favour of the opinion that this represents the plesiomorphic character state in the Bilateria. The systematic position of the gnathostomulids is uncertain; I have chosen to regard them as specialized annelids (Chapter 17) and accordingly to interpret their monociliarity as a secondary reversal to the ancestral condition. In light of the present knowledge about ciliogenesis in multiciliate cells (Chapter 9) this possibility cannot be rejected, but more information about this group must be obtained before a firm stand can be taken.

In the Gastrotricha, the monociliate condition occurs scattered within a number of families and genera which in several cases comprise other species with multiciliate epidermis (Rieger 1976). If the generally accepted systematics of the Gastrotricha is correct it follows that either the monociliate or the multiciliate condition has evolved several times within the phylum. Rieger (1976) regarded the monociliate condition as ancestral and concluded that the multiciliate condition had evolved convergently several times. The present knowledge of the ciliogenesis in some types of multiciliate cells (Chapter 9) gives us an alternative choice, namely that the monociliate condition is a 'return' to the ancestral character state through an abbreviation in the ciliogenesis, and this appears to be a much more 'parsimonious' explanation for the gastrotrichs.

Characters of mesoderm and coelom have played a major role in phylogenetic discussions. It has often been stated that one of the major differences between protostomes and deuterostomes is that the protostomes form coelomic cavities through schizocoely and the deuterostomes through enterocoely. There is of course some truth in the statement, but when the variation within the two groups is considered it turns out that the situation is much more complex, and the origin of the mesoderm, which by definition surrounds the coelomic cavities, is perhaps more significant. The mesoderm and coelom of the deuterostomes are discussed in Chapter 43, and here it should suffice to mention that the mesoderm in all cases originates from the walls of the archenteron (and from the neural crest in the chordates); the coelomic pouches (usually three pairs) are often formed through enterocoely, but even among for example enteropneusts there is a variation ranging from typical schizocoely to typical enterocoely (Chapter 50; see also Fig. 43.2).

In the protostomes, the mesoderm originates either from the blastopore lips or as ectomesoderm, and the coelomic cavities originate through schizocoely; only the chaetognaths (Chapter 31) form an exception (see below). The protostome sister groups, Spiralia and Aschelminthes, are very different with respect to the origin of the mesoderm and are discussed in detail in Chapters 11 and 28, respectively.

The four main characters discussed above, viz. the lateral blastopore closure, the nervous system with the apical brain and ventral nerve cord(s), the trochophora larval type, and the mesoderm formation from the blastopore rim, are well-defined and none of them have been observed in the phyla which are here classified as deuterostomes. Their occurrence in the protostome phyla is scattered, but together (and especially when the spiral cleavage is taken into consideration too) these characters define the protostomes unequivocally (the only problematic group being the ectoproct bryozoans; Chapter 24). They can be interpreted as the results of the series of linked evolutionary steps envisaged by the trochaea theory (Chapter 2), leading from the holoplanktonic, radially symmetrical trochaea with a pouch-shaped archenteron to the protostome ancestor, gastroneuron, which had a tube-shaped gut and a pelago-benthic life cycle. Mesoderm occurs in all the phyla, but its origin cannot be used to characterize the protostomes. It should be stressed that the hypothetical protostome ancestor, gastroneuron, was without a coelom.

The Protostomia can therefore be characterized as a monophyletic group, whose origin from simple, gastrula-like ancestors can be explained in a series of steps which each have been adaptations to changes in life cycles: from holoplanktonic, planktotrophic organisms to pelago-benthic organisms with planktotrophic trochophora larvae and deposit-feeding, benthic adults.

References

Anderson, D.T. 1973. Embryology and Phylogeny in Annelids and Arthropods. – Pergamon Press, Oxford.

Apel, W. 1885. Beitrag zur Anatomie und Histologie des *Priapulus caudatus* (Lam.) und des *Halicryptus spinulosus* (v. Sieb.). – Z. wiss. Zool. **42**: 459-529, pls 15-17.

Ax, P. 1987. The Phylogenetic System. – John Wiley, Chichester.

Ax, P. 1989. Basic phylogenetic systematization of the Metazoa. – *In* B. Fernholm, K. Bremer & H. Jörnvall (eds): The Hierarchy of Life, pp 229-245. Elsevier Science, Amsterdam.

Bartolomaeus, T. 1989a. Larvale Nierenorgane bei *Lepidochiton cinereus* (Polyplacophora) und *Aeolidia papillosa* (Gastropoda). – Zoomorphology **108**: 297-307.

Bartolomaeus, T. 1989b. Ultrastructure and development of the nephridia in *Anaitides mucosa* (Annelida, Polychaeta). – Zoomorphology **109**: 15-32.

Cuénot, L. 1949. Les Tardigrades. – Traité de Zoologie 6: 39-59. Masson, Paris.

Davis, C.W.C. 1967. A comparative study of larval embryogenesis in the mosquito *Culex fatigans* Wiedemann (Diptera: Culicidae) and the sheep-fly *Lucilia sericata* Meigen (Diptera: Calliphoridae). – Aust. J. Zool. **15**: 547-579.

Dawydoff, C. 1959. Ontogenèse des Annélides. – Traité de Zoologie 5(1): 594-686. Masson, Paris.

Dohle, W. 1964. Die Embryonalentwicklung von *Glomeris marginata* (Villers) im Vergleich zur Entwicklung anderer Diplopoden. – Zool. Jb., Anat. **81**: 241-310.

Drew, G.A. 1899. Some observations on the habits, anatomy and embryology of members of the Protobranchia. – Anat. Anz. **15**: 493-519.

Goto, T. & M. Yoshida 1986. Nervous system in Chaetognatha. – *In* M.A. Ali (ed.): Nervous Systems in Invertebrates (NATO ASI, Ser. A 141), pp 461-481. Plenum Press, New York.

Hanström, B. 1928. Vergleichende Anatomie des Nervensystems der wirbellosen Tiere. – Springer, Berlin.

Hatschek, B. 1878. Studien über Entwicklungsgeschichte der Anneliden. – Arb. zool. Inst. Univ. Wien **1**: 277-404, pls 23-30.

Hatschek, B. 1883. Über Entwicklung von *Sipunculus nudus*. – Arb. zool. Inst. Univ. Wien **5**: 61-140, pls 4-9.

Hatschek, B. 1891. Lehrbuch der Zoologie, 3. Lieferung (pp 305-432). – Gustav Fischer, Jena.

Heath, H. 1905. The morphology of a solenogastre. – Zool. Jb., Anat. **21**: 703-734, pls 42-43.

Hennig, W. 1984. Taschenbuch der Zoologie, Band 2, Wirbellose I. – Gustav Fischer, Jena.

Herrmann, K. 1986. *Polygordius appendiculatus* (Archiannelida) – Metamorphose. – Publ. wiss. Film., Sekt. Biol., Ser. 18, No. **36**/E2716: 1-15.

Ivanova-Kazas, O.M. 1986. Analysis of larval development in Tentaculata. 1. Larvae in Phoronida and Brachiopoda. – Zool. Zh. **65**: 757-770 (In Russian, English summary; English translation available from: Library, Canadian Museum of Nature, P.O. Box 3443, Stn.D, Ottawa, Ontario, Canada K1P 6P4).

Malakhov, V.V. 1986. Nematodes. Anatomy, Development, Systematics and Phylogeny. – Nauka, Moskva (In Russian).

Manton, S.F. 1949. Studies on the Onychophora VII. The early embryonic stages of *Peripatopsis*, and some general considerations concerning the morphology and phylogeny of the Arthropoda. – Phil. Trans. R. Soc. B **233**: 483-580, pls 31-41.

Metalnikoff, S. 1900. *Sipunculus nudus*. – Z. wiss. Zool. **68**: 261-322, pls 17-22.

Nielsen, C. 1985. Animal phylogeny in the light of the trochaea theory. – Biol. J. Linn. Soc. **25**: 243-299.

Nielsen, C. 1987. Structure and function of metazoan ciliary bands and their phylogenetic significance. – Acta zool. (Stockh.) **68**: 205.262.

Reisinger, E. 1925. Untersuchungen am Nervensystem der *Bothrioplana semperi* Braun. – Z. Morph. Ökol. Tiere **5**: 119-149.

Reisinger, E. 1972. Die Evolution des Orthogons der Spiralier und das Archicölomatenproblem. – Z. zool. Syst. Evolutionsforsch. **10**: 1-43.

Rieger, R.M. 1976. Monociliated epidermal cells in Gastrotricha: significance for concepts of early metazoan evolution. – Z. zool. Syst. Evolutionsforsch. **14**: 198-226.

Salvini-Plawen, L.v. 1972. Zur Morphologie und Phylogenie der Mollusken: die Beziehungen der Caudofoveata und der Solenogastres als Aculifera, als Mollusca und als Spiralia. – Z. wiss. Zool. **184**: 205-394.

Salvini-Plawen, L.v. 1980. Was ist eine Trochophore? Eine Analyse der Larventypen mariner Protostomier. – Zool. Jb., Anat. **103**: 389-423.

Snodgrass, R.E. 1938. Evolution of the Annelida, Onychophora, and Arthropoda. – Smithsonian misc. Coll. **97**(6): 1-159.

Storch, V. & U. Schlötzer-Schrehardt 1988. Sensory structures. – Microfauna mar. **4**: 121-133.

Storch, V. & U. Welsch 1991. Systematische Zoologie (4. ed.). – Gustav Fischer, Stuttgart.

Teuchert, G. 1977. The ultrastructure of the marine gastrotrich *Turbanella cornuta* Remane (Macrodasyoidea) and its functional and phylogenetic importance. – Zoomorphologie **88**: 189-246.

Thompson, T.E. 1960. The development of *Neomenia carinata* Tullberg (Mollusca Aplacophora). – Proc. R. Soc. Lond. B **153**: 263-278.

Weygoldt, P. 1964. Vergleichend-embryologische Untersuchungen an Pseudoscorpionen (Chelineti). – Z. Morph. Ökol. Tiere **54**: 1-106.

Wilson, D.P. 1932. On the mitraria larva of *Owenia fusiformis* Delle Chiaje. – Phil. Trans. R. Soc. B **221**: 231-334, pls 29-32.

Woltereck, R. 1902. Trochophora-Studien I. Histologie der Larve und die Entstehung des Annelids bei den *Polygordius*-Arten der Nordsee. – Zoologica (Stuttg.) **13**(34): 1-71, 11 pls.

Woltereck, R. 1904. Wurm'kopf', Wurmrumpf und Trochophora. – Zool. Anz. **28**: 273-322.

11

SPIRALIA

The spiral cleavage is a highly characteristic developmental pattern (Fig. 11.1), which exhibits not only very conspicuous blastomere patterns but usually also strongly determined blastomere fates. The generally accepted notation for the blastomeres (first used by Conklin (1897) for the cleavage of the prosobranch *Crepidula*) can be seen from Table 11.1. This cleavage type has been reported from most of the phyla included here in the group Spiralia (Table 10.1), but many modifications have been described and the spiral pattern cannot be recognized at all for example in insects and cephalopods, which for other reasons are included in the group.

The cleavage pattern is not the only apomorphy which unites the phyla. Also the morphology and fate of the apical organ and its relation to the adult brain are highly characteristic.

The spiral cleavage pattern is not necessarily coupled with a high degree of determination; the pattern and its modifications will be discussed first.

The main (apical-blastoporal) axis of the embryo becomes fixed during oogenesis (Huebner & Anderson 1976). The orientation is known to be related to the position of the oocyte in the ovary with the blastoporal (vegetal) pole facing the ovary wall, for example in several molluscs which have intraovarian oogenesis (Raven 1976) and in nemertines (Wilson 1903), but for example in many polychaetes the oocytes mature floating in the coelomic fluid so that this relationship cannot be recognized (Eckelbarger 1988). The polar bodies are given off at the apical (animal) pole, and since fertilization usually takes place before the meiotic divisions have been completed, the polar bodies are retained inside the fertilization membrane and can be used as markers for the orientation of the embryos.

Fertilization may take place anywhere on the egg, as in some polychaetes, but the entry point of the spermatozoon is fixed in other forms, such as many arthropods which have an egg envelope with a micropyle. The entry point of the spermatozoon determines the secondary main axis of the embryo (the dorso-ventral axis) in many species because the first cleavage furrow forms through this point and the apical pole (Guerrier 1970, van den Biggelaar & Guerrier 1983). The axes of the embryo and the adult can therefore be recognized already at the first cleavage stage.

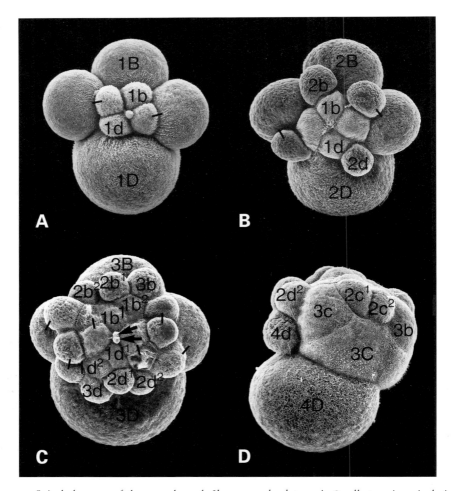

Fig. 11.1. Spiral cleavage of the prosobranch *Ilyanassa obsoleta.* – A, 8-cell stage in apical view. – B, 12-cell stage in apical view , – C, 24-cell stage in apical view. – D, 28-cell stage in right view. In A-C, the cells of the B and D quadrants are numbered, and the daughter blastomeres of the last cleavage are united by short lines in the A and C quadrants; the polar bodies are indicated by arrows. (SEM courtesy of Dr M.M. Craig, Southwest Missouri State Univ., Springfield, MO, USA; see Craig & Morrill 1986).

The sizes of the blastomeres may be equal so that it is difficult to identify specific blastomeres at the early stages, but size differences or other characters make it possible to name the individual blastomeres as early as at the 2-cell stage of many species. When size differences are apparent, it is usually the D-quadrant which is larger than the others.

The 'typical' spiral cleavage with four quartets and mesoderm-formation from the 4d-cell has been found in several phyla and is described in most text-books. The cleavage and the normal fate of the resulting blastomeres can be correlated in detail

Table 11.1. Spiral cleavage in an annelid *(Arenicola)* and a mollusc *(Trochus)*. Based on Siewing (1969).

Arenicola — *Trochus*

Arenicola lineage	Fate	Trochus lineage
$1a^{111}$, $1a^{112}$	apical rosette cells	$1a^{111}$, $1a^{112}$
$1a^{12}$	ectoderm of episphere	$1a^{12}$
$1a^2$	primary trochoblasts	$1a^2$
$2a$	ectoderm	$2a$
$3a$	ectomesoderm	$3a$
$3A$	endoderm	$3A$
B, C	development as A	B, C
$1d$	ectoderm of episphere	$1d$
$2d$	somatoblast	$2d$
$3d$	ectoderm	$3d$
$4d$	mesoblast	$4d$
$4D$	endoderm	$4D$

(Groupings in the original dendrogram: Arenicola — $Z\{AB\{A\{1A,2A\},B\},C,CD\{D\{1D,2D,3D\}\}\}$; with $1a\{1a^1\{1a^{11}\{1a^{111},1a^{112}\},1a^{12}\},1a^2\}$. Trochus shown as the mirror image.)

in embryos of annelids and molluscs (Table 11.1); the cleavages of most of the other spiralian phyla have not been studied in sufficient detail to allow similar comparisons. A detail not included in the table could be added here: the prototroch is usually formed from descendants of the primary trochoblasts, $1a^2$-$1d^2$, often supplemented by some secondary trochoblasts, $2a^{11}$-$2c^{11}$, $2a^{121}$-$2c^{121}$ or $1a^{1222}$-$1c^{1222}$ (Anderson 1973), leaving a small temporary break at the dorsal side of the prototroch corresponding to the break postulated in the ancestral trochophore larva (Chapter 10).

A sometimes quite conspicuous, cross-shaped pattern of blastomeres around the apical pole has been observed in several molluscs, annelids and sipunculans, but the differences in the fate of the cross cells between the phyla (sometimes characterized as annelid cross versus molluscan cross) only reflect that the crosses are not formed by identical cells as defined by the cleavage pattern (Siewing 1969). These differences appear to be of minor phylogenetic importance.

The spiral pattern is easily recognized in species with small eggs (typically 50-300 µm in diameter), but many types of modifications occur for example in species with large, yolky eggs and in species with placentally nourished embryos. There is, however, no direct correlation between amount of yolk and cleavage pattern. The eggs of the prosobranch *Busycon carica* are about 1.7 mm in diameter, but the cleavage is nevertheless holoblastic and the spiral pattern can easily be followed (Conklin 1907). The very large eggs of the cephalopods (0.6-17 mm in diameter) on the other hand all show discoidal cleavage (Fioroni 1978). Large systematic groups such as classes may have uniform developmental features, such as the discoidal cleavage of the cephalopods or the superficial cleavage of the insects (Anderson 1973), but in other groups considerable variation may occur even within genera. An example of this is the the hardly recognizable spiral pattern in the embryos of the entoproct *Loxosomella vivipara*, which has very small eggs and placentally nourished embryos, whereas many other species of the same genus have a normal spiral pattern (Chapter 23). It is generally recognized that spiral cleavage has been lost in many groups under influence of large amounts of yolk or placental nourishment of the embryos, so it should not be controversial to include classes, or even phyla, without spiral cleavage, such as cephalopods, in the Spiralia.

Spiral cleavage with only two 'quartets' – usually called duet cleavage – appears to be characteristic of the acoel turbellarians (Chapter 26). The blastomere pattern resembles a normal quartet cleavage in which the first (or the second) cleavage does not take place. The fates of the blastomeres correspond well with that of the normal quartet cleavage, except that the acoels do not have a well defined gut. This cleavage type can hardly be interpreted as anything but a specialized spiral cleavage.

Costello & Henley (1976) reinterpreted earlier studies on cirripedes and rotifers and proposed that these groups have a spiral cleavage with only one quartet – a monet cleavage (or in other words a spiral cleavage in which the two first cleavages of the normal spiral pattern have disappeared). A scrutiny of the original descriptions of cirripede embryology (Chapter 20, Table 20.2) shows, however, that the cleavage pattern of *Balanus* can better be interpreted as a quartet spiral cleavage because mesoderm originates equally from the three smaller cells formed by the two first cleavages, whereas mesoderm is apparently never formed from the first quartet of micromeres in the normal spiral cleavage. Also, the second polar body lies in the cross between the cells in the 4-cell stage and is retained in a position indicating the apical pole (Bigelow 1902). Anderson's (1969) studies of several cirripedes show well defined spiral patterns. The mesoderm originates exclusively from the A-C-quadrants according to Anderson (1973), but Bigelow (1902) claimed that mesoderm was formed also from the D-quadrant.

All reports of spiral cleavage in aschelminths (rotifers, see above and Chapter 29, and gastrotrichs, Chapter 33) and deuterostomes (phoronids, Chapter 44) have turned out to be erroneous. The presence of oblique spindles in the two first cleavages is definitely not enough to characterize a cleavage pattern as spiral.

Cleavage patterns of the spiral type can be recognized in sipunculans, molluscs, annelids (see above), arthropods, entoprocts, platyhelminths, nemertines, and perhaps in ectoprocts (Fig. 11.2), while there are no reliable reports of this pattern in onychophorans and tardigrades, which for other reasons have been included in the Spiralia. There are many deviations from the 'normal' spiral pattern with quartets, but they can all be interpreted as modified spiral cleavages.

In sipunculans, only *Golfingia vulgare* has been studied in detail (Chapter 13). It has a very typical spiral cleavage pattern (Fig. 11.2), with the apical cells giving rise to a sensory organ with a tuft of stiff cilia and the cells $1a^2$-$1d^2$ being the primary trochoblasts; the cells $1a^{122}$-$1c^{122}$ are supposedly secondary trochoblasts; the 4d-cell gives rise to mesoderm.

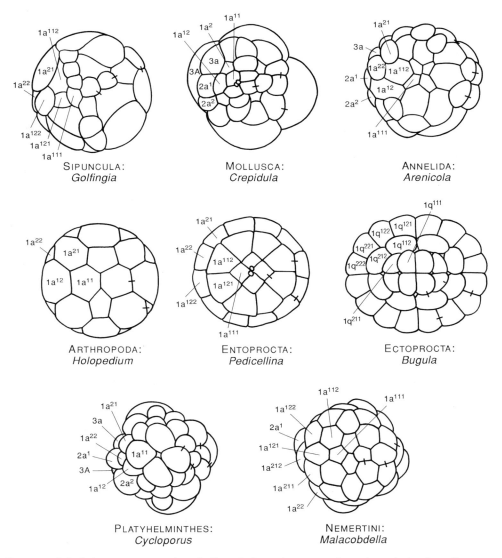

Fig. 11.2. Spiral cleavage patterns in spiralian phyla; embryos seen from the apical pole. – Sipuncula: 48-cell stage of *Golfingia vulgare* (redrawn after Gerould 1906). – Mollusca: 29-cell stage of *Crepidula fornicata* (redrawn from Conklin 1897). – Annelida: 48-cell stage of *Arenicola marina* (redrawn from Child 1900). – Arthropoda: 31-cell stage of *Holopedium gibberum* (redrawn from Baldass 1937). – Entoprocta: 48-cell stage of *Pedicellina cernua* (redrawn after Marcus 1939). – Ectoprocta: 64-cell stage of *Bugula neritina* (redrawn from Corrêa 1948). – Platyhelminthes: 32-cell stage of *Cycloporus papillosus* (redrawn from Bresslau 1928-31). – Nemertini: 44-cell stage of *Malacobdella grossa* (redrawn from Hammarsten 1918). The blastomeres of the A-quadrant are numbered and the blastomere pairs of the latest cleavage are indicated by small lines in the C-quadrant. The quadrants cannot be identified in the ectoproct embryos and in the early stages of the nemertine embryos.

Many studies on cleavage patterns of molluscs and annelids demonstrate how the 'typical' spiral pattern, shown for example in Table 11.1, can be modified in many ways (see also Chapters 15 and 17; review in Anderson 1973).

A cleavage in which the endoderm originates exclusively from the D-quadrant has been observed in the clitellate *Chaetogaster* (Anderson 1973, p. 59). Another cleavage type, resembling a real 'monet' type, has been described from the earthworm *Eisenia* (Devries 1968): its 4-cell stage has a D-cell which is much larger than the other three blastomeres and the three small cells remain undivided during the following stages; the D-cell develops much like the D-cell in a normal spiral cleavage and forms the usual cell groups. Further deviations from the spiral pattern have been described for example in the oligochaete *Stylaria*, in which the whole embryo develops from the cells 2B and 2C while the micromeres of the first quartet and the macromeres 1A and 1B together develop into a protective envelope surrounding the embryo (Dawydoff 1941). A complete lack of 4d-cell mesoderm has been reported for the prosobranch *Paludina* (Chapter 15), where all the mesodermal organs originate from ectomesoderm. This somewhat unexpected finding has been met with some doubt by various authors, but it appears well documented.

Among the arthropods, only a few crustaceans show spiral cleavage (Chapter 20). The cirripedes (Fig. 20.2) were mentioned above, but also some copepods and cladocerans (Fig. 11.2) show the spiral pattern. The cladoceran *Holopedium* (Baldass 1937) has a rather small egg in which the nucleus divides three times before the cell divisions begin, and the first cleavage stage is an '8-cell' stage, but with undivided centre. The next two cleavages show the normal spiral pattern, but the following stages are modified; the 2D-cell becomes a primordial germ cell surrounded by the 2d-cell which becomes mesoderm and a horseshoe of anterior and lateral blastomeres which become endoderm. This developmental type forms a transition to the more derived, entolecithal types in malacostracans and insects.

All entoprocts studied so far show spiral cleavage, and the early embryology of most species resembles that described for *Pedicellina cernua* (Chapter 23 and Fig. 11.2). The apical rosette cells give rise to an apical organ with a tuft of stiff cilia and the cells $1a^2$-$1d^2$ and $1a^{122}$-$1d^{122}$ give rise to the prototroch; the 4d-cell forms mesoderm.

There are only very few studies on the early development of ectoprocts (Chapter 24), but the cell-lineage study of *Bugula* by Corrêa (1948) shows that the cells of the corona or prototroch develops from the cells $1q^{12}$ and $1q^{22}$ (Fig. 11.2), which is in agreement with the general spiralian cell lineage; the cleavage is sometimes described as biradial and there is no sign of the spiral pattern, but similar bilateral stages are known from older embryos of both annelids and crustaceans, so the ectoproct cleavage may be a modified spiral cleavage after all (Chapter 24).

Among the platyhelminths, the acoels have the duet cleavage described above, while a normal spiral cleavage with quartets has been observed in most of the groups with entolecithal eggs (Chapter 26 and Fig. 11.2). The cleavage pattern is the normal one with alternating laeotrophic and dexiotrophic cleavages and the most apical cells develop an apical organ with stiff cilia; some of the larvae develop a more conspicuous band of cilia at a later stage, but the fates of the blastomeres have not been followed so far. The mesoderm develops from descendants of the 4d-cell and from the cells 2a-d.

All nemertines show spiral cleavage (Chapter 27 and Fig. 11.2). The apical cells form an apical organ with a few cells with mono- or multiciliate cells; the planktotrophic pilidium larvae develop a band of compound cilia around the equatorial part of the embryo, and some experiments with marking of blastomeres have shown that the large ciliary band originates from descendants of the cells $1a^2$-$1d^2$ and 2a-d as in normal trochophores. Mesoderm is apparently formed from different cells in the various species: from the 4d-cell (with additional cells from the invaginating endoderm) in *Lineus*, from the $2a^{III}$-$2d^{III}$ in *Malacobdella* (ectomesoderm), and from the 3A-C cells in *Tubulanus*, which has only three micromere quartets (Chapter 27).

Anderson (1973) has used fate maps for comparisons of the embryology of annelids and arthropods. These very informative maps have made it possible to compare developmental types without holoblastic cleavage with the more primitive forms. The results are discussed in some detail in Chapters 17 and 20. It is possible that such maps could be used to contrast spiralians and aschelminths, but the embryology of most of the aschelminth phyla is so poorly known that this can only be pointed out as an interesting possibility in the future.

The high degree of determination in the development is demonstrated by the fact that isolated blastomeres or blastomere groups are not able to regulate and form 'normal' embryos. This does not mean that the spiralian embryo lacks regulative powers, as shown in annelids by Dorresteijn, Bornewasser & Fischer (1987) and molluscs by van den Biggelaar & Guerrier (1979). Separations of blastomeres of 2- and 4-cell stages of the nemertine *Cerebratulus* have given at least partially normal embryos (Chapter 27), but the small larvae were not followed to metamorphosis. It is clear that a certain degree of regulative powers is present, but this is still far from the totipotentiality of the blastomeres of the 4-cell stage of the sea-urchin embryo (Chapter 46).

Apical organs are found in larvae of almost all eumetazoans, but they have a special morphology in the spiralians, and this is probably another apomorphy of the group (Fig. 11.3). The spiralian apical organ is often onion-shaped with a pair of anterior nerves (or a single, median nerve) extending to the prototroch and further to the ventral nervous system, and a pair of muscles to the region of the mouth. The molluscs have very small apical organs with the cerebral ganglia developing from cells of the episphere just lateral to the apical cells; ventral extensions from these ganglia go to the anterior part of the foot where a pair of statocysts are formed (Conklin 1897). Muscles are not developed until at a later stage. The apical organs of platyhelminths and nemertines are of a similar shape but lack the nerves to a ventral nervous system, but the parenchymians apparently lack the ventral nervous system in all stages (Chapter 25). Among the spiralians, only the arthropods lack apical organs. The apical organs of cnidarian and deuterostome larvae are usually a rather thin ectodermal cell group with a tuft of long cilia and a basiepithelial nerve plexus; only the tornaria larva (Chapter 50) has a more onion-shaped apical organ, in some species even with a pair of eyes, but it lacks nerves extending to a ventral ganglion and also muscles to the mouth region. The aschelminths apparently lack apical organs altogether (Chapter 28).

The very special type of cleavage and possibly also the type of apical organ characterize the spiralians as a monophyletic group.

The spiralian pattern of mesoderm formation – the 4d-cell mesoderm – could perhaps be seen as a specialization from the pattern of mesoderm formation found in the aschelminths, viz. from cells around the blastopore; this would make the aschelminths a paraphyletic group. However, other characters, such as the direct development and the lack of apical organs, could indicate that the aschelminths are the specialized group (Chapter 28). At the present stage it appears best to consider the two groups as sister groups (Fig. 9.1).

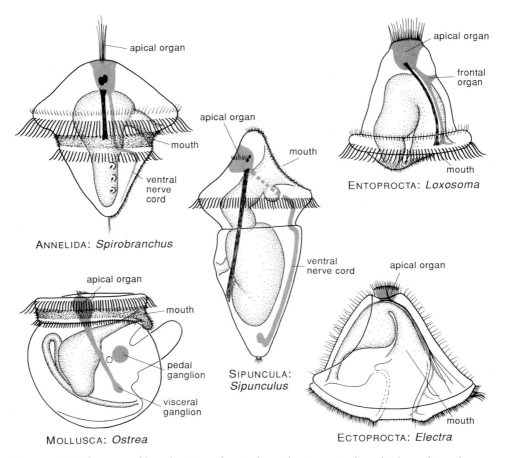

Fig. 11.3. Apical organs of larval or juvenile spiralians showing apical cerebral ganglia and nerves and muscles to the ventral side; in entoprocts and ectoprocts the nerves and muscles extend only to the prototroch zone; all specimens are seen from the right side. – Sipuncula: *Sipunculus nudus* larva ready for metamorphosis (redrawn from Hatschek 1883). – Mollusca: larva of *Ostrea edulis* (based on Erdmann 1935). – Annelida: early metatrochophore of *Spirobranchus polycerus* (based on Lacalli 1984). – Entoprocta: larva of *Loxosoma pectinaricola* (based on Nielsen 1971). – Ectoprocta: larva of *Electra pilosa* (based on Nielsen 1971).

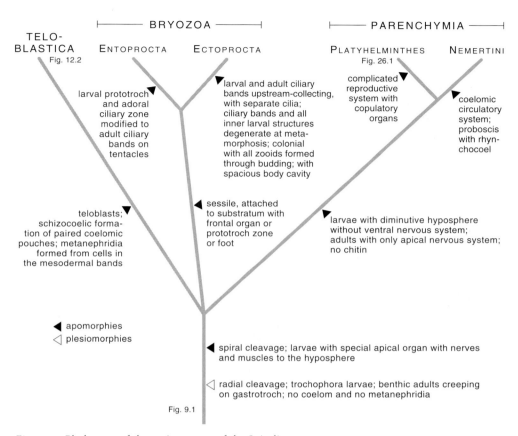

TELO-
BLASTICA
Fig. 12.2

ENTOPROCTA

⊢———— BRYOZOA ————⊣

ECTOPROCTA

⊢——— PARENCHYMIA ———⊣

PLATYHELMINTHES
Fig. 26.1

NEMERTINI

larval prototroch▼
and adoral
ciliary zone
modified to
adult ciliary
bands on
tentacles

▼larval and adult ciliary
bands upstream-collecting,
with separate cilia;
ciliary bands and all
inner larval structures
degenerate at meta-
morphosis; colonial
with all zooids formed
through budding; with
spacious body cavity

complicated▼
reproductive
system with
copulatory
organs

▼coelomic
circulatory
system;
proboscis
with rhyn-
chocoel

teloblasts;▼
schizocoelic forma-
tion of paired coelomic
pouches; metanephridia
formed from cells in
the mesodermal bands

◄ sessile, attached
to substratum with
frontal organ or
prototroch zone
or foot

▼larvae with diminutive hyposphere
without ventral nervous system;
adults with only apical nervous system;
no chitin

◄ apomorphies
◁ plesiomorphies

◄ spiral cleavage; larvae with special apical organ with nerves
and muscles to the hyposphere

◁ radial cleavage; trochophora larvae; benthic adults creeping
on gastrotroch; no coelom and no metanephridia

Fig. 9.1

Fig. 11.4. Phylogeny of the main groups of the Spiralia.

References

Anderson, D.T. 1969. On the embryology of the cirripede crustaceans *Tetraclita rosea* (Krauss), *Tetraclita purpurascens* (Wood), *Chthamalus antennatus* (Darwin) and *Chamaesipho columna* (Spengler) and some considerations of crustacean phylogenetic relationships. – Phil. Trans. R. Soc. B **256**: 183-235.

Anderson, D.T. 1973. Embryology and phylogeny in annelids and arthropods. – Pergamon Press, Oxford.

Baldass, F.v. 1937. Entwicklung von *Holopedium gibberum*. – Zool. Jb., Anat. **63**: 399-545.

Bigelow, M.A. 1902. The early development of *Lepas*. A study of cell-lineage and germ-layers. – Bull. Mus. comp. Zool. Harv. **40**: 61-144, 12 pls.

Bresslau, E. 1928-31. Turbellaria. – Handbuch der Zoologie, 2. Band, 1. Hälfte (2), pp 52-320. Walter de Gruyter, Berlin.

Child, C.M. 1900. The early development of *Arenicola* and *Sternaspis*. – Arch. Entwicklungsmech. Org. **9**: 587-723, 22-25.

Conklin, E.G. 1897. The embryology of *Crepidula*. – J. Morph. **13**: 1-226, pls 1-9.

Conklin, E.G. 1907. The embryology of *Fulgur*: a study of the influence of yolk on development. – Proc. Acad. nat. Sci. Philad. **59**: 320-359, pls 23-28.

Corrêa, D.D. 1948. A embryologia de *Bugula flabellata* (J.V. Thompson) (Bryozoa Ectoprocta). – Bolm Fac. Filos. Ciênc. Univ. S. Paulo, Zool. **13**: 7-54, pls 1-8.

Costello, D.P. & C. Henley 1976. Spiralian development: a perspective. – Am. Zool. 16: 277-291.

Craig, M.M. & J.B. Morrill 1986. Cellular arrangements and surface topography during early development in embryos of *Ilyanassa obsoleta*. – Int. J. Invert. Reprod. Dev. 9: 209-228.

Dawydoff, C. 1941. Études sur l'embryologie des Naididae indochinoises. – Archs Zool. exp. gén. 81 (Notes et Revue): 173-194.

Devries, J. 1968. Les premières étapes de la segmentation (formation de la jeune blastule) chez le Lombricien *Eisenia foetida*. – Bull. Soc. zool. Fr. 93: 87-97.

Dorresteijn, A.W.C., H. Bornewasser & A. Fischer 1987. A correlative study of experimentally changed first cleavage and Janus development in the trunk of *Platynereis dumerilii* (Annelida, Polychaeta). – Roux's Arch. dev. Biol. 196: 51-58.

Eckelbarger, K.J. 1988. Oogenesis and female gametes. – Microfauna mar. 4: 281-307.

Erdmann, W. 1935. Untersuchungen über die Lebensgeschichte der Auster. Nr. 5. Über die Entwicklung und die Anatomie det 'ansatzreifen' Larve von *Ostrea edulis* mit Bemerkungen über die Lebensgeschichte der Auster. – Wiss. Meeresunters., N.F. Helgoland 19(6): 1-25, 8 pls.

Fioroni, P. 1978. Cephalopoda, Tintenfische. – *In* F. Seidel (ed.): Morphogenese der Tiere, Deskriptive Morphogenese, 2. Lieferung, pp 1-181. VEB Gustav Fischer, Jena.

Gerould, J.H. 1906. The development of *Phascolosoma*. – Zool. Jb., Anat. 23: 77-162, pls 4-11.

Guerrier, P. 1970. Les caractères de la segmentation et la détermination de la polarité dorsoventrale dans le développement de quelques Spiralia. – J. Embryol. exp. Morph. 23: 667-692.

Hammarsten, O.D. 1918. Beitrag zur Embryonalentwicklung der *Malacobdella grossa* (Müll.). – Arb. zootom. Inst. Univ. Stockh. 1: 1-96, 10 pls.

Hatschek, B. 1883. Über Entwicklung von *Sipunculus nudus*. – Arb. zool. Inst. Univ. Wien 5: 61-140, pls 4-9.

Huebner, E. & E. Anderson 1976. Comparative spiralian oogenesis – structural aspects: an overview. – Am. Zool. 16: 315-343.

Lacalli, T.C. 1984. Structure and organization of the nervous system in the trochophore larva of *Spirobranchus*. – Phil. Trans. R. Soc. B 306: 79-135, 19 pls.

Marcus, E. 1939. Briozoários marinhos brasileiros III. – Bolm Fac. Filos. Ciênc. Univ. S. Paulo, Zool. 3: 111-299, pls 5-31.

Nielsen, C. 1971. Entoproct life-cycles and the entoproct/ectoproct relationship. – Ophelia 9: 209-341.

Raven, C. P. 1976. Morphogenetic analysis of spiralian development. – Am. Zool. 16: 395-403.

Siewing, R. 1969. Lehrbuch der vergleichenden Entwicklungsgeschichte der Tiere. – Paul Parey, Hamburg.

van den Biggelaar, J.A.M. & P. Guerrier 1983. Origin of spatial organization. – *In* K.M. Wilbur (ed.): The Mollusca, vol. 3, pp 197-213. Academic Press, New York.

Wilson, E. B. 1903. Experiments on cleavage and localization in the nemertine-egg. – Arch. Entwicklungsmech. Org. 16: 411-460.

12

TELOBLASTICA

The metameric spiralians (annelids and panarthropods) have often been united in a group called articulates, which has been considered as rather closely related to the molluscs and the sipunculans, although these two phyla have mostly been interpreted as unsegmented. The following discussion of the molluscs (Chapter 14) concludes that the ancestor of this phylum must have had eight segments, but that these segments most probably lacked spacious coelomic cavities functioning as a hydrostatic skeleton. The sipunculans show no traces of segmentation.

There is, however, a complex of fundamental characters which unite these groups. The spiral cleavage gives rise to a pair of mesoblasts which are formed from the 4d-cell, and these two cells are teloblasts which give off smaller cells anteriorly, so that a pair of mesodermal bands are formed (Fig. 12.1). The coelomic cavities are formed as slits in these compact mesodermal masses, the method of coelom formation called schizocoely. The teloblasts are more or less easily recognized in sipunculans, molluscs and annelids, whereas mesoderm of the arthropods generally develops from segmental mesoblasts; only some crustaceans have a modified type of

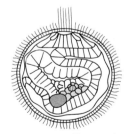 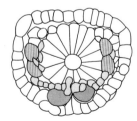 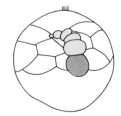

SIPUNCULA: *Sipunculus*   MOLLUSCA: *Physa*   ANNELIDA: *Amphitrite*

*Fig. 12.1. Formation of mesodermal bands from teloblasts (shaded) in the sipunculid *Sipunculus nudus* (redrawn from Hatschek 1883), the mollusc *Physa fontinalis* (redrawn from Wierzejski 1905), and the annelid *Amphitrite ornata* (redrawn from Mead 1897); the nephroblasts are hatched; teloblasts in oligochaete annelids are seen in Fig. 17.7.

mesoteloblasts (Chapter 20). None of the other spiralian phyla have mesoteloblasts. The schizocoelic development of coelomic cavities is well-documented in all the phyla except the tardigrades, where the origin of the coelomic cavities is uncertain (Chapter 21). The coelom of the ectoprocts develops in a most unusual way and its homology is very uncertain (Chapter 24). The blood vessel system and the rhynchocoel of the nemerteans are coeloms according to the usual definition of the word, but they are probably not homologous with those found in the Teloblastica (Chapter 27). The adult excretory organs, which also function as gonoducts, are metanephridia formed from cells in the mesodermal bands. None of the other spiralians have metanephridia (see discussion of the coelomoducts/gonoducts of the ectoprocts in Chapter 24).

The mesoteloblasts and the coelomic cavities formed through schizocoely in the mesodermal bands are unique and define the Teloblastica as a monophyletic group.

The segmentation is not the only synapomorphy of molluscs and euarticulates (annelids + panarthropods); also the haemal system with a dorsal contractile heart with the muscular walls formed by the median walls of pericardial coelomic cavities (Chapters 14, 15) appears to be an advanced character at this stage. The sipunculans have no traces of any of these characters (Chapter 13). This indicates a phylogenetic tree with the Sipuncula as the sister group of the Articulata, a group which comprises

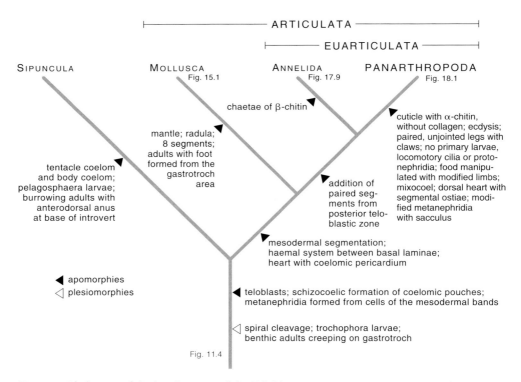

Fig. 12.2. Phylogeny of the basal groups of the Teloblastica.

Mollusca and Euarticulata (Fig. 12.2).

The ectoderm is a monolayered epithelium (only known exception is the polychaete *Travisia*, Chapter 17) resting on a basement membrane. In the non-arthropod groups, the apical side of the cells has numerous microvilli, usually with an extracellular meshwork of fibrils in the subdistal zone; the fibrils are known to consist of collagen in sipunculans and annelids (Chapters 13, 17). The panarthropods have an epithelium without microvilli and collagen, but with a chitinous cuticle (Chapter 18).

References

Hatschek, B. 1883. Über Entwicklung von *Sipunculus nudus*. – Arb. zool. Inst. Univ. Wien 5: 61-140, pls 4-9.

Mead, A.D. 1897. The early development of marine annelids. – J. Morph. 13: 227-326, pls 10-19.

Wierzejski, A. 1905. Embryologie von *Physa fontinalis* L. – Z. wiss. Zool. 83: 503-706, pls 18-27.

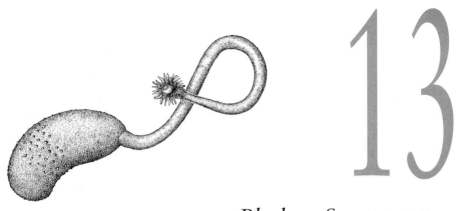

Phylum SIPUNCULA

Sipunculans form a small, well-defined phylum comprising about 320 described species, which are all marine. It is usually divided into four or six families. There is no reliable fossil record.

The more or less cylindrical organisms are divided into a body and a more slender, anterior, retractile introvert. The mouth is terminal and usually surrounded by a ring of ciliated tentacles. The gut forms a loop and the anus is situated dorsally at the base of the introvert.

The epidermis of the tentacles is covered by a thin cuticle of cross-layered collagen fibres with interspaced microvilli; the cuticle of the body is of the same structure, but much thicker (Pilger 1982, Storch 1984). Chitin appears to be completely absent (Jeuniaux 1982).

The nervous system (Fig. 10.2) comprises a dorsal, bilobed brain which is connected with a ventral nerve cord by a pair of circumoesophageal connectives. The ventral nerve cord is surrounded by a thin peritoneum and lies free in the body cavity only attached in the ventral mid-line by a number of fine nerves covered by the peritoneum; the nerve cord is accompanied by a pair of lateral muscles which cover the whole periphery in some zones (Harms 1920).

There are two body cavities, a small, anterior tentacle coelom and a large, posterior body coelom. Both cavities contain various types of coelomocytes, which are formed from the mesothelia (Ohuye 1942), and which under certain circumstances may move between the cavities (Metalnikoff 1900); nevertheless, the two cavities are normally isolated from each other, as shown by the different properties of the haemerythrins in the two cavities (Manwell 1963).

The tentacle coelom is circumoesophageal and sends canals into the branched tentacles and one or two median canals (compensation sacs) along the dorsal and ventral sides of the gut. The coelom is lined by a peritoneum which is ciliated especially in the tentacle canals, where the cilia are presumed to create a circulation of the coelomic fluid with haemocytes through a median and two lateral canals in each ten-

---

Chapter vignette: *Phascolion strombi*. (Based on Théel 1905.)

tacle (Pilger 1982). Some types have two rather short compensation sacs without diverticula, for example *Sipunculus*, while in others, such as *Themiste* (Fisher 1952), the dorsal sac is greatly expanded posteriorly (and called the contractile vessel) with numerous long, thin diverticula. This system clearly functions both as a hydrostatic skeleton expanding the tentacles and as a respiratory system which can transport oxygen from the expanded tentacles to the body coelom. It has been described as a system of blood vessels, but the ciliated coelomic walls show that it is a true coelom.

The main body coelom is a spacious cavity lined by a peritoneum overlying longitudinal and circular muscles and containing a fluid with haemocytes. The muscle layers are more or less continuous in the genera considered primitive and divided into separate muscles in the more advanced genera (Cutler 1986). The coelomic cavity extends into longitudinal canals or sacs between the muscles in the advanced genera. The body coelom functions as a hydrostatic skeleton both in eversion of the introvert and in burrowing.

A haemal system is absent, its functions apparently being carried out by the tentacle coelom and the coelomic canals of the body wall.

A pair of large metanephridia is found in the body coelom with the nephridiopores situated near the anus. The funnel is very large and ciliated and has a special function in separating the ripe eggs from the several other cell types in the coelomic fluid. In most metazoans, the primary urine is filtered from a haemal system to the coelom, but as also pointed out by Ruppert & Smith (1988) the sipunculan metanephridium is exceptional in that there is no haemal system from where the primary urine can be filtered. The suggestion that the filtration should be from the compensation sac of the tentacle coelom is not supported by experiments and appears unlikely since the tentacle coelom is so restricted and the body coelom is in direct contact with most of the muscles through the coelomic canals of the body wall. Podocytes have been observed in the trunk peritoneum of the contractile vessel (Pilger & Rice 1987), but not in the apposed peritoneum of the tentacle coelom, and podocytes obviously without a function in ultrafiltration have been reported from crustaceans (Wägele & Walter 1990) and enteropneusts (Chapter 50), so the mere presence of podocytes is not a proof of production of primary urine.

The gonad is a ventral, lobed organ surrounded by peritoneum and suspended in the body coelom by a mesentery. The oocytes develop to the first meiotic prophase in the ovary and are then released into the coelom where ripening takes place. The ripe egg is surrounded by a thick envelope with many pores. Some species have spherical eggs while others have spindle-shaped to flattened eggs with a shallow depression at the apical pole, indicating that the polarity of the egg is determined before spawning. The sperm may penetrate the egg envelope everywhere except at the apical pole (Rice 1989). Most species are free spawners, and the polar bodies are given off at the apical pole soon after fertilization (Rice 1989).

Cleavage is spiral (Fig. 11.2) with a cell lineage closely resembling that of annelids and molluscs. An apical organ is formed at the position of the polar bodies. The early prototroch is formed by descendants of the four primary trochoblasts, $1a^2$-$1d^2$, and three intermediate cells, $1a^{122}$-$1c^{122}$, so that there is a narrow dorsal gap, which closes at a later stage when the two ends of the band fuse (Gerould

1906). Gastrulation is embolic to epibolic according to the amount of yolk. Mesoderm is formed from the 4d-cell and a pair of mesoteloblasts (Fig. 12.1) have been observed in *Sipunculus* (Hatschek 1883), *Phascolopsis* (Gerould 1906) and *Phascolosoma* (Rice 1973). Gerould described and illustrated 3-4 small coelomic pouches in early trochophores of *Phascolopsis*, but according to Hyman (1959, p. 657) he later changed his interpretation and explained that the apparent metamerism was caused by contraction and buckling. Coelomic cavities arise through schizocoely in the paired mesodermal bands, but the two coelomic sacs fuse completely at a later stage. Hatscheck's (1883) drawings of an early metamorphosis stage of *Sipunculus* shows an undivided coelomic cavity extending to the anterior end of the larva in front of the mouth, and the tentacles are described as developing from the rim of the mouth; the origin of the tentacle coelom is not mentioned, but his observations indicate that it becomes pinched off from the body coelom after metamorphosis.

A few species have direct development but most species have lecithotrophic trochophores which swim with the cilia of the prototroch protruding through pores in the egg envelope. In some of these species, the trochophores metamorphose directly into the juveniles, but most species go through a planktotrophic or lecithotrophic stage called pelagosphaera (Rice 1981; Fig. 13.1). This larval type is characterized by a prominent ring of compound cilia behind the mouth and a prototroch which has become overgrown more or less completely by neighbouring ectodermal cells (Gerould 1906). The fully developed pelagosphaera larva has an extended ciliated lower lip with a buccal organ which can be protruded from a deep, transverse ectodermal fold and a lip gland (Rice 1973). A further characteristic is a complicated retractile terminal organ with both sensory and secretory cells (Ruppert & Rice 1983). Both of these organs appear to represent sipunculan apomorphies.

Sipunculus has a special, pericalymma-like larval stage with the prototrochal epithelium extended posteriorly, covering mouth and hyposphere; this thin extension, called serosa, is shed at hatching and the larva becomes a normal pelagosphaera (Hatschek 1883).

Gerould (1906) observed the development of circular muscles from ectomesodermal cells in the zone just behind the prototroch, and also the retractor muscles of the introvert are believed to be ectomesodermal (Rice 1973). The other muscles are believed to originate from the 4d-mesoderm.

---

*Fig. 13.1. Pelagic developmental types of sipunculans. – *Golfingia vulgare* has a completely lecithotrophic development; the first stage is a roundish trochophore which swims with a wide band of cilia anterior to and around the mouth; the second stage is more elongate, the prototroch cells have become infolded and degenerate and the larva swims with a ring of compound cilia behind the mouth; the third stage is cylindrical and the gut is developing (based on Gerould 1906). – *Golfingia misakiensis* goes through a similar trochophora stage, but the gut becomes functional at an early stage; the full-grown pelagosphaera larva swims with the postoral ring of compound cilia and is able to retract the anterior part of the body, including the ciliary ring, into the posterior part (redrawn from Rice 1978). – *Sipunculus nudus* has a pericalymma larva (type 1) with the hyposphere completely covered by an extension of the prototrochal area; the hyposphere breaks out through the posterior end of the serosa, which is for a short time carried as a helmet over the episphere and then cast off; the full-grown larva is a normal pelagosphaera (based on Hatschek 1883).*

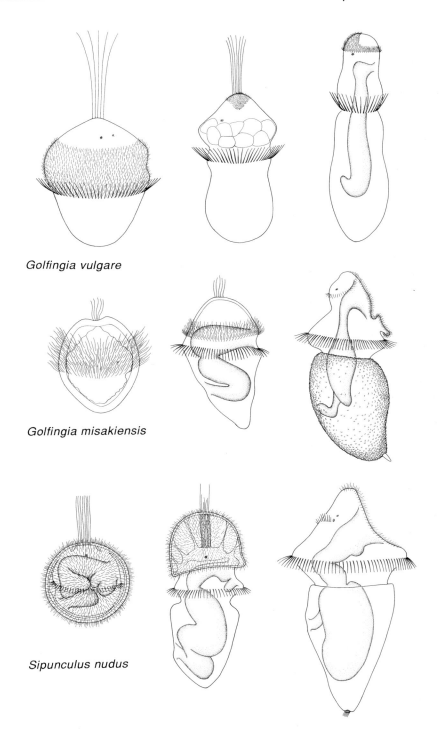

Golfingia vulgare

Golfingia misakiensis

Sipunculus nudus

The brain develops from the apical organ (Åkesson 1961). In *Sipunculus*, the ventral nerve cord develops as a median longitudinal thickening of the ectoderm which later splits off from the ectoderm and sinks into the body cavity covered by the mesothelium; the circumenteric connectives to the brain arise later as similar thickenings of the ectoderm (Hatschek 1883). In *Phascolosoma*, Rice (1973) observed a pair of ventral nervous cords in recently metamorphosed specimens and a fusing of the two cords during the following development; she also observed the circumoesophageal connectives at a stage where the ventral nerves were not yet developed, and suggested that the connectives develop from the brain.

Protonephridia have not been observed at any stage, but a pair of metanephridia develops in the early pelagosphaera stage. In *Sipunculus*, Hatschek (1883) observed a yellowish cell completely embedded in each of the lateral groups of mesodermal cells. These cells developed into U-shaped cell groups with a narrow canal, which came into contact with the ectoderm and formed the nephridiopore, and which broke through the peritoneum forming the ciliated funnel of a typical metanephridium. Rice (1973) observed a similar development in *Phascolosoma*, with ectodermal cells giving rise to the pore region. However, Gerould (1906) believed that the main part of the metanephridia in *Golfingia* develops from the ectoderm, with only the ciliated funnel originating from the mesoderm. It cannot be excluded that variations occur in the development of metanephridia, but Gerould's report appears less well documented.

The planktotrophic pelagosphaera larvae do not use the ciliary bands in filter feeding, but the nature of the feeding mechanism is unknown. Jägersten (1963) observed swallowing of large particles, such as fragments of other larvae, and found copepods in the gut of freshly caught larvae, so the larvae may be carnivorous.

The presence of a typical spiral cleavage with mesoderm originating from the 4d-cell and the development and morphology of the central nervous system clearly place the Sipuncula in the spiralian line of the Protostomia. The teloblastic proliferation of the mesoblasts and the schizocoelic formation of paired coelomic pouches indicate a close relationship with the mollusc-articulate group. The complete absence of segmentation and of blood vessels indicates that the Sipuncula are descendants of unsegmented metazoans without a haemal system and that they are the sister group of Mollusca + Articulata, which have both these characteristics. It is possible that segmentation can be lost during evolution but it appears most unlikely that an organ system as important as the haemal system should disappear and its functions be taken over by an apparently less efficient system like that of the tentacle coelom. A similar conclusion about the position of the sipunculans was reached by Rice (1985). The pelagosphaera larva is a quite aberrant trochophore; its prominent ring of compound cilia behind the mouth is usually interpreted as a metatroch, but the direction of the effective stroke is opposite to that of metatrochs of other spiralians. Jägersten (1972) considered the ciliated lower lip with the lip gland as a creeping organ, which had been the main locomotory organ in a creeping ancestor, i.e. a foot derived from the gastrotroch like the foot of the molluscs. This supports the interpretation of the large ciliary band as an additional band like those occurring just behind the mouth in larvae of polychaetes such as *Chaetopterus* or mollusc larvae such as those of pteropods (Nielsen 1987).

Interesting subjects for future research

1. Development of the ventral cord.
2. Origin of the tentacle coelom.

References

Åkesson, B. 1961. The development of *Golfingia elongata* Keferstein (Sipunculidea) with some remarks on the development of neurosecretory cells in sipunculids. – Ark. Zool., 2. ser., **13**: 511-531.

Cutler, E.B. 1986. The family Sipunculidae (Sipuncula): body wall structure and phylogenetic relationships. – Bull. mar. Sci. **38**: 488-497.

Fisher, W.K. 1952. The sipunculid worms of California and Baja California. – Proc. U.S. natn. Mus. **102**: 371-450, pls 18-39.

Gerould, J.H. 1906. The development of *Phascolosoma*. – Zool. Jb., Anat. **23**: 77-162, pls 4-11.

Harms, W. 1920. Bauchnervenstrang und Spindelmuskel von *Physcosoma* in Anpassung an die Formveränderungen dieses Tieres. – Zool. Anz. **52**: 67-76.

Hatschek, B. 1883. Über Entwicklung von *Sipunculus nudus*. – Arb. zool. Inst. Univ. Wien **5**: 61-140, pls 4-9.

Hyman, L.H. 1959. The Invertebrates, vol. 5. – McGraw-Hill, New York.

Jägersten, G. 1963. On the morphology and behaviour of pelagosphaera larvae (Sipunculoidea). – Zool. Bidr. Uppsala **36**: 27-35.

Jägersten, G. 1972. Evolution of the Metazoan Life Cycle. – Academic Press, London.

Jeuniaux, C. 1982. La chitine dans le règne animal. – Bull. Soc. zool. Fr. **107**: 363-386.

Manwell, C. 1963. Genetic control of hemerythrin specificity in a marine worm. – Science **139**: 755-758.

Metalnikoff, S. 1900. *Sipunculus nudus*. – Z. wiss. Zool. **68**: 261-322, pls 17-22.

Nielsen, C. 1987. Structure and function of metazoan ciliary bands and their phylogenetic significance. – Acta zool. (Stockh.) **68**: 205-262.

Ohuye, T. 1942. On the blood corpuscules and the hemopoiesis of a nemertean, *Lineus fuscoviridis*, and a sipunculan, *Dendrostoma minor*. – Sci. Rep. Tôhoku imp. Univ., Biol. **17**: 187-196.

Pilger, J.F. 1982. Ultrastructure of the tentacles of *Themiste lageniformis* (Sipuncula). – Zoomorphology **100**: 143-156.

Pilger, J.F. & M.E. Rice 1987. Ultrastructural evidence for the retractile vessel of sipunculans as a possible ultrafiltration site. – Am. Zool. **27**: 152A.

Rice, M.E. 1973. Morphology, behavior, and histogenesis of the pelagosphera larva of *Phascolosoma agassizii* (Sipuncula). – Smithson. Contr. Zool. **132**: 1-51.

Rice, M.E. 1978. Morphological and behavioral changes at metamorphosis in the Sipuncula. – In F.-S. Chia & M.E. Rice (eds): Settlement and Metamorphosis of Marine Invertebrate Larvae, pp 83-102. Elsevier, New York.

Rice, M.E. 1981. Larvae adrift: patterns and problems in life histories of sipunculans. – Am. Zool. **21**: 605-619.

Rice, M.E. 1985. Sipuncula: developmental evidence for phylogenetic inference. – In S. Conway Morris, J.D. George, R. Gibson & H.M. Platt (eds): The Origins and Relationships of Lower Invertebrates, pp 274-296. Oxford Univ. Press, Oxford.

Rice, M.E. 1989. Comparative observations of gametes, fertilization, and maturation in sipunculans. – In J.S. Ryland & P.A. Tyler (eds): Reproduction, Genetics and Distribution of Marine Organisms, pp 167-182. Olsen & Olsen, Fredensborg.

Ruppert, E.E. & M.E. Rice 1983. Structure, ultrastructure, and function of the terminal organ of a pelagosphaera larva (Sipuncula). – Zoomorphology **102**: 143-163.

Ruppert, E.E. & P.R. Smith 1988. The functional organization of filtration nephridia. – Biol. Rev. **63**: 231-258.

Storch, V. 1984. Echiura and Sipuncula. – *In* J. Bereiter-Hahn, A.G. Matoltsy & K.S. Richards (eds): Biology of the Integument, vol. 1, pp 368-375. Springer, Berlin.

Theel, H. 1905. Northern and Arctic invertebrates in the collection of the Swedish State Museum. I. Sipunculids. – K. svenska Vetenskapsakad. Handl. 39(1): 1-130, 15 pls.

Wägele, J.-W. & U. Walter 1990. Discovery of extranephridial podocytes in isopods. – J. crust. Res. 10: 400-405.

14

ARTICULATA

The segmented protostomes have often been treated under the collective name Articulata, but there has been much discussion about the status of the molluscs. The discussion of the molluscs (Chapter 15) has led me to the conclusion that the ancestor of this phylum had eight segments, and I have chosen to use the term Articulata in a wide sense, namely for the molluscs plus the euarticulates (annelids and panarthropods). These phyla share the teloblastic mesoderm formation and the development of the coelomic sacs through schizocoely with the sipunculans (Chapter 13), but the apomorphies of the Articulata comprise not only the presence of segments but also the presence of a haemal system (a circulatory system with blood flowing in extra-coelomic spaces; Chapter 9); the sipunculans have apparently never had a circulatory system of this type, and have developed special extensions from the tentacle coelom and in some species additional extensions from the body coelom into the body wall for the transport of substances (Chapter 13). The haemal system is surrounded by basement membran of mesoderm or endoderm, and the circulation is set up by muscle cells in the bordering epithelia. There is usually a dorsal heart surrounded by a pericardium, the pericardial sac being a small coelomic cavity.

The phylogeny of the main groups of articulates is shown in Fig. 12.2.

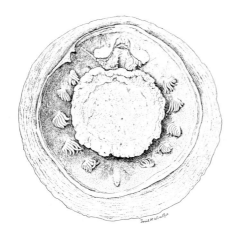

Phylum MOLLUSCA

Molluscs are one of the largest animal phyla with about 100 000 known living species and a very extensive fossil record. The number of known living species is somewhat uncertain – the numbers given by various authors lie between 50 000 and 150 000 – because malacologists from amateur shell collectors to professional biologists during the last centuries have given names to the same species several times often with rather chaotic results. Early Cambrian faunas of small, calcareous-shelled forms show considerable diversity, including representatives of several living and extinct classes and orders (Runnegar & Pojeta 1985). These well-preserved shells with distinct muscle scars have given the study of the early radiation of the shelled groups, especially the Conchifera (see below), a firm basis. However, the unshelled forms (such as the Aplacophora) are not known as fossils, so the early phylogeny must be based strongly on soft-part anatomy and embryology of living forms.

All the living classes are predominantly marine; only snails and mussels have spread to freshwater and only the snails have entered the terrestrial habitat. Some families of snails and mussels comprise commensal species and one family of snails contains entoparasitic species, which can only be recognized as molluscs by their larvae.

The phylum is very clearly delimited, but a wide variation in morphology of both adults and larvae makes it difficult to give a short definition of a mollusc. There are, however, a few features which can be recognized in almost all molluscs and which can therefore be used as defining characters:

1. The mantle: A large area of the dorsal epithelium which has a thick cuticle with calcareous spicules or one or more calcareous shells. The mantle is usually expanded in peripheral folds which protect the lateral sides of the body, sometimes covering it completely. The mantle is easily recognized in representatives of all the eight living classes.
2. The foot: A flat, ciliated, postoral, ventral expansion used in creeping or modified for other types of locomotion. A series of muscles from the foot to the mantle can pull the protective mantle towards the substratum or in other ways bring the soft

Chapter vignette: The monoplacophoran *Neopilina galatheae*. (From Lemche & Wingstrand 1959.)

parts into the protection of the mantle. The foot is easily identified in most classes, but it is reduced to a narrow keel in the solenogasters and has disappeared altogether in the caudofoveates.

3. The radula: A cuticular band with teeth formed in a pocket of the ventral epithelium of the oesophagus and used in feeding (Fig. 15.2). Radulae occur in all the classes except in the bivalves.

4. The central nervous system, including the cerebral ganglia, is concentrated around the oesophagus and there are two pairs of longitudinal main nerves. The nervous system is highly modified in several classes (see below).

The relationships of the eight classes of living molluscs have been discussed by a number of recent authors, and there appears to be some agreement about a phylogeny as in Fig. 15.1 (Lauterbach 1984, Wingstrand 1985, Scheltema 1988). The adult ancestral mollusc is believed to have been creeping on a ciliated foot and to have had a mantle with a mucous lining with calcareous spicules. Two evolutionary lines can be traced from this ancestor: 1) The aplacophoran line in which the foot lost most of the locomotive function and became narrow (Solenogastres) or completely reduced (Caudofoveata). In connection with the reduction or loss of the foot, the lateral parts of the mantle came close to each other or fused along the ventral mid-line. Locomotion be-

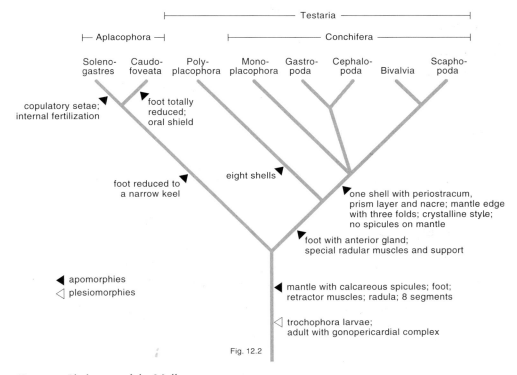

Fig. 15.1. Phylogeny of the Mollusca.

came worm-like, and the mantle retained the rather soft character with only small cal-careous spicules. 2) The testarian line in which the foot with the ciliary creeping sole was retained and the mantle expanded laterally and secreted calcareous shell plates.

Runnegar & Pojeta (1985) have presented a slightly different view of the evolu-tion within the Conchifera: They believe that the Monoplacophora represent a stem group which gave rise to gastropods, cephalopods and the now extinct rostroconchs (which in turn gave rise to bivalves and scaphopods); this concept fits equally well with the following discussion, which is mostly concerned with the evolution of the 'lower' molluscs.

A somewhat different phylogeny with the Caudofoveata as the sister group of 'Adenopoda' (= Solenogastres + Testaria) has been advocated by Salvini-Plawen (1972, 1985). It is based on the assumption that the molluscan ancestor was a turbel-lariomorph organism with the mouth situated in the anterior part of the ciliated creeping sole (foot). The anterior part of its foot was innervated by the cerebral gang-lia while the posterior part of the foot was innervated by the pedal nerves (Salvini-Plawen 1972, figure 45A). The oral shield (called pedal shield) of the caudofoveates was thus considered to represent the anterior part of the foot with scattered glandu-lar cells and to be homologous with the ciliated, glandular pedal pit of the soleno-gasters and the anterior foot gland of the testarians (specialized as the byssus gland in the bivalves). The glandular pit/foot gland was therefore interpreted as a synapo-morphy uniting the Solenogastres and the Testaria, with the 'foot shield' with scat-tered gland cells of the Caudofoveata representing the plesiomorphic character state. However, Scheltema (1988) showed that the oral shield cuticle is continuous with the cuticle of the pharynx in several solenogasters and further pointed out that also the innervation from the cerebral ganglia points to an association with the mouth rather than with the foot. Ontogenetically, the molluscan foot develops as a specialization of the gastrotroch area (clearly seen in polyplacophorans, gastropods and bivalves) and is thus postoral and not circumoral (although the gastrotroch is continuous with the adoral ciliary zone in the larva). The interpretation of the oral shield as the an-terior part of the foot must therefore be rejected, and Scheltema's (1988) interpretation of the oral shield as a specialized oral zone and thus as an apomorphy of the Solenogastres accepted. The ciliated pedal pit/gland of the solenogasters may repre-sent the plesiomorphic character state, and its disappearance in the caudofoveates and concentration as a foot gland in testarians can be seen as two divergent apomor-phies. This agrees with the phylogeny in Fig. 15.1.

The pelago-benthic life cycle occurs in all living classes except the cephalopods and was probably characteristic of the ancestral mollusc. The larva of the ancestor was probably a trochophore (see below). Comparative studies of living molluscs in-dicate that the adult of the ancestral mollusc was a creeping benthic organism re-sembling gastroneuron but having the mantle, foot, radula, and nervous system mentioned above.

Most reconstructions of a somewhat more advanced ancestor (see for example Yonge & Thompson 1976) include a number of additional characters: a mantle cav-ity with pectinate gills which developed in connection with the restriction of the res-piratory surface caused by the thickening of the mantle cuticle and the expansion of the foot. A circulatory system with a median aorta leading forwards from a posterior heart and paired vessels each with an atrium receiving blood from the pectinate gills. The muscular walls of the heart were formed by the median walls of one or more pairs of coelomic sacs (pericardia), which each had a ciliated duct functioning as gonoduct and metanephridium. The remaining part of the mesoderm was a more or less compact filling between the ectoderm and the gut. These characters are discussed in some detail below.

As mentioned above, the mantle can be recognized in almost all molluscs. It is covered by a layer of mucopolysaccharides with chitin and calcareous spicules secreted by single cells in the aplacophorans, and the perinotum surrounding the shells of the polyplacophorans has a similar structure (Haas 1981). The eight shells of the polyplacophorans have been interpreted as fused spicules, but this is not substantiated by direct observations, and Kniprath (1980) directly observed the formation of uninterrupted transverse shell plates in larvae of two species; larvae reared at raised temperatures formed isolated calcareous granules which fused to abnormal plates. The shells lack a periostracum and differ from the shells of the Conchifera both in deposition and structure (Haas 1981, Scheltema 1988). The shells of the Conchifera consist of periostracum, prisms and nacre and are secreted by the mantle, which has a characteristic edge with three folds separating different areas of secretory epithelium (Beedham & Trueman 1967). The slight modification of this pattern with addition of extra folds in some arcoid bivalves is interpreted as a specialization within this group (Waller 1980). The periostracum consists of quinone-tanned protein sometimes with β-chitin, and chitin occurs in the organic matrix of the calcareous shells as well (Watabe 1984).

The foot and the retractor muscles originating at the mantle and fanning out in the sole of the foot can be recognized in almost all molluscs. The caudofoveates have lost the foot completely by the fusion of the lateral mantle edges, but a narrow, midventral seam can be recognized in the primitive *Scutopus* (Salvini-Plawen 1972).

The radula (Fig. 15.2) is a band of thickened, toothed cuticle secreted by the apposed epithelia of a deep, posterior fold of the ventral side of the buccal cavity, the radular gland or sac. It consists of α-chitin and quinone-tanned proteins and may be impregnated with iron and silicon salts (Bubel 1984, Lowenstam & Weiner 1989). It can be protruded through the mouth and pulled back and forth over the tips of a pair of elongate cartilaginous structures and used to scrape particles from the substratum. Its presence in all classes, except the bivalves, indicates that the adult molluscan ancestor was a benthic deposit feeder or scraper, since a radula appears to be without function in ciliary filter feeders (a radula is present in the species which feed on plankton by various types of mucus nets).

Cuticular thickenings in the shape of teeth or jaws occur in several protostomes, for example rotifers and annelids, but a band with many similar transverse rows of chitinous cuticular teeth is only found in molluscs.

Wingstrand (1985) identified not only identical radular structures but also several homologous structures associated with the radulae of polyplacophorans and testarians, including hollow radula vesicles, cartilages of the odontophore and several sets of radular muscles, and these characters strongly support the monophyletic nature of the Conchifera.

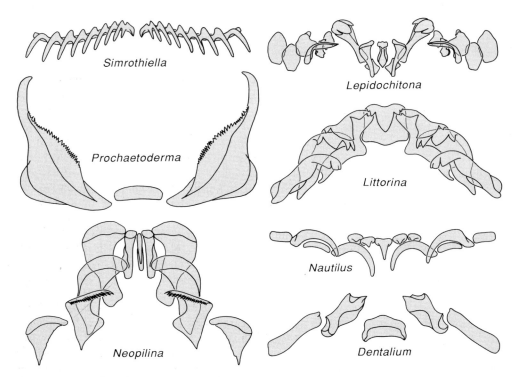

Fig. 15.2. Examples of radulae of the molluscan classes; the bivalves lack a radula. – Solenogastres: *Simrothiella* sp. (redrawn from Scheltema 1988). – Caudofoveata: *Prochaetoderma* sp. (redrawn from Scheltema 1981). – Polyplacophora: *Lepidochitona cinerea* (redrawn from Kaas & Van Belle 1985). – Monoplacophora: *Neopilina galatheae* (redrawn from Lemche & Wingstrand (1959). – Gastropoda: *Littorina littorea* (redrawn from Ankel 1936). – Cephalopoda: *Nautilus pompilius* (redrawn from Naef 1921). – Scaphopoda: *Dentalium entale* (redrawn from Lacaze-Duthiers 1858/59).

The general epidermis of the body is a monolayered ectodermal epithelium with microvilli and a subterminal web of extracellular fibrils (Bubel 1984).

The gut is straight in aplacophorans and more or less coiled in the testarians; only the conchiferans have a crystalline style (Wingstrand 1985). The foregut with the radula is formed from the stomodaeum, while the mid- and hindgut are usually formed from endoderm (Raven 1966).

The nervous system consists of a ring with ganglia surrounding the oesophagus; paired cerebral, pleural and pedal ganglia can usually be recognized. Two pairs of prominent nerves extend from the the cerebral ganglia to the two last-mentioned pairs of ganglia, the ventral (pedal) and lateral (pleural) nerves. These nerves are generally in the shape of nerve cords in aplacophorans and polyplacophorans, but well-defined ganglia connected by nerves without cell bodies are found in the solenogaster *Genitoconia* (Salvini-Plawen 1967), and ganglia and 'cell-free' nerves are the rule in the testarians.

Coelomic cavities functioning as hydrostatic skeletons, used for example in burrowing, are known in in many phyla, but this function is carried out by blood sinuses in molluscs (Trueman & Clarke 1988).

The circulatory system comprises a median, dorsal vessel with a heart; a pair of more or less fused posterior atria are found in most phyla, but two completely separate pairs of atria are found in some polyplacophorans, and in *Neopilina* and *Nautilus* (Wingstrand 1985); seven or eight pairs of gills are found in the monoplacophorans (see the chapter vignette), and two pairs in *Nautilus*. The peripheral part of the system comprises distinct capillaries in some organs but large lacunae are found, for example in the foot of burrowing bivalves (Trueman & Clarke 1988). The enormous swelling of the foot in some burrowing naticid gastropods is accomplished through intake of water into a complex sinus which is completely isolated from the circulatory system (Bernard 1968). The musculature of the heart is formed by the walls of the pericardial (coelomic) sac(s) and the blood spaces are clearly located between basement membranes as in most invertebrates. Only the cephalopods have vessels with endothelia, but these are incomplete in the capillaries; the presence of an endothelium can probably be ascribed to the high level of activity. It is sometimes stated that the cephalopods have a closed circulatory system while the other molluscs have open systems, but this distinction appears to be rather useless (Trueman & Clarke 1988).

The excretory organs of adult molluscs are paired metanephridia which drain the pericardial sac. Primary urine is filtered from blood vessels/spaces to special pericardial expansions called auricles through areas with podocytes (Andrews 1988). The primary urine becomes modified during passage through the metanephridial ducts which open in the mantle cavity. The proximal part of the metanephridium is a small ciliated canal, the renopericardial canal, leading to the usually quite voluminous kidney which is responsible for both osmoregulation and excretion. Most molluscs have one pair of nephridia (or only one for example in many gastropods), but *Nautilus* has two pairs of kidneys, which are not connected with the pericardium. The monoplacophorans have 6-7 pairs of nephridiopores in the mantle groove, but some of the nephridia on each side are fused and the numbers are difficult to ascertain because of the state of fixation of the material studied; two of the posterior pairs of nephridia are possibly connected to the pericardium (Wingstrand 1985).

The metanephridia develop from mesodermal cells, but an ectodermal invagination gives rise to the distal part of the excretory canal, the ureter (Raven 1966).

The gonads, which originate from the mesoderm, are connected with the nephridia in most groups and the gametes are spawned through the nephridiopores, but the reproductive system is in many species so specialized that the original structure is hard to recognize. Many species spawn small eggs freely in the water, but intricate egg masses are constructed in many species with internal fertilization.

The primary axis of the embryo can be recognized already in the mature eggs, which have the apical pole facing away from the attachment of the egg to the ovary (van den Biggelaar & Guerrier 1983). The dorso-ventral axis and the entrance point of the sperm are apparently correlated, and experiments with eggs of *Spisula* and *Pholas* (Guerrier 1970) indicate that the entrance of the sperm determines the position of the first cleavage furrow.

Spiral cleavage can be recognized in all the classes except the cephalopods, which have very large eggs and discoidal cleavage, and the holoblastic, spiral cleavage type is undoubtedly ancestral in the phylum. A small number of cells at the apical pole develop cilia forming an apical tuft, and a group of cells on each side of the apical pole form the cephalic plates from which eyes, tentacles and the cerebral ganglia develop (Verdonk & van den Biggelaar 1983). In the several species of Polyplacophora, Gastropoda, Bivalvia and Scaphopoda which have been studied, the prototroch, which may consist of one to three rows of cells, originates mainly from the primary trochoblasts ($1a^2$-$1d^2$). The gaps between these four groups of cells become closed anteriorly and laterally by the secondary trochoblasts ($2a^{11}$-$2c^{11}$). The dorsal gap in the prototroch usually closes by a fusion of the posterior tips of the prototroch (Verdonk & van den Biggelaar 1983).

Gastrulation is through invagination in species with small eggs and a coeloblastula, and through epiboly in species with large, yolky eggs and a sterroblastula (Verdonk & van den Biggelaar 1983). The blastopore becomes the definitive mouth in many species where it partially closes from the posterior side but more conspicuously becomes shifted anteriorly by a curving of the embryo (Verdonk & van den Biggelaar 1983). The mouth opening may remain open while the stomodaeal invagination is formed but there is a temporal closure at this point in many species. The anus is formed as a secondary opening from the proctodaeum. Species with larger eggs have epibolic gastrulation and mouth and anus are formed as separate invaginations. The only known exception to this general type of blastopore fate is found in *Viviparus*, in which the blastopore becomes the anus and the mouth is formed through a stomodaeal invagination (Dautert 1929).

Both endomesoderm and ectomesoderm can be recognized in the development of most species. The endomesoderm originates from the 4d-cell, which divides into a right and a left cell. These cells usually give off some enteroblasts and then become true mesoteloblasts which produce a pair of lateral mesoderm bands (observed for example in the gastropod *Physa* (Wierzejski 1905; Fig. 12.1) and the bivalve *Sphaerium* (Okada 1939)). Larval protonephridia, primordial gonocytes and much of the larval musculature develop from these bands, and this is assumed also for the adult heart, kidneys and gonads. Ectomesoderm is formed from the second and third micromere quartets and soon becomes so intermingled with the endomesoderm that a separation has not been possible (Verdonk & van den Biggelaar 1983). Dautert (1929) reported a complete absence of endomesoderm in the embryos of *Viviparus*; his study was based on serial sections of many embryos and the illustrations appear to support his interpretation, but it has nevertheless been questioned. The compact mesodermal bands give rise to the body musculature of the adults, and further differentiate into small coelomic sacs, which become the pericardium with the connected metanephridia and the gonads; it is sometimes stated that these organs can be traced back to the 4d-cell, but the possibility of ectomesodermal participation cannot be excluded.

The larval cerebral ganglia become incorporated in the brain of the adults, whereas the origin of the pedal and lateral nerves is more uncertain. In aplacophorans and polyplacophorans the development is poorly known, and the descriptions partly contradictory. Kowalevsky (1883), who studied *Lepidochitona*, reported that two lateroventral pairs of longitudinal ectodermal thickenings move in and form the nerves, just as in the annelids, and his drawings of transverse sections of a number of stages support this interpretation. However, Hammarsten & Runnström (1925), who studied *Acanthochitona*, claimed that the pedal and lateral nerves originate from cells proliferating from the cerebral ganglion, and a similar origin was reported in *Epimenia* by Baba (1940), but none of these accounts show transverse sections of the relevant stages and their conclusions appear uncertain. In testarians the pleural, pedal and visceral ganglia arise as ectodermal thickenings; the origin of their connecting nerves is uncertain and is described either as outgrowths from the ganglia or as proliferation of cell strands from the ectoderm (Raven 1966).

Trochophora larvae with all the characteristic ciliary bands are not known in living molluscs, but the species with planktotrophic development have the characteristic trochophore prototroch, adoral ciliary zone and metatroch and the larvae of some solenogasters and the caudofoveate *Chaetoderma* have a telotroch (Fig. 15.3). Many species with lecithotrophic development have larvae which pass through stages where the prototroch is the swimming organ (Fig. 15.3). Later developmental stages are often more or less modified trochophores, which have certain areas of the body modified for various functions. Many snails and mussels have the well known veliger larvae, which have the prototroch, adoral ciliary zone and metatroch pulled out in lateral loops along large, thin, hemicircular or lobed expansions, the velum (Fig. 15.3).

Characteristic pericalymma larvae (Chapter 10; Fig. 10.5) are found in a few groups. Lecithotrophic larvae with the area of the prototroch or the area immediately below it expanded into a thin, ciliated sheet, the serosa, covering the mouth and the hyposphere (Type 1 larvae) are known from solenogasters and bivalves. In the solenogaster *Neomenia* the serosa covers the episphere only partially, leaving the posterior end of the larva with the anus surrounded by the telotroch exposed (Fig. 15.3); in the bivalve *Solemya* (Gustafson & Reid 1986) and some of the protobranchs (Fig. 15.3) the serosa covers the episphere completely and there is no telotroch. At metamorphosis, the serosa folds over anteriorly and becomes invaginated with the whole episphere in *Neomenia* while it becomes cast off in the bivalves. Another type of pericalymma larva is found in brooding teredinid bivalves of the genus *Lyrodus* (Fig. 10.5) where the adoral ciliary zone below the mouth is greatly expanded posteriorly so that it covers the valves almost completely (Type 2 larva); this serosa retracts at a later stage and the larvae are released as pediveligers (Fig. 15.3). Various types of planktotrophic and lecithotrophic trochophore larvae are known from the classes Gastropoda and Bivalvia, and lecithotrophic, trochophore-like larvae are known also from Caudofoveata, Solenogastres, Polyplacophora, and Scaphopoda (the larvae of monoplacophorans are unknown and cephalopods have direct development) so it must be concluded that the trochophore is the ancestral larval type of the Mollusca, and that the pericalymma larvae are specializations which have evolved independently in solenogasters and bivalves. The veliger larva, which is often considered characteristic of the Mollusca, is only found in gastropods and bivalves; it may represent a parallel specialization in the two groups increasing the length of the ciliary bands used in swimming and feeding. The veliger larva can definitely not be regarded as an apomorphy of the whole phylum.

The foot and the mantle, with shell glands in the testarians, develop mainly from the 2d and 3d-cells, but in some gastropods, cells from the C-quadrant are involved too (Verdonk & van den Biggelaar 1983).

Paired protonephridia have been observed in larvae or embryos of polyplacophorans, gastropods and bivalves (Brandenburg 1966, Bartolomaeus 1989). Their origin has been claimed to be from the ectoderm, from the mesoderm or mixed (Raven 1966); the ultrastructure of *Lepidochitona* (Bartolomaeus 1989) supports the latter interpretation.

The presence of segmentation in molluscs has been questioned by many authors over time (see for example Salvini-Plawen 1985), but independent characters indicate that the body of the ancestral mollusc had eight segments (see for example Wingstrand 1985).

Eight (or seven) transverse rows of calcareous spicules or plates are found on the dorsal side of larvae of solenogasters, caudofoveates and polyplacophorans and are characteristic of the adult polyplacophorans (Fig. 15.4). The foot retractor muscles of polyplacophorans and monoplacophorans are grouped in eight pairs and several of the individual muscles can be homologized between the two groups (Wingstrand 1985; Fig. 15.4). The presence of eight pairs of foot retractor muscles in the mid-Ordovician bivalve *Babinka* can be inferred from the well-preserved muscle scars (Fig. 15.4), and the same number can with some uncertainty be recognized in the homologous foot and byssus retractor muscles of living bivalves such as *Mytilus* and *Yoldia* (Fig. 15.4).

Segmentation of other organ systems is not well documented, but Wingstrand (1985, p. 43) demonstrated that serial repetition of nerve connectives, nephridiopores, gills, gonoducts, and atria (in order of decreasing numbers) correlated well with the pedal retractors in both *Neopilina* and *Vema*. Two pairs of gills, atria and nephridia are found in *Nautilus*. The musculature unequivocally demonstrates that the testarian ancestor must have had eight segments, and the transverse rows of spicules in the aplacophoran larvae indicate that this was also the case in the com-

---

Fig. 15.3. Larval types of the molluscan classes; Monoplacophora and Cephalopoda are omitted, the development of the former is unknown and the latter has direct development. – Solenogastres: *Epimenia verrucosa* (redrawn from Baba 1940); *Neomenia carinata* (redrawn from Thompson 1960). – Caudofoveata: *Chaetoderma nitidulum* Lovén (after an unpublished drawing by the late Dr Gunnar Gustafsson, Kristineberg Marine Biological Station, Sweden). – Polyplacophora: *Mopalia muscosa* (Gould) (larva reared at Friday Harbor Laboratories, WA, USA, June 1992). – Gastropoda: *Crepidula fornicata* (redrawn from Werner 1955); *Lottia pelta* (Rathke)(early larva reared at Friday Harbor Laboratories, WA, USA, June 1992; a later stage is a lecithotrophic veliger). – Scaphopoda: *Dentalium entale* (redrawn from Lacaze-Duthiers 1858/59). – Bivalvia: *Barnea candida* (redrawn from Nielsen 1987); *Lyrodus pedicellatus* (Quatrefages) (pediveliger redrawn after a scanning micrograph by Drs C.B. Calloway and R.D. Turner, Museum of Comparative Zoology, Harvard Univ., MA, USA; the early stage is a trochophore of a structure similar to that of *Lottia*, the intermediate stage is of the pericalymma-type (Fig. 10.6)); *Yoldia limatula* (after Drew 1899). The presence of compound cilia in the larvae of *Epimenia*, *Neomenia* and *Dentalium* is inferred from the descriptions.

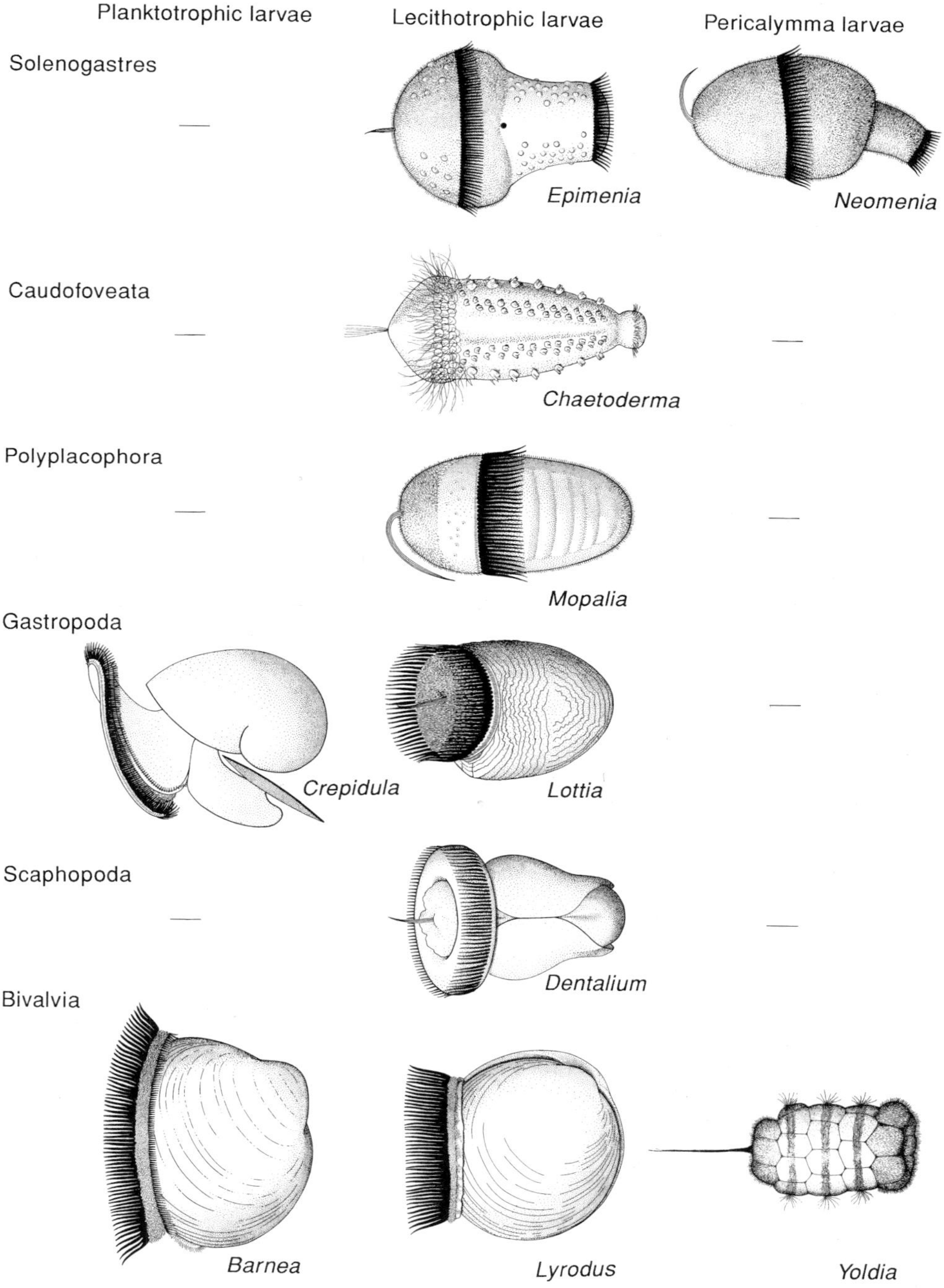

Planktotrophic larvae   Lecithotrophic larvae   Pericalymma larvae

Solenogastres

Epimenia   *Neomenia*

Caudofoveata

Chaetoderma

Polyplacophora

Mopalia

Gastropoda

Crepidula   *Lottia*

Scaphopoda

Dentalium

Bivalvia

Barnea   *Lyrodus*   *Yoldia*

119

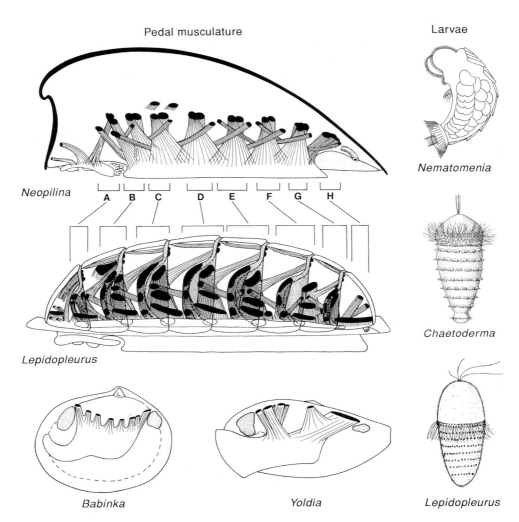

Pedal musculature

Larvae

Neopilina

A B C D E F G H

Lepidopleurus

Nematomenia

Chaetoderma

Babinka

Yoldia

Lepidopleurus

Fig. 15.4. Indications of segmentation in molluscs. – Shell plates of larvae and juveniles: newly metamorphosed larva of *Nematomenia banyulensis* (redrawn from Prouvot 1890); young larva of *Chaetoderma nitidulum* Lovén (an unpublished drawing by the late Dr Gunnar Gustafsson, Kristineberg Marine Biological Station, Sweden); larva of *Lepidopleurus asellus* (from Christiansen 1954). – Pedal musculature: *Lepidopleurus* sp. and *Neopilina galatheae*: reconstructions based on serial sections (after Wingstrand 1985); *Babinka prima*: reconstruction of the pedal muscles of the mid-Ordovician fossil (after McAlester 1965); *Yoldia limatula*: diagram of the pedal muscles (redrawn from Heath 1937).

mon molluscan ancestor. The alternative interpretation, viz. that the segmentation originated in the testarian ancestor and that the indications of eight segments in aplacophorans are without importance, appears less likely, but cannot be ruled out completely as long as the development of the aplacophorans is so poorly known. The aplacophorans have several plesiomorphic characters, but the cylindrical body shape is definitely an apomorphy, and the specialization of locomotion has obviously influenced the musculature and very likely resulted in the development of a high number of smaller muscles. Again, an investigation of the embryology of one or more of the aplacophorans is needed before firm conclusions can be drawn.

There is nothing to indicate that the molluscan ancestor had large coelomic cavities functioning as a hydrostatic skeleton in connection with burrowing. Many living molluscs burrow, for example with the foot, but it is always the haemal system which functions as hydrostatic skeleton. All the more 'primitive' molluscs have quite narrow coelomic cavities (pericardia) and in the classes where the embryology has been studied there are no signs of larger coelomic pouches at any stage. This could indicate that small pericardial cavities, possibly connected with gonads, was the first step in the evolution of the schizocoelous articulates. At a later stage, burrowing habits could have favoured the enlargement of such cavities and their specialization as the hydrostatic skeleton seen today in euarticulates.

There can apparently be no doubt about the monophyly of the Mollusca, and the indications of eight segments in the ancestral mollusc make it logical to regard them as the sister group of the euarticulates, which have numbers of segments increasing during ontogeny (Chapter 16). A further synapomorphy of the two groups may be the gonopericardial complex, i.e. the small pericardial coelomic sacs which contain the gonads and which are drained by metanephridia with specialized 'nephridial' sections of the duct (Wingstrand 1985).

Interesting subjects for future research

1. Larval development of Solenogastres, especially the the differentiation of the mesoderm and the formation of the calcareous spicules.
2. Embryology and larval development of Caudofoveata.
3. Differentiation of the nervous system and of the mesoderm in relation to the shells in Polyplacophora.
4. Embryology and larval development of Monoplacophora, especially the differentiation of the mesoderm.

References

Andrews, E.B. 1988. Excretory systems of molluscs. – In K.M. Wilbur (ed.): The Mollusca, vol. 11, pp 381-448. Academic Press, San Diego.

Ankel, W.E. 1936. Die Frassspuren von Helcion und Littorina und die Funktion der Radula. – Verh. dt. zool. Ges. 38: 174-182.

Baba, K. 1940. The early development of a solenogastre, Epimenia verrucosa (Nierstrasz). – Annotnes zool. Jap. 19: 107-113.

Bartolomaeus, T. 1989. Larvale Nierenorgane bei Lepidochiton cinereus (Polyplacophora) und Aeolidia papillosa (Gastropoda). – Zoomorphology 108: 297-307.

Beedham, G.E. & E.R. Trueman 1967. The relationship of the mantle and shell of the Polyplacophora in comparison with that of other Mollusca. – J. Zool. (Lond.) 151: 215-231.

Bernard, F.R. 1968. The aquiferous system of Polynices lewisi (Gastropoda, Prosobranchiata). – J. Fish. Res. Bd Can. 25: 541-546.

Brandenburg, J. 1966. Die Reusenformen der Cyrtocyten. – Zool. Beitr. **12**: 345-417.

Bubel, A. 1984. Mollusca: epidermal cells. – *In* J. Bereiter-Hahn, A.G. Matoltsy & K.S. Richards (eds): Biology of the Integument, vol. 1, pp 400-447. Springer, Berlin.

Christiansen, M.E. 1954. The life history of *Lepidopleurus asellus* (Spengler) (Placophora). – Nyt Mag. Zool. **2**: 52-72.

Dautert, E. 1929. Die Bildung der Keimblätter von *Paludina vivipara*. – Zool. Jb., Anat. **50**: 433-496.

Drew, G.A. 1899. Some observations on the habits, anatomy and embryology of members of the Protobranchia. – Anat. Anz. **15**: 493-519.

Guerrier, P. 1970. Les caractères de la segmentation et de la détermination de la polarité dorsoventrale dans le développement de quelques Spiralia. III. *Pholas dactylus* et *Spisula subtruncata* (Mollusques, Lamellibranches). – J. Embryol. exp. Morphol. **23**: 667-692.

Gustafson, R.G. & R.G.B. Reid 1986. Development of the pericalymma larva of *Solemya reidi* (Bivalvia: Cryptodonta: Solemyidae) as revealed by light and electron microscopy. – Mar. Biol. (Berl.) **93**: 411-427.

Haas, W. 1981. Evolution of calcareous hardparts in primitive molluscs. – Malacologia **21**: 403-418.

Hammarsten, O.D. & J. Runnström 1925. Zur Embryologie von *Acanthochiton discrepans* Brown. – Zool. Jb., Anat. **47**: 261-318.

Heath, H. 1937. The anatomy of some protobranch molluscs. – Mém. Mus. Hist. nat. Belg., 2. sér. **10**: 1-26, 10 pls.

Kaas, P. & R.A. Van Belle 1985. Monograph of Living Chitons, vol. 1. – Brill/Backhuys, Leiden.

Kniprath, E. 1980. Ontogenetic plate and plate field development in two chitons, *Middendorfia* and *Ischnochiton*. – Roux's Arch. dev. Biol. **189**: 97-106.

Kowalevsky, M.A. 1883. Embryogénie du *Chiton polii* (Philippi). – Annls Mus. Hist. nat. Marseille, Zool. **1**(5): 1-46, 8 pls.

Lacaze-Duthiers, H. 1858/59. Histoire de l'organisation et du développement du *Dentale*. – Annls Sci. nat., 4. sér., Zool. **6**: 225-281 & 319-385, pls 8-13, and **7**: 5-51 & 171-255, pls 2-49.

Lauterbach, K.-E. 1984. Das phylogenetische System der Mollusca. – Mitt. dt. malakozool. Ges. **37**: 66-81.

Lemche, H. & K.G. Wingstrand 1959. The anatomy of *Neopilina galatheae* Lemche, 1957. – Galathea Rep. **3**: 9-71, 56 pls.

Lowenstam, H.A. & S. Weiner 1989. On Biomineralization. – Oxford Univ. Press, New York.

McAlester, A.L. 1965. Systematics, affinities, and life habits of *Babinka*, a transitional Ordovician lucinoid bivalve. – Palaeontology **8**: 231-246.

Naef, A. 1921. Die Cephalopoden. – Fauna Flora Golf. Neapel **35**: 1-148.

Nielsen, C. 1987. Structure and function of metazoan ciliary bands and their phylogenetic significance. – Acta zool. (Stockh.) **68**: 205-262.

Okada, K. 1939. The development of the primary mesoderm in *Sphaerium japonicum biwaense* Mori. – Sci. Rep. Tohoku imp. Univ., Biol. **14**: 25-48, pls 1-2.

Prouvot, G. 1890. Sur le développement d'un Solenogastre. – C.r. hebd. Séanc. Acad. Sci., Paris **111**: 689-695.

Raven, C.P. 1966. Morphogenesis: The Analysis of Molluscan Development (2nd ed.). – Pergamon Press, Oxford.

Runnegar, B. & J. Pojeta, Jr. 1985. Origin and diversification of the Mollusca. – *In* K.M. Wilbur (ed.): The Mollusca, vol. 10, pp 1-57. Academic Press, Orlando.

Salvini-Plawen, L.v. 1967. Neue scandinavische Aplacophora (Mollusca, Aculifera). – Sarsia **27**: 1-63.

Salvini-Plawen, L.v. 1972. Zur Morphologie und Phylogenie der Mollusken: die Beziehungen der Caudofoveata und der Solenogastres als Aculifera, als Mollusca und als Spiralia. – Z. wiss. Zool. **184**: 205-394.

Salvini-Plawen, L.v. 1985. Early evolution and the primitive groups. – *In* K.M. Wilbur (ed.): The Mollusca, vol. 10, pp 59-150. Academic Press, Orlando.

Scheltema, A.H. 1981. Comparative morphology of the radulae and alimentary tracts in the Aplacophora. – Malacologia **20**: 361-383.

Scheltema, A.H. 1988. Ancestors and descendents: relationships of the Aplacophora and Polyplacophora. – Am. malac. Bull. **6**: 57-68.

Thompson, T.E. 1960. The development of *Neomenia carinata* Tullberg (Mollusca Aplacophora). – Proc. R. Soc. Lond. B **153**: 263-278.

Trueman, E.R. & M.R. Clarke 1988. Introduction. – *In* K.M. Wilbur (ed.): The Mollusca, vol 11, pp 1-9. Academic Press, San Diego.

van den Biggelaar, J.A.M. & P. Guerrier 1983. Origin of spatial organization. – *In* K.M. Wilbur (ed.): The Mollusca, vol. 3, pp 179-213. Academic Press, New York.

Verdonk, N.H. & J.A.M. van den Biggelaar 1983. Early development and the formation of the germ layers. – *In* K.M. Wilbur (ed.): The Mollusca, vol. 3, pp 91-122. Academic Press, New York.

Waller, T.R. 1980. Scanning electron microscopy of shell and mantle in the order Arcoida (Mollusca: Bivalvia). – Smithsonian Contr. Zool. **313**: 1-58.

Watabe, N. 1984. Mollusca: Shell. – *In* J. Bereiter-Hahn, A.G. Matoltsy & K.S. Richards (eds): Biology of the Integument, vol. 1, pp 448-485. Springer, Berlin.

Werner, B. 1955. Über die Anatomie, die Entwicklung und Biologie des Veligers und der Veliconcha von *Crepidula fornicata* L. (Gastropoda, Prosobranchia). – Helgoländer wiss. Meeresunters. **5**: 169-217.

Wierzejski, A. 1905. Embryologie von *Physa fontinalis* L. – Z. wiss. Zool. **83**: 502-706, pls 18-27.

Wingstrand, K.G. 1985. On the anatomy and relationships of recent Monoplacophora. – Galathea Rep. **16**: 7-94, 12 pls.

Yonge, C.M. & T.E. Thompson 1976. Living Marine Molluscs. – Collins, London.

16

EUARTICULATA

The typical euarticulate has an elongate body with a number of more or less identical segments. Each segment contains a pair of coelomic compartments with associated organs, for example nephridia and gonads, and a pair of ventral ganglia formed from the ectoderm. The mesoderm originates from the 4d-cell which divides to form a pair of teloblasts situated in the anal area; these two cells give off a number of stem cells from the anterior side, corresponding to the number of segments, and each of these cells divides to form the mesodermal elements of one half segment. In some types, for example in leeches, the teloblasts give off a finite number of stem cells during the early embryological stages, corresponding to the adult number of segments, while new stem cells are given off both during the early development and during the individual's later life, for example in many 'polychaetes' (Chapter 17), which continue adding segments.

This pattern is clearly seen in several annelids and some of the panarthropod groups (onychophorans and arthropods), whereas especially the insects show highly modified types of development with superficial cleavage and blastoderm formation which make it impossible to trace cell lineages.

The cell lineage of several polychaetes was studied already around the turn of the century, and the origin of the mesodermal bands from the two mesodermal teloblasts has been documented in many genera (Anderson 1966). The lateral mesodermal bands are compact at first, but small coelomic cavities arise when splits develop between the cells (schizocoely). The mesodermal bands subsequently break up so that each coelomic cavity with its surrounding mesodermal cells becomes one coelomic sac. The mesodermal bands and the coelomic sacs fill the whole blastocoel in some polychaete larvae, while they form flat bands on the ventral side of a spacious blastocoel in other species where the blastocoel becomes obliterated after metamorphosis (Chapter 17).

The panarthropods lack primary larvae and their embryology is modified accordingly. In many onychophorans the cleavage is modified as a result of large amounts of yolk or the development of a placenta; in species with small eggs the cleavage is irregular, but results in the formation of a gastrula with a group of mesodermal cells at the posterior side of the blastopore; this median group of mesodermal cells proliferates a pair of lateral mesodermal bands which become organized into coelomic sacs (Chapter 19). The mesoderm and coelom formation in tardigrades is probably completely misunderstood (Chapter 21). In the arthropods, mesoteloblasts have been observed in certain crustaceans whereas strongly modified cleavage patterns with a blastoderm surrounding a large mass of yolk is characteristic of most of the phylum (Chapter 20).

Teloblastic proliferation from the mesoblasts occurs also in molluscs (see Chapter 15) and sipunculans (Chapter 13), but the arrangement of the mesoderm as a paired series of coelomic sacs which define the segments and the addition of segments from the teloblastic zone appear to be good synapomorphies of the euarticulates (Weygoldt 1986).

References

Anderson, D.T. 1966. The comparative embryology of the Polychaeta. – Acta zool. (Stockh.) **47**: 1-42.
Weygoldt, P. 1986. Arthropod interrelationships – the phylogenetic-systematic approach. – Z. zool. Syst. Evolutionsforsch. **24**: 19-35.

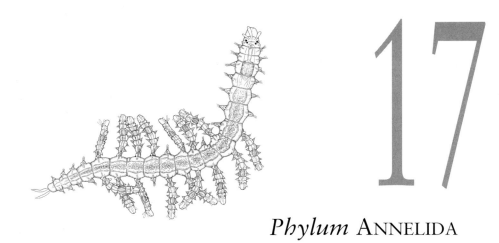

Phylum ANNELIDA

Annelids are an ecologically and systematically important phylum of aquatic or terrestrial animals, comprising at least 15 000 living species. The fossil record is meagre; the records of Precambrian annelids are uncertain, and the first unquestionable remains of annelids appear to be from the Lower Cambrian (Brasier 1979). A diverse fauna of quite large species is known from the Middle Cambrian Burgess Shale (Conway Morris 1979).

The phylogeny of the annelids is completely unresolved. The phylum is traditionally divided into the classes Polychaeta and Clitellata, or Polychaeta, Oligochaeta and Hirudinea, sometimes with Myzostomida as an additional class, without any discussion of sister-group relationships. Recent authors, such as Fauchald (1974), Pettibone (1982) and George & Hartmann-Schröder (1985), regard the polychaetes as a more 'primitive' stem group from which both clitellates and myzostomids have evolved.

The archiannelids, which were earlier considered as a separate class, are now regarded as specialized interstitial forms and are integrated in various polychaete orders with larger forms or as separate orders (George & Hartmann-Schröder 1985, Westheide 1990).

Other taxa, such as Pogonophora, Vestimentifera, Lobatocerebridae, Gnathostomulida, Echiura, and Sipuncula, have sometimes been included in the Annelida, so before the characters of an ancestor common to all annelids can be traced – and its relationships with the other spiralians discussed – it appears necessary to evaluate the relationships of all these groups so that the phylum Annelida and its constituent groups can be defined.

The Polychaeta, as defined by Fauchald (1977) and George & Hartmann-Schröder (1985), clearly includes all the taxa regarded as most 'primitive' and it appears practical first to identify the ancestral characters of this group as a basis for the discussion of the relationships with the other groups mentioned above.

The polychaetes are segmented worms with groups of chaetae (setae) on the

Chapter vignette: The polychaete *Exogone gemmifera* with attached juveniles. (From Rasmussen 1973.)

sides of each segment. The segmentation is almost complete with septa and mesenteries separating a row of paired coelomic sacs in many 'errant' and tubicolous forms such as nereidids, spionids and sabellids, but both septa and mesenteries are lacking in the anterior portion of the body in many forms with a large, eversible pharynx such as glycerids, and the inner partitions are also strongly reduced in burrowing forms such as arenicolids and scalibregmatids, in tubiculous forms such as pectinariids, and in the pelagic *Poeobius* (Robbins 1965). On the other hand, the coelomic cavities are completely absent in some interstitial forms, such as *Protodrilus* and *Psammodriloides* (Fransen 1980).

The head region consists of a preoral region, the prostomium, which is definitely presegmental, and a perioral region, the peristomium, which originates from the area between prototroch and metatroch in the trochophora larva (Fig. 17.1); the ectoderm of the peristomium originates behind the prototroch but in front of the ectoteloblast ring which proliferates the ectoderm of the body segments and it seems that the mouth region ectoderm is topographically and phylogenetically presegmental (Åkesson 1967, Anderson 1973). The mesoderm of prostomium and peristomium is less well described, but mesoderm in the shape of one or more pairs of coelomic sacs may be present; it appears that the mesoderm initially originates from ectomesodermal cells and that additional mesoderm may be formed by proliferation from the anterior coelomic sacs formed from the mesoteloblasts (Anderson 1966a, 1973).

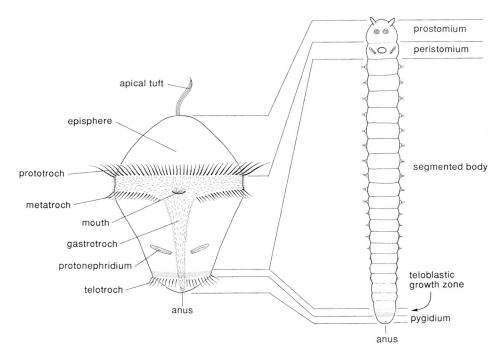

Fig. 17.1. The correlation between body regions in a larval and an adult annelid. (Based on Schroeder & Hermans 1975.)

The ectoderm is a monolayered epithelium except in *Travisia*, which has a strat-ified epithelium with extensive intercellular spaces (Storch 1988). Myoepithelial cells have been observed in tentacles and cirri of nereidids (Boilly-Marer 1972).

The ciliated epithelia of annelids consist of multiciliate cells, but *Owenia* in the family Oweniidae forms an exception, having only monociliate cells both as adults and larvae – even the prototroch and metatroch arise from monociliate cells (Gar-diner 1978, Nielsen 1987). This has been interpreted as a plesiomorphic feature and the oweniids have accordingly been regarded as located at the base of the polychaetes (Gardiner 1978, Smith, Ruppert & Gardiner 1987), but in the larger context it ap-pears more plausible to regard it as an advanced character – a reversal to the original monociliate stage through loss of the 'additional cilia' (see Chapter 10); this is further supported by the fact that the larvae of *Myriochele*, also belonging to the Oweniidae, have multiciliate cells (Smith, Ruppert & Gardiner 1987). Monociliate cells with a very short cilium with an accessory centriole occur also in developing coelomic cells of *Magelona* (Turbeville 1986), so monociliary, non-sensory cells are known from other epithelia of species which are generally multiciliate.

The cuticle consists of several layers of parallel collagen fibrillae with alternat-ing orientation and with microvilli extending between the fibrillae to the surface. The microvilli often terminate in a small knob and there is in many cases an electron-dense epicuticle at the surface (Richards 1984, Storch 1988). Chitin is generally ab-sent in the cuticle, but Bubel *et al.* (1983) have demonstrated the presence of α-chitin in the opercular filament cuticle of the serpulid *Pomatoceros*.

The chaetae consist mainly of β-chitin associated with protein, each chaeta being formed by a chaetoblast with long microvilli; the chaetae have characteristic longitudinal channels corresponding to these microvilli (O'Clair & Cloney 1974, Schroeder 1984). Some polychaetes, such as capitellids and oweniids, have the chaetae projecting directly from the cylindrical body and are burrowing or tubi-colous and have segments with longitudinal and circular muscles functioning as hy-drostatic units (Clark 1964); others are creeping on ciliary fields, for example many of the interstitial types. However, most polychaetes have protruding muscular ap-pendages, parapodia, with chaetae on a dorsal and a ventral branch and with an elaborate musculature which makes the parapodia suited for various types of creep-ing or swimming; the longitudinal and circular segmental muscles may be rather weak. It has been customary to regard types with large parapodia, such as *Nereis*, as the typical polychaetes, but Fauchald (1974) proposed that the ancestral polychaete was a burrowing form with chaetae but without parapodia, superficially resembling a capitellid. The parapodia should then be seen as locomotory appendages which en-abled the more advanced polychaetes to crawl in soft, flocculent substrates such as the rich detritus layer at the surface of the sediment. The swimming and tube-build-ing types are clearly more advanced. This interpretation fits well with the general image of the ancestral polychaete.

The pharynx represents the stomodeal invagination and accordingly has a cu-ticle similar to that of the outer body wall. Most species of the orders Phyllodocida and Eunicida have a pharynx with jaws which are heavily sclerotized parts of the cu-ticle (Purschke 1988). Collagen is an important constituent and quinone tanning has been demonstrated in some species; chitin has not been found. The basal layer of most jaws shows short canals with microvilli.

Many polychaetes secrete tubes with or without incorporated mud, shells, sand grains or other foreign objects; the tubes of serpulids are heavily calcified. The composition of the organic material is not well known, but both carbohydrates and proteins, in some species in a keratin-like form, are present; chitin has not been found (Gaill & Hunt 1988).

The central nervous system consists of a paired cerebral ganglion, connectives on each side of the pharynx and a pair of ventral nerves (Fig. 10.2). These nerves are situated within the epithelium in early developmental stages and also in some adult forms, for example *Polygordius* and *Protodrilus* (Westheide 1990), but in many forms the cords sink in from the epithelium during ontogeny (Fig. 17.2) and the perikarya become arranged in paired ganglia, connected by transverse and paired longitudinal nerves; there is usually one pair of ganglia per segment, but the ganglia are sometimes less well defined, and two or three pairs of ganglion-like swellings with lateral nerves are observed in each parapodial segment, for example in *Pectinaria* (Nilsson 1912).

Special parapodial ganglia usually connected by lateral nerves are found in species with well developed parapodia.

The main part of the mesoderm develops from the two mesoteloblasts derived from the 4d-cell, but some of the mesoderm in the head region is ectomesoderm (see below). The coelomic cavities are covered by a monolayered peritoneum, often with multiciliate cells, and monociliate muscle cells have been reported from *Owenia* (Gardiner & Rieger 1980).

Most annelids have a haemal system which consists of more or less well-defined vessels surrounded by basement membranes of the various epithelia; there is no endothelium (Ruppert & Carle 1983). Some families, such as capitellids and glycerids,

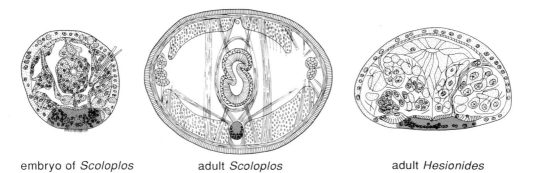

embryo of *Scoloplos*   adult *Scoloplos*   adult *Hesionides*

Fig. 17.2. Transverse sections of polychaetes showing the position of the ventral longitudinal nerve cords (dark shading). – Two-chaetiger embryo of *Scoloplos armiger* with intraepithelial nerve cords (redrawn from Anderson 1959). – Adult *Scoloplos armiger* with completely internalized and fused nerve cords (redrawn from Mau 1881). – Adult *Hesionides arenaria* with intraepithelial nerve cords (redrawn after Westheide 1967).

lack blood vessels, and the coelomic fluid, which may contain respiratory pigment, functions as a circulatory system.

The excretory organs of polychaetes show enormous variation. Protonephridia of several types are found in both larvae and adults, and metanephridia occur in adults of many families (Bartolomaeus 1989). The protonephridia are clearly of ectodermal origin and are generally known to be surrounded by a basement membrane (Hay-Schmidt 1987). The nephridial sacs of the mitraria larva of *Owenia* have complicated, podocyte-like fenestrated areas, but are in principle like protonephridia (Smith, Ruppert & Gardiner 1987). Metanephridia are usually thought to be modified coelomoducts and to originate from the mesoderm (and this has been shown to be the case in oligochaetes, see below), but Bartolomaeus (1989) and Bartolomaeus & Ax (1992) have shown that the metanephridia of a number of polychaetes develop from protonephridia which open into the coelom and are thus ectodermal.

Gonads of mesodermal origin are found in a large number of segments in many families, but for example some capitellids have the gonads restricted to a small number of segments. The gametes are usually liberated to the coelom, where the final maturation takes place, and spawning is through the ciliated metanephridia or gonoducts.

Polychaetes exhibit a wide variation in developmental types. Many forms spawn sperm and small eggs directly in the water and the zygote develops into a planktotrophic larva which metamorphoses into a benthic adult; this is considered the ancestral developmental type. Other forms have large yolky eggs which develop into lecithotrophic larvae or the development may be direct without a larval stage (see the chapter vignette).

Meiosis is usually halted in the prophase of the first division and becomes reactivated at fertilization. The apical-blastoporal axis is fixed already during maturation, and the entrance of the spermatozoon determines the position of the first cleavage and thereby the orientation of the anterior-posterior axis (Dorresteijn & Fischer 1988). The fertilized egg is surrounded by a thin fertilization membrane, which in some species becomes incorporated in the larval cuticle through which the cilia penetrate (Eckelbarger & Chia 1978); in other species this membrane forms a protecting envelope from which the larva hatches.

Cleavage is total and spiral, and since the polar bodies are always situated at the apical pole and the D-cell is often larger than the other three, it has been possible to follow the fate of many of the important cells from the 2-cell stage (Table 11.1; review in Anderson 1973). The 'typical' spiral pattern can be observed in all species, and fate-maps of 64-cell stages with the positions of the cells indicated in the notation of the spiral cleavage have been constructed for a number of species (Anderson 1973); the map of *Podarke* can be taken as an example (Fig. 17.3). The apical cells give rise to the apical sense organ with a tuft of long cilia. Just over the equator lies a horseshoe of cells which give rise to the prototroch; the number of cells which give rise to the prototroch varies somewhat between species, but the prototroch cells are always descendants of the primary trochoblasts, $2a-c^{11}$, and the secondary trochoblasts, $1a-d^2$ and $2a-c^{121}$ (Anderson 1973); the posterior break closes at a later stage (the posterior break should be expected if the prototroch is the anterior part of the archaeotroch; Chapter 2). The cells between the apical area and the prototroch

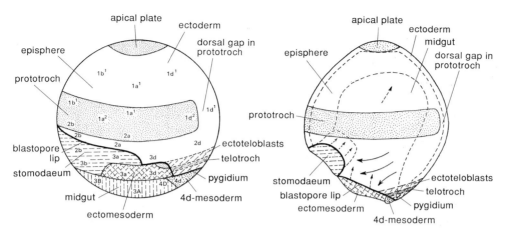

Fig. 17.3. Fate maps of blastula (left) and gastrula (right) stages of *Podarke obscura* seen from the left side. The heavy lines (marked blastopore lip) separate ectoderm from mesoderm + endoderm. (Modified from Anderson 1973.)

are presumptive ectoderm cells, as are the cells in the posterior break of the prototroch and a zone just behind the prototroch; the anterior part of this zone will form the stomodaeum. There is a narrow posterior strip of ectodermal cells, the so-called ectoteloblasts (descendants of 2d-, (3a-?), 3c- and 3d-cells in species with small eggs and of 2d only in species with very large D-cells), which will proliferate the ectoderm of the body region of the worm, and just behind those a narrow zone of cells which will form the telotroch (with a ventral gap, as foreseen in the trochaea theory). A small area on each side is the presumptive gastrotroch; the larva is lecithotrophic and has no metatroch. The blastoporal pole is occupied by the presumptive endoderm cells and by the mesoderm originating from the 4d-cell. At each side there is a narrow area of presumptive ectomesoderm just above the endoderm.

After gastrulation the cell areas lie in the positions characteristic of the trochophore (Fig. 17.3). The 4d-cell divides into a right and a left cell which give off a few small cells to the endoderm before they become the mesoteloblasts.

There is a good deal of variation in the extent and shape of the areas in the fate maps and the areas are not strictly related to identical cells (reviews in Anderson 1973 and Korn 1982), but the maps can all be seen as modifications of the general pattern described above: the apical organ is not developed in several types, for example *Scoloplos*; the prototroch cells form a complete ring, for example in *Eunice* (the drawings of the gastrula of *Eunice* (Åkesson 1967) seem to indicate that the mesoderm is exclusively ectomesodermal, but this should be studied further).

Gastrulation is embolic in species with small eggs and a blastocoel, such as *Podarke*, *Eupomatus* and *Polygordius*. There is a whole series of modifications with increasing amounts of yolk; in species with considerable amounts of rather evenly distributed yolk, for example *Arenicola*, the endodermal cells immigrate into the narrow blastocoel, which finally becomes filled completely; in types with large amounts

131

of yolk located in the prospective endoderm cells the embryos are solid and the gastrulation epibolic, for example *Nereis* and *Neanthes* (review in Korn 1982).

The blastopore of forms with embolic gastrulation may become laterally compressed leaving the adult mouth and anus and a tube-shaped gut (as in *Podarke*, see Treadwell 1901); in *Polygordius* (Woltereck 1904; Fig 10.1) the formation of a deep stomodaeum makes the mouth sink into a deep funnel and the anus closes temporarily, but reopens in the same region later on. More commonly, the blastopore closes from behind so that only the mouth remains while the anus breaks through to the sac-shaped archenteron at a later stage (as in *Eupomatus*, see Hatschek 1885). In *Eunice*, gastrulation is embolic and the blastopore constricts completely in the area where the anus develops at a later stage; the stomodaeum develops from an area isolated from the blastopore by a wide band of ectoderm (Åkesson 1967). In types with a solid endoderm the lumen of the gut forms as a slit between the cells while the two ectodermal invaginations stomodaeum and proctodaeum break through to the gut to form mouth and anus, respectively (as in *Arenicola*, see Child 1900).

There is thus an enormous variation in the developmental patterns including gastrulation and blastopore fate among the polychaetes, but the diversity can be interpreted as variations over the pattern of spiral cleavage followed by embolic gastrulation and partial blastopore closure which leaves mouth and anus, as shown by *Polygordius*.

Two types of mesoderm give rise to different structures in most forms, the ectomesoderm developing into muscles traversing the blastocoel in the episphere of the larva and musculature in the prostomium-peristomium of the adult (see for example Åkesson 1968 and Anderson 1973), and the 4d-mesoderm developing into the mesoderm of the true segments. The development of the paired lateral series of coelomic sacs from the pygidial growth zone has been documented for several species (for example *Owenia*, Wilson 1932 and *Scoloplos*, Anderson 1959). As mentioned above, the whole mesoderm appears to originate from ectomesoderm in *Eunice*, and further studies may well reveal more variation.

In the trochophore of *Polygordius*, a pair of protonephridia develop from descendants of the cells 3c and 3d, and the metatroch develops from other descendants of the same cells (Woltereck 1904).

The planktonic larvae show much variation, but the typical trochophore with an apical tuft, a prototroch and a metatroch of compound cilia functioning in a downstream-collecting system with an adoral zone of single cilia transporting the captured particles to the mouth, a gastrotroch of single cilia and a telotroch of compound cilia around the anus (as in *Polygordius*, which lacks the gastrotroch; Fig. 10.3) is the type from which all the other types may have developed through losses of one or more of the ciliary bands (Nielsen 1987). The telotroch is absent in many planktotrophic larvae, which thus swim only by means of the prototroch (for example *Serpula*, see Fig. 17.4). The metatroch is absent in all lecithotrophic larvae, which may have a telotroch (as in the larvae of spionids and terebellids, see Fig. 17.4) or which may have only the prototroch (as in many phyllodocids). Accessory rings of single or compound cilia are found in many species, and various specialized types of feeding in the plankton are also observed.

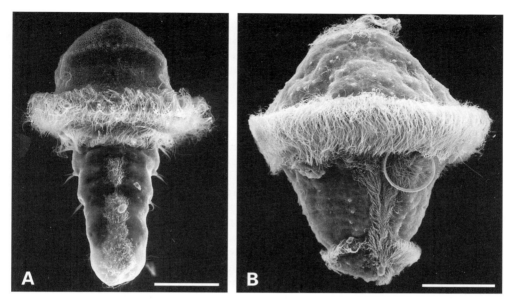

Fig. 17.4. Metatrochophora larvae of polychaetes. – A, 3-pair chaetiger larva of *Serpula vermicularis* (Friday Harbor Laboratories, WA, USA, July 1980). – B, an unidentified terebellid larva (earlier believed to be a *Pectinaria*-larva) (from Nielsen 1987.) – Scale bars: 50 μm.

The trochophore larvae have developing coelomic sacs and chaetae already in later planktonic stages, sometimes with long, special larval chaetae functioning as protection towards predators. Metamorphosis may be rather gradual, as in many lecithotrophic *Nereis* larvae, or more abrupt, as in sabellariid larvae which shift from planktotrophic larvae to sessile adults with a new feeding apparatus; metamorphosis may even be 'catastrophic', as in *Owenia* and *Polygordius*, where the larval organs used in feeding are cast off (see below).

Cerebral ganglia, with eyes in several species, develop at the lateral sides of the apical organ, and this larval brain is connected to nerves along the bases of the prototroch and metatroch cells through a pair of connectives (Lacalli 1981, 1984). These connectives extend further behind the mouth and become connected to the paired ventral chain of ganglia (Lacalli 1988). The larval brain with its connectives becomes incorporated directly into the adult nervous system (Segrove 1941; Korn 1958, 1960; Lacalli 1984); this can be followed even in species with a catastrophic metamorphosis in which major parts of the larval episphere are cast off (for example *Owenia*, see Wilson 1932). The ventral nervous chain develops from the ectoderm along the zone of the fused blastopore lips (Anderson 1959). It remains intraepithelial for example in many of the small, interstitial species (Westheide 1990), but sinks in and becomes situated along the ventral attachment of the mesentery, surrounded by its basement membrane in many of the larger forms (Fig. 17.2).

Special larval types called pericalymma (or serosa) larvae (Fig. 10.5) are found in *Polygordius* and oweniids. These larvae have more or less 'blown-up' bodies with the normal trochophoran prototroch and metatroch at the equator. *Poly-gordius neapolitanus* has larvae which develop an elongate segmented body and change gradually into the adult, while *P. appendiculatus* and *P. lacteus* have larvae in which the segmented body develops strongly retracted (like an accordion) and covered by a circular fold of the region behind the metatroch (type 2 pericalymma larvae; Figs 10.5, 17.5); at metamorphosis the body stretches out and the larger part of the spherical larval body with the ciliary feeding apparatus is shed (Woltereck 1902, Herrmann 1986). Oweniid larvae (usually called mitraria larvae) have ciliary bands which form wide lobes and the chaetae of the first segment develop early and become very long; the following segments have short chaetae and are pulled up into a deep circular fold behind the long chaetae (type 3 pericalymma larvae; Fig. 10.5). At metamorphosis, the parts of the hyposphere carrying the ciliary feeding structures are cast off together with the long larval chaetae and the body stretches out so that a small worm resembling an adult emerges in less than an hour (Wilson 1932).

The variation among the *Polygordius* larvae shows that these pericalymma larvae can be considered as specialized polychaete trochophores, and no special phylogenetic importance needs to be placed on them at the phylum level.

The ancestral polychaete has apparently been a pelago-benthic organism with a planktotrophic trochophora larva and a segmented, burrowing adult. New segments were added from a posterior growth zone with a pair of mesoteloblasts which gave rise to new coelomic sacs. Each segment had a pair of coelomic sacs and a group of chaetae on each side; all segments had gonads and gonoducts (which per definition were also coelomoducts). Segmentation may have evolved in connection with burrowing, and the chaetae may have aided this mode of locomotion.

The ectoparasitic group Myzostomida has no obvious sister taxon within the Polychaeta, but their pelagic larvae (Jägersten 1939; Fig. 17.6) are typical polychaete nectochaetes with larval chaetae whereas the parasitic adults not unexpectedly are quite modified; the adults have paired protonephridia associated with the short parapodia like many of the more usual polychaetes (Pietsch & Westheide 1987). In agreement with Pettibone (1982) they are here regarded not as a sister group of the polychaetes, but as a specialized group of polychaetes.

The clitellates (Oligochaeta and Hirudinea) are characterized by the reduction of the number of segments with reproductive organs and by development of special structures of the fertile region (the clitellum) connected with copulation and formation of protective cocoons for the eggs. Other characteristics include hermaphroditism, direct development, and in the leeches restriction of the coelomic cavities and absence of chaetae. The clitellates are usually treated as a group of equal rank with the polychaetes, but it could perhaps be more useful to regard the polychaete family Capitellidae as the sister group instead of the whole polychaete 'class'. Many species of this family are described as earthworm-like with 'reduced' parapodia and small chaetae, but as mentioned above this condition may actually be ancestral. The number of segments with reproductive organs is restricted in all genera and there is only one pair of genital openings in *Capitella* and *Capitomastus* with special copulatory organs on the sexual and one neighbouring segment. Males of *Capitella* may under

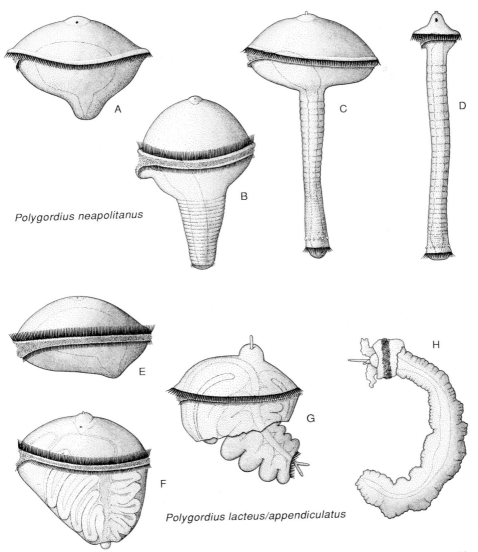

Polygordius neapolitanus

Polygordius lacteus/appendiculatus

Fig. 17.5. Larval types and metamorphoses in *Polygordius*. The trochophora larvae have a blown-up body with a spacious blastocoel in both main types, but the segmented body of the metatrochophores develops along two main lines: the exolarva, which has the body as a posterior appendage (*P. neapolitanus*, redrawn from Hatschek 1878), or the endolarva (a type 2 pericalymma), in which the segmented body is contracted like an accordion and retracted into an extension of the post-metatrochal zone (serosa) of the trochophore (*P. lacteus* and *P. appendiculatus*, redrawn from Woltereck 1902, 1926 and Herrmann 1986). – A, a young trochophore. – B-C, the segmented body develops. – D, the larval body has contracted strongly and the larva is ready for settling. – E, a young trochophore. – F, the segmented body develops inside the serosa. – G, the serosa is ruptured when the segmented body stretches. – H, most of the larval organs degenerate and the juvenile is ready for the benthic life.

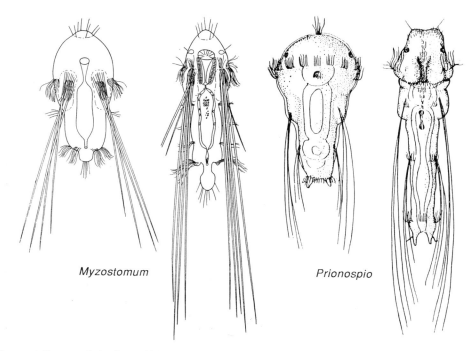

Myzostomum   Prionospio

Fig. 17.6. Four- and six-days-old larvae of *Myzostomum parasiticum* and young and late 2-segment larval stage of the spionid *Prionospio malmgreni*; both types of larvae have chaetae which are shed in later stages. (From Jägersten 1939 and Hannerz 1956.)

certain circumstances develop a pair of ovaries and become simultaneous herma-phrodites, just like the clitellates (Holbrook & Grassle 1984).

The embryology of the clitellates is characterized by the presence of consider-able amounts of yolk, and development is direct in all species; cleavage is highly un-equal (Anderson 1966b). Gastrulation is epibolic; the ectodermal and mesodermal cells form a micromere cap which spreads ventrally over the large endodermal cells. The cell lineage of the leech *Helobdella* (Fernández 1980, Weisblat *et al.* 1980, Weis-blat, Kim & Stent 1984, see Fig. 17.7) has been studied in fine detail: at a certain stage of the cleavage, the D-macromere divides into two cells of nearly equal size: one cell, called DM, corresponds to the 4d-cell in the usual terminology of the spiral cleavage, and the other, called DNOPQ, gives rise to the ectoteloblasts. DM divides into a pair of large mesoteloblasts, which give off a number of cells, each of which develops into the mesoderm of one side of a segment. Also DNOPQ divides laterally and each of the resulting cells divides to form four large ectoteloblasts: N, O, P, and Q (see Fig. 17.7); these teloblasts each give off two cells for each mesodermal cell. The bands of ectodermal cells (n, o, p, and q cells) cover the bands of mesodermal cells (m) and these bands now migrate over the macromeres, finally meeting at the ventral mid-line where the two rows of n-cells come into contact, leaving only an an-terior and a posterior opening which become mouth and anus, respectively. The ec-

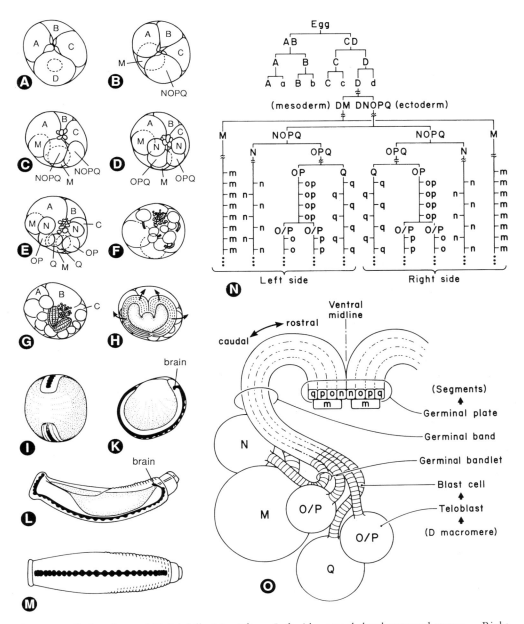

Fig. 17.7. Embryology of *Helobdella triserialis.* – Left side: actual developmental stages. – Right side: cell-lineage of the D (or 1D)-cell; the table shows the pedigrees of the teloblasts and the diagram shows the development of the germinal bands of the left side of an embryo. The M-cells give rise to mesoderm and the N,O,P,Q-cells give rise to ectoderm and the ventral chain of ganglia. (Combined from Weisblat et al. 1980 and Weisblat, Kim & Stent 1984.)

toderm of the whole body as well as the chain of ventral ganglia originates from the ectoteloblasts. The supraoesophageal ganglion (brain) originates from the A, B, and C-cells, probably from the apical micromeres.

Other oligochaetes and hirudineans have more aberrant cleavage and cell-lineage (Anderson 1966b). The mid-gut originates solely from the D-quadrant in *Erpobdella*, and in *Stylaria*, descendants of the first micromeres and the macromeres 1A and 1B become an embryonic envelope, which is shed at a later stage (Dawydoff 1941).

Within the clitellates the two groups Oligochaeta and Hirudinea may not be sister groups; Erséus (1987) suggested that the leeches could be included in the Tubificida, which would make the oligochaetes another example of a paraphyletic stem group.

From the above, it appears that the clitellates resemble one of the polychaete families more than they resemble the hypothetical polychaete ancestor, and that they are therefore not a sister group of the polychaetes but rather a group within the 'Polychaeta', which is thus paraphyletic (Fig. 17.9), as also suggested for example by Fauchald (1974) and George & Hartmann-Schröder (1985).

The systematic position of the pogonophorans has been much debated. (The nomenclature adopted here, viz. Pogonophora = Frenulata + Vestimentifera, cannot be described as elegant, but it appears to be the only way in which the well-known names Pogonophora and Vestimentifera can be retained without introduction of new names.) However, recent descriptions of newly settled larvae of the vestimentiferan *Ridgeia* (Southward 1988, Jones & Gardiner 1988, 1989) unequivocally settle the discussions about the dorsal/ventral orientation of the adults and about the origin of the trophosome with the symbiotic bacteria; these results (see below) have removed the last doubt about the annelid character of the pogonophorans.

The youngest known stages of *Ridgeia*, about 1/4 mm long, are tube-dwelling and have a normal, tubular gut and a pair of tentacles developing on the dorsal side (Southward 1988); the central nervous system consists of supra- and suboesophageal ganglia connected by commissures encircling the oesophagus and a ventral main longitudinal nerve trunk (Fig. 17.8). The gut contains bacteria in the lumen and in vacuoles in the epithelial cells. There is an anterior coelomic cavity surrounding the mouth region and separated from a main coelom by a normal septum. The main coelom appears to be partially divided into a pair of cavities by a dorsal and a ventral mesentery. The first opisthosomal coelom is bordered anteriorly by a normal septum, but the preservation made it difficult to make out the further details. The fourteen-tentacle stage has a ciliated gut in which some of the endodermal cells contain vacuoles with bacteria and have lost contact with the lumen of the gut. It is clear that the trophosome develops from the gut. The opisthosome now has five flat coelomic segments, and rows of chaetae are associated with each segment. Somewhat older specimens have the mouth situated on the tip of a peculiar proboscis, but this structure disappears in the adults, which lack a gut completely.

Although the embryology of the vestimentiferans is totally unknown, the anatomy of the young *Ridgeia* demonstrates beyond doubt that the vestimentiferans are annelids with an extremely elongate postoral segment and a gut which is modified as a storage organ for symbiotic bacteria.

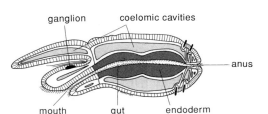

Fig. 17.8. Diagrammatic median section of a juvenile *Ridgeia piscesae*. (Redrawn from Southward 1988, with addition of the dorsal ganglion from an unpublished drawing by Dr Eve Southward, Plymouth, UK.)

The frenulate pogonophorans share a number of important characters with the vestimentiferans: the general regionation of the body with a small opisthosome with rings of chaetae, the presence of an elongate mass of internal cells (the trophosome or bacterial cylinder) with symbiotic sulphur bacteria instead of a normal gut (although the trophosome is actually a tube which quite possibly represents the gut, see Southward 1982), the shape and position of the nervous system, and the general morphology of the larvae/juveniles. The main differences between the two groups may be related to the much larger size of the vestimentiferans: an example is that the pinnules of the tentacles are multicellular in vestimentiferans (van der Land & Nørrevang 1977) and unicellular in frenulates (Ivanov 1955). The embryology has been studied in a few species. The small species of *Siboglinum* have elongate eggs with a large amount of yolk and a cleavage pattern with strong spiralian traits (Bakke 1980, Ivanov 1988); the formation of the mesoderm has been described in a number of papers, but there has not been agreement about the interpretation. The somewhat larger *Oligobrachia* have more rounded eggs with less yolk, which makes it easier to observe the development of the mesoderm; Gureeva & Ivanov (1986; see also Ivanov 1988) reported an embryological stage with a pair of long lateral coelomic pockets in open connection with the anterior end of the archenteron. This may indicate an enterocoelic mode of mesoderm formation, a feature otherwise almost unknown in the spiralians, but further studies based on serial sections of various stages are needed before a full understanding of the mesoderm formation can be reached. There are, however, so many detailed similarities between the adults of frenulates and vestimentiferans that the two groups must be regarded as closely related groups, perhaps as sister groups.

The precise systematic position of the pogonophorans cannot be ascertained at present, but some of the features considered characteristic of pogonophorans can be found in various polychaete families, for example long papillate tentacles in *Magelona*. The tubes of some chaetopterids (*Spiochaetopterus* and *Phyllochaetopterus*) are superficially very similar to pogonophore tubes, but the chemical composition is different, the tubes of *Siboglinum* containing about 33% β-chitin (Southward 1971) while chitin has not been found in 'normal' polychaete tubes. Symbiotic sulphur bacteria are known from other polychaetes from the hydrothermal vents, from oligochaetes (Giere & Langheld 1987) and from various molluscs (Southward 1986), but in all these organisms the bacteria are associated with the epidermis; the endosym-

139

bionts in the modified gut appear so far to be a unique feature of the pogono-phorans. At present the pogonophorans must thus be regarded as a specialized poly-chaete group and not as a sister group of the Annelida.

The family Lobatocerebridae (with the only genus *Lobatocerebrum*) comprises a few small, interstitial, unsegmented, completely ciliated 'worms', which are usually regarded as very specialized annelids (Rieger 1980, 1981, 1988). At first sight, the animals resemble small turbellarians, but some of the interstitial polychaetes are also superficially quite similar. Studies of the ultrastructure have shown that there is no sign of a segmentation of the mesoderm, which has no coelomic cavities. *Lobato-cerebrum* is thus technically acoelomate, but similar conditions have been described from small species belonging to a number of polychaete families which show a seg-mentation (Fransen 1980). Segmentation is one of the most constant characters of annelids, but the tiny (evidently neotenic) dwarf male of the polychaete *Dinophilus gyrociliatus* is unsegmented (Westheide 1988). As discussed by Rieger (1988), it is difficult to find convincing synapomorphies which can link the lobatocerebrids with either platyhelminths or annelids. Many characters are generalized spiralian, but there seem to be a few characteristics connected with reproduction which may give some indications. The platyhelminths have copulation and internal fertilization and the male gonads are connected with characteristic copulatory organs with intercellu-lar stylets of proteinaceous/collagenous material (Brüggemann 1985); their sperma-tozoa are highly modified with a pair of cilia with only one central tubule in the Acoela and in the more advanced groups (the Trepaxonemata (Ehlers 1985), which comprises all platyhelminths except Catenulida, Nemertodermatida, Acoela and Macrostomida); the catenulids and the macrostomids have non-ciliated spermatozoa (Ehlers 1985) and only the nemertodermatoids have spermatozoa with a (short) nor-mal cilium (Hendelberg 1986). Although the sperm type with a long undulatory cil-ium is believed to represent the primitive type among the metazoans, it appears that the specialized sperm type connected with internal fertilization evolved at the very stem of the platyhelminths (Chapter 26). *Lobatocerebrum* lacks copulatory struc-tures and has a ciliated male duct and spermatozoa with a filiform head and a long cilium; these characters are common among polychaetes. The gut is complete with rectum and anus as in the polychaetes. Reproduction and development are com-pletely unknown.

As already stated by Rieger (1988), it is not possible to find any convincing synapomorphies between *Lobatocerebrum* and the Platyhelminthes, whereas the presence of an anus and a number of characteristics of the male reproductive organs resemble those of certain annelids (perhaps especially oligochaetes). Until further ob-servations become available, it appears necessary to regard the Lobatocerebridae as specialized annelids.

Another group of small, totally ciliated, interstitial 'worms' which have puzzled zoologists is the Gnathostomulida. The group has generally been treated as a separ-ate phylum of uncertain affinities, often regarded as closely related to the Platy-helminthes, but sometimes considered closer to annelids or gastrotrichs. Ax (1987) has presented a lengthy discussion of gnathostomulids and platyhelminths as sister groups, without mentioning that other possibilities exist. As with *Lobatocerebrum* the gnathostomulids show many characters which must be characterized as general protostomian or spiralian, for example the structure of the epithelium and the cleavage, but it is difficult to pick out characters which can be used to identify a sister group.

The gnathostomulid epithelium consists of monociliate cells and the cilia beat in the usual metachronal pattern (Rieger & Mainitz 1977; Ax 1964, 1987); similar epithelia are found in several gastrotrichs (see Chapter 33), but since a complete reversal from the multiciliary state to a monociliary state is a definite possibility (discussed in Chapter 10), this shared character can hardly be interpreted as a synapomorphy. The ontogeny has not been studied in detail, but Riedl (1969) described the cleavage of *Gnathostomula* as spiral and identified two possible mesoblasts; this indicates that the gnathostomulids are not related to gastrotrichs. A further difference is found in the structure of the pharynx, which has mesodermal muscles in gnathostomulids (Kristensen & Nørrevang 1977), whereas that of gastrotrichs has an ectodermal myoepithelium (Ruppert 1982). An anus is generally absent, but Knauss (1979) observed a close connection and absence of the basement membrane between endoderm and ectoderm in a posterodorsal area in *Haplognathia* and suggested that it may function as an anus.

The choice between platyhelminths and annelids as sister group (or most closely related group) of the gnathostomulids can only be based on few characters, and none of them can be taken as definitive. The lack of an anus seems an uncertain character. The lack of segmentation and coelom could indicate affinity with the platyhelminths, but the Lobatocerebridae and the dwarf male of *Dinophilus* are similar in this respect. The gnathostomulid jaws are cuticular structures on a ventral pharyngeal bulb like that of several polychaetes (Kristensen & Nørrevang 1977), while the jaws, copulatory stylets and spines of platyhelminths are either reinforced areas of the basement membrane or intraepithelial structures (Doe 1976, Brüggemann 1985). The muscle cells of platyhelminths are of the smooth type (Ehlers 1985) while the gnathostomulids have cross-striated muscles (see for example Kristensen & Nørrevang 1977). The two last-mentioned characters strongly indicate annelid affinities.

The nature of the relationship between the gnathostomulids and the annelids is apparently not that of sister groups, and at present I am inclined to regard the gnathostomulids as another highly specialized polychaete group.

The Echiura have been regarded as closely related to or even included in the Annelida ever since Hatschek's (1880) remarkably detailed observations on the larva of *Echiurus* and its metamorphosis. The chaetae have exactly the same structure and chemical composition as those of the polychaetes (Storch 1984) – and of the brachiopods (Chapter 45). The only uncertainty appears to be whether the echiurans should be regarded as primarily unsegmented – and therefore a sister group to the articulates – or as secondarily unsegmented – and therefore as specialized annelids.

The development shows a spiral cleavage almost identical to that of several polychaetes with the 4d-cell giving rise to the two mesoblasts (for example *Urechis*; Newby 1940). The larva of *Echiurus* is an almost schematic trochophore (Hatschek 1880, Baltzer 1917) with all the ciliary bands and nerves as proposed by the trochaea theory (see Chapter 2). Hatschek (1880) reported that the mesodermal bands are

141

proliferated from a pair of large teloblasts, but this was not confirmed by the investigations of Baltzer (1917), Torrey (1903, on *Thalassema*) and Newby (1940, on *Urechis*). A coelomic cavity is formed as a slit in the anterior end of each mesodermal band; the two coelomic sacs spread and meet dorsally and ventrally where they fuse and form mesenteries which later disappear almost totally (Newby 1940). There is no indication of a segmentation in the mesoderm.

The nervous system is of a generalized gastroneuralian type. The brain develops from the apical plate and the ventral nerve cord from the ectoderm between mouth and anus. In early stages of *Echiurus*, the paired ventral cord shows an arrangement of cells resembling a segmentation, but segmentation has not been observed in the adults; the larval ventral cords become strongly transversely folded, which may have been the reason for earlier reports of ventral ganglia (Korn 1982). Juveniles of *Urechis* have one row of ill-defined ventral ganglia, but the number of ganglia increases by division of already formed ganglia instead of through development of new ganglia from a posterior growth zone (Newby 1940).

At present it seems impossible to form a well founded opinion about the precise phylogenetic position of the echiurans. The presence of blood vessels and heart indicates a molluscan-articulate relationship, and the chaetae point specifically to relationships with annelids, but similar structures are found in other phyla (see below), and the evidence about segmentation is inconclusive. I have chosen to put emphasis on the chaetae and to include the echiurans in the Annelida, but new information may well make it necessary to change opinion. A clear sister-group relationship with a special annelid group cannot be recognized, and it is not possible to regard the echiurans as sister group of all the annelids.

A number of authors have also incorporated the Sipuncula in the Annelida; the sipunculans have many spiralian characters, but since they have neither chaetae nor any signs of a segmentation or blood vessels (Chapter 13), there is no synapomorphy which can be used to unite the two groups.

The chaetae are here considered the important synapomorphy of the Annelida although more or less similar structures are known from other phyla. Very similar chaetae occur at the mantle edge of brachiopods (Chapter 45); some molluscs have chitinous structures consisting of tubular secretions from microvilli of one cell: the girdle of some chitons has hairs with sensory cells wrapped in a layer of a structure very similar to that of the annelid chaetae (Leise & Cloney 1982), and most of the epithelia of newly hatched cephalopods have tufts of chitinous tubules (Kölliker's tufts) which resemble annelid setae but which are not fused, so that the tubules can spread out in tufts (Brocco, O'Clair & Cloney 1974). The gizzard teeth of some ctenostome bryozoans are of a similar structure with channels and microvilli but of a quite different shape and situated in the pharynx (Chapter 24). The homology of these structures has been discussed by several authors, for example George & Southward (1973), and it has been considered possible that chaetae or similar structures were present in the common ancestor of these groups, which in consequence means in the early bilaterians. This can be questioned for a number of reasons. Chitin is present in several protists, in cnidarians, in the cuticle of most protostomians and has also been found in tunicates (Jeuniaux 1982), which shows that the ability to synthesize chitin is widespread. The structure of the chaetae and the more tooth-like structures appears to be identical, but the position on the body varies: the annelid chaetae are formed in lateral rows along the body; the larval brachiopod setae (Chapter 45) occur in a similar pattern, while those of the adults are situated along the mantle edge; the girdle hairs of chitons and the Kölliker's tufts of juvenile cephalopods occur in larger areas of the epidermis, and the gizzard teeth of the bryozoans are located in the pharynx. It appears that Remane's homology criterion of position is not fulfilled, and it is therefore not probable that the various chaeta-like structures are homologous.

A consequence of the above discussion of the various groups considered as annelids is that the Polychaeta have become a paraphyletic stem group (just like the Turbellaria; see Chapter 26); the word 'polychaetes' can be used as a colloquial term for the less specialized annelids, but it has no place in the systematic vocabulary. The

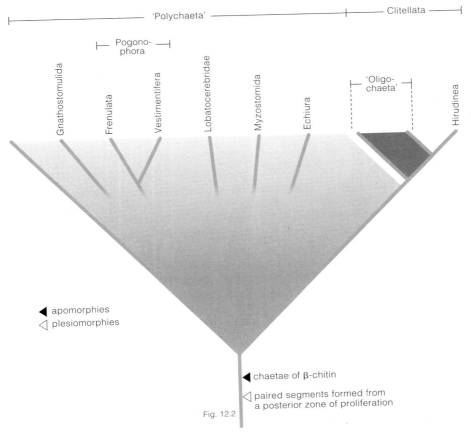

Fig. 17.9. A 'phylogenetic tree' of the Annelida with indications of the special groups discussed in the text. The two traditional groups 'Polychaeta' and 'Oligochaeta' are considered paraphyletic.

phylogenetic relationships within the phylum are completely unresolved, and it is only possible to recognize certain classes, orders or families emerging from a bush (Fig. 17.9).

The ancestral annelid can now be characterized as follows: it had a trochophora larva and the adult was small, benthic, probably burrowing, with segments with a pair of coelomic sacs and chaetae, but without parapodia. It could be distinguished from the ancestral panarthropod by the presence of chaetae and of the 'normal' spiralian cuticle with collagen (Chapter 12).

The paired coelomic sacs clearly characterize the Annelida as articulates. Cleavage is of the spiralian type, although much variation occurs. The ontogeny of the nervous systems both in the planktotrophic 'polychaete' larvae and in the lecithotrophic leeches clearly demonstrates that the brain originates from apical cells and the ventral chain of ganglia from cells along the fused blastopore lips, as considered diagnostic of the protostomes (gastroneuralians).

Interesting subjects for future research

1. Cleavage and early larval development of the vestimentiferans – is the cleavage spiral and are the larvae planktotrophic ?
2. Echiuran development – do they have teloblasts? – are there any unquestionable signs of segmentation in mesoderm, ectoderm or nervous system ?

References

Åkesson, B. 1967. The embryology of the polychaete *Eunice kobiensis*. – Acta zool. (Stockh.) 48: 141-192.

Åkesson, B. 1968. The ontogeny of the glycerid prostomium (Annelida; Polychaeta). – Acta zool. (Stockh.) 49: 203-217.

Anderson, D.T. 1959. The embryology of the polychaete *Scoloplos armiger*. – Q. Jl microsc. Sci. 100: 89-166.

Anderson, D.T. 1966a. The comparative embryology of the Polychaeta. – Acta zool. (Stockh.) 47: 1-42.

Anderson, D.T. 1966b. The comparative embryology of the Oligochaeta, Hirudinea and Onychophora. – Proc. Linn. Soc. N.S.W. 91: 10-43.

Anderson, D.T. 1973. Embryology and Phylogeny of Annelids and Arthropods. (Int. Ser. Monogr. pure appl. Biol., Zool. 50) – Pergamon Press, Oxford.

Ax, P. 1964. Das Hautgeisselepithel der Gnathostomulida. – Zool. Anz., Suppl 27: 452-461.

Ax, P. 1987. The Phylogenetic System. – John Wiley & Sons, Chichester.

Bakke, T. 1980. Embryonic and post-embryonic development in the Pogonophora. – Zool. Jb., Anat. 103: 276-284.

Baltzer, F. 1917. Echiuriden 1. Teil: *Echiurus abyssalis*. – Fauna Flora Golf. Neapel 34: 1-234, 12 pls.

Bartolomaeus, T. 1989. Ultrastructure and development of the nephridia in *Anaitides mucosa* (Annelida, Polychaeta). – Zoomorphology 109: 15-32

Bartolomaeus, T. & P. Ax 1992. Protonephridia and metanephridia – their relation within the Bilateria. – Z. zool. Syst. Evolutionsforsch. 30: 21-45.

Boilly-Marer, Y. 1972. Présence de cellules de type myoépitélial chez les Nereidae (Annélides Polychètes). – J. Microscopie 15: 253-277.

Brasier, M.D. 1979. The Cambrian radiation event. – *In* M.R. House (ed.): The Origin of Major Invertebrate Groups, pp 103-159. Academic Press, London.

Brocco, S.L., R. O'Clair & R.A. Cloney 1974. Cephalopod integument: the ultrastructure of Köl-liker's organs and their relationships to setae. – Cell Tissue Res. 151: 293-308.

Brüggemann, J. 1985. Ultrastruktur und Bildungsweise penialer Hartstrukturen bei freilebenden Plathelminthen. – Zoomorphology 105: 143-189.

Bubel, A., R.M. Stephens, R.H. Fenn & P. Fieth 1983. An electron microscope, X-ray diffraction and amino acid analysis study of the opercular filament cuticle, calcareous opercular plate and habitation tube of Pomatoceros lamarckii Quatrefages (Polychaeta: Serpulidae). – Comp. Bio-chem. Physiol. 74B: 837-850.

Child, C.M. 1900. The early development of Arenicola and Sternaspis. – Arch. Entwicklungsmech. Org. 9: 587-723, pls 21-25.

Clark, R.B. 1964. Dynamics in Metazoan Evolution. – Clarendon Press, Oxford.

Conway Morris, S. 1979. Middle Cambrian polychaetes from the Burgess Shale of British Colum-bia. – Phil. Trans. R. Soc. B 285: 227-274, 9 pls.

Dawydoff, C. 1941. Études sur l'embryologie des Naïdidae Indochinoises. – Archs Zool. exp. gén. 81 (Notes et Revue): 173-194.

Doe, D.A. 1976. The proboscis hooks in Karkinorhynchidae and Gnathorhynchidae (Turbellaria, Kalyptorhynchidae) as basement membrane or intracellular specializations. – Zool. Scr. 5: 105-115.

Dorresteijn, A.W.C. & A. Fischer 1988. The process of early development. – Microfauna mar. 4: 335-352.

Eckelbarger, K.J. & F.-S. Chia 1978. Morphogenesis of larval cuticle in the polychaete Phrag-matopoma lapidosa. – Cell Tissue Res. 186: 187-201.

Ehlers, U. 1985. Das phylogenetische System der Plathelminthes. – Gustav Fischer, Stuttgart.

Erséus, C. 1987. Phylogenetic analysis of the aquatic Oligochaeta under the principle of parsimony. – Hydrobiologia 155: 75-89

Fauchald, K. 1974. Polychaete phylogeny: a problem in protostome evolution. – Syst. Zool. 24: 493-506.

Fauchald, K. 1977. The polychaete worms: definitions and keys to the orders, families and genera. – Nat. Hist. Mus. Los Angeles Co., Sci. Ser. 28: 1-190.

Fernández, J. 1980. Embryonic development of the glossiphoniid leech Theromyzon rude: charac-terization of developmental stages. – Dev. Biol. 76: 245-262.

Fransen, M.E. 1980. Ultrastructure of coelomic organization in annelids. I. Archiannelids and other small polychaetes. – Zoomorphology 95: 235-249.

Gaill, F. & S. Hunt 1988. Tubes. – Microfauna mar. 4: 61-70.

Gardiner, S.L. 1978. Fine structure of the ciliated epidermis on the tentacles of Owenia fusiformis (Polychaeta, Oweniidae). – Zoomorphologie 91: 37-48.

Gardiner, S.L. & R.M. Rieger 1980. Rudimentary cilia in muscle cells of annelids and echinoderms. – Cell Tissue Res. 213: 247-252.

George, J.D. & G. Hartmann-Schröder 1985. Polychaetes: British Amphinomida, Spintherida and Eunicida. – Synopses Br. Fauna, N.S. 32: 1-221.

George, J.D. & E.C. Southward 1973. A comparative study of the setae of Pogonophora and poly-chaetous Annelida. – J. mar. biol. Ass. U.K. 53: 403-424.

Giere, O. & C. Langheld 1987. Structural organisation, transfer and biological fate of endosymbio-tic bacteria in gutless oligochaetes. – Mar. Biol. (Berl.) 93: 641-650.

Gureeva, M.A. & A.V. Ivanov 1986. On the coelomic mesoderm formation in embryos of Oligo-brachia mashikoi (Pogonophora). – Zool. Zh. 780-788 (In Russian, English summary; trans-lation obtained from the Translation Bureau of the Canadian Secretary of State).

Hannerz, L. 1956. Larval development of the polychaete families Spionidae Sars, Disomidae Mes-nil, and Poecilochaetidae n.fam. in the Gullmar Fjord. – Zool. Bidr. Upps. 31: 1-204.

Hatschek, B. 1878. Studien über Entwicklungsgeschichte der Anneliden. – Arb. zool. Inst. Univ. Wien 1: 277-404, pls 23-30.

Hatschek, B. 1880. Ueber Entwicklungsgeschichte von Echiurus und die systematische Stellung der Echiuridae (Gephyrei chaetiferi). – Arb. zool. Inst. Univ. Wien 3: 45-78, pls 4-6.

Hatschek, B. 1885. Entwicklung der Trochophora von Eupomatus uncinatus, Philippi (Serpula un-cinatus). – Arb. zool. Inst. Univ. Wien 6: 121-148, pls 8-13.

Hay-Schmidt, A. 1987. The ultrastructure of the protonephridium of the actinotroch larva (Phoronida). – Acta zool. (Stockh.) **68**: 35-47.

Hendelberg, J. 1986. The phylogenetic significance of sperm morphology in the Plathelminthes. – Hydrobiologia **132**: 53-58.

Herrmann, K. 1986. *Polygordius appendiculatus* (Archiannelida) – Metamorphose. – Publ. wiss. Film, Biol., **18**. ser. 36/E2716: 1-15.

Holbrook, M.J.L. & J.P. Grassle 1984. The effect of low density on the development of simultaneous hermaphroditism in male *Capitella* species I (Polychaeta). – Biol. Bull. Woods Hole **166**: 103-109.

Ivanov, A.V. 1955. On external digestion in Pogonophora. – Syst. Zool. **4**: 174-176.

Ivanov, A.V. 1988. Analysis of the embryonic development of Pogonophora in connection with the problems of phylogenetics. – Z. zool. Syst. Evolutionsforsch. **26**: 161-185.

Jägersten, G. 1939. Zur Kenntniss der Larvenentwicklung bei *Myzostomum*. – Ark. Zool. 31A(11): 1-21.

Jeuniaux, C. 1982. La chitine dans le règne Animal. – Bull. Soc. zool. Fr. **107**: 363-386

Jones, M.L. & S.L. Gardiner 1988. Evidence for a transient digestive tract in Vestimentifera. – Proc. biol. Soc. Wash. **101**: 423-433.

Jones, M.L. & S.L. Gardiner 1989. On the early development of the vestimentiferan tube worm *Ridgeia* sp. and observations on the nervous system and trophosome of *Ridgeia* sp. and *Riftia pachyptila*. – Biol. Bull. Woods Hole **177**: 254-276.

Knauss, E.B. 1979. Indication of an anal pore in Gnathostomulida. – Zool. Scr. **8**: 181-186.

Korn, H. 1958. Vergleichend-embryologische Untersuchungen an *Harmothoe* Kinberg, 1857 (Polychaeta, Annelida). Organogenese und Neurosekretion. – Z. wiss. Zool. **161**: 346-443.

Korn, H. 1960. Das larvale Nervensystem von *Pectinaria* Lamarck und *Nephthys* Cuvier (Annelida, Polychaeta). – Zool. Jb., Anat. **78**: 427-456.

Korn, H. 1982. Annelida (einschliesslich Echiurida und Sipunculida). – *In* F. Seidel (ed.): Morphogenese der Tiere, Erste Reihe, Lief. 5, pp 1-599. Gustav Fischer, Stuttgart.

Kristensen, R.M. & A. Nørrevang 1977. On the fine structure of *Rastrognathia macrostoma* gen. et sp. n. placed in Rastrognathiidae fam. n. (Gnathostomulida). – Zool. Scr. **6**: 27-41.

Lacalli, T.C. 1981. Structure and development of the apical organ in trochophores of *Spirobranchus polycerus*, *Phyllodoce maculata* and *Phyllodoce mucosa* (Polychaeta). – Proc. R. Soc. Lond. B **212**: 381-402, 7 pls.

Lacalli, T.C. 1984. Structure and organization of the nervous system in the trochophore larva of *Spirobranchus*. – Phil. Trans. R. Soc. B **306**: 79-135, 19 pls.

Lacalli, T.C. 1988. The larval reticulum in *Phyllodoce* (Polychaeta, Phyllodocida). – Zoomorphology **108**: 61-68.

Leise, E.M. & R.A. Cloney 1982. Chiton integument: ultrastructure of the sensory hairs of *Mopalia muscosa* (Mollusca: Polyplacophora). – Cell Tissue Res. **223**: 43-59.

Mau, W. 1881. Über *Scoloplos armiger* O.F. Müller. – Z. wiss. Zool. **36**: 389-432, pls 26-27.

Newby, W.W. 1940. The embryology of the echiuroid worm *Urechis caupo*. – Mem. Am. phil. Soc. **16**: 1-219.

Nielsen, C. 1987. Structure and function of metazoan ciliary bands and their phylogenetic significance. – Acta zool. (Stockh.) **68**: 205-262.

Nilsson, D. 1912. Beiträge zur Kenntniss des Nervensystems der Polychaeten. – Zool. Bidr. Upps. **1**: 85-161, pls 3-5.

O'Clair, R.M. & R.A. Cloney 1974. Patterns of morphogenesis mediated by dynamic microvilli: chaetogenesis in *Nereis vexillosa*. – Cell Tissue Res. **151**: 141-157.

Pettibone, M.H. 1982. Annelida, Polychaeta. – *In* S.P. Parker (ed.): Synopsis and Classification of Living Organisms, vol. 2, pp 1-43. McGraw-Hill, New York.

Pietsch, A. & W. Westheide 1987. Protonephridial organs in *Myzostoma cirriferum* (Myzostomida). – Acta zool. (Stockh.) **68**: 195-203.

Purschke, G. 1988. Pharynx. – Microfauna mar. **4**: 177-197.

Rasmussen, E. 1973. Systematics and ecology of the Isefjord marine fauna. – Ophelia **11**: 1-495.

Richards, K.S. 1984. Cuticle. – *In* J. Bereiter-Hahn, A.G. Matoltsy & K.S. Richards (eds): Biology of the Integument, vol. 1, pp 310-322. Springer, Berlin.

Riedl, R.J. 1969. Gnathostomulida from America. – Science **163**: 445-462.

Rieger, R.M. 1980. A new group of interstitial worms, Lobatocerebridae nov. fam. (Annelida) and its significance for metazoan phylogeny. – Zoomorphologie **95**: 41-84.

Rieger, R.M. 1981. Fine structure of the body wall, nervous system, and digestive tract in the Lobatocerebridae Rieger and the organization of the gliointerstitial system in Annelida. – J. Morph. **167**: 139-165.

Rieger, R.M. 1988. Comparative ultrastructure and the Lobatocerebridae: keys to understand the phylogenetic relationship of Annelida and the Acoelomates. – Microfauna mar. **4**: 373-382.

Rieger, R.M. & M. Mainitz 1977. Comparative fine structure study of the body wall in Gnathostomulida and their phylogenetic position between Platyhelminthes and Aschelminthes. – Z. zool. Syst. Evolutionsforsch. **15**: 9-35.

Robbins, D.E. 1965. The biology and morphology of the pelagic annelid *Poeobius meseres* Heath. – J. Zool. (Lond.) **146**: 197-212.

Ruppert, E.E. 1982. Comparative ultrastructure of the gastrotrich pharynx and the evolution of myoepithelial foreguts in Aschelminthes. – Zoomorphology **99**: 181-220.

Ruppert, E.E. & K.J. Carle 1983. Morphology of metazoan circulatory systems. – Zoomorphology **103**: 193-208.

Schroeder, P.C. 1984. Chaetae. – In J. Bereiter-Hahn, A.G. Matoltsy & K.S. Richards (eds): Biology of the Integument, vol. 1, pp 297-309. Springer, Berlin.

Schroeder, P.C. & C.O. Hermans 1975. Annelida: Polychaeta. – In A.C. Giese & J.S. Pearse (eds): Reproduction of Marine Invertebrates, vol. 3, pp 1-213. Academic Press, New York.

Segrove, F. 1941. The development of the serpulid *Pomatoceros triqueter* L. – Q. Jl microsc. Sci. **82**: 467-540.

Smith, P.R., E.E. Ruppert & S.L. Gardiner 1987. A deuterostome-like nephridium in the mitraria larva of *Owenia fusiformis* (Polychaeta, Annelida). – Biol. Bull. Woods Hole **172**: 315-323.

Southward, E.C. 1971. Recent researches on the Pogonophora. – Oceanogr. mar. Biol. ann. Rev. **9**: 193-220.

Southward, E.C. 1982. Bacterial symbionts in Pogonophora. – J. mar. biol. Ass. U.K. **62**: 889-906.

Southward, E.C. 1986. Gill symbionts in thyasirids and other bivalve molluscs. – J. mar. biol. Ass. U.K. **66**: 889-914.

Southward, E.C. 1988. Development of the gut and segmentation of newly settled stages of *Ridgeia* (Vestimentifera): implications for relationships between Vestimentifera and Pogonophora. – J. mar. Biol. Ass. U.K. **68**: 465-487.

Storch, V. 1984. Echiura and Sipuncula. – In J. Bereiter-Hahn, A.G. Matoltsy & K.S. Richards (eds): Biology of the Integument, vol.1, pp 368-375. Springer, Berlin.

Storch, V. 1988. Integument. – Microfauna mar. **4**: 13-36.

Torrey, J.C. 1903. The early embryology of *Thalassema mellita* (Conn). – Ann. N.Y. Acad. Sci. **14**: 165-246, pls 1-2.

Treadwell, A.L. 1901. Cytogeny of *Podarke obscura* Verrill. – J. Morph. **17**: 399-486, pls 36-40.

Turbeville, J.M. 1986. An ultrastructural analysis of coelomogenesis in the hoplonemertine *Prosorhochmus americanus* and the polychaete *Magelona* sp. – J. Morph. **187**: 51-60.

van der Land, J. & A. Nørrevang 1977. Structure and relationships of *Lamellibrachia* (Annelida, Vestimentifera). – Biol. Skr. Danm. Vid. Selsk. **21**(3): 1-102, pls 1-26.

Weisblat, D.A., G. Harper, G.S. Stent & R.T. Sawyer 1980. Embryonic cell lineages in the nervous system of the glossiphoniid leech *Helobdella triserialis*. – Dev. Biol. **76**: 58-78.

Weisblat, D.A., S.Y. Kim & G.S. Stent 1984. Embryonic origins of cells in the leech *Helobdella triserialis*. – Dev. Biol. **104**: 65-85.

Westheide, W. 1967. Monographie der Gattungen *Hesionides* Friedrich und *Microphthalmus* Mecznikow (Polychaeta, Hesionidae). – Z. Morph. Ökol. Tiere **61**: 1-159.

Westheide, W. 1988. The nervous system of the male *Dinophilus gyrociliatus* (Annelida: Polychaeta). I. Number, types and distribution pattern of sensory cells. – Acta zool. (Stockh.) **69**: 55-64.

Westheide, W. 1990. Polychaetes: interstitial families. – Synopses Br. Fauna, N.S. **44**: 1-152.

Wilson, D.P. 1932. On the mitraria larva of *Owenia fusiformis* Delle Chiaje. – Phil. Trans. R. Soc. B **221**: 231-334, pls 29-32.

Woltereck, R. 1902. Trochophora-Studien I. Histologie der Larve und die Entstehung des Annelids bei den *Polygordius*-Arten der Nordsee. – Zoologica (Stuttg.) **13**(34): 1-71, 11 pls.

Woltereck, R. 1904. Beiträge zur praktischen Analyse der *Polygordius*-Entwicklung nach dem 'Nordsee-' und dem 'Mittelmeer-Typus'. – Arch. Entwicklungsmech. Org. **18**: 377-403, pls 22-23.

Woltereck, R. 1926. Neue und alte Beobachtungen zur Metamorphose der Endolarve von *Polygordius*. – Zool. Anz. **65**: 49-60.

18

PANARTHROPODA

Arthropoda, Onychophora and Tardigrada are often regarded as closely related, and the two latter phyla have sometimes been united in a group called Proarthropoda; this is not in accordance with the conclusions reached below. There seems to be no generally accepted collective name for the three phyla; Weygoldt (1986) and Ax (1987) used Arthropoda in a wide sense comprising Onychophora and Euarthropoda, but I have chosen to introduce the term Panarthropoda to maintain the accustomed sense of the term Arthropoda. The pentastomids are regarded as parasitic crustaceans (probably closely related to Branchiura, see Chapter 20). Very lucid discussions of the phylogeny of onychophorans and arthropods are given by Weygoldt (1986) and Ax (1987), and I agree with most of their conclusions.

This is a group in which the fossils play an important role in the phylogenetic discussions. The rich Middle-Cambrian fauna of the Burgess Shale shows variations over the 'arthropod' theme which are unknown today (see also Chapter 20). One member of this fauna, *Aysheaia*, was originally interpreted as an annelid, but more recent studies usually emphasize its general likeness to onychophorans, and relationships with tardigrades have also been suggested. The thorough study by Whittington (1978) concludes, however, that it may be regarded as an example of the early panarthropod groups from which the living phyla are derived. In a treatise of the living animal phyla it may suffice to say that onychophorans (Chapter 19), tardigrades (Chapter 21) and the several Lower-Cambrian lobopodian panarthropods, such as *Aysheaia*, *Luolishania* and perhaps *Hallucigenia* (see Ramsköld & Hou 1991) can be interpreted as representatives of a diverse stem group from which the living panarthropod phyla originated.

The panarthropods are articulates with a cuticle containing α-chitin and protein, but lacking collagen (the type of chitin in the tardigrade cuticle has not been identified). There are essential similarities between amino acid compositions in the non-sclerotized cuticles of crustaceans, insects, merostomes and onychophorans (Hackman & Goldberg 1976). The cuticle is in most types rather thick and functions as an exoskeleton; it has limited capability of expansion, so growth is made possible through a series of moultings (ecdysis). The segments have lobopodial appendages with muscles and one or more terminal claws; there are no chaetae (as those of the annelids). The food is typically manipulated with modified anterior limbs. The development is not through a ciliary-feeding, primary larval stage, and locomotory cilia are lacking totally in all stages.

A whole series of additional synapomorphies can be recognized (Weygoldt 1986, Ax 1987): coelomic sacs develop during ontogeny, but their walls disintegrate partially later on so that the body cavity is a combination of the blastocoel and the coelomic cavities, i.e. a mixocoel or haemocoel. There is a dorsal heart which is a tubular structure consisting of circular muscles and having a pair of ostia in each segment; the blood enters the heart through the ostia from the haemocoel, is pumped anteriorly, and leaves the heart through its anterior end. Protonephridia are lacking, but modified metanephridia are found in most segments in the onychophorans and traces have been observed in a few segments of most arthropods. These metanephridia develop from small pockets from the embryonic coelomic sacs and become differentiated into a sacculus with podocytes and a duct which modifies the primary urine from the sacculus. The onychophorans have a ciliated funnel as the beginning of the duct, whereas cilia are lacking in the arthropods; the ciliated funnel resembles that of the annelids, and the structures are obviously homologous. There is a complex brain comprising a protocerebrum, innervating the lateral eyes, and one or two ganglia innervating antennal appendages; the homology between the brain regions of onychophorans and arthropods is uncertain, but the tardigrades have brain regions corresponding closely to those of the arthropods (Chapter 20). Small sensory organs, sensilla, consisting of one or a few primary sense cells with a modified cilium and surrounded by three types of cells, are found in all three phyla and should probably be regarded as a synapomorphy.

The tardigrades lack metanephridia and heart, probably as a function of their small size, and their ontogeny is practically unknown (see discussion in Chapter 21).

The development with a serial arrangement of coelomic sacs (which break up in later stages) formed from a post-anal zone and extending anteriorly to the mouth (and often further anteriorly on both sides of the oesophagus) clearly demonstrates that the panarthropods are euarticulates. Like annelids, many panarthropods add posterior segments during ontogeny, and teloblasts have been observed in a number of arthropods; the proliferation pattern of both onychophorans and arthropods resembles that of the annelids in many details. Spiral cleavage is the exception in the panarthropods, but a cleavage pattern with many spiral traits has been observed in some crustaceans (Chapter 20).

The arthropod ancestor was probably a marine, planktotrophic, crustacean-like organism with a number of similar, phyllopodous/biramous limbs, but benthic, crawling forms evolved already in the Cambrian, as shown for example by *Aysheaia*.

It seems clear that the panarthropods form a monophyletic unit, and that they are closely related to the annelids; a sister-group relationship is usually accepted. Proposals for deriving the panarthropods from various annelid groups (deriving the onychophorans and the tracheates together as the group 'Uniramia') are discussed in Chapter 19 and found not to be tenable. The lobopodia are most often regarded as modified parapodia, but I believe that the ventrally directed panarthropodean limbs used in swimming, walking or food collecting are not homologous with the laterally

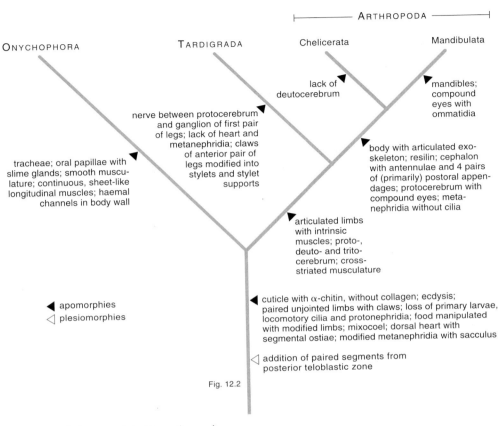

Fig. 12.2

Fig. 18.1. Phylogeny of the Panarthropoda.

directed parapodia, which probably evolved in some 'polychaetes' as adaptations to crawling or swimming; parapodia were probably not present in the annelid ancestor (Chapter 17).

References

Ax, P. 1987. The Phylogenetic System. The Systematization of Organisms on the Basis of their Phylogenies. – John Wiley, Chichester.

Hackman, R.H. & M. Goldberg 1976. Comparative chemistry of arthropod cuticular proteins. – Comp. Biochem. Physiol. 55B: 201-206.

Ramsköld, L. & X. Hou 1991. New early Cambrian animal and onychophoran affinities of enigmatic metazoans. – Nature **352**: 225-228.

Weygoldt, P. 1986. Arthropod interrelationships – the phylogenetic-systematic approach. – Z. zool. Syst. Evolutionsforsch. **24**: 19-35.

Whittington, H.B. 1978. The lobopod animal *Aysheaia pedunculata* Walcott, Middle Cambrian, Burgess Shale, British Columbia. – Phil. Trans. R. Soc. B **284**: 166-197, 14 pls.

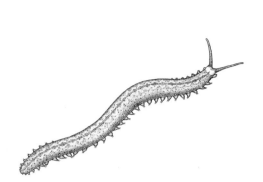

Phylum ONYCHOPHORA

Onychophorans are a small phylum of terrestrial animals found in humid tropical and southern temperate regions; only about 80 species are recognized, representing two families. The onychophoran-like *Aysheaia* and other early panarthropods from the Cambrian are discussed in Chapter 18.

The body is cylindrical with a pair of long anterior antennae, a pair of oral papillae, and a number of segments with a pair of lobopods, i.e. sac-shaped, unarticulated legs with internal muscles, terminating in claws. Locomotion is essentially annelid-like with the body cavity functioning as a hydrostatic skeleton.

The ectoderm (Storch 1984) is monolayered, unciliated and covered by a leathery, non-articulated cuticle with α-chitin and protein but without collagen; cuticular pore canals are lacking. The cuticle is shed at regular intervals (ecdysis), and the new cuticle is secreted underneath the old one by the ectodermal cells which develop microvilli, but these are subsequently withdrawn; the whole process is very similar to that of the arthropods (Robson 1964). It is often stated that ecdysis is induced by ecdysteroids, but I have been unable to find the source of this information. The epithelium forms numerous small papillae with scales and small sensory organs (see below). There are no unicellular glands, but various types of larger, multicellular glands are present: crural glands are ectodermal invaginations of the underside of the legs near the claws. The very large slime glands opening on the oral papillae are modified crural glands (Ruhberg & Storch 1977), as are the accessory genital glands found in varying numbers in the males (Ruhberg & Storch 1978).

Each segment has a high number of small spiracles which are the openings of the respiratory organs. The spiracles, which may be situated at the bottom of a wide ectodermal invagination, the atrium, are the openings for numerous narrow, tubular tracheae extending to the organs they supply (Pflugfelder 1968).

The mouth is heavily cuticularized with a pair of lateral jaws, which represent the claws of modified legs. The gut comprises a muscular pharynx and a thin-walled

Chapter vignette: *Macroperipatus geayi.* (Redrawn from Pearse *et al.* 1987.)

oesophagus, both with a cuticle, a wide, cylindrical gut with two cell types, one of which secretes a peritrophic membrane surrounding the gut content, and a rectum with a thin cuticle.

The central nervous system consists of a brain above the pharynx, a pair of connectives around the pharynx and a pair of ventral, longitudinal nerve cords.

The brain shows a structure generally resembling that of annelids and arthropods. The antennae are innervated from the posterior part of the brain, which has a preoral commissure, but it is difficult to homologize the posterior part of the brain with deutocerebrum or tritocerebrum of the arthropods (Schürmann 1987). A pair of eyes are situated at the base of the antennae; they develop from an epithelial invagination and consist of a domed lens covered by ectoderm with a transparent cuticle and a circular, almost flat basal retina with sensory and pigment cells (Dakin 1921). The ultrastructure of the eyes has been described by Eakin & Westfall (1965). The exterior side of the lens is covered by a thin cell layer which is an extension of the retina and which probably secretes the lens. The sensory cells have a long distal process with numerous orderly arranged microvilli at the periphery and a small cilium situated in a narrow extracellular space at the base of the process. Embryos of *Macroperipatus* (Eakin 1966) showed that the sensory cells first develop a cilium surrounded by microvilli, and that the cell surface with microvilli overgrows the cilium, so that the structure of the fully developed cell is of the rhabdomeric type.

The two nerve cords are situated lateroventrally and are connected with several commissures per segment; the cells are not arranged in ganglia, but each cord has one swelling per segment, and two nerves from each swelling innervate the leg muscles.

The skin bears several types of sensory organs (Storch & Ruhberg 1977); some are apparently chemoreceptors whereas others obviously are mechanoreceptors. They all contain various types of bipolar sensory cells with more or less strongly modified cilia often surrounded by microvilli. Some of the sense organs have been reported to have an opening through the cuticle, but this has not been substantiated by newer investigations using electron microscopy.

The spacious body cavity, haemocoel, arises through confluence of coelomic sacs and the primary body cavity, i.e. it is a mixocoel (see below). There is dorsal, longitudinal heart, which is open at both ends and has a pair of lateral ostia in each segment; the lumen of the heart is surrounded by a basement membrane surrounded by muscle cells (Nylund *et al.* 1988); this is the normal structure of the heart of an articulate, but the openings between the primary blood space surrounded by a basement membrane and the coelomic pericardial cavity is an panarthropod apomorphy. The 'blood' enters the heart through the ostia, is pumped anteriorly and leaves the heart through the anterior opening. The body wall is covered with a complicated arrangement of several layers of smooth muscles (Birket-Smith 1974), which extend into the legs; the whole body as well as the limbs functions as parts of a hydrostatic unit (Manton 1977).

The excretory organs are paired, segmental metanephridia with a ciliated funnel which is surrounded by a small coelomic vesicle (sacculus), and a duct which opens through a small pore at the ventral side near the base of the legs (Storch, Ruhberg & Alberti 1978). The wall of the sacculus consists of podocytes resting on an outer basement membrane, obviously the site of the ultrafiltration of primary urine from the mixocoel. The multiciliate cells of the funnel and the various cell types of the duct show many structures associated with absorption and secretion. Also the cells of the mid-gut have excretory functions. A pair of salivary glands, which open into the foregut through a common duct, are modified nephridia with a small sacculus with podocytes (Storch, Alberti & Ruhberg 1979).

The gonads develop through fusion of dorsal series of coelomic compartments, and the gonoducts are modified nephridia. The spermatozoa have a long head with a helically coiled nuclear surface and a long tail consisting of a cilium with accessory tubules (Storch & Ruhberg 1990). The sperm is transferred in a spermatophore which in some species becomes deposited in the genital opening of the young female; in other species the males attach the spermatophore to the female in a random area, and the spermatozoa wander through the epidermis to the haemocoel to reach the ovary (Ruhberg 1990). A few species deposit large (up to 2 mm in diameter), yolk-rich eggs with chitinous shells, but many more species are ovoviparous with medium-sized eggs, or viviparous with small eggs (down to 40 μm in diameter) nourished by secretions from the uterus. The ovoviparous type of development is generally regarded as the more primitive and the viviparous type without or with a placenta as the more specialized (Manton 1949), although the embryology of the species with a placenta is seemingly the more primitive.

The oviparous and the ovoviviparous species have superficial cleavage resembling that of the insects, but their development is incompletely described (Sheldon 1889, Anderson 1973). The development of the viviparous species (Manton 1949, Anderson 1966) shows a wide variation, from large eggs with much yolk and no development of placental structures to very small eggs which develop large placental structures, which in the most developed form resemble the amnion of higher vertebrates. The placental structures are enlarged anterodorsal areas of the embryo, which in the least developed types have the shape of a dorsal sac of ectoderm surrounding the yolk cells; this dorsal structure becomes increasingly larger with decreasing amounts of yolk, and includes mesoderm and endoderm in the most specialized types with an 'amnion'. The larger eggs show superficial cleavage and formation of a blastoderm, while species with very small eggs have total, almost equal cleavage which leads to the formation of a coeloblastula or a sterroblastula (Sclater 1888, Kennel 1885). The further development of these species appears very specialized, with the embryo enclosed in an amniotic cavity and connected to the amniotic wall through a dorsal 'umbilical cord' (Kennel 1885, Anderson & Manton 1972).

The development of species with rather small amounts of yolk, as for example *Peripatopsis capensis* (Sedgwick 1885, 1886, 1887, 1888, Manton 1949), is better described and perhaps the most easy to follow. The cleavage leads to the formation of a plate of cells on one side of the elongate embryo; somewhat later stages become spherical and show one half of the embryo covered by a layer of smaller cells with small amounts of yolk covering a mass of larger, yolky cells; the smaller cells become the ectoderm and the larger ones the endoderm. A gastrulation takes place, resulting in a gastrula with a compressed blastopore, but a wide opening, called mouth-anus or stomo-proctodaeum, soon opens in the same area. Mesoderm is now being produced from a small area at the posterior side of the anus, and compact mesodermal bands grow forwards along the sides of the mouth-anus; teloblasts have not been reported. Schizocoelic spaces are formed in these bands, which divide to form a row of coelomic sacs on each side. The mouth-anus divides by lateral fusion of the mouth-anus lips (Fig. 19.1). Other species of *Peripatopsis* show modifications to this pattern, probably induced by larger amounts of yolk (Manton 1949). Each coelomic sac divides into a dorsal and a ventral compartment, but the fates of the two sacs have been documented with different certainty (Kennel 1888, Sedgwick 1887, Evans 1901). The ventral sac apparently differentiates into the nephridium with a thin-walled sacculus and a thick-walled, ciliated funnel, which becomes connected to the exterior through a short ectodermal duct. The fate of the dorsal sacs is less well known. Kennel (1888; *Epiperipatus edwardsii* and *E. torquatus*) was of the opinion that the walls of the sacs disintegrate so that the coelomic and pseudocoelomis spaces become confluent, whereas Sedgwick (1887; *Peripatopsis capensis*) and Evans (1901; *Eoperipatus weldoni*) reported that the sacs collapse and that the cells become incorporated into the ventral wall of the heart. Manton (1949) pointed out that the methods of these earlier investigations were so imperfect that a number of details

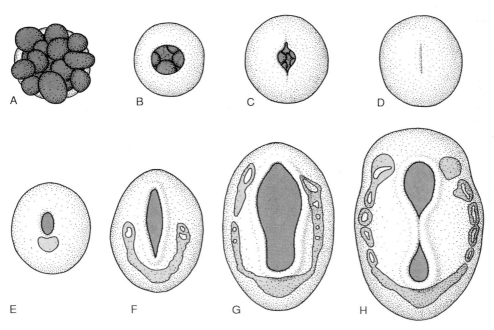

Fig. 19.1. Early development of *Peripatopsis capensis*; embryos seen from the blastoporal side; the endoderm is indicated in dark shading and the mesoderm, which is seen through transparency, in light shading. – A, 'blastula'. – B, 'gastrula'. – C, lateral blastopore closure. – D, blastopore lips in contact. – E, blastopore reopens; the mesoderm is recognized behind the blastopore. – F, blastopore elongates; the mesoderm extends laterally along the blastopore. – G, the mesoderm has reached the anterior part of the embryo. – H, the blastopore closes laterally, leaving mouth and anus. (Redrawn from Manton 1949.)

were overlooked or misinterpreted; it seems clear that the differentiation of the coelomic sacs of a number of species should be studied. The pseudocoelomic/mixo-coelomic nature of the body cavity is thus somewhat uncertain, but most authors state that it is a mixocoel (for example Anderson 1973). The gonads develop from the dorsal sacs, which connect to modified nephridia.

The onychophorans show several specializations which indicate that the group originated in a terrestrial environment. The tracheae and slime glands could not function in water, and the nephridia are of a structure found in several terrestrial groups. The phylum must be seen as a specialized, terrestrial offshoot from the diverse Cambrian fauna of early aquatic panarthropods.

Several apomorphies characterize the onychophorans as a monophyletic group: second pair of limbs transformed into a pair of knobs with very large claws, the oral hooks or mandibles. Third pair of limbs transformed into oral papillae with the slime glands; the corresponding pair of metanephridia transformed into salivary glands. Ventral nerve cords widely separated and with several commissures in each segment. Large number of stigmata with tracheal bundles on each segment.

Some authors, notably Manton (1973, 1977) and Anderson (1973) emphasized similarities between the onychophorans and the tracheates (myriapods + insects) and united these groups in the phylum Uniramia and regarded the taxa Chelicerata, Crustacea and Trilobita as equally separate phyla (Chapter 20). Emphasis was on the functional morphology of the limbs and on embryology, and the basic assumption was that the limbs of the four groups were of so different structure/function that they had to have evolved independently from lobopodan ancestors (annelids). The chitinous cuticle and ecdysis were not considered. The embryological evidence is less convincing, because considerable deviation from the primitive spiral cleavage development is to be expected in groups which lack feeding primary larvae. Manton's main argument for regarding her 'Uniramia' as separate from the other groups was that the onychophorans and tracheates have uniramous limbs (including uniramous mandibles) whereas the other groups have biramous legs, and that the uniramians manipulate the food with the tips of the mandibles, whereas the other groups use the gnathobase of the mandibles. This is discussed further in Chapter 20.

It is difficult to separate the discussion of the position of the onychophorans from the general discussion of the 'arthropods', but the point I wish to make here is that the onychophorans have one set of apomorphies (see above) and the arthropods have another set of apomorphies (Chapter 20), which in my view make it unavoidable to regard the two phyla as separate, monophyletic groups. The synapomorphies of the panarthropods (Chapter 18), especially the cuticle with α-chitin, the sense organs, and the division of the embryonic coelomic sacs into pericardium, nephridial sacculus plus duct and a larger cavity which partially loses its identity by fusing with the blastocoel to form a mixocoel (haemocoel), clearly set the onychophorans apart from the annelids. The ciliated funnel of the onychophoran nephridia resembles that of the annelids, but the sacculus with podocytes is only found in panarthropods. The different types of nephridia can perhaps be interpreted as a transformation series from the primitive annelidan metanephridium opening into a spacious coelom through the onychophoran ciliated funnel with a sacculus with podocytes to the advanced arthropod antennal gland with a sacculus with podocytes but without cilia.

A direct link between onychophorans and clitellate annelids was proposed by Jamieson (1986) on the basis of some similarities between the spermatozoa of the two groups. Some of the embryological similarities between clitellates and 'uniramians' stressed by the Manton-Anderson school were also mentioned, but the mono-phyletic view was almost totally ignored. Also Sawyer (1984) saw evidence for a derivation of the 'uniramians' from clitellates, especially leeches. He focused on the similarity between the mouthparts of some terrestrial leeches and onychophorans, but no evidence was presented indicating that the jaws of the leeches are modified limbs like those of the onychophorans; the similarities must be considered completely superficial. Also the likeness between the lateral respiratory structures of two glos-siphoniid leeches and the onychophoran lobopodia must be superficial.

It appears unquestionable that the onychophorans are closer to the arthropods than to the annelids. The unarticulated appendages, the ciliated funnel of the nephridia, and the sperm with a long cilium are characters shared with the annelids, but these characters must be interpreted as symplesiomorphies. The development of the eyes shows how the rhabdomeric type characteristic of the panarthropods may have evolved from the ciliary type. Taken together the just-mentioned characters support the interrelationships between the phyla shown in Fig. 18.1.

The position of the tardigrades is discussed in Chapter 21.

Interesting subjects for future research

1. Ontogeny of nephridia and heart.
2. Ultrastructure of tracheae.
3. Physiology of moulting.

References

Anderson, D.T. 1966. The comparative early embryology of the Oligochaeta, Hirudinea and Ony-chophora. – Proc. Linn. Soc. N.S.W. **91**: 11-43.
Anderson, D.T. 1973. Embryology and Phylogeny in Annelids and Arthropods. – Pergamon Press, Oxford.
Anderson, D.T. & S.M. Manton 1972. Studies on the Onychophora VIII. The relationship between the embryos and the oviduct in the viviparous placental onychophorans *Epiperipatus trinidadensis* Bouvier and *Macroperipatus torquatus* (Kennel) from Trinidad. – Phil. Trans. R. Soc. B **264**: 161-189, pls 31-32.
Birket-Smith, S.J.R. 1974. The anatomy of the body wall of Onychophora. – Zool. Jb., Anat. **93**: 123-154.
Dakin, W.J. 1921. The eye of *Peripatus*. – Q. Jl microsc. Sci., N.S. **65**: 163-172, pl. 7.
Eakin, R.E. 1966. Differentiation in the embryonic eye of *Peripatus* (Onychophora). – 6th int. Congr. Elect. Microsc. Kyoto, pp 507-508. Maruzen, Tokyo.
Eakin, R.E. & J.A. Westfall 1965. Fine structure of the eye of *Peripatus* (Onychophora). – Z. Zell-forsch. **68**: 278-300.
Evans, R. 1901. On the Malayan species of Onychophora. Part II. – The development of *Eoperipa-tus weldoni*. – Q. Jl microsc. Sci., N.S. **45**: 41-88, pls 5-9.

Jamieson, B.G.M. 1986. Onychophoran-euclitellate relationships: evidence from spermatozoal ultrastructure. – Zool. Scr. 15: 141-155.

Kennel, J. 1885. Entwicklungsgeschichte von *Peripatus edwardsii* Blanch. und *Peripatus torquatus* n.sp. – Arb. zool.-zootom. Inst. Würzburg 7: 95-299, pls 5-11.

Kennel, J. 1888. Entwicklungsgeschichte von *Peripatus edwardsii* Blanch. und *Peripatus torquatus* n. sp. II Theil. – Arb. zool.-zootom. Inst. Würzburg 8: 1-93, pls 1-6.

Manton, S.M. 1949. Studies on the Onychophora VII. The early embryonic stages of *Peripatopsis*, and some general considerations concerning the morphology and phylogeny of the Arthropoda. – Phil. Trans. R. Soc. B 233: 483-580.

Manton, S.M. 1973. Arthropod phylogeny – a modern synthesis. – J. Zool. (Lond.) 171: 111-130.

Manton, S.M. 1977. The Arthropoda. Habits, Functional Morphology, and Evolution. – Oxford Univ. Press, Oxford.

Nylund, A., H. Ruhberg, A. Tjønneland & B. Meidell 1988. Heart ultrastructure in four species of Onychophora (Peripatopsidae and Peripatidae) and phylogenetic implications. – Zool. Beitr., N.F. 32: 17-30.

Pearse, V., J. Pearse, M. Buchsbaum & R. Buchsbaum 1987. Living Invertebrates. – Blackwell, Palo Alto, CA.

Pflugfelder, O. 1968. Onychophora. – In G. Czihak (ed.): Grosses zoologisches Praktikum, Heft 13a, pp 1-42. Gustav Fischer, Stuttgart.

Robson, E.A. 1964. The cuticle of *Peripatopsis moseleyi*. – Q. Jl microsc. Sci. 105: 281-299.

Ruhberg, H. 1990. Onychophora. – In K.G. Adiyodi & R.G. Adiyodi (eds): Reproductive Biology of Invertebrates, vol. 4B, pp 61-76. Oxford & IBH Publ. Co., New Delhi.

Ruhberg, H. & V. Storch 1977. Über Wehrdrüsen und Wehrsekret von *Peripatopsis moseleyi* (Onychophora). – Zool. Anz. 198: 9-19.

Ruhberg, H. & V. Storch 1978. Zur Ultrastruktur der accessorischen Genitaldrüsen von *Opisthopatus cinctipes* (Onychophora, Peripatopsidae). – Zool. Anz. 199: 289-299.

Sawyer, R.T. 1984. Arthropodization in the Hirudinea: evidence for a phylogenetic link with insects and other Uniramia? – Zool. J. Linn. Soc. 80: 303-322.

Schürmann, F.-W. 1987. Histology and ultrastructure of the onychophoran brain. – In A.P. Gupta (ed.): Arthropod Brain, its Evolution, Development, Structure and Functions, pp 159-180. Wiley-Interscience, New York.

Sclater, W.L. 1888. – On the early stages of the development of a South American species of *Peripatus*. – Q. Jl microsc. Sci., N.S. 28: 343-363, pl. 24.

Sedgwick, A. 1885. The development of *Peripatus capensis*. Part I. – Q. Jl microsc. Sci., N.S. 25: 449-466, pls 31-32.

Sedgwick, A. 1886. The development of the Cape species of *Peripatus*. Part II. – Q. Jl microsc. Sci., N.S. 26: 175-212, pls 12-14.

Sedgwick, A. 1887. The development of the Cape species of *Peripatus*. Part III. On the changes from stage A to stage F. – Q. Jl microsc. Sci., N.S. 27: 467-550, pls 34-37.

Sedgwick, A. 1888. The development of the Cape species of *Peripatus*. Part IV. The changes from stage G to birth. – Q. Jl microsc. Sci., N.S. 28: 373-396, pls 26-29.

Sheldon, L. 1889. On the development of *Peripatus novæ-Zelandiæ*. – Q. Jl microsc. Sci. 29: 283-294, pls 25-26.

Storch, V. 1984. Onychophora. – In J. Bereiter-Hahn, A.G. Matoltsy & K.S. Richards (eds): Biology of the Integument, vol. 1, pp 703-708. Springer, Berlin.

Storch, V., G. Alberti & H. Ruhberg 1979. Light and electron microscopical investigations on the salivary glands of *Opisthopatus cinctipes* and *Peripatopsis moseleyi* (Onychophora: Peripatopsidae). – Zool. Anz. 203: 35-47.

Storch, V. & H. Ruhberg 1977. Fine structure of the sensilla of *Peripatopsis moseleyi* (Onychophora). – Cell Tissue Res. 177: 539-553.

Storch, V. & H. Ruhberg 1990. Electron microscopic observations on the male genital tract and sperm development in *Peripatus sedgwicki* (Peripatidae, Onychophora). – Invert. Reprod. Dev. 17: 47-56.

Storch, V., H. Ruhberg & G. Alberti 1978. Zur Ultrastruktur der Segmentalorgane der Peripatopsidae (Onychophora). – Zool. Jb., Anat. 100: 47-63.

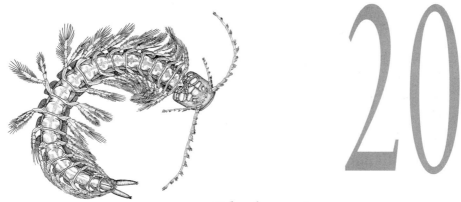

Phylum ARThropoda

Arthropoda, with the three living groups Chelicerata, Crustacea (including the Pentastomida) and Tracheata (see discussion below), is one of the largest animal phyla; the insects alone are now believed to comprise more than a million living species, while the other arthropods number more than 100,000. They inhabit almost every conceivable habitat from the deep sea to the deserts, and some of the parasitic forms are so modified that their arthropod nature can only be recognized in certain developmental stages. There is a very substantial fossil record going back to the Early Cambrian, and the fossils are important for our understanding of the evolution of the phylum.

Arthropods can usually be recognized by their articulated chitinous exoskeleton which has thicker areas corresponding to the segments and thin rings 'between' segments; each segment typically carries a pair of articulated legs. A number of anterior segments are fused to a cephalon which usually carries compound eyes, antennules, antennae, mandibles, and two pairs of maxillae (Ax 1987).

Nevertheless, the monophyletic nature of the phylum has been questioned, for example by Manton (1972, 1977), who united the onychophorans with the tracheates (Myriapoda + Insecta) in the group Uniramia and considered Crustacea, Chelicerata and Trilobita as separate phyla which had reached the 'arthropod grade of organization' independently. (The term Uniramia is unfortunately used both for tracheates + onychophorans, as proposed by Manton, and as synonymous with the term Tracheata; I will therefore avoid the use of the term). Anderson (1973) carried Manton's thought even further by proposing that the 'uniramians' were derived from annelids with a clitellate type of development. The position of the onychophorans has been discussed in Chapter 19, where it was concluded that they cannot be derived from the clitellates and must be regarded as a separate phylum with a sister-group relationship to the arthropods. (The position of the tardigrades is discussed in Chapter 21). Anderson (1973) further stressed the differences he had observed between the fate maps of crustaceans and the other arthropods (and annelids), where the crustacean mesoderm should be situated mid-ventrally between the areas of the

Chapter vignette: The remipede *Speleonectes ondinae*. (Redrawn from Schram 1986.)

endoderm and the stomodaeum and originate from the A-C-quadrants; the fate maps of other arthropod groups and the annelids show the mesoderm originating (mainly) from an area behind the endoderm, i.e. from the D-quadrant. His generalized fate map of the cirripedes was based on his earlier observations on the development of four species (belonging to the genera *Tetraclita*, *Chthamalus* and *Chamaesipho*), and this was compared with fate maps of other crustacean groups constructed on the base of information from the literature (Anderson 1969, fig. 28). However, the two very well-documented, older studies of cirripede embryology by Bigelow (1902, on *Lepas*) and Delsman (1917, on *Balanus*) emphasize that a major part of the mesoderm originates from the D-quadrant (see below and Table 20.2), but these studies were simply dismissed (Anderson 1969, p. 224) because their results were not in accordance with his results. I find it more reasonable to accept the differences observed within the cirripedes, and to use them as a key to the understanding of the origin of the mesoderm in the crustaceans (see below). A separation of the Crustacea as a phylum distinct from other 'arthropod' phyla, on the basis of the embryology, is definitely not warranted.

The whole concept of 'polyphyly' of the panarthropods has been questioned by a number of authors, and Weygoldt (1986, p. 22) expressed the central critique very clearly: 'The only method by which a polyphyletic origin of the arthropods could be made probable would be the demonstration of different relationships of the various subgroups with different non-arthropods, for example with annelids, molluscs, or plathelminths. This has never been done.' Most authors now support the 'monophyletic' opinion, but it seems necessary to discuss the phylogeny of the whole group before an opinion about the characters of the ancestral arthropod can be formed. It appears practical first to discuss the relationships of the living groups, and then to consider the additional information provided by the extinct forms.

The living arthropods are usually arranged in three groups, Chelicerata (comprising Pantopoda, Merostomata and Arachnida), Crustacea, and Tracheata (Myriapoda and Insecta). The last-mentioned two groups are often united in the group Mandibulata, and Weygoldt (1986) stressed the following synapomorphies: 1) third head appendage transformed into the most important feeding organ with the strong coxal endite developed as a mandible; and 2) compound eyes with ommatidia consisting of two corneagene cells, a tetrapartite eucone crystalline cone, and a retinula consisting of eight cells (Paulus 1979, see also Ax 1987). The sister group, Chelicerata, lacks mandibles and has compound eyes with a higher, variable number of cells in the subunits.

The homology of the mouth parts was questioned by the 'polyphyletic' theory, which homologized the mandibles of tracheates and onychophorans and rejected the homology of crustacean and tracheate mandibles (Manton 1977). This was founded on analyses of the musculature and movement patterns of the various groups, but since the mandibles of crustaceans and tracheates have homologous innervation (from the first ventral ganglion, see below) it can hardly be denied that the mandibles of the tracheates could be modified crustacean coxae lacking both endopod and exopod. Gnathobases are found on anterior limbs of members of almost all arthropod groups, even on the antennae of some crustacean larvae (Fig. 20.3; Table 20.1). It is

Table 20.1. Somites and their ganglia and eyes or appendages in four crustacean groups. * The appendages marked with an asterisk may have gnathobases. The position of the mouth is marked by the broken line. The tritocerebrum is marked as postoral because it is situated lateral to the mouth in primitive crustaceans and in many embryos and because it has a postoral commissure. The brain and the nerves of the trilobitomorphs are unknown, but they have been interpreted as lacking the deutocerebrum as in the chelicerates. (Modified from Schram 1986 and Brusca & Brusca 1990.)

Somite	Brain region or ventral ganglion	Crustacea	Tracheata	Trilobitomorpha	Chelicerata
0	protocerebrum	compound eyes	compound eyes	compound eyes	compound eyes
1	deutocerebrum	antennules	antennae	absent	absent
2	tritocerebrum	antennae*	(embryonic)	antennae	chelicerae
3	1. ventral g.	mandibles*	mandibles*	1. legs*	pedipalps or 1. legs*
4	2. ventral g.	maxillae 1*	maxillae 1	2. legs*	1. or 2. legs*
5	3. ventral g.	maxillae 2*	maxillae 2	3. legs*	2. or 3. legs*

possible that the 'mandibles' of the onychophorans and tracheates are formed from the same pair of legs, but this only underlines the common descent of all the panarthropods.

The ommatidia with the characteristic number and arrangement of cells are found in crustaceans and insects, while the myriapod eyes appear to be secondarily modified (Paulus 1979). The ommatidia of the two groups consist of a constant number of cells with complex structure/function and constant spatial relationships. Each ommatidium develops through specialization of a number of initially similar epithelial cells (Such 1975). The ommatidium must be regarded as a phylogenetic character of high importance.

A third characteristic of the mandibulates may be found in the appendages of the cephalic region as indicated by their innervation (Table 20.1): the compound eyes are innervated from the protocerebrum in both mandibulates and chelicerates; in the mandibulates, the first pair of appendages is the preoral antennules innervated from the deutocerebrum, whereas this region of the brain and the corresponding limbs appear to be absent in the chelicerates; the antennae of the mandibulates and the chelicerae of the chelicerates are innervated from the tritocerebrum, which has a postoral commissure. The mandibles are innervated from the first ventral pair of ganglia and the maxillae from the following two pairs. The presence of deutocerebrum and antennules can be interpreted either as a plesiomorphy or an apomorphy, but several of the characters discussed below indicate the primitiveness of the crustaceans, so the presence of a deutocerebrum is probably the plesiomorphic character state.

I think that the monophyletic nature of the group Mandibulata cannot be questioned, but the sister-group relationship between Crustacea and Tracheata is more uncertain since apomorphies of the Crustacea are difficult to identify. Weygoldt (1986) mentioned the fusion of four median eyes to a nauplius eye as a possible apomorphy, but pointed out that most other characters must be characterized as plesiomorphies.

The discussion of the interrelationships between tracheates and crustaceans has been centred around the structure of the limbs, the crustaceans having biramous limbs (at least in the more primitive groups) and the tracheates uniramous limbs (see also Kukalová-Peck 1987). The living chelicerates have uniramous legs, so the biramous legs of the crustaceans could be interpreted as an apomorphy, but the fossils contradict this (see below). The derivation of the uniramous leg from a biramous one has been favoured by most authors, and this theory is supported by the fact that several of the 'higher' crustaceans, such as brachyurans, have biramous legs in larval or juvenile stages and uniramous legs as adults. A derivation of the biramous legs from the uniramous was proposed by Schram & Emerson (1991), who offered the surprising idea that the biramian limb should have evolved by the fusion of limbs from two segments. The background for this theory is the Carboniferous enantiopod remipede *Tesnusocaris*, which apparently had wide segments with a pair of lateral limbs alternating with quite narrow segments situated in a posterior notch in the wide segments and carrying a median pair of limbs; a fusion of these paired segments followed by the development of a common base or coxa for the inner and outer limb of each side should then have resulted in the formation of the biramous limb. An important difficulty for this idea is that *Tesnusocaris* is of Carboniferous age whereas all the main biramian groups are known already from the Cambrian, and the Recent sister group of the enantiopods, the nectipods, are biramous too. A number of observations on other panarthropods were taken to support the theory, but none of them appear to stand a closer scrutiny: two (or three) pairs of ganglia per segment are found not only in crustaceans, but also in some polychaetes (Chapter 17), and onychophorans (Chapter 19) have one pair of swellings of the ventral nerve cords, but two nerves to each leg and several commissures per segment; several of the myriapod groups have two (or three) pairs of legs on each fused segment, but I know of no example of two alternating types of legs.

I find it most probable that the tracheates are a specialized group of terrestrial crustaceans. The tracheae are obviously an adaptation to the terrestrial habitat (but they are probably not homologous in the various tracheate groups, see below), and the exopodite of the biramous legs, which is usually associated with respiration in water, has been lost completely during the same transition. This is in accordance with the comparatively late occurrence of the tracheates in the fossil record, the oldest reliable representative recorded being from the Silurian (Almond 1985, Shear & Kukalová-Peck 1990); the Middle-Cambrian *Cambropodus* was interpreted as an early, marine myriapod (Robison 1990), but its position seems very uncertain.

This interpretation leaves the crustaceans as a paraphyletic group, but I will for practical reasons use the term in its accustomed sense in the following.

The fossil record of arthropods is very considerable, and there is a diverse fauna of very well-preserved specimens in a number of strata, notably the famous Middle-Cambrian Burgess Shale and the Upper-Cambrian 'Orsten' deposits. Gould (1989) stressed the high diversity of the Burgess Shale fauna, but other authors have put more emphasis on the similarities between the fossil and living faunas. Briggs & Fortey (1989) analysed the Cambrian and a few living arthropods (the two living crustacean genera believed to exhibit the most plesiomorphic characters: the remipede *Speleonectes* and the cephalocarid *Hutchinsoniella*) and obtained a cladogram (Fig. 20.1) with the Burgess Shale forms *Marella* and *Branchiocaris* at the base of the tree together with the living crustaceans, and most of the other Cambrian fossils and the trilobites 'on the way' to *Limulus* (representing the living chelicerates). This would make the Crustacea the paraphyletic stem group and the trilobite-chelicerate line the advanced crown group, but Briggs (1990) pointed out that the taxa bracketed as crustaceans in the cladogram (Fig. 20.1) would come out as a monophyletic group if each of these taxa were coded for the presence of mandibles. Walossek & Müller (1990) analysed a number of the astonishingly well-preserved Orsten arthropods, which they mostly interpreted as representatives of the stem group preceding the crustaceans as we know them today (Fig. 20.1). The position of *Agnostus* was found to be very uncertain, but Fortey & Owens (1990) placed this group at the base of the trilobites. The earliest known larval stages of the Upper-Cambrian 'stem-group crustaceans' (Müller & Walossek 1986, Walossek & Müller 1990) and *Bredocaris* (Müller & Walossek 1988) are nauplii or metanauplii with more or less well developed first maxillae, and their adults had a cephalon with the characteristic five pairs of appendages (the second maxillar segment appears to have been free in some of the forms, so the earliest type may have had one segment less incorporated in the

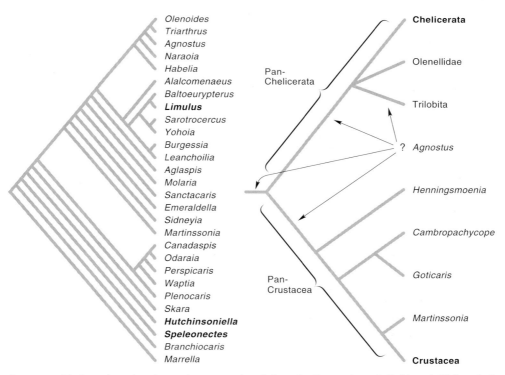

Fig. 20.1. Phylogenies of arthropod groups after Briggs & Fortey (1989) (left) and Walossek & Müller (1990) (right). Living groups are in boldface.

cephalon); the nauplii were non-feeding, but some of the metanauplii had mouth and appendages with gnathobases and were probably feeding. Both the larvae and the adults of these early forms were remarkably similar to many of the living crustaceans.

The earliest larval or juvenile stages of trilobites reconstructed by Lauterbach (1980) had a cephalon with four pairs of appendages and these could be homologous with the appendages on the cephalon of the living chelicerates, which lack the antennules (Table 20.1).

It is clear that the fossils have brought very important information to the cladistic analysis, and it appears that almost all the information supports the phylogenetic branching pattern deduced on the base of the living forms.

Taken together, all these pieces of information strongly indicate that the primitive 'crustaceans' are the living forms which come closest to the ancestral arthropod, and this is supported by the fact that the most crustaceans are aquatic, whereas the tracheates and chelicerates are mainly terrestrial with all the anatomical and embryological specializations related to the changed habitat. In the considerations about the phylogenetic position of the arthropods I have therefore chosen to concentrate on characters of the crustaceans, and to emphasize the apparently 'primitive' remipedes (*Speleonectes* (see the chapter vignette) and a few other genera; Schram, Yager & Emerson 1986) and cephalocarids (*Hutchinsoniella* and a few other genera; Schram 1986), although both groups show several specializations, for example lack of eyes (Elofsson & Hessler 1990).

The position of the parasitic group Pentastomida has been a matter for discussion for a long time, and it is usually placed as a separate 'proarthropod' phylum. The origin of both endoderm and mesoderm is undescribed and the regions of the central nervous system are difficult to homologize with those of the other arthropods. Four pairs of coelomic pouches and seven pairs of ganglionic cell groups can be recognized during the embryonic development; the four anterior ganglia are connected with the coelomic pouches. The three anterior pairs of ganglia fuse to form a 'brain' and the three posterior ones fuse too (Böckeler 1984). These observations demonstrate the articulate character of the group, but give no certain hints about its more precise position. Investigations on *Raillietiella* and the branchiuran crustacean *Argulus* have shown that both genera have highly specialized sperm of very similar type (Wingstrand 1972; see also Storch & Jamieson 1992), and this together with observations on embryology and structure and moulting of the cuticle (Riley, Banaja & James 1978) has led to the inclusion of the group in the Crustacea. The report by Karuppaswamy (1977) of β-chitin in the cuticle instead of α-chitin seems to contradict the inclusion of the group in the Panarthropoda, and if this observation turns out to be correct, a reevaluation of the systematic position is definitely needed.

A generalized crustacean has a head or cephalon with a pair of compound eyes and five pairs of articulated appendages: antennules, antennae, mandibles, and two pairs of maxillae (or only one pair of maxillae, see above). The limbs on the following segments, some of which may have fused with the cephalon, may be similar, but different types may be found in different body regions. A preoral extension, the labrum, partially covers the mouth, which is situated on the ventral side, facing the anterior end of the food groove between the limbs. The pharynx-oesophagus-stomach or foregut is a stomodaeum with cuticular lining. The endodermal midgut is a shorter or longer intestine with a pair of digestive caeca (*Speleonectes* has serially arranged diverticula, see Yager 1981) lacking motile cilia, but biciliate cells with cilia lacking the central tubules and dynein-arms have been reported from the midgut of a phyllopod (Rieder & Schlecht 1978). The rectum or hindgut is lined with a cuticle.

The antennules are apparently not involved in feeding, but all the following appendages may at some stage be engaged in creating food currents or collecting and handling particles. The remipedes are probably carnivores or scavengers (Schram 1986) and have specialized appendages on the cephalon; the posterior limbs are all laterally oriented, biramous swimming legs, which move in a metachronal pattern giving the animal a quite polychaete-like appearance, but this is probably a secondary specialization (see below). A number of the cephalocarids, branchiopods, maxillopods, and the primitive malacostracan group Phyllocarida have biramous limbs with endopodites used in filter feeding and coxae with gnathobases used in handling of food particles. Antennae, mandibles and maxillae of the larval stages are in many cases of this type too, and the ancestral crustacean was most probably a swimming, filter-feeding organism with a series of similar, biramous limbs.

The ectoderm (reviews in Bereiter-Hahn, Matoltsy & Richards 1984) is simple, cuboid and covered with a cuticle consisting of α-chitin with non-collagenous protein; locomotory cilia are completely absent, but modified cilia are found in many of the sensory organs called sensillae. The cuticle is often heavily sclerotized and/or calcified in the stiff plates or sclerites, while the articulation membranes connecting the sclerites are thin and flexible. The special protein resilin is believed to occur in all arthropod groups, and is probably an arthropod apomorphy. The cuticle can only expand slightly, so growth is restricted to moultings, where the old cuticle breaks open along preformed sutures and a folded, thin, soft, unsclerotized cuticle preformed below the old cuticle becomes stretched out. Experiments have shown that moulting is controlled by ecdysones in crustaceans, myriapods and insects; in chelicerates, the ecdysones are known to control regeneration, while it is only presumed that they control moulting (Käuser 1989). The process is best known in crustaceans and insects, and there are unexpected differences between the organs involved in secretion of the ecdysteroids and in their ways of control (Watson, Spaziani & Bollenbacher 1989): in crustaceans, the ecdysteroids are secreted by the Y-organs located at the base of the eyestalks, while they come from the prothoracic glands in insects. Both glands are controlled by neuropeptide hormones, but in different ways: in crustaceans, a moult-inhibiting hormone from the X-glands in the eyestalks suppresses the production of the ecdysial hormone, while in insects, the prothoracicotrophic hormone produced in the corpora cardiaca in the brain stimulates the production of the ecdysial hormone. These differences and the widespread occurrence and several presumed functions of the ecdysial hormones make it difficult to use these hormones in phylogenetic argumentation at this point.

Respiration is through the thin areas of the body wall in small, aquatic crustaceans, but special gill structures are developed in the larger forms. The terrestrial arachnids and tracheates have developed tracheae with paired, segmentally arranged stigmata, but these organs appear to have evolved independently several times, as indicated by their scattered occurrence in the groups and their variable position (Dohle 1980).

The central nervous system of the crustaceans is of the usual articulate type. There is a non-segmental protocerebrum, which innervates the compound eyes (Table 20.1), and a deutocerebrum, which belongs to the first segment and which innervates the antennules; these two parts are fused into an anterodorsal brain. The second segmental pair of ganglia innervates the antennae and is situated lateral to the oesophagus, on the connectives to the first ventral ganglion in nauplius larvae (Walley 1969) and in adults of primitive forms, such as *Hutchinsoniella* (Elofsson & Hessler 1990) and the phyllopod *Triops* (Henry 1948), but it is fused with the deutocerebrum in the more advanced forms. The more primitive, elongate arthropods have a chain of paired ventral ganglia (Fig. 10.2), but forms with a short body have the ganglia fused into a single mass. The chelicerates lack the deutocerebrum and the corresponding appendage.

Arthropod eyes are of two types, ocelli and compound eyes, both innervated from the protocerebrum. The ocelli show much variation between the groups, but Paulus (1979) proposed that the ancestral arthropod had four pairs of medial photoreceptors, which then have differentiated in the various groups, one example being the ocelli in the nauplius eye of the crustaceans, another the ventral and dorsal frontal organs of crustaceans and tracheates. The compound or faceted eyes consisting of many ommatidia, i.e. groups of cells with characteristic structure and function (lens cells, retinula cells etc.), are found in mandibulates, merostomes, and were probably found also in trilobites. All the compound eyes have been regarded as homologous, but Paulus (1979) concluded that a parallel evolution of the eyes of mandibulates and chelicerates cannot be excluded. The ommatidia of the mandibulate eyes have been discussed above.

Other sense organs are sensilla of many different types (reviews in Schmidt & Gnatzy 1984 and Bereiter-Hahn, Matoltsy & Richards 1984), i.e. one or a few primary receptor cells usually with a strongly modified cilium surrounded by three cell types: tormogen, trichogen and thecogen cells, the two last-mentioned cells secreting a cuticular structure. All the cells of a sensillum develop from one epithelial cell through a fixed sequence of differential cell divisions. Both the mechanoreceptive sensory hairs and the chemoreceptive sensilla have very thin cuticular areas or small pores through the cuticle.

The mesoderm forms small segmental, coelomic cavities in early developmental stages, but later divides into a number of different structures (see below). The main part of the coelom loses the epithelial character and splits up into muscles and other organs so that the coelomic cavities fuse with extracoelomic spaces forming a mixocoel or haemocoel; small sacs remain as nephridial sacculi and gonadal walls. The musculature is striated and the locomotory muscles are usually attached to the exoskeleton of neighbouring sclerites, often on long, internal extensions called apodemes; the locomotion is thus typically not of the hydrostatic-skeleton type. The appendages have intrinsic musculature, i.e. muscles between the joints.

Modified metanephridia without cilia, each originating in a small coelomic compartment, the sacculus, are found in crustaceans, merostomes, chelicerates, and insects. Crustaceans have antennal glands (at the base of the antennae) or maxillary glands (at the base of the second maxillae); many types have antennal glands in the larval stage and maxillary glands in the adult stage, but some retain the antennal glands as adults; and a few types, such as mysids, have both glands. The sacculi consist of podocytes resting on a basement membrane and resemble those of the nephridia of onychophorans and further of the glomeruli of the vertebrate kidney (Taylor & Harris 1986, White & Walker 1981). The formation of primary urine through ultrafiltration from the haemocoel to the sacculus and its modification during the passage through the duct are well documented. *Limulus* has a pair of coxal glands which develop from the ventral somites of the six prosomal segments with the nephridial funnel and duct originating from segment number five (Patten & Hazen 1900). The arachnids have one or two pairs of coxal glands which develop from tubular outgrowths of the splanchnic mesoderm (Moritz 1959). Insects have a pair of labial glands which are excretory in collembolans (Feustel 1958). In terrestrial forms, the excretion is usually taken over more or less completely by Malpighian tubules, which arise from the posterior part of the endodermal midgut in arachnids and from the zone between midgut and hindgut in tracheates (Anderson 1973).

The heart is a dorsal tube with a pair of ostia in each segment. The haemocoelic fluid enters through the ostia and is pumped anteriorly by the contractions of the circular muscles which constitute the wall of the heart (Clarke 1979, Tjønneland, Økland & Nylund 1987).

The gonads (Clarke 1979) are derived from the coelomic pouches, but their cavity is normally formed secondarily; they often occupy various regions of the trunk, but they are situated in the head region in the cirripedes. There is one pair of gonoducts with gonopores located on different segments in different groups.

Almost all arthropods have internal fertilization, and the fertilized eggs become surrounded by various types of protecting membranes before deposition. In some forms, for example insects, the eggs become surrounded by the membranes already before fertilization, which takes place through special openings in the membrane, indicating an early determination of embryonic axes. Only the cirripedes and *Limulus* have external fertilization. The cirripedes shed the eggs in a gelatinous mass in the mantle cavity and there is a pseudocopulation where the 'male' deposits the sperm in the mantle cavity of the 'female' and the spermatozoa swim to the eggs (Klepal 1990). The females of *Limulus* deposit the eggs in a small cavity in the sand, and the males shed the sperm over the eggs (Shuster 1950).

Arthropod embryology shows tremendous variation. The embryology of remipedes and cephalocarids has unfortunately not been described, but Yager (1991) believes that the development of the remipedes is direct without a pelagic stage. Total cleavage is found not only in species with small eggs, but also in a few species with large eggs, such as *Limulus*; superficial cleavage is observed in many groups. The total cleavage in crustaceans with small eggs makes it possible to follow the cleavage pattern and the origin of the germ layers, but it is difficult to account for the origin of the various organs.

Anderson (1969, 1973) recognized the spiral-cleavage traits in a number of cirripedes and found that many older descriptions of several crustacean types could be reinterpreted as more or less strongly modified spiral cleavages. However, he dismissed parts of the detailed studies by Bigelow (1902) and Delsman (1917) on cirripedes, which both demonstrated close similarities between cleavage patterns and cell-lineages of cirripedes and annelids. His documentation was concentrated on the fate maps, but he did not construct cell-lineage diagrams. The paper by Delsman (1917) on the development of *Balanus balanoides* is especially well documented, showing the fate of the polar body and a close series of developmental stages with indications of cell divisions. The notation is difficult to follow, but the description and drawings make it possible to 'translate' the cell-lineage chart to the usual spiral-cleavage notation (Table 20.2). I will take Delsman's study as the starting point for the discussion of arthropod development.

The apical-blastoporal orientation of crustacean embryos cannot be seen directly, because a ciliated apical organ is lacking, but the second polar body is situated inside the fertilization membrane for example in *Balanus*, and this makes it possible to follow the position of the apical pole through the early embryonic stages.

The first two cleavages are almost perpendicular and go through the main axis of the egg with the polar body situated at the apical pole. At the end of the second cleavage, the A and C cells are in wide contact apically and the B and D cells blasto-

Table 20.2. Cell-lineage of *Balanus balanoides* after Delsman (1917); the notation is the normal spiral cleavage notation followed by the original notation of Delsman. l-left, r-right.

$$
\begin{array}{l}
Z \left\{
\begin{array}{l}
AB \left\{
\begin{array}{l}
A = a^3 \left\{
\begin{array}{l}
1a = a^{4.2} \ - \ \text{ectoderm} \\
1A = a^{4.1} \left\{
\begin{array}{l}
2a = a^{5.1} \ - \ \text{ectoderm} \\
2A = a^{5.2} \left\{
\begin{array}{l}
3a = a^{6.4} \ - \ \text{ectoderm} \\
3A = a^{6.3} \left\{
\begin{array}{l}
4a = a^{7.6} \ - \ \text{ectoderm} \\
4A = a^{7.5} \ - \ \text{mesoderm}
\end{array}\right.
\end{array}\right.
\end{array}\right.
\end{array}\right. \\
B \ \} \ \text{as A}
\end{array}\right. \\
CD \left\{
\begin{array}{l}
C \ \} \ \text{as A} \\
D = d^3 \left\{
\begin{array}{l}
1d = d^{4.2} \left\{
\begin{array}{l}
1d^l = d^{5.4} \ - \ \text{ectoderm + mesoderm} \\
1d^r = d^{5.3} \ - \ \text{ectoderm + mesoderm}
\end{array}\right. \\
1D = d^{4.1} \left\{
\begin{array}{l}
2d = d^{5.2} \ - \ \text{mesoderm} \\
2D = d^{5.1} \left\{
\begin{array}{l}
2D^l = d^{6.2} \left\{
\begin{array}{l}
2D^{lr} = d^{7.4} \ - \ \text{mesoderm} \\
2D^{l2} = d^{7.3} \ - \ \text{endoderm}
\end{array}\right. \\
2D^r = d^{6.1} \left\{
\begin{array}{l}
2D^{rr} = d^{7.2} \ - \ \text{mesoderm} \\
2D^{r2} = d^{7.1} \ - \ \text{endoderm}
\end{array}\right.
\end{array}\right.
\end{array}\right.
\end{array}\right.
\end{array}\right.
\end{array}
$$

porally; the polar body becomes situated where the cells A, B and C are in contact (Fig. 20.2 A). The D-cell is larger than the others and contains most of the yolk; it becomes situated in the posterior end of the ellipsoidal egg with the A-C blastomeres at the anterior end. The 16-cell stage shows an anterior cap of smaller cells and a large posterior 2D-cell (Fig. 20.2 B). The polar body becomes invaginated between the three apical cells of the A-C-quadrants and disappears soon after (Fig. 20.2 C). The anterior cells gradually spread posteriorly in an epibolic gastrulation. The fates of the three anterior quadrants are quite similar, with the blastomeres closest to the apical pole becoming ectoderm and those closest to the blastopore becoming mesoderm (Fig. 20.2 D-E). The D-quadrant develops differently with the 2d-cell becoming mesoderm and the 2D-cell dividing horizontally; each of the daughter cells of 2D di-

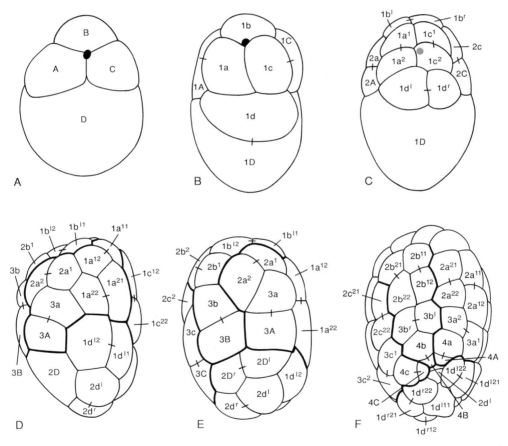

Fig. 20.2. The cleavage of *Balanus balanoides* redrawn from Delsman (1917) with notation changed to the spiral cleavage notation used for annelids and molluscs. – A, 4-cell stage in dorsal view. – B, 8-cell stage in dorsal view. – C, 16-cell stage in dorsal view. – D, 31-cell stage seen from the left side. – E, 32-cell stage in somewhat oblique ventral view. – F, stage close to 64 cells in ventral view.

169

vide into a small mesodermal cell and a large endodermal cell. The mesodermal cells move into the blastopore and spread as a thin sheet between ectoderm and endoderm. Delsman was unable to link the position of stomodaeum and proctodaeum to the cell lineage.

The early cleavage stages of the cirripedes studied by Anderson (1969) initially showed a similar cleavage pattern, but the cell-lineage was not followed in detail beyond the 33-cell stage. It was stated that the mesoderm derives exclusively from the A-C-quadrants.

Cladocerans such as *Holopedium* and *Polyphemus* have more modified spiral cleavage patterns (Anderson 1973). The fertilized egg of *Holopedium* (Baldass 1937) begins the development with three divisions of the nucleus without division of the cytoplasm; cleavage furrows develop at the '8-cell' stage, but do not reach the centre of the embryo. At the 31-cell stage, the blastomeres are completely separate and the primordial germ cell can be recognized at the blastoporal pole, surrounded by presumptive mesoderm cells from the 2d-cell and a horseshoe of cells giving rise to endoderm. Other crustaceans show a whole spectrum of cleavage types ending with intralecithal cleavage and blastoderm formation in several of the malacostracans. The other arthropod groups show variations over the same themes (Anderson 1973).

The cleavage of *Balanus* deviates from the typical spiral pattern in that endoderm originates only from the D-quadrant, but a similar pattern is found in the annelids *Chaetogaster* and *Erpobdella* (Chapter 17). The mesoderm of the cirripedes studied by Anderson (1969) apparently comes exclusively from the A-C-quadrants, and this is perhaps a parallel to the conditions in those annelids which have only ectomesoderm (Chapter 17).

The blastopore closes completely and mouth and anus are formed at a later stage from stomodaeal and proctodaeal invaginations. None of the two new openings are parts of the blastopore, but the general pattern seems to be that the stomodaeum forms in front of the blastopore while the proctodaeum forms in the region of the closed blastopore, and this is often observed in other articulates too (Manton 1949).

The mesoderm proliferates laterally along the endodermal cells which form the midgut and fill the ectodermal evaginations which become the appendages of the nauplius larva (Delsman 1917, Anderson 1969). The mesodermal cells differentiate into the muscles of the body and appendages of the larva without passing a stage resembling coelomic sacs, with the exception of the small coelomic sacs from which the antennal organs develop. A small caudal papilla at the posterior tip of the embryo contains a compact mass of mesodermal cells, and a pair of these cells grow larger than the others; these cells each divide twice and become a group of four teloblasts on each side, giving rise to the mesoderm of the following segments after the hatching of the larva (Anderson 1969). Similar development of the mesoderm with two posterior groups of mesoteloblasts has been reported from a number of crustacean groups (Anderson 1973).

The ganglia develop from ectodermal thickenings which eventually separate from the epithelium. The protocerebral and deutocerebral ganglia arise from preoral areas, while the tritocerebral ganglia and the ganglia of the following segments develop from postoral areas and have postoral commissures (Walley 1969).

The first larval stage of *Balanus balanoides* is a planktotrophic nauplius, and this stage has been studied for example by Walley (1969). The larva has three pairs of appendages: uniramous antennules and biramous antennae and mandibles with well developed gnathobases; all three pairs of appendages are used in swimming, and the food particles are handled with the gnathobases. Similar nauplius stages (or slightly more developed metanauplius stages with elongated posterior papillae with signs of segmentation but without limbs) are found in *Hutchinsoniella* and in representatives of all major crustacean groups (Fig. 20.3). Not all nauplius larvae feed, and the non-feeding types usually have more laterally situated appendages without gnathobases. The nauplius of *Balanus* has a pair of small antennal glands, but these excretory organs degenerate during the last naupliar stages and their function is taken over by the maxillary glands in the cypris stage.

The further development is through different larval stages characteristic of the various crustacean groups, but the most important part of the development is the addition of posterior segments with characteristic appendages: mandibles and two pairs of maxillae (all biramous) belonging to the cephalon, and further limbs which all are biramous locomotory limbs in most of the more 'primitive' types. The mesoderm of the body behind the three naupliar segments develops from the paired groups of mesoteloblasts; two rows of compact cell groups are given off anteriorly and a coelomic cavity develops in each cell mass in many groups (Anderson 1973). In each segment, the mesoderm spreads around the gut towards the dorsal side (Weygoldt 1958). The mesodermal cell groups split up and the coelomic cavities fuse with the primary body cavity forming a mixocoel or haemocoel. Some cell groups become muscles between the segments and in the limbs, others become arranged around a longitudinal haemocoelic cavity forming the heart with a pair of ostia in each segment, and still others form the pericardial septum, connective tissue and segmental organs. Coelomic cavities are retained only in the nephridia. Nephridial vestiges develop in a number of segments in some crustaceans (Benesch 1969), but only those of the antennal or the maxillary glands become functional. Each segmental organ develops from a small group of mesodermal cells which hollows out to form the sac, while another group of cells forms the duct; only the short exit duct is of ectodermal origin (Anderson 1982).

The arthropods share a number of obvious apomorphies, such as the articulated exoskeleton, articulated appendages with intrinsic musculature, cephalon of fused anterior segments, typically with compound eyes (questionable), antennules, antennae, mandibles and (one or) two pairs of maxillae; the presence of the special protein resilin is probably related to the structure of the exoskeleton. The mouth is directed posteriorly and partially covered by a labrum; in the filter-feeding crustaceans, the particles caught by the setae of the appendages are brought to the mouth along the food groove between the coxae, and this is believed to represent the ancestral functional morphology of the arthropods. The internal structure shows metanephridial funnels without cilia. There is no primary, ciliary filter-feeding larval stage, the earliest free stage (in the crustaceans) being a planktotrophic nauplius larva with three

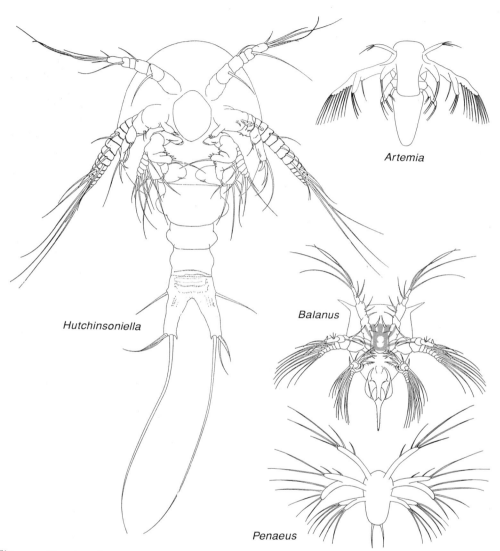

Artemia

Hutchinsoniella

Balanus

Penaeus

Fig. 20.3. First larval stage of selected crustacean groups; the first three larvae are planktotrophic and the antennae and mandibles are directed ventrally and have gnathobases; the krill larva is lecithotrophic and its appendages are directed laterally and lack gnathobases. – Cephalocarida: metanauplius of *Hutchinsoniella macracantha* (redrawn from Sanders 1963). – Branchiopoda: Anostraca: nauplius of *Artemia salina* (redrawn from Heath 1924). – Maxillopoda: Cirripedia: nauplius of *Balanus balanoides*, the nervous system is indicated in grey (redrawn from Sanders 1963 and Walley 1969). – Malacostraca: Decapoda: nauplius of *Penaeus setiferus* (redrawn from Pearson 1939).

pairs of appendages: uniramous antennules without gnathobases, and biramous antennae and mandibles with gnathobases. All these apomorphies clearly demonstrate that the Arthropoda are a monophyletic group.

Total cleavage with a spiral pattern is found in several of the crustacean groups considered most 'primitive', and these groups are all aquatic, mostly marine, which is obviously the ancestral habitat of the phylum. The cirripedes appear to have the most primitive type of cleavage and this group also has external fertilization, which must be considered ancestral. The possibility of a convergent evolution of total, spiral cleavage from the superficial type in several crustacean groups is apparently not indicated by any direct observations, so the spiral type must be considered a plesiomorphy, which fits the overall phylogenetic pattern advocated here. The origin and development of the mesoderm with teloblasts and segmental pairs of coelomic sacs (in early developmental stages) and the addition of segments from a pygidial growth zone are in good agreement with the development of for example annelids and molluscs. These characters immediately indicate that the arthropods belong to the group Teloblastica within the Spiralia.

A sister-group relationship with the onychophorans is indicated by the presence of an exoskeleton with α-chitin without collagen, which is moulted at regular intervals, and of a mixocoel, metanephridia originating in a sacculus with podocytes and a dorsal longitudinal heart with segmental pairs of ostia.

The relationships with the Tardigrada are discussed in Chapter 21.

Interesting subjects for future research

1. Anatomy and development of remipedes.
2. Embryology of cephalocarids.

References

Almond, J.E. 1985. The Silurian-Devonian fossil record of the Myriapoda. – Phil. Trans. R. Soc. B 309: 227-237, 1 pl.
Anderson, D.T. 1969. On the embryology of the cirripede crustaceans *Tetraclita rosea* (Krauss), *Tetraclita purpurascens* (Wood), *Chthamalus antennatus* (Darwin) and *Chamaesipho columna* (Spengler) and some considerations of crustacean phylogenetic relationships. – Phil. Trans. R. Soc. B 256: 183-235.
Anderson, D.T. 1973. Embryology and Phylogeny in Annelids and Arthropods. – Pergamon Press, Oxford.
Anderson, D.T. 1982. Embryology. – *In* D.E. Bliss (ed.): The Biology of Crustacea, vol. 2, pp 1-41. Academic Press, New York.
Ax, P. 1987. The Phylogenetic System. The Systematization of Organisms on the Basis of their Phylogenies. – John Wiley, Chichester.
Baldass, F.v. 1937. Entwicklung von *Holopedium gibberum*. – Zool. Jb., Anat. 63: 399-454.
Benesch, R. 1969. Zur Ontogenie und Morphologie von *Artemia salina*. – Zool. Jb., Anat. 86: 307-458.
Bereiter-Hahn, J., A.G. Matoltsy & K.S. Richards (eds) 1984. Biology of the Integument, vol. 1. – Springer, Berlin.
Bigelow, M.A. 1902. The early development of *Lepas*. A study of cell-lineage and germ-layers. – Bull. Mus. comp. Zool. Harv. 40: 61-144, 12 pls.

Böckeler, W. 1984. Embryogenese und ZNS-Differenzierung bei *Reighardia sternae*. Licht- und elektronenmikroskopische Untersuchungen zur Tagmosis und systematischen Stellung der Pentastomiden. – Zool. Jb., Anat. 111: 297-342.

Briggs, D.E.G. 1990. Early arthropods: dampening the Cambrian explosion. – Short Courses Paleontol. 3: 24-43.

Briggs, D.E.G. & R.A. Fortey 1989. The early radiation and relationships of the major arthropod groups. – Science 246: 241-243.

Brusca, R.C. & G.J. Brusca 1990. Invertebrates. – Sinauer Associates, Sunderland.

Clarke, K.U. 1979. Visceral anatomy and arthropod phylogeny. – *In* A.P. Gupta (ed.): Arthropod Phylogeny, pp 467-549. Van Nostrand Reinhold Co., New York.

Delsman, H.C. 1917. Die Embryonalentwicklung von *Balanus balanoides* Linn. – Tijdschr. ned. dierk. Vereen., 2. ser., 15: 419-520, 12 pls.

Dohle, W. 1980. Sind die Myriapoden eine monophyletische Gruppe? – Abh. naturwiss. Ver. Hamburg, N.F. 23: 45-104.

Elofsson, R. & R.R. Hessler 1990. Central nervous system of *Hutchinsoniella macracantha* (Cephalocarida). – J. crust. Biol. 10: 423-439.

Feustel, H. 1958. Untersuchungen über die Exkretion bei Collembolen. – Z. wiss. Zool. 161: 209-238.

Fortey, R.A. & R.M. Owens 1990. Evolutionary radiation in the Trilobita. – *In* P.D. Taylor & G.P. Larwood (eds): Major Evolutionary Radiations, pp 139-164. Oxford Univ. Press, Oxford.

Gould, S.J. 1989. Wonderful Life. The Burgess Shale and the Nature of History. – Hutchinson Radius, London.

Heath, H. 1924. The external development of certain phyllopods. – J. Morph. 38: 453-483.

Henry, L.M. 1948. The nervous system and the segmentation of the head in the Annulata. – Microentomology 13: 1-26.

Käuser, G. 1989. On the evolution of ecdysteroid hormones. – *In* J. Koolman (ed.): Ecdysone, pp 327-336. Georg Thieme, Stuttgart.

Karuppaswamy, S.A. 1977. Occurrence of β-chitin in the cuticle of a pentastomid *Railletiella gowrii*. – Experientia 33: 735-736.

Klepal, W. 1990. The fundamentals of insemination in cirripedes. – Oceanogr. mar. Biol. ann. Rev. 28: 353-379.

Kukalová-Peck, J. 1987. New Carboniferous Diplura, Monura, and Thysanura, the hexapod ground plan, and the role of thoracic side lobes in the origin of wings (Insecta). – Can. J. Zool. 65: 2327-2345.

Lauterbach, K.-E. 1980. Schlüsselereignisse in der Evolution des Grundplans der Arachnata (Arthropoda). – Abh. naturwiss. Ver. Hamburg, N.F. 23: 163-327.

Manton, S.F. 1949. Studies on the Onychophora VII. The early embryonic stages of *Peripatopsis*, and some general considerations concerning the morphology and phylogeny of the Arthropoda. – Phil. Trans. R. Soc. B 233: 483-580, pls 31-41.

Manton, S.F. 1972. The evolution of arthropodan locomotory mechanisms. Part 10. Locomotory habits, morphology and evolution of the hexapod classes. – Zool. J. Linn. Soc. 51: 203-400.

Manton, S.F. 1977. The Arthropoda. Habits, Functional Morphology, and Evolution. – Oxford Univ. Press, Oxford.

Moritz, M. 1959. Zur Embryonalentwicklung der Phalangiidae (Opiliones, Palpatores) II. Die Anlage und Entwicklung der Coxaldrüse bei *Phalangium opilio* L. – Zool. Jb., Anat. 77: 229-240.

Müller, K.J. & D. Walossek 1986. *Martinssonia elongata* gen. et sp. n., a crustacean-like euarthropod from the Upper Cambrian 'Orsten' of Sweden. – Zool. Scr. 15: 73-92.

Müller, K.J. & D. Walossek 1988. External morphology and larval development of the Upper Cambrian maxillopod *Bredocaris admirabilis*. – Fossils Strata 23: 1-70.

Paulus, H.F. 1979. Eye structure and the monophyly af the Arthropoda. – *In* A.P Gupta (ed.): Arthropod Phylogeny, pp 299-383. Van Nostrand Reinhold Co., New York.

Patten, W. & A.P. Hazen 1900. The development of the coxal gland, branchial cartilages, and genital ducts of *Limulus polyphemus*. – J. Morph. 16: 459-502, pls 22-28.

Pearson, J.C. 1939. The early life histories of some American Penaeidae, chiefly the commercial shrimp *Penaeus setiferus* (Linn.). – Bull. Bur. Fish. **49**(30): 1-73.

Rieder, N. & F. Schlecht 1978. Erster Nachweis von freien Cilien im Mitteldarm von Arthropoden. – Z. Naturforsch. **33**c: 598-599.

Riley, J., A.A. Banaja & J.L. James 1978. The phylogenetic relationships of the Pentastomida: the case for their inclusion within the Crustacea. – Int. J. Parasitol. **8**: 245-254.

Robison, R.A. 1990. Earliest-known uniramian arthropod. – Nature **343**: 163-164.

Sanders, H.L. 1963. The Cephalocarida. Functional morphology, larval development, comparative external anatomy. – Mem. Conn. Acad. Arts Sci. **15**: 1-80.

Schmidt, M. & W. Gnatzy 1984. Are the funnel-canal organs the 'campaniform sensilla' of the shore crab, *Carcinus maenas* (Decapoda, Crustacea)? II. Ultrastructure. – Cell Tiss. Res. **237**: 81-93.

Schram, F.R. 1986. Crustacea. – Oxford Univ. Press, New York.

Schram, F.R. & M.J. Emerson 1991. Arthropod pattern theory: a new approach to arthropod phylogeny. – Mem. Qd Mus. **31**: 1-18.

Schram, F.R., J. Yager & M.J. Emerson 1986. Remipedia. Part I. Systematics. – Mem. San Diego Soc. nat. Hist. **15**: 1-60.

Shear, W.A. & J. Kukalová-Peck 1990. The ecology of Paleozoic terrestrial arthropods: the fossil evidence. – Can. J. Zool. **68**: 1807-1834.

Shuster, C.N., Jr. 1950. Observations on the natural history of the American horseshoe crab, *Limulus polyphemus*. – Contr. Woods Hole oceanogr. Inst. **564**: 18-23.

Storch, V. & B.G.M. Jamieson 1992. Further spermatological evidence for including the Pentastomida (Tongue worms) in the Crustacea. – Int. J. Parasit. **22**: 95-108.

Such, J. 1975. Analyse ultrastructurale de la morphogenèse ommatidienne au cours du développement embryonnaire de l'œil composé, chez le Phasme *Carausius morosus* Br. – C. r. hebd. Séanc. Acad. Sci. Paris **281**: 67-70.

Taylor, P.M. & R.R. Harris 1986. Osmoregulation in *Corophium curvispinum* (Crustacea: Amphipoda), a recent coloniser of freshwater. – J. comp. Physiol. B **156**: 331-337.

Tjønneland, A., S. Økland & A. Nylund 1987. Evolutionary aspects of the arthropod heart. – Zool. Scr. **16**: 167-175.

Walley, L.J. 1969. Studies on the larval structure and metamorphosis of *Balanus balanoides* (L.). – Phil. Trans. R. Soc. B **256**: 237-280.

Walossek, D. & K.J. Müller 1990. Upper Cambrian stem-lineage crustaceans and their bearing upon the monophyletic origin of Crustacea and the position of *Agnostus*. – Lethaia **23**: 409-427.

Watson, R.D., E. Spaziani & W.E. Bollenbacher 1989. Regulation of ecdysone biosynthesis in insects and crustaceans: a comparison. – *In* J. Koolman (ed.): Ecdysone, pp 188-203. Georg Thieme, Stuttgart.

Weygoldt, P. 1958. Die Embryonalentwicklung des Amphipoden *Gammarus pulex pulex* (L.). – Zool. Jb., Anat. **77**: 51-110.

Weygoldt, P. 1986. Arthropod interrelationships – the phylogenetic-systematic approach. – Z. zool. Syst. Evolutionsforsch. **24**: 19-35.

White, K.N. & G. Walker 1981. The barnacle excretory organ. – J. mar. biol. Ass. U.K. **61**: 529-547.

Wingstrand, K.G. 1972. Comparative spermatology of a pentastomid, *Raillietiella hemidactyli*, and a branchiuran crustacean, *Argulus foliaceus*, with a discussion of pentastomid relationships. – Biol. Skr. Dan. Vid. Selsk. **19**(4): 1-72, 23 pls.

Yager, J. 1981. Remipedia, a new class of Crustacea from a marine cave in the Bahamas. – J. crust. Biol. **1**: 328-333.

Yager, J. 1991. The reproductive biology of two species of remipedes. – *In* R.T. Bauer & J.W. Martin (eds): Crustacean Sexual Biology, pp 271-289. Columbia Univ. Press, New York.

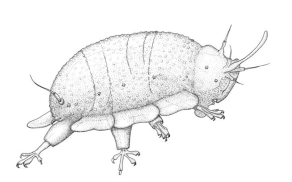

21

Phylum TARDIGRADA

The tardigrades or water bears are a small, easily recognized phylum of microscopic, often charmingly clumsy animals occurring in almost all types of habitats which are permanently or periodically moist – from the deep sea to soil, mosses, hot springs, and glaciers. So far about 600 species have been described, but new species, both terrestrial and marine, are constantly being added. Three orders are usually recognized: Heterotardigrada (with mainly marine suborder Arthrotardigrada and the terrestrial or limnic suborder Echiniscoidea), Mesotardigrada (represented only by *Thermozodium* from hot sulfur springs in Japan) and the mainly limnic-terrestrial Eutardigrada (Ramazzotti & Maucci 1983, Kristensen & Higgins 1984, Nelson & Higgins 1990). The extraordinary abilities of the terrestrial species to withstand extreme conditions, cryptobiosis, must be seen as a specialization to special habitats and are probably not of importance for the understanding of the position of the phylum. The knowledge of the morphology, development, physiology, and ecology, especially of the limnic-terrestrial species, was summarized by Marcus (1929b), which is still an important key to the older literature, although some of the conclusions must be treated with care (see below); a more up to date review was given by Greven (1980). This chapter includes a lot of new information from an unpublished manuscript on morphology and ultrastructure of marine tardigrades, which Dr R.M. Kristensen (Univ. Copenhagen) has generously permitted me to use; information from this source is marked RMK.

The body always has a head, with or without various spines or other appendages, and four segments, each with a pair of legs with a terminal group of sucking discs or claws (Fig. 21.1). The anterior or anteroventral mouth is a small circular opening through which a peculiar telescoping mouth cone with stylets can be protruded. The complicated buccal tube, pharynx and oesophagus are described in more detail below. The food seems to be bacteria or parts of larger plants or animals, live or dead. The food becomes surrounded by a peritrophic membrane and passes through the simple, endodermal midgut, consisting of cells with microvilli (Greven 1976), to the short, ectodermal intestine. The legs have a proximal joint, sometimes

---

Chapter vignette: *Wingstrandarctus corallinus*. (Redrawn from Kristensen 1984.)

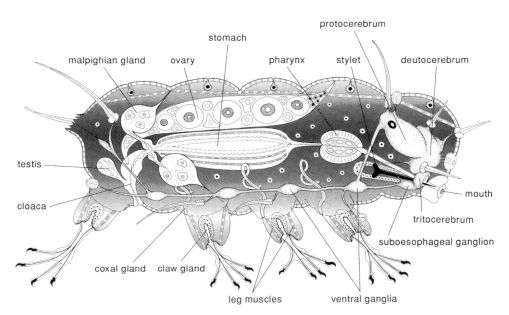

Fig. 21.1. A diagram of the ancestral tardigrade (RMK).

called coxa, and a second joint, sometimes called femur, which carries a group of cuticular claws or toes (see below); these cuticular structures are divided into a proximal shaft and distal swelling with toes in the arthrotardigrades, and these two parts
are sometimes called tibia and tarsus. The legs have short, intrinsic muscles (RMK).

The ectoderm secretes a complex chitinous cuticle (Jeuniaux 1975, Greven &
Peters 1986, RMK) with lipid-containing layers (Wright 1988); the type of the chitin
is not known. The cuticle is moulted periodically; it is heavily sclerotized in many
terrestrial forms. The claws, the cuticle of the pharynx with stylets and stylet supports and the cuticle of the rectum are all shed at moulting. The pharynx bulb is Y-
shaped in cross section with radiating muscle cells, comprising a single sarcomere of
a cross-striated muscle, in a constant pattern (Marcus 1929b, RMK). The pharyngeal muscles are usually considered to be ectodermal, i.e. myoepithelial (Dewel &
Clark 1973b, Walz 1973), but it now appears that the cuticle is secreted by very thin
extensions of ectodermal cells along the bottom of the three grooves; the radiating
muscles of the pharyngeal bulb must therefore be mesodermal (RMK).

Both the claws and the complicated toes, with cuticular 'tendons' connecting the
suction pads and terminal claws at the tips of the toes with the muscle attachment at
the base, are cuticular structures secreted by the so-called foot glands (Kristensen
1976); these ectodermal structures have no glandular function and have nothing to
do with the crural glands of the onychophorans.

The more or less strongly calcified stylets, their chitinous supports and the claws
with their complex 'tendons' are all secreted by ectodermal 'glands' (Wenck 1914,
RMK). The glands secreting the stylets and stylet supports are usually called salivary

glands, and they contain a secretion of unknown nature in the intermoult stage (Dewel & Clark 1973a), but there is apparently no evidence for the secretion of digestive products. Stylets and stylet supports are formed as separate structures in separate lobes of the 'salivary glands' in *Batillipes* (Kristensen 1976), and this may apply to the other forms too. Both the secretion of the stylets and their supports and the innervation of the muscles to the stylets (see below) indicate that they are modified legs (RMK).

The ectoderm is one layer of cells, which is very thin on the body, where a fixed number of cells are arranged in a symmetrical pattern which can be followed from species to species even between heterotardigrades and eutardigrades (Marcus 1928).

The brain (RMK, Fig. 21.1) comprises three main regions: 1) a large, anterodorsal protocerebrum with a pair of lateroposterior extensions, each with a small eye; each eye comprises a pigment cell, 1-2 cells with a cilium and a microvillar cell (Kristensen 1982); a pair of conspicuous nerves connect the posterior side of the two extensions to the ganglion of the first pair of legs; 2) a more ventral, rather compact deutocerebrum; 3) a circumoesophageal tritocerebrum, which has a postoral commissure, and which innervates the muscles of the stylets. All the three regions innervate one or two pairs of sensillae. A small suboesophageal ganglion is connected to the tritocerebrum and the ganglion of the first pair of legs through paired connectives and sends nerves to a number of buccal sense organs and perhaps to the stylet supports (Walz 1978, RMK). The four ganglia of the body segments are connected through paired connectives and innervate the legs and one or two pairs of sensillae on each segment.

Sense organs of the sensilla type are found on each segment in the arthrotardigrades. They consist of a few sensory cells with a modified cilium surrounded by a trichogen cell and a tormogen cell (Kristensen 1981).

There is a spacious body cavity which is usually described as a mixocoel or haemocoel, but its ontogenetic development is not documented. The musculature consists of separate muscles often consisting of few, large cells arranged in a metameric pattern. All the muscles are cross-striated in the arthrotardigrades, whereas the eutardigrades have oblique-cross-striated or smooth muscles except for the stylet muscles and the muscles of the pharyngeal bulb (Kristensen 1978). The muscles are attached to the cuticle through ectodermal cells with microfilaments and the usual hemidesmosomes (RMK).

Malpighian tubules are found in heterotardigrades and *Thermozodium*, where they are reported to originate from the anterior part of the hindgut or from the pylorus (Dewel & Dewel 1979); however, it appears that the tubules always open into the microvillar zone of the gut, so there may be some confusion about the terminology (RMK).

An unpaired testis or ovarium is situated dorsal to the gut with paired or unpaired gonoducts ending either in a separate gonopore in front of the anus (in the heterotardigrades) or in the rectum (in the eutardigrades). Sperm is transferred to the female during a copulation.

Tardigrade embryology is not well known. The development of some eutardigrades was studied by Erlanger (1895), Wenck (1914) and Marcus (1929a), but important parts of their descriptions have been questioned. Erlanger's drawings are so schematic that they have no documentary value. The cleavage is total and the endoderm is probably formed through delamination. The reports of four pairs of pouches from the embryonic gut have been taken as indicative of an enterocoelic mode of coelom formation, but the fate of these pouches has not been documented. I feel that the uncertainty about the development is so strong that I cannot use it in the phylogenetic discussion.

It seems probable that the marine arthrotardigrades are the group with most plesiomorphies, and that for example the malpighian tubules and the ability to go into cryptobiosis are adaptations to the terrestrial habitat (Renaud-Mornant 1982, Kristensen & Higgins 1984).

The monophyly of the Tardigrada seems to be unquestioned; the nerve from the lateral protocerebral lobes to the first ventral ganglion and the structure of the eyes are unique. The relationships of the tardigrades to other groups are much debated. Two main views are held: the tardigrades are panarthropods or proarthropods, or they are related to the aschelminths.

The 'aschelminth-theory' has focused on tardigrade characters such as the structure of the pharynx, the constant number of cells in certain organs (eutely), structure of sensillae, the cuticle of the heterotardigrades, and the ability to go into cryptobiosis (Crowe, Newell & Thomson 1970, Dewell & Clark 1973b). Also characters such as lack of haemal system and coelom have been interpreted as synapomorphies of tardigrades and aschelminths, but the lack of a character is a very weak phylogenetic argument, and the character of the tardigrade body cavity is a matter for conjecture since its ontogenetic origin is uncertain. Eutely must be treated as a character of the same nature as for example eyes; the mere presence of eyes is not a phylogenetic character, but presence of eyes which can be regarded as homologous is important. The constant number and patterns of epithelial cells in the tardigrade orders are characters indicating the monophyly of the phylum, but the occurrence of eutely for example in certain organs of nematodes and rotifers does not indicate relationships. The arguments about the gut structure are of course much dependent on the correct interpretation of the anatomy of the tardigrade pharynx, and the comparison with the nematodes falls if the above-mentioned report of a thin ectodermal cell layer below the cuticle is accepted. The tardigrade sensillae resemble both those of nematodes (Chapter 35) and loriciferans (Chapter 40) and those of arthropods (Chapter 20) so this character cannot be used.

On the other hand, there are very strong arguments in favour of the 'arthropod-theory'. The articulated legs with intrinsic muscles are without counterpart among the aschelminths, and they resemble those of the arthropods both in structure and function. The moveable spines in kinorhynchs and some nematodes are the only structures which could be taken as having similar function, but their structure is completely different and can at the most be compared to the claws of the tardigrade legs. Even more convincing similarities are found in the central nervous system, which has the same brain regions, protocerebrum, deutocerebrum and tritocerebrum, in tardigrades and arthropods; the tritocerebrum apparently innervates a pair of limbs in both groups. Also the ventral chain of ganglia with paired connectives is

179

very similar in the two groups. The central nervous system of nematodes, kinorhynchs and chaetognaths comprises a brain which may be divided into more or
less well-defined regions but each region surrounds the gut; a completely fused ventral nervous cord without ganglia is found in most of the phyla, and only the kinorhynchs have a paired system with ganglion-like swellings (Chapter 39).

To me, there is no doubt that the tardigrades are panarthropods closely related
to the arthropods, and the tardigrade apomorphies mentioned above make it reasonable to regard the two phyla as sister groups with the onychophorans as the first outgroup (Fig. 18.1).

Interesting subjects for future research

1. Embryology of all major groups.
2. Type of the chitin in the cuticle.

References

Erlanger, R.v. 1895. Beiträge zur Morphologie der Tardigraden. I. Zur Embryologie eines Tardigraden: *Macrobiotus macronyx* Dujardin. – Morph. Jb. **22**: 491-513, pls 21-22.

Dewel, R.A. & W.H. Clark, Jr. 1973a. Studies on the tardigrades. I. Fine structure of the anterior
foregut of *Milnesium tardigradum* Doyère. – Tissue Cell **5**: 133-146.

Dewel, R.A. & W.H. Clark, Jr. 1973b. Studies on the tardigrades. II. Fine structure of the pharynx
of *Milnesium tardigradum* Doyère. – Tissue Cell **5**: 147-159.

Dewel, R.A. & W.C. Dewel 1979. Studies on the tardigrades. IV. Fine structure of the hindgut of
Milnesium tardigradum Doyère. – J. Morph. **161**: 79-110.

Greven, H. 1976. Some ultrastructural observations on the midgut epithelium of *Isohypsibius au-
gusti* (Murray, 1907)(Eutardigrada). – Cell Tiss. Res. **166**: 339-351.

Greven, H. 1980. Die Bärtierchen. Tardigrada. (Neue Brehm-Bücherei 537). – A. Ziemsen Verlag,
Wittenberg.

Greven, H. & W. Peters 1986. Localization of chitin in the cuticle of Tardigrada using wheat germ
agglutinin-gold conjugate as a specific electron-dense marker. – Tissue Cell **18**: 297-304.

Jeuniaux, C. 1975. Principes de systématique biochimique et application à quelques problèmes particuliers concernant les Aschelminthes, les Polychètes et les Tardigrades. – Cah. Biol. mar. **16**:
597-612.

Kristensen, R.M. 1976. On the fine structure of *Batillipes noerrevangi* Kristensen 1976. 1. Tegument and moulting cycle. – Zool. Anz. **197**: 129-150.

Kristensen, R.M. 1978. On the structure of *Batillipes noerrevangi* Kristensen 1978. 2. The muscle-
attachments and the true cross-striated muscles. – Zool. Anz. **200**: 173-184.

Kristensen, R.M. 1981. Sense organs of two marine arthrotardigrades (Heterotardigrada, Tardigrada). – Acta zool. (Stockh.) **62**: 27-41.

Kristensen, R.M. 1982. The first record of cyclomorphosis in Tardigrada based on a new genus and
species from Arctic meiobenthos. – Z. zool. Syst. Evolutionsforsch. **20**: 249-270.

Kristensen, R.M. 1984. On the biology of *Wingstrandarctus corallinus* nov. gen. et sp., with notes
on the symbiontic bacteria in the subfamily Florarctinae (Arthrotardigrada). – Vidensk. Meddr
dansk naturh. Foren. **145**: 201-218.

Kristensen, R.M. & R.P. Higgins 1984. A new family of Arthrotardigrada (Tardigrada: Heterotardigrada) from the Atlantic coast of Florida, U.S.A. – Trans. Am. microsc. Soc. **103**: 295-311.

Marcus, E. 1928. Zur vergleichenden Anatomie und Histologie der Tardigraden. – Zool. Jb., Allg.
Zool. **45**: 99-158, pls 6-9.

Marcus, E. 1929a. Zur Embryologie der Tardigraden. – Zool. Jb., Anat. **50**: 333-384, pl. 8.

Marcus, E. 1929b. Tardigrada. – Bronn's Klassen und Ordnungen des Tierreichs, 5. Band, 4. Abt., 3. Buch, pp 1-608. Akademische Verlagsgesellschaft, Leipzig.

Nelson, D.R. & R.P. Higgins 1990. Tardigrada. – In D.L. Dindal (ed.): Soil Biology Guide, pp 393-419. John Wiley, New York.

Ramazzotti, G. & W. Maucci 1983. Il Philum Tardigrada. – Mem. Ist. ital. Idrobiol. 41: 1-1012.

Renaud-Mornant, J. 1982. Species diversity in marine tardigrades. - In D.R. Neslon (ed.): Proceedings of the Third International Symposium on the Tardigrada, pp 149-177. East Tennessee State Univ. Press, Johnson City.

Walz, B. 1973. Zur Feinstruktur der Muskelzellen des Pharynx-Bulbus von Tardigraden. – Z. Zellforsch. 140: 389-399.

Walz, B. 1978. Electron microscopic investigation of cephalic sense organs of the tardigrade Macrobiotus hufelandi C.A.S. Schultze. – Zoomorphologie 89: 1-19.

Wenck, W.v. 1914. Entwicklungsgeschichtliche Untersuchungen an Tardigraden (Macrobiotus lacustris Duj.). – Zool. Jb., Anat. 37: 465-514, pls 35-38.

Wright, J.C. 1988. The tardigrade cuticle I. Fine structure and the distribution of lipids. – Tissue Cell 20: 745-758.

22

BRYOZOA

Originally, *Pedicellina* and *Loxosoma* were placed in the phylum Bryozoa, but Nitsche (1869) pointed out that they lack a coelom whereas the other bryozoans have a body cavity lined by a peritoneum; he created the names Entoprocta and Ectoprocta and placed the two groups far from each other in his system. I have earlier argued in favour of reuniting the two groups (Nielsen 1971, 1985, 1987), but it must be admitted that my opinion has been based on only few positive facts.

The spiralian nature of the Entoprocta appears unquestionable (Chapter 23). The spiral cleavage is well-documented and the larvae are typical trochophores, often with a specialization of the gastrotroch as a foot resembling that of the molluscs. There is no trace of coelomic cavities, but spaces between the mesodermal cells have been interpreted as a 'pseudocoel'. This has led some authors to place the phylum in a group called pseudocoelomates, together with most of the phyla here classified as aschelminths (Hyman 1951, Brusca & Brusca 1990); this 'monothetic' method of classification which places all the weight on one character and disregards all others is incompatible with the cladistic argumentation used here. Some textbooks, for example Barnes (1986), discuss the entoprocts together with the ectoprocts, phoronids and brachiopods, i.e. the 'lophophorates', but this is quite confusing, because the entoprocts clearly belong with the spiralians.

The position of the Ectoprocta is more uncertain (Chapter 24). They are traditionally placed together with the phoronids and brachiopods in the group Tentaculata (Hatschek 1891) or Lophophorata (Hyman 1959) because they have a crown of ciliated tentacles on a 'lophophore' around the mouth. The deuterostome nature of phoronids and brachiopods is clearly demonstrated by their archimeric (trimeric) regionation with the mesosome carrying the tentacles, which have ciliary bands of the structure and function found in other deuterostomes, and with a dorsal central nervous system which develops independently of the larval apical organ (see Chapters 44 and 45). The ectoprocts show a cleavage pattern with some spiralian traits and a preoral ciliary band which resembles a prototroch both in cell-lineage and in some structural details, and the embryology shows no trace of enterocoely or archimery which seems to characterize most deuterostomes (Figs 43.1, 43.2); the origin of their adult body cavity after metamorphosis is very unusual and is not in any way reminiscent of enterocoely; all the larval organs are lost at metamorphosis and the polypides develop through a budding process so that it is not possible to ascertain the dorsal-ventral orientation of the polypides; the tentacles have bands of multiciliate cells and this is unknown among the phyla here classified as deuterostomes (Chapter 24). It seems impossible to identify a single synapomorphy uniting the ectoprocts with the phoronids and brachiopods.

In the search for a sister group of the ectoprocts, the entoprocts come in as a candidate because some of the colonial species show a metamorphosis which resembles that of some of the ectoprocts in several details (Fig. 22.1). This may of course be a case of convergence, but to my knowledge no other phylogenetic position of the ectoprocts has been discussed in a serious cladistic context.

It is difficult to identify synapomorphies of the two phyla, but the presence of myoepithelial cells in the apical organ of both groups may be one.

Another specific similarity between entoprocts and ectoprocts has been observed in the structure of the larval eyes. The presumed photoreceptors of larvae of both groups comprise one or more pigment cells and a photoreceptor cell with numerous almost unmodified cilia oriented at right angles to the direction of the incom-

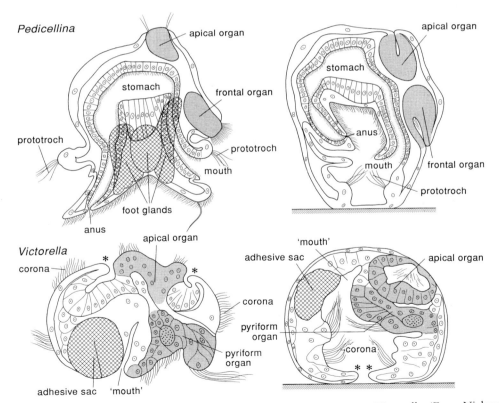

Fig. 22.1. Metamorphosis of the entoproct *Pedicellina* and the ectoproct *Victorella*. (From Nielsen 1971.)

183

ing light, and this is apparently a unique structure of photoreceptor cells. The only difference reported between the two cells is that the cilia of the entoproct sensory cell lack the dynein arms on the microtubules. On the other hand, a homology of the organs in the two groups is not indicated by their position.

A more detailed discussion of the position of the ectoprocts is given in Chapter 24.

References

Barnes, R.D. 1986. Invertebrate Zoology, 5. ed. – Saunders College Publishing, Philadelphia.

Brusca, R.C. & G.J. Brusca 1990. Invertebrates. – Sinauer Associates, Sunderland.

Hatschek, B. 1891. Lehrbuch der Zoologie, 3. Lieferung (pp 305-432). – Gustav Fischer, Jena.

Hyman, L.H. 1951. Acanthocephala, Aschelminthes, and Entoprocta. The pseudocoelomate Bilateria. The Invertebrates, vol. 3. – McGraw-Hill, New York.

Hyman, L.H. 1959. Smaller Coelomate groups. The Invertebrates, vol. 5. – McGraw-Hill, New York.

Nielsen, C. 1971. Entoproct life-cycles and the entoproct/ectoproct relationship. – Ophelia 9: 209-341.

Nielsen, C. 1985. Animal phylogeny in the light of the trochaea theory. – Biol. J. Linn. Soc. 25: 243-299.

Nielsen, C. 1987. Structure and function of metazoan ciliary bands and their phylogenetic significance. – Acta zool. (Stockh.) 68: 205-262.

Nitsche, H. 1869. Beiträge zur Kentniss der Bryozoen I-II. – Z. wiss. Zool. 20: 1-36, pls 1-3.

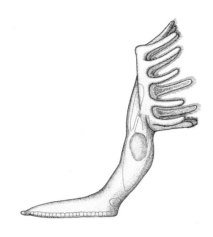

23

Phylum ENTOPROCTA

Entoprocta is a small phylum comprising about 150 described species (Nielsen 1989); the fossil record consists of Upper Jurassic, bioimmured colonies of a species of *Barentsia* very similar to the Recent *B. matsushimana* (Todd & Taylor 1992) (the Burgess Shale *Dinomiscus* is clearly not an entoproct, see Briggs & Conway Morris 1986 and Todd & Taylor 1992). All species are benthic with trochophora-type larvae. The family Loxosomatidae comprises solitary, usually commensal species, whereas the other three families are colonial; most species are marine, a few enter the brackish zone and one occurs in freshwater.

Each individual or zooid consists of a more or less globular body with a horseshoe of ciliated tentacles surrounding the depressed ventral side of the body, called the atrium, and a shorter or longer cylindrical stalk. There is no sign of a secondary body cavity, but spaces between ectodermal, endodermal and mesodermal elements form a narrow primary body cavity. The fluid of this cavity can be moved to and fro between body and stalk in the pedicellinids and barentsiids by the action of a small organ consisting of a stack of star-shaped cells with contractile peripheral rays (Emschermann 1969).

Most areas of the ectoderm are covered by a cuticle of crossing fibrils between branching microvilli with swollen tips, but there is only a thin glycocalyx without fibres at the ciliated faces of the tentacles (Nielsen & Rostgaard 1976, Emschermann 1982). The stiff portions of the stalks of the barentsiids and the thick cuticle around the characteristic resting bodies of some barentsiid species have a very thick layer of fibrils and an aschelminth-like surface layer without microvilli (Emschermann 1972b, 1982), but the illustrations suggest that the surface layer has been secreted by microvilli which have then retracted or degenerated. The cuticle contains chitin (Jeuniaux 1982). The ciliated cells of both ectodermal and endodermal epithelia as well as in the protonephridia are all multiciliate (Nielsen 1987).

The tentacles (Nielsen & Rostgaard 1976, Emschermann 1982) are almost cylindrical with a frontal row of cells (i.e. at the side of the tentacle facing the atrium)

Chapter vignette: *Loxosomella elegans*. (Redrawn from Nielsen 1964.)

with separate cilia beating towards the base of the tentacle, where they join a similar band, the adoral ciliary zone, which beats along the tentacle bases to the mouth. A row of cells along the lateral sides of the tentacles carries a band of long compound cilia; these compound cilia function as a downstream-collecting system creating a water current flowing between the tentacles away from the atrium and straining particles, which are then taken over by the frontal ciliary band leading to the mouth. A row of cells with separate cilia, the laterofrontal cells, lies between these two types of bands; their function is unknown. The lateral bands of compound cilia on the abfrontal pair of tentacles turn back and follow the adoral ciliary zone along the atrium. The laterofrontal ciliary cells are myoepithelial, and a row of mesodermal muscle cells is found lateral to the lateral ciliary cells.

There is a U-shaped gut with the mouth surrounded by the ciliary system of the tentacles and the anus situated near the aboral opening in the horseshoe of tentacles.

The nervous system comprises a dumbbell-shaped ganglion situated ventrally at the bottom of the atrium. Fine nerves go to sensory organs on the abfrontal side of the tentacles and in some species to lateral sense organs on the body (Harmer 1885). The lateral ciliated cells are innervated (Nielsen & Rostgaard 1976), and it is probable that this innervation controls the beat of the lateral cilia. A pair of lateral nerves is situated in the stalk of *Pedicellina*, and the longitudinal muscles send cytoplasmic strands to the nerves where synapses are formed (Emschermann 1985). Other parts of the peripheral nervous system including the nerves probably connecting the zooids in the colonies are poorly known; Hilton (1923) described a nervous network with numerous sensory pits staining with methylene blue in *Barentsia*, but the nature of these cells is uncertain.

There is no coelomic cavity and no haemal system.

There is a pair of protonephridia each consisting of a few cells and with a common nephridiopore at the bottom of the atrium (Kümmel 1965, Emschermann 1965b, Brandenburg 1966). Only the freshwater species *Urnatella gracilis* has several protonephridia on branched ducts in the body and several protonephridia in each joint of the stalk; all the protonephridia develop from ectodermal cells (Emschermann 1965b).

The gonads are simple, sac-shaped structures opening at the bottom of the atrium. Fertilization has not been observed directly, but Marcus (1939) found spermatozoa in eggs in the ovaries of *Pedicellina*. The fertilized eggs usually become enveloped in a secretion from glands in the gonoduct, and this secretion is pulled out into a string which attaches the egg to a special area of the atrial epithelium. When the egg membrane bursts open and the larva starts feeding, a ring around the apical organ remains, and the larvae are retained in the atrium for a period. *Loxosomella vivipara* has very small eggs which are taken up in a narrow invagination from the atrial epithelium where a placenta develops, nourishing the embryo through the apical area; the embryo increases enormously in size, and the fully grown larva has a large internal bud (Nielsen 1971). *Urnatella gracilis* is viviparous (Emschermann 1965a).

Cleavage is spiral with quartets (Marcus 1939: *Pedicellina*; Malakhov 1990: *Barentsia*). Only *Loxosomella vivipara* shows a cleavage in which the spiral pattern can be recognized only at the 8-cell stage (Nielsen 1971). The cleavage follows the normal spiral pattern, and Marcus (1939) observed a 56-cell stage with the cells $1a^{111}$-$1d^{111}$ forming the apical rosette, a ring of trochoblasts consisting of the daughter cells of the primary trochoblast cells $1a^2$-$1d^2$ and the secondary trochoblasts $1a^{122}$-$1d^{122}$, the mesoblast 4d, and the endoderm cells 4a-c, 5a-d and 5A-D (Fig. 11.2).

All entoproct larvae (Nielsen 1971) have the typical protostomian downstream-collecting ciliary system with a large prototroch of compound cilia, a smaller metatroch of compound cilia and an adoral ciliary zone. The prototroch comprises two rows of cells in *Barentsia discreta* (Malakhov 1990). The cilia are used in filter-feeding both in the older larvae still in the maternal atrium and after liberation. The gastrotroch is a small ventral band in some species, but most species have a conspicuous expansion of the ventral area, resembling the molluscan foot. The nervous system comprises a large apical organ with a pair of thin nerves to a paired frontal ganglion and further to a ring nerve at the base of the large prototroch cells (Nielsen 1971; Fig. 11.3); some larvae have a pair of lateral sense organs innervated from the frontal ganglion. The apical organ consists of a circle of multiciliate cells, a number of monociliate cells, myoepithelial cells and vacuolated cells (Sensenbaugh 1987). Mariscal (1965) described a large ventral ganglion in the larva of *Barentsia*, but this has not been seen by other authors. Most larvae have a large, retractable, ciliated frontal organ in contact with the frontal ganglion, but the organ is very small and lacks cilia in others (Nielsen 1971). Many species of *Loxosomella* have a ring of gland cells around the organ (see below). Larvae of most loxosomatids have a pair of eyes in the frontal organ. Each eye consists of a cup-shaped pigment cell, a lens cell and a photoreceptor cell with a bundle of cilia oriented perpendicular to the incoming light. The cilia have a normal $9 \times 2 + 2$ axoneme, but the dynein-arms are lacking (Woollacott & Eakin 1973). Many of the loxosomatid larvae and all the pedicellinid and barentsiid larvae have a large ciliated foot with a transverse row of long compound cilia at the anterior end; the larvae of the colonial species have three pairs of large glands with different types of secretion granules in the foot. There is a pair of protonephridia, each consisting of three cells and opening into a groove surrounding the foot.

Most larvae have a free period of only few hours before they settle, but some of the loxosomatid larvae apparently stay in the plankton for weeks; their development is mostly unknown.

Solitary and colonial entoprocts show considerable differences in the life cycles (Nielsen 1971; Fig. 23.1), but the variations within both types make it possible to interpret the types as variations over one common theme.

When the larvae with a foot are ready to settle, they creep on the substratum on the foot and test it with the frontal organ. Some species of *Loxosomella* have a very straightforward metamorphosis: the larva settles on the frontal organ, which apparently becomes glued to the substratum by a secretion from the gland cells around the ciliated sensory cells, with the hyposphere retracted and the muscles along the prototroch constricted so that all the larval organs are enclosed. During the following metamorphosis the apical and frontal organs disintegrate and the gut rotates only slightly. The larval ciliary bands disintegrate, but similar bands develop from neighbouring areas on tentacle buds at the frontal side of the closed atrium. A new gan-

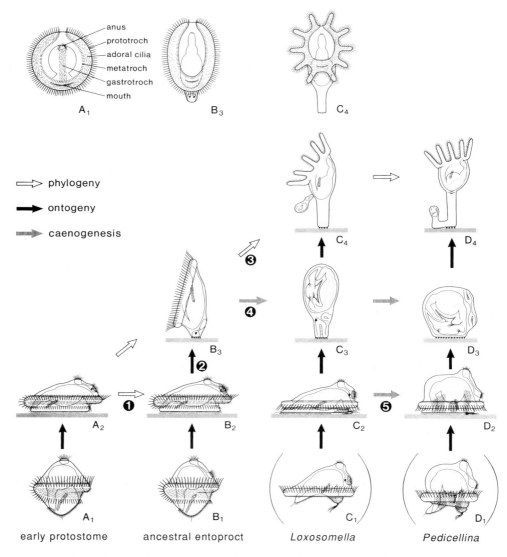

Fig. 23.1. The evolution of entoproct types from a hypothetic ancestor. The early ancestor had a pelagic trochophora larva and a benthic adult creeping on the enlarged gastrotroch. – The later ancestor had a trochophora larva with the gastrotroch extended as a creeping sole and a frontal sense organ surrounded by gland cells (1); at settling, the larva cemented the frontal organ to the substratum by secretion from the gland cells (2); the adult had the same general structure as the larva. – *Loxosomella* evolved from this ancestor probably in two steps: 1) the prototroch became extended onto a horseshoe of tentacles (3), and 2) the metamorphosis involved a temporary closure of the atrium by constriction of the ring of cells at the apical side of the prototroch (4). – *Pedicellina* probably evolved through a specialization of the settling mechanism (5), which comprised the evolution of a set of attachment glands in the larval foot and the attachment of the settling larva to the substratum by the area above the contracted prototroch. (Modified from Nielsen 1971.)

glion forms from an invagination of the ventral epithelium, and the atrium reopens exposing the short adult tentacles. The larval protonephridia may be retained, but this has not been studied.

Other species of *Loxosomella* and probably all species of *Loxosoma* have precocious budding from areas of the episphere, corresponding to the laterofrontal budding zones of the adult, and the larval body disintegrates after having given off the buds. In some species of both genera, the budding points are situated at the bottom of ectodermal invaginations, and the buds appear to be internal; these larvae become disrupted when the buds are liberated.

When the larvae of pedicellinids and barentsiids settle, the large glands in the foot apparently give off their secretion and the larva then retracts the hyposphere and contracts the prototroch muscle so that the contracted larva becomes glued to the substratum with the ring-shaped zone above the constricted prototroch. The frontal organ is obviously only acting as a sensory organ testing potential settling spots. The retracted apical and frontal organs disintegrate and the gut rotates about 180° with the mouth in front; the degeneration of the larval ciliary bands and the development of the adult bands, the formation of a new brain and the reopening of the atrium take place as in the loxosomatids.

The adult loxosomatids form buds from laterofrontal areas of the body and the buds detach after having reached the shape of a small adult, while the buds in the colonial species are formed at the base of the stalk (*Loxokalypus*, Emschermann 1972a) or from the growing tips of stolons (pedicellinids and barentsiids). The buds develop from small thickened ectodermal areas, which form a gut through invagination. Thick-walled resting buds are formed from the stolons of some barentsiids; at germination they develop new zooids through the same budding process.

The entoprocts are clearly a monophyletic group. Their larvae are more or less modified trochophores, but the adults do not resemble any of the teloblastic or parenchymian phyla. Some comprehensive texts have placed the phylum together with the 'acoelomates', but this has been founded on lack of characters rather than on synapomorphies.

The entoprocts are undoubtedly spiralians. The cleavage pattern is spiralian with quartets and with the mesoderm originating from the 4d-cell (although its further development has not been followed). The larvae have the ciliary bands characteristic of trochophores, both with regard to cell-lineage, structure and function, and the shape of their apical organ and the presence of a pair of protonephridia conform with the generalized protostome larva. Salvini-Plawen (1980) gave a special name to the entoproct larva and rejected the homology of its ciliary bands with those of the annelid trochophore; his view was based on the idea that the pericalymma larva is ancestral to the trochophore; this has been discussed and judged untenable in Chapter 10. The ciliated foot of most entoproct larvae is a specialization of the gastrotroch corresponding to the creeping area of gastroneuron; it resembles the molluscan foot, but a homology is not indicated. The foot is lacking in some larvae of *Loxosomella* and *Loxosoma*, which then resemble polychaete trochophores, but this may be a secondary reduction; the following development of these larvae is unknown. The central nervous system of the adults deviates from the protostomian pattern in

189

lacking the apical brain and the longitudinal ventral nervous system originating from the fused blastopore lips, but strongly concentrated, quite aberrant central nervous systems are found in many sessile forms, and the ventral nervous system is absent in the parenchymians (Chapter 25).

The relationships with other spiralian phyla are not obvious. There is no sign of teloblasts or coelomic cavities, so an inclusion in the Teloblastica is not indicated. The larvae are typical trochophores usually with the gastrotroch on a prominent foot, whereas the Parenchymia have larvae which are interpreted as strongly modified trochophores lacking the hyposphere almost completely. The adult entoprocts have a ventral ganglion, whereas the ventral nervous system appears to be lacking altogether in the parenchymians. The sessile habits of the entoprocts have obviously influenced their structure fundamentally, and there is no special indication of a closer relationship with the Parenchymia. At present there seems to be no better alternative than to accept a trichotomy with the Entoprocta in a separate group, Bryozoa, parallel to the two other major groups Articulata and Parenchymia. The phylum Ectoprocta is here placed together with the Entoprocta, and this will be discussed in the following chapter.

Interesting subjects for future research

1. The nervous system of stalk and stolon in the colonial species.
2. Structure of the nervous system of the larvae and the origin of the adult ganglion at metamorphosis.

References

Atkins, D. 1932. The ciliary feeding mechanism of the entoproct Polyzoa, and a comparison with that of the ectoproct Polyzoa. – Q. Jl microsc. Sci. 75: 393-423.

Brandenburg, J. 1966. Die Reusenformen der Cyrtocyten. – Zool. Beitr., N.F. 12: 345-417.

Briggs, D.E.G. & S. Conway Morris 1986. Problematica from the Middle Cambrian Burgess Shale of British Columbia. – *In* A. Hoffman & M.H. Nitecki (eds): Problematic Fossil Taxa, pp 167-183. Oxford Univ. Press, Oxford.

Emschermann, P. 1961. Über Brutkörper bei dem Kamptozoon *Barentsia gracilis* Sars. – Zool. Jb., Allg. Zool. 69: 333-338.

Emschermann, P. 1965a. Über die sexuelle Fortpflanzung und die Larve von *Urnatella gracilis* Leidy (Kamptozoa). – Z. Morph. Ökol. Tiere 55: 100-114.

Emschermann, P. 1965b. Das Protonephridiensystem von *Urnatella gracilis* Leidy (Kamptozoa). Bau, Entwicklung und Funktion. – Z. Morph. Ökol. Tiere 55: 859-914.

Emschermann, P. 1969. Ein Kreislauforgan bei Kamptozoen. – Z. Zellforsch. 97: 576-607.

Emschermann, P. 1972a. *Loxokalypus socialis* gen. et sp. nov. (Kamptozoa, Loxokalypodidae fam. nov.), ein neuer Kamptozoentyp aus dem nördlichen Pazifischen Ozean. Ein Vorschlag zur Neufassung der Kamptozoensystematik. – Mar. Biol. (Berl.) 12: 237-254.

Emschermann, P. 1972b. Cuticular pores and spines in the Pedicellinidae and Barentsiidae (Entoprocta), their relationship, ultrastructure, and suggested function, and their phylogenetic evidence. – Sarsia 51: 7-16.

Emschermann, P. 1982. Les Kamptozoaires. État actuel de nos connaissances sur leur anatomie, leur développement, leur biologie et leur position phylogénétique. – Bull. Soc. zool. Fr. 107: 317-344, 3 pls.

Emschermann, P. 1985. Cladus Kamptozoa = Entoprocta, Kelchwürmer, Nicktiere. – *In* R. Siewing (ed.): H. Wurmbach's Lehrbuch der Zoologie, vol. 2, Systematik, pp 576-586. Gustav Fischer, Stuttgart.

Harmer, S. F. 1885. On the structure and development of *Loxosoma*. – Q. Jl microsc. Sci. **25**: 261-337, pls 19-21.

Hilton, W.A. 1923. A study of the movements of entoproctan bryozoans. – Trans. Am. microsc. Soc. **42**: 135-143.

Kümmel, G. 1962. Zwei neue formen von Cyrtocyten. Vergleich der bisher bekannten Cyrtocyten und Erörterung des Begriffes 'Zelltyp'. – Z. Zellforsch. **57**: 172-201.

Malakhov, V.V. 1990. Description of the development of *Ascopodaria discreta* (Coloniales, Barentsiidae) and discussion of the Kamptozoa status in the animal kingdom. – Zool. Zh. **69**(10): 20-30 (In Russian, English summary. English translation available from: Library, Canadian Museum of Nature, P.O. Box 3443, Stn. D, Ottawa, Ontario, Canada K1P 6P4).

Marcus, E. 1939. Bryozoarios marinhos brasileiros III. – Bolm Fac. Filos. Ciênc. Univ. S Paulo, Zool. **3**: 111-354.

Mariscal, R.N. 1965. The adult and larval morphology and life history of the entoproct *Barentsia gracilis* (M. Sars, 1835). – J. Morph. **116**: 311-338.

Nielsen, C. 1964. Studies on Danish Entoprocta. – Ophelia **1**: 1-76.

Nielsen, C. 1971. Entoproct life-cycles and the entoproct/ectoproct relationship. – Ophelia **9**: 209-341.

Nielsen, C. 1989. Entoprocta. – Synopses Br. Fauna, N.S. **41**: 1-131.

Nielsen, C. & J. Rostgaard 1976. Structure and function of an entoproct tentacle with discussion of ciliary feeding types. – Ophelia **15**:115-140.

Salvini-Plawen, L.v. 1980. Was ist eine Trochophora? Eine Analyse der Larventypen mariner Protostomier. – Zool. Jb., Anat. **103**: 389-423.

Sensenbaugh, T. 1987. Ultrastructural observations on the larva of *Loxosoma pectinaricola* Franzén (Entoprocta, Loxosomatidae). – Acta zool. (Stockh.) **68**: 135-145.

Todd, J.A. & P.D. Taylor 1992. The first fossil entoproct. – Naturwissenschaften **79**: 311-314.

Woollacott, R.M. & R.M. Eakin 1973. Ultrastructure of a potential photoreceptor organ in the larva of an entoproct. – J. Ultrastruct. Res. **43**: 412-425.

Phylum ECTOPROCTA

Ectoprocts or moss-animals constitute a quite isolated phylum of sessile, colonial, aquatic organisms; about 4000 living species are known and there is an extensive fossil record (Boardman & Cheetham 1987). Three classes are recognized: Gymnolaemata, which are marine, brackish or limnic, with calcified or non-calcified body wall with a normal peritoneum, the earliest fossils from the Upper Ordovician; Stenolaemata, which are marine, with calcified body wall and a detached peritoneum forming the membranous sac, the earliest fossils from the Lower Ordovician; and Phylactolaemata, which are limnic, with non-calcified body wall, the earliest fossils from the Upper Tertiary. The gymnolaemates are often divided into the calcified, operculate Cheilostomata and the non-calcified, non-operculate Ctenostomata (which are only occasionally found as fossils, the oldest fossils being borings in calcareous shells), but the classification is somewhat uncertain. The stenolaemates flourished with five orders in the Palaeozoic, but only one order, Cyclostomata (also called Tubuliporata to avoid the synonymy with the Agnatha), has survived.

The colonies consist of individuals or zooids which arise by budding and which remain in more or less open contact. Some types, especially the cheilostomes, show polymorphism with zooids specialized for defence, cleaning, reproduction, anchoring or other functions. Some of these special zooids feed, but several types lack feeding structures and are nourished by the neighbouring zooids. A generalized feeding zooid has a box- or tube-shaped, mostly rather stiff body wall called the cystid and a moveable polypide consisting of the gut and a ring of ciliated tentacles around the mouth. The tentacles or lophophore can be retracted into the cystid, which closes either through constriction or with a small operculum. Retraction is caused by strong retractor muscles extending from the basal part of the cystid to the thickened basement membrane at the base of the lophophore. Protrusion is caused by contraction of various muscles of the cystid wall with the body cavity acting as a hydrostatic skeleton; the muscles may either constrict the whole cystid, as in the ctenostomes; special, non-calcified parts of the cystid wall, as in the cheilostomes; or only the detached peritoneum

---

Chapter vignette: A branch of the ctenostome *Farrella repens*. (Redrawn after Marcus 1926a.)

or membranous sac of the cystid wall, as in the cyclostomes (Taylor 1981). The first zooid in a colony, the ancestrula, arises from a metamorphosed larva through rearrangement and differentiation of larval tissues and blastemas, and the following polypides develop from invaginations of ectoderm and mesoderm of the body wall. The body cavity remains continuous in phylactolaemate colonies, where only incomplete walls separate the zooids (Brien 1953). The developing gymnolaemate zooids form almost complete septa between neighbouring zooids, but the ectoderm remains continuous around openings in pore plates, which are plugged by special, mesodermal rosette cells (Bobin 1958). The cyclostome colonies have a continuous primary body cavity but the zooids have individual coelomic cavities (Nielsen & Pedersen 1979).

The lophophore is a ridge with a circle of tentacles around the ciliated mouth; most of the phylactolaemates have the lateral sides of the lophophore extended posteriorly (in the direction of the anus) so that the tentacle crown becomes horseshoe-shaped; they also have a lip originating from the posterior side of the mouth. The tentacles show characteristic patterns of ciliated cells (Fig. 24.1). Each tentacle has two rows of ciliated ectodermal cells along the lateral sides; in gymnolaemates and phylactolaemates these lateral cells have numerous cilia forming a wide band (Lutaud 1973, Brien 1960), whereas the stenolaemates have two rows of closely set lateral cilia (Nielsen 1987). The frontal side of the tentacles has one row of multiciliate cells in gymnolaemates (Lutaud 1973), whereas a wide band of multiciliated cells with interspersed sensory cells is found in the phylactolaemates (Brien 1960); the stenolaemates have a row of unciliated, apparently secretory frontal cells (Nielsen 1987). A basiepithelial nerve runs beneath the frontal cells in all three groups. A row of monociliate laterofrontal sensory cells lies along the frontal cells, and a basiepithelial nerve follows each row of cells (not reported in phylactolaemates). The function of the ciliary bands has been studied by Strathmann (1973, 1982), who described the cilia of the tentacles as an upstream-collecting system with particle capture through local ciliary reversal, and by Gilmour (1978), who proposed a combination of impingement and direct filtration. The rather passive type of particle capture is in almost all species complicated by various types of tentacle movements, which result in capture of particles of different characters (Winston 1978). The differences in ciliation between the three classes, especially the lack of frontal cilia in the stenolaemates, the conflicting descriptions of the filtration method, and also the different types of behaviour indicate that the feeding mechanism of representatives of all three classes should be studied in more detail.

The lophophore has a monolayered ectoderm with a thin cuticle between branched microvilli. The ectoderm of the cystid secretes a cuticle but lacks microvilli; the ctenostomes and phylactolaemates have a sometimes quite thick cuticle with some chitin whereas the cheilostomes and cyclostomes have calcified areas where the cuticle consists of an outer, organic periostracum and an inner, calcified layer (Tavener-Smith & Williams 1972, Jeuniaux 1982).

All ciliated epithelial cells are multiciliate (Nielsen 1987).

The gut is a U-shaped tube with a number of ciliated regions; it develops from the ectoderm during budding, so it is not possible to make direct statements about ecto- and endodermal regions. The pharynx is triradial and consists of myoepithelial

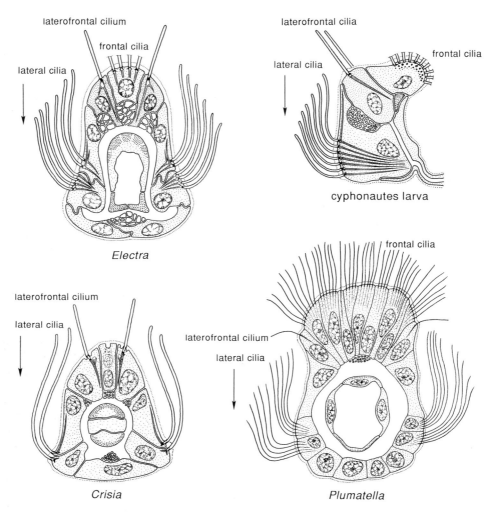

Fig. 24.1. Cross-sections of ectoproct tentacles and of the ciliated ridge of a cyphonautes larva. – Gymnolaemata: *Electra pilosa* (redrawn from Lutaud 1973); cyphonautes of *Membranipora* sp. (based on Nielsen 1987 and additional TEM observations). – Stenolaemata: *Crisia eburnea* (based on Nielsen 1987 and additional TEM observations). – Phylactolaemata: *Plumatella fungosa* (redrawn from Brien 1960). The direction of the current created by the lateral cilia is indicated by the large arrows.

cells each with a large vacuole; the contraction of the radial, cross-striated myofilaments shortens and widens the cells thereby expanding the pharynx (Bullivant & Bils 1968). Some gymnolaemates have a gizzard at the entrance to the stomach. Each gizzard tooth is secreted by an epithelial cell with microvilli; the teeth have a honeycomb-structure with cylindrical canals, with the microvilli extending into the basal part of the canals (Gordon 1975). The basal point of the stomach is attached to the cystid by a tissue strand called the funiculus. It is simple, comprising only muscle cells and peritoneal cells (with testes) in the cyclostomes (Carle & Ruppert 1983); in the phylactolaemates it additionally contains an extension of the ectoderm which secretes the cuticle of the resting bodies called statoblasts (Brien 1953). The gymnolaemates have a more complicated funicular system, comprising additional, hollow, branching mesodermal strands, which attach to the cystid wall in connection with the interzooidal pores (Lutaud 1982, Carle & Ruppert 1983). The funicular system apparently functions as a haemal system, capable of transporting substances within a zooid, and the rosette cells are obviously polarized and have been shown to transport organic molecules across the pores (Lutaud 1982). Carle & Ruppert (1983) interpreted the funiculi and funicular systems of all ectoprocts as homologous to the haemal systems of brachiopods and phoronids, based on their structure, i.e. blastocoelic cavities surrounded by basement membrane, and their function, i.e. transport. However, there is nothing to indicate that the small, unbranched funiculi of phylactolaemates and cyclostomes function as circulatory organs, and the ground plan of the gymnolaemate funicular system bears no resemblance to the haemal systems of phoronids and brachiopods. If the homology is followed further, it leads to the conclusion that all blastocoels are homologous, and the information about homology of specialized haemal systems is lost.

Each polypide has a ganglion at the posterior (anal) side of the oral opening with lateral extensions following the lophophore base around the mouth (Lutaud 1977). The peripheral nervous system is delicate, and most of the studies are based on vital staining supplemented by a few electron microscopical observations (Gymnolaemata: Lutaud 1969, 1973, 1977, 1979; Phylactolaemata: Gerwerzhagen 1913, Marcus 1934; the peripheral nervous system of the Cyclostomata is poorly known). A few nerves connect the ganglion with a fine nerve net which connects the zooids. The connection is through special cells in the rosettes of the pore plates in the gymnolaemates, but is a more uncomplicated net in the phylactolaemates, which lack walls between the zooids.

The colonies have species-specific growth patterns, and the growth areas vary from wide zones along the edge of the colonies, as in phylactolaemates and *Membranipora*, to narrow points at the tips of stolons, as in stolonate ctenostomes, or to certain points of the cystid, as in *Electra* and *Crisia*. Ectoderm and mesoderm are difficult to distinguish in these areas (Brien & Huysmans 1938, Borg 1926), but the two layers become distinct a short distance from the growth zone. The peritoneum surrounds a spacious coelom, which shows some important differences between the classes.

Gymnolaemate zooids each have a well-delimited coelom with the cystid wall consisting of ectoderm, muscles and peritoneum. There is a ring-shaped lophophore coelom with extensions into the tentacles and a dorsal opening to the main cystid cavity (Brien 1960).

Stenolaemates have very unusual body cavities (Nielsen & Pedersen 1979). The peritoneum has the usual close connection with the gut and with the ectoderm of the polypide, but the peritoneum of the cystid is detached from the ectoderm and forms the membranous sac, which consists of the very thin peritoneum with its basement membrane and a series of very thin, annular muscles. The coelomic cavity of each polypide is completely separated from that of the neighbouring zooids, whereas the spacious primary body cavity surrounding the membranous sac is in open connection with that of the neighbouring zooids through communication pores or via the common extrazooidal cavity at the surface of the colonies. The lophophoral coelom is narrow and continuous with the body coelom through a pore above the ganglion (Borg 1926).

Phylactolaemates have a somewhat complicated coelom in the protrusible part of the polypide, but the description by Brien (1960), based on *Plumatella*, *Fredericella* and *Cristatella*, settles much of the uncertainty found in the older literature. There is a coelomic canal in the lophophore with branches extending to the tips of the tentacles. A pair of strongly ciliated canals along the posterior (anal) side of the buccal cavity and the ganglion connect the posterior part of the lophophoral coelom to the main coelom. In *Cristatella*, the median part of the lophophoral canal (at the upper part of the ciliated canals) is expanded into a small posterior bladder, in which amoebocytes with excretory products accumulate. The amoebocytes may become expelled from the bladder, but there is no permanent excretory pore. A narrow canal from the main coelom extends between the two ciliated canals to a more spacious cavity in the lip. The cilia of the peritoneum create a circulation of the coelomic fluid. It is clear that there is one, rather complicated, coelomic cavity in the polypide, and that archimery and metanephridia are not present.

Nephridia have not been observed in any ectoproct. Waste products accumulate in the cells of the gut, and the whole polypide degenerates periodically and a new polypide forms by budding from the cystid. The degenerated polypide is either taken into the gut of the new polypide and expelled as the first faeces, or it remains in the basal part of the cystid as a brown body (Gordon 1977).

The gonads are special areas of the peritoneum; the testes are usually situated on the funiculus and the ovaries on the body wall. The ripe gametes float in the coelom, and the spermatozoa are liberated through a small, transitory pore at the tentacle tips in both gymnolaemates and stenolaemates (the phylactolaemates have not been studied) (Silén 1966, 1972). The eggs are shed through a median supraneural pore at the posterior side of the lophophore (Silén 1945). During the reproductive periods, species of *Alcyonidium*, *Bowerbankia* and *Membranipora* have a temporary gutter with lateral rows of compound cilia on the posterior side of the pharynx, transporting the ripe eggs to the supraneural pore (Reed 1988b). The pore is simple in many species, but during the reproductive period it is extended into a ciliated funnel, the intertentacular organ, for example in *Electra* and *Membranipora* (Marcus 1926a).

The eggs are fertilized in the coelom in some of the brooding species (Marcus 1938, Dyrynda & King 1982, 1983), but Silén (1966) observed that the eggs of the free-spawning *Electra posidoniae* become fertilized when leaving the intertentacular organ.

Gymnolaemate reproduction, development and metamorphosis have been studied by a number of authors, but there are still large gaps in our knowledge. The few free spawners have planktotrophic, shelled cyphonautes larvae, but the majority brood the embryos in one of a bewildering variety of ways, and their larvae are lecithotrophic or placentally nourished, and are usually without shells. Simple retention of the fertilized eggs attached to the tentacle sheath is seen for example in the cteno-stome *Triticella* (Ström 1969), and brooding in the retracted tentacle sheath of zooids with partially degenerated polypide is known for example in *Bowerbankia* (Reed 1988b). Brooding in special ovicels formed by the zooid distal to the maternal zooid is found in many cheilostomes (Nielsen 1981). The eggs of the brooding species are usually quite large and the development lecithotrophic, but a few species, such as *Bugula neritina* (Woollacott & Zimmer 1972a, 1975) deposit the very small egg in an ovicel closed by an extension of the maternal epithelium, which becomes a placenta nourishing the developing embryo. *Epistomia* is viviparous with a single, tiny egg nourished in the maternal cystid after the degeneration of the polypide (Dyrynda & King 1982).

The regulative powers of the blastomeres are almost unstudied, but Zimmer (1973) isolated the blastomeres of 2-cell stages of *Membranipora isabelleana* and found complete regulation.

The small, freely spawned eggs of *Electra, Alcyonidium albidum* and *Hypophorella* (Prouho 1892) cleave equally so that a radially symmetrical 8-cell stage is formed, but the two planes of the following cleavage are lateral to the primary, api-cal-blastoporal axis, so that the 16-cell stage is transversely elongate consisting of two apical and two blastoporal rows of four cells each. The polar bodies remain vis-ible near the apical pole until gastrulation. The two following cleavages result in a coeloblastula, and four or eight cells at the blastoporal pole move into the blastocoel. These cells form the endoderm, and possibly also the mesoderm, but the differentia-tion was not followed. The blastopore closes, the endoderm rearranges as a small archenteron, and an anterior stomodeal invagination becomes the vestibule. A pair of mesodermal cells were observed at the anterior side of the vestibule, but their ori-gin was not observed; they appeared to give rise to a row of cells from the pyriform organ to the apical organ and possibly to become differentiated into muscle cells. The embryos develop cilia on the apical cells and on a ring of large coronal cells (which can be interpreted as prototrochal cells, see below), and the larvae break the egg membrane. The larvae are planktotrophic and spend weeks in the plankton be-fore settling; they soon become laterally compressed, the pallial epithelium, i.e. the epithelium of the episphere, secretes a pair of triangular shells, and the larvae be-come the well-known cyphonautes larvae. This larval type has been reported from the cheilostomes *Membranipora, Conopeum* and *Electra* and the ctenostomes *Alcy-onidium albidum, Hypophorella* and *Farrella* (although shelled stages have not actu-ally been observed in any of the ctenostome species; Nielsen 1971).

Also species with non-planktotrophic development, such as *Siniopelta* and *Bu-gula flabellata* (Marcus 1938, Corrêa 1948), show a biradial cleavage pattern (Fig. 24.2). All species studied so far appear to have the same cleavage pattern, but the two just-mentioned species have been followed in more detail, and it has been poss-ible to follow the cell lineage to the 64-cell stage. The cleavages do not show the al-ternating oblique cleavage furrows of the spiral cleavage, but if the notation of the spiral cleavage is used (Table 24.1), it turns out that the corona cells which give rise to the large ciliary band are descendants of the cells $1q^{12}$ and $1q^{22}$, which is practi-cally identical to the pattern seen in other spiralians, where the primary prototroch

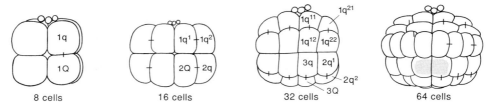

Fig. 24.2. Embryology of *Bugula flabellata*. All embryos are in lateral view. Stages: 8 cells, 16 cells, 32 cells, 64 cells (the 4q-cells inside the blastocoel are indicated by shading). (Based on Corrêa 1948.)

Table 24.1. Cell lineage of one quadrant of *Bugula flabellata* based on Corrêa (1948); the four quadrants show biradial symmetry so the individual quadrants cannot be identified and the letter q is therefore used. The first notation for each blastomere is given in accordance with the usual spiralian pattern with the lowest number being given to the cells closest to the apical pole along the meridian of the embryo; the numbers in parentheses are the notation given by Corrêa.

4	8	16	32 cells	64 cells	cell fates
Q	1q	1q^{I}	1q^{II} (2)	1q^{III} (2), 1q$^{\mathrm{II2}}$ (10)	apical plate
			1q$^{\mathrm{I2}}$ (10)	1q$^{\mathrm{I2I}}$ (18), 1q$^{\mathrm{I22}}$ (22)	prototroch
		1q^2	1q$^{\mathrm{2I}}$ (6)	1q$^{\mathrm{2II}}$ (6), 1q$^{\mathrm{2I2}}$ (14)	
			1q$^{\mathrm{22}}$ (14)	1q$^{\mathrm{22I}}$ (26), 1q$^{\mathrm{222}}$ (30)	prototroch
	1Q	2q	2q^{I} (22)	2q^{II} (38), 2q$^{\mathrm{I2}}$ (54)	
			2q^2 (26)	2q$^{\mathrm{2I}}$ (42), 2q$^{\mathrm{22}}$ (58)	
		2Q	3q (18)	3q^{I} (34), 3q^2 (50)	
			3Q (30)	4q (46), 4Q (62)	mesoderm + endoderm

cells derive from the cells 1q^2 and various secondary prototroch cells from 1q$^{\mathrm{I2}}$ and 2q (Chapter 11). The corona can therefore be interpreted as a prototroch. The eight coronal cells divide twice horizontally, and the 32 cells become the very large cells of the corona. This number of prototroch cells has been observed in many gymnolaemate larvae, but higher numbers are observed in other species. The four cells at the blastoporal pole of the 32-cell stage divide horizontally so that four cells become situated inside the blastula; these four upper cells migrate towards the periphery of the blastocoel, and it appears that these four internal cells can be recognized in embryos of many species (Corrêa 1948, d'Hondt 1983). The following cell divisions are more difficult to follow, and it appears unclear whether more cells enter the blastocoel. Pace (1906) reported that also the four lower cells at the blastoporal pole enter, but the following stages of differentiation of the internal cells is unknown. The labelling of the eight lower cells is therefore tentative. The development of the apical organ, the adhesive sac and the other larval organs has not been studied in detail, but all reports point to processes resembling those in the developing cyphonautes described above.

The structure of the cyphonautes larvae of *Electra* (Kupelwieser 1905) and especially those of *Membranipora* (Atkins 1955, Stricker 1987, Stricker, Reed & Zimmer 1988a,b) has been studied in detail. The apical organ consists of concentric rings of unciliated and monociliate cells and myoepithelial cells. A cord of neurons and muscle cells connects the apical organ and the pyriform organ. This organ is a strongly ciliated cleft of thick epithelium at the anterior part of the prototroch (corona); it contains both gland cells and nerve cells and is obviously active in the explorative phase of the settling. The prototroch is situated along the lower edge of the shells and consists of two rows of large, multiciliate cells with a row of smaller biciliate (Kupelwieser 1905) or monociliate cells on the apical side (supracoronal cells; Stricker, Reed & Zimmer 1988a). The coronal cilia are not organized as compound cilia; they show metachronal waves, and the effective stroke makes the larva swim with the apical organ in front, but the stroke can be reversed for short periods (Nielsen 1987). The spacious vestibule is divided into an anterior inhalant and a posterior exhalant chamber by a U-shaped ciliated ridge. The ridge bears three rows of ciliated cells (Fig. 24.1): a row of multiciliate lateral cells creating the current through the gap between the ridges, a row of biciliate laterofrontal cells, and a row of multiciliate frontal cells. It has generally been believed that the lateral cilia function as an upstream-collecting band, like the cilia of one side of a tentacle, but observations by Strathmann & McEdward (1986) indicate that the laterofrontal cilia are stiff and function as a mechanical filter. This appears to be a unique ciliary mechanism among the metazoans. The frontal cilia and cilia in the upper part of the vestibule carry the food particles to the mouth. The median epithelium of the exhalant chamber has formed a large, complicated, glandular invagination, the adhesive sac. The gut consists of a ciliated oesophagus and stomach and an unciliated rectum.

The non-planktotrophic larvae are of several types (Zimmer & Woollacott 1977a). Both shelled larvae with a non-functioning but almost complete gut and cyphonautes-like larvae without shells are known. Various types of a more spherical or elongate larvae with the prototroch expanded to cover the entire surface of the larva, except the apical plate, so-called coronate larvae, are found in many genera. The well-described larva of *Bugula neritina* has a large apical organ with a number of cell types, covering a thick blastema of undifferentiated cells (Woollacott & Zimmer 1971), and similar, although less conspicuous structures have been reported in most other gymnolaemate larvae. The nervous and muscular systems of the lecithotrophic larvae are generally less complex than those of the cyphonautes (Reed 1988a). Many of the coronate larvae have pigment spots, which are presumably photosensory, associated with the corona; each spot consists of a number of pigment cells surrounding a depression with a putative sensory cell (which may have pigment too) with a bundle of almost unmodified cilia (Woollacott & Zimmer 1972b, Hughes & Woollacott 1979).

Settling behaviour and the first phases of the metamorphosis are similar in all the gymnolaemates, but the later stages show considerable variation (Fig. 24.3). The larvae creep on the substratum, exploring with the pyriform organ, and when a suitable spot has been found, the adhesive sac everts and gives off its secretion. Settling cyphonautes larvae expand the adhesive sac over the substratum and contract muscles between the adhesive sac and the apical organ so that the adductor muscles of the shells rupture and the larva changes from a laterally compressed to a dorso-ventrally compressed shape. The pallial epithelium releases the shells partially and extends over the upper side of the metamorphosing larva. The periphery of the pallial epithelium fuses with the periphery of the adhesive sac, enclosing the prototroch and all the larval organs which degenerate; the apical organ becomes invaginated (Kupelwieser 1905, Stricker 1988, 1989). The just metamorphosed larva, the primary disc or preancestrula, is completely covered by ectoderm, but its internal morphology is rather chaotic, consisting of a ring-shaped portion of the former exterior (at first surrounded by the infolded prototroch) and various partially degenerating larval organs. A layer of mesodermal cells along the body wall has not been reported. A few other gymnolaemates, such as *Flustrellidra* (Prouho 1890) and *Watersipora* (Lyke, Reed & Woollacott 1983), have a similar metamorphosis, but two opposite lines can be followed from the cyphonautes type (Fig. 24.3): the *Bugula*-type, characterized by an extreme expansion of the adhesive sac, which finally covers also the upper side of the primary disc (Woollacott & Zimmer 1971, Reed & Woollacott 1983, 1984), and the *Bowerbankia*-type, which goes to the opposite extreme by withdrawing the adhesive sac and expanding the pallial epithelium along the lower side of the primary disc (Reed & Cloney 1982a,b, Reed 1984).

The polypide of the ancestrula develops from the blastema at the underside of the apical organ or from an infracoronal ring (Stricker 1989, Reed & Cloney 1982a, review in Zimmer & Woollacott 1989).

The cystid wall expands in various patterns, and new polypides develop from the cystid wall through small invaginations of the body wall (review in Nielsen 1971).

Fig. 24.3. Types of metamorphosis in ectoproct bryozoans showing variations in the origin of the cystid epithelium. – Gymnolaemata: *Electra pilosa*: free-swimming larva and just formed primary disc; the cystid epithelium originates from the adhesive sac and the pallial epithelium (mostly underlying the shells) (based on Nielsen 1971). – *Bugula neritina*: free-swimming larva, newly settled larva, and just formed primary disc; the cystid epithelium originates from the adhesive sac (based on Reed & Woollacott 1982, 1983). – *Bowerbankia gracilis*: free-swimming and just settled larva with everted adhesive sac, and young primary disc; the cystid epithelium originates from the pallial epithelium (based on Reed & Cloney 1982a,b). – Stenolaemata: *Crisia eburnea*: free-swimming larva, settling larva, and just formed primary disc; the cystid epithelium originates from the adhesive sac and the cuticle-lined epithelium of the apical invagination (based on Nielsen 1970). – Phylactolaemata: *Plumatella fungosa*: larva with precociously developed polypides at the bottom of the apical invagination, free-swimming stage with everted polypides, and fully metamorphosed stage; the cystid epithelium originates from the epithelium of the apical invagination, i.e. the pallial epithelium (based on Brien 1953).

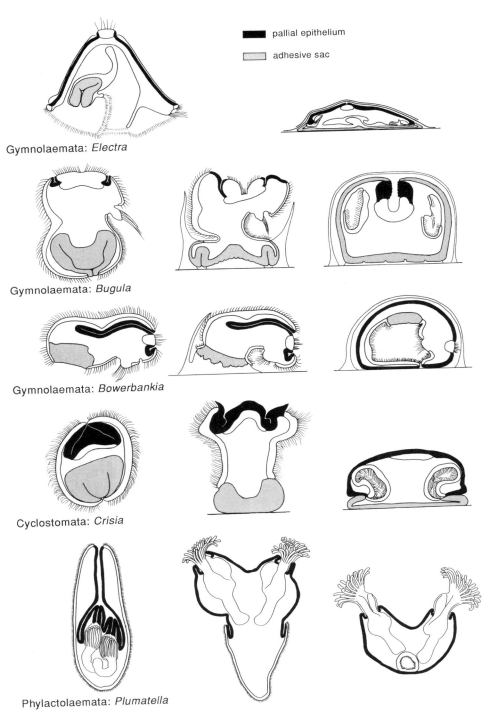

pallial epithelium

adhesive sac

Gymnolaemata: *Electra*

Gymnolaemata: *Bugula*

Gymnolaemata: *Bowerbankia*

Cyclostomata: *Crisia*

Phylactolaemata: *Plumatella*

There is some uncertainty about the origin of the various parts of the gut, but most accounts describe a posterior invagination which differentiates into rectum–intestine–stomach and an anterior invagination which becomes the oesophagus; a secondary opening is then formed between oesophagus and stomach. The ganglion develops from an ectodermal invagination at the posterior side of the area of the mouth.

Cyclostomes have a highly specialized type of reproduction with polyembryony (Harmer 1893). Only a few zooids in special positions in the colonies are female, and when their single egg has been fertilized, the polypide degenerates and the zooid becomes a large brooding structure, a gonozooid, nourished by the neighbouring zooids. The embryo becomes irregular, and secondary and even tertiary embryos are given off. The embryos finally differentiate into almost spherical larvae without any trace of apical organ, pyriform organ or gut. The ectoderm forms a large invagination of the apical side and secretes a cuticle and a corresponding invagination of the opposite side becomes a large adhesive sac. The whole external part of the ectoderm is ciliated. The liberated larvae (Nielsen 1970) swim for a few hours and settle by everting the adhesive sac on the substratum; the apical invagination everts at the same time, and the ciliated ectoderm becomes internalized as a ring-shaped cavity with all the cilia and degenerates (Fig. 24.3). The first polypide develops from a layer of cells below the ectoderm of the apical area. The origin and differentiation of the mesoderm including the membranous sac are in need of further studies.

The polyembryony and lack of an apical organ make it difficult to ascertain the axes of the cyclostome larva, but the cuticle is probably secreted by the epithelial area corresponding to the pallial area which secretes the shells in the cyphonautes larva, and the adhesive organs of the larvae are probably homologous. The ciliated epithelium is probably homologous with the corona, although it consists of many small cells instead of one ring of large cells. The primary disc is thus covered by pallial and adhesive sac epithelia like the primary disc of *Electra* (Fig. 24.3).

The subsequent buds are formed through a process which resembles that of the gymnolaemates, but the origin of the various cell layers has not been studied.

The phylactolaemates are viviparous, but the fertilization has never been described. Species of most of the eight genera have been studied, but the most detailed information comes from *Plumatella* (Brien 1953, review in Nielsen 1971). The small fertilized egg enters an embryo sac, which is an invagination of the body wall. The early development shows no definite patterns of cleavage or germ layer formation, but an elongate, two-layered embryo without any indication of a gut has been observed in all species investigated. The embryos become nourished by placental structures situated either in an equatorial zone, for example in *Plumatella*, or an 'apical' zone (see below), for example in *Fredericella*. One or more polypide buds develop from invaginations of the body wall above the annular placenta or lateral to the apical placenta, and the polypides become fully developed before the 'larva' is liberated. There is one polypide in *Fredericella*, usually two in *Plumatella* and from four to several in *Cristatella* and *Pectinatella*. The 'larva' hatches from the maternal zooid and swims with the pole opposite the invagination with polypides in front; Marcus (1926b) noted a concentration of nerve cells at this 'anterior' pole, but there is no well-defined epithelial thickening as in other spiralian apical organs. Franzén &

Sensenbaugh (1983) studied the ultrastructure of the anterior pole and found a number of different cell types, including ciliated cells, sensory cells, nerve cells, and gland cells, but not a single cell type showing specific similarities with cells of the apical organ of the gymnolaemate larvae could be pointed out. On the other hand, it was stated that the glandular cells resemble the cells of the adhesive sac of the gymnolaemate larvae and that their secretion is released at settling. My conclusion is that the anterior pole of the phylactolaemate larva is homologous with the adhesive sac epithelium of the gymnolaemate larvae and that the larva therefore swims 'backwards' with the apical end trailing (Nielsen 1971, Fig. 24.3). This is in no way contradicted by the beat direction of the cilia of the phylactolaemate larvae because the gymnolaemate larvae are known to be able to reverse the beat of the coronal cilia (Reed & Cloney 1982b, Nielsen 1987). This orientation makes the position of the precociously formed polypide buds agree with the position of the twin buds in *Membranipora*, which develop from an area of the pallial epithelium. The free-swimming colonies settle after a short pelagic phase and the whole ciliated epithelium, representing the extended prototroch and the adhesive sac, becomes invaginated.

The polypides of the 'larva' are normal polypide buds, as are the buds from the germinating statoblasts, and their development from the two-layered body wall has been studied in a number of species (Brien 1953, review in Nielsen 1971). The development of the polypides resembles that of the gymnolaemates. The statoblasts are formed by ectodermal cells in the funiculus which secrete chitinous shells around a mass of cells rich in stored nutrients.

The monophyly of the Ectoprocta is almost unquestioned. The structure and development both of the colonies and the individual polypides show many unique characters: the development of the polypide of the ancestrula from a blastema at the lower side of the apical organ or from an infracoronal ring, the division of the body into cystid and polypide, the lophophore with an 'upstream-collecting' ciliary system with multiciliate cells, the filter system of the ciliated ridge of the cyphonautes larva. The origin of the three classes has been much debated, and the conclusions have sometimes been heavily influenced by preconceived ideas about the relationships to other phyla.

Gymnolaemates have a 'normal' embryology and some have planktotrophic larvae, but their metamorphoses show an astonishing variation in the origin of the ectoderm of the primary disc (Fig. 24.3). The morphology of the adult body wall is 'normal', with ectoderm and peritoneum, although the origin of the peritoneum is unusual in all ectoprocts.

The living stenolaemates apparently all have the very aberrant polyembryony, and the larvae lack any trace of a gut; their metamorphosis can be interpreted as a modification of the *Membranipora*-type seen in several gymnolaemates (Fig. 24.3). The body wall is very unusual with the detached peritoneum forming the membranous sac.

The phylactolaemates generally have larger zooids than the other groups and their lophophores are usually horseshoe-shaped instead of circular, but this may be a simple adaptation to the larger size. Both some large sabellid polychaetes and the larger phoronids show curved to spirally coiled tentacle crowns, and even an unusually large, undescribed loxosomatid entoproct from Thailand shows a curved lophophore. The embryology and the 'larva' appear highly specialized too, and this is probably related to the limnic habitat.

The fossil evidence is difficult to interpret, because some of the groups are heavily calcified and therefore well-represented whereas other groups are only slightly calcified or non-calcified and therefore only preserved very occasionally, for example by bioimmuration. The earliest known fossils are stenolaemates, but that does not prove that this group represents the most ancestral type. The soft-parts of the early stenolaemates can, of course, only be inferred from observations of the living species, but well-preserved specimens of several Ordovician stenolaemates show traces of the membranous sac, indicating that also the early stenolaemates had the very unusual detached mesoderm; structures interpreted as gonozooids have been found in a few genera of two extinct stenolaemate orders too, indicating brooding like that of the living cyclostomes (Boardman 1983). Other Palaeozoic stenolaemates had small brood-chambers, indicating brooding of single larvae as in the cheilostomes (Schäfer 1991). Some of the very early (mid-Ordovician) cyclostomes, such as *Corynotrypa*, resemble the ctenostome *Arachnidium*, and this could indicate that both groups have evolved from an ancestor of this type (Taylor 1985). The cheilostomes may have originated in the Jurassic from other *Arachnidium*-like ctenostomes; the newly described genus *Cardoarachnidium* (Taylor 1990), which has been preserved through bioimmuration, lacks calcified body walls but has an operculum and could represent a 'missing link' between ctenostomes and cheilostomes. The late occurrence of fossil traces of phylactolaemates and their limnic habitat may indicate a rather late origin, perhaps from brackish-water ctenostomes.

Judging on the above evidence, I can only conclude that the 'primitive' gymnolaemates with planktotrophic larvae show the most plesiomorphic characters among the ectoprocts, and that both stenolaemates and phylactolaemates are derived groups; a cladistic analysis of the phylum will not be attempted here.

The relationships with other phyla is much debated. The early authors just added *Pedicellina* and *Loxosoma* to the Bryozoa, but Nitsche (1869) noted some of the important differences between the two groups and created the names Entoprocta and Ectoprocta. Hatschek (1891) stressed the similarity between the entoproct larvae and rotifers and placed the entoprocts in the 'Scolecida' next to the rotifers, whereas the ectoprocts were united with the phoronids in the group Tentaculata (see also Hatschek 1911). Hyman (1959) introduced the name Lophophorata for the same group, and this name is now in common use in most textbooks. Jägersten (1972) compared the cyphonautes and actinotrocha larvae and concluded that the differences between the two types are differences in proportions, and Farmer (1977) built this idea into an 'adaptive' model for the evolution of colonial ectoprocts from a semi-colonial phoronid. These ideas completely disregard both the differences in the structure between the prototroch of the cyphonautes and the ciliated edge of the epistome of the actinotrocha (Chapter 44) and the structural and functional differences between the adults. I (Nielsen 1971, 1985, 1987) have argued for a reunification of the ectoprocts and entoprocts in the supraphyletic group Bryozoa, placed within the Spiralia, inspired by new investigations of entoproct ontogeny.

Two positions of the Ectoprocta therefore need to be discussed: in the Lophophorata within the Deuterostomia, with Phoronida and Brachiopoda as the closest sister groups, and in the Bryozoa within the Spiralia in the Protostomia, with the Entoprocta as the sister group.

Lophophorates are usually defined as trimeric animals with a lophophore and a U-shaped gut; this definition fits the pterobranchs equally well, but this is usually ignored. Trimeric (or archimeric) means that the body consists of three regions: prosome, mesosome and metasome, each with a paired or unpaired coelomic compartment: protocoel, mesocoel and metacoel. A lophophore is defined as a mesosomal extension with ciliated tentacles containing mesocoelomic canals. Archimery and ciliated tentacles containing mesocoelomic extensions are here regarded as deuterostome synapomorphies, and phoronids, brachiopods and pterobranchs (Chapter 44, 45 and 47) unquestionably belong to the deuterostome line (Chapter 43). The discussion of ectoproct relationship could best begin with a discussion of their coelomic compartments.

The postulated trimery of the ectoprocts is not based on embryological evidence: there is no trace of coelomic sacs or of a tripartition of mesoderm during the larval development. The adult polypides comprise a large body cavity in the cystid connected through a wide opening with a smaller, ring-shaped lophophoral cavity around the mouth with extensions into the tentacles; the phylactolaemates have a lip at the posterior side of the mouth with a median extension from the body coelom. The continuity of the coelomic cavities is also seen by the fact that the ripe sperm moves to the lophophoral cavity and is shed through the tips of some of the tentacles. Thus, trimery is not indicated in the polypides. The phoronids and brachiopods have large metacoelomic metanephridia which also function as gonoducts; metanephridia are not found in ectoprocts and their male and female gametes or embryos are shed through different openings.

The gizzard teeth have a structure similar to that of both annelid and brachiopod setae (Chapters 17, 45), so this character must be polyphyletic.

The structure and function of larval and adult ciliary bands of the ectoprocts show some similarities with those of the deuterostomes, but the new information about the ciliary ridge of the cyphonautes larva indicates that this constitutes a different filtering system. The lack of frontal cilia on the tentacles of the stenolaemates indicates that this system is something special too, and it may turn out that the upstream-collecting ciliary system on the ectoproct tentacles has a function which is different from that of the deuterostome ciliary bands. This is supported by the fact that the ectoproct tentacles have multiciliate cells whereas the deuterostome upstream-collecting bands consist of monociliate cells and also by the complicated behaviour of most ectoproct polypides as opposed to the rather passive filtering process of phoronids and brachiopods.

Deuterostomes have a dorsal central nervous system developed from an area behind the apical organ, which is lost at metamorphosis (Chapter 43). Ectoprocts have a ganglion at the posterior side of the oesophagus, but the dorsal/ventral orientation of the polypides cannot be ascertained, because all the polypides develop through budding. All statements about a dorsal ganglion in ectoprocts are based on circular argumentation (Nielsen 1971).

It should be clear that I find the arguments for uniting the ectoprocts with phoronids and brachiopods (and pterobranchs) unconvincing – some of the arguments, such as the 'trimery' and the 'metanephridia' of the ectoprocts, are without embryological or morphological foundation, and the statements about their presence have the character of incantations.

On the other hand, there are several ectoproct characters which point to relationships with the spiralians: cleavage with spiralian traits in the cell lineage, such as the origin of the prototroch cells and the meso- and endoderm; the regulative powers of the 2-cell stage are similar to those of the nemertines (Chapter 27). A small pretroch and a prototroch with two rows of large, multiciliate cells in the cyphonautes; a pretroch has been pictured in trochophores of entoprocts, annelids and molluscs (Nielsen 1987) and the double row of prototrochal cells has been reported from entoprocts (Malakhov 1990), annelids (Hatschek 1878, Holborow, Laverack & Barber 1969), and molluscs (Patten 1885, Erdmann 1935). Extension of the prototroch area over most of the surface of the larvae has been observed in a number of spiralian larvae (pericalymma larvae of type 1; Fig. 10.5), and the coronate larvae may be interpreted as a special type of pericalymma larvae. Metamorphosis with infolding and degeneration of the prototroch (corona) inside the body like that of all the ectoproct larvae has also been observed in entoprocts (Chapter 23) and molluscs (Thompson 1960). Also the multiciliarity of the epithelial cells indicates relationships with the spiralians rather than with the 'lower' deuterostomes (Fig. 9.2).

The arguments for regarding ectoprocts and entoprocts as sister groups come from their ontogeny. The adhesive glands secreting the attachment substance in the ectoproct larvae are located in the same area and have the same function as the foot glands of the entoproct larvae, and the whole metamorphosis of the *Bowerbankia*-type ectoproct larvae strongly resembles that of the pedicellinid and barentsiid entoprocts. The apparently unique type of photoreceptor cells found in eye spots of both ectoproct and entoproct larvae (Chapter 23), may indicate close relationships, but the homology of the eyespots seems uncertain.

The frontal organ of the entoproct larvae and the pyriform organ of the ectoproct larvae have sometimes been considered as homologous, but their different position relative to the prototroch contradicts this.

My conclusion is that the ectoprocts are a very specialized group, and that their phylogenetic position cannot at present be stated with certainty. However, I find that the arguments for uniting them with phoronids and brachiopods in the group 'lophophorates' are mainly circumstantial, whereas a number of observations on embryology, metamorphosis and ultrastructure indicate a position within the protostomes with a sister-group relationship with the entoprocts. It should be stressed that I have never proposed to include the entoprocts in the lophophorates.

Interesting subjects for future research

1. Embryology of both planktotrophic and lecithotrophic gymnolaemates.
2. Polypide development in gymnolaemate ancestrulae of a number of groups.

3. Budding of cyclostomes.
4. Function of the ciliary bands on the ciliated ridge of the cyphonautes larva and on the tentacles of all three classes.

References

Atkins, D. 1955. The ciliary feeding mechanism of the cyphonautes larva (Polyzoa Ectoprocta). – J. mar. biol. Ass. U.K. **34**: 451-466.

Boardman, R.S. 1983. General features of the class Stenolaemata. – Treatise of Invertebrate Paleontology, Part G (revised), vol. 1, pp 49-137. Geological Society of America, Boulder.

Boardman, R.S. & A.H. Cheetham 1987. Phylum Bryozoa. – In R.S. Boardman, A.H. Cheetham & A.J. Rowell (eds): Fossil Invertebrates, pp 497-549. Blackwell Scientific Publications, Palo Alto.

Bobin, G. 1958. Structure et genèse des diaphragmes autozoéciaux chez *Bowerbankia imbricata* (Adams). – Archs Zool. exp. gén. **96**: 53-99.

Borg, F. 1926. Studies on Recent cyclostomatous Bryozoa. – Zool. Bidr. Upps. **10**: 181-507, 14 pls.

Brien, P. 1953. Étude sur les Phylactolémates. – Annls Soc. r. zool. Belg. **84**: 301-444.

Brien, P. 1960. Classe des Bryozoaires. – Traité de Zoologie, vol. 5(2): 1053-1355. Masson, Paris.

Brien, P. & G. Huysmans 1938. La croissance et le bourgeonnement du stolon chez les Stolonifera (*Bowerbankia* (Farre)). – Annls Soc. r. zool. Belg. **68**: 13-40.

Bullivant, J.S. & R.F. Bils 1968. The pharyngeal cells of *Zoobotryon verticillatum* (delle Chiaje) a gymnolaemate bryozoan. – N.Z. J. mar. Freshwat. Res. **3**: 438-446.

Carle, K.J. & E.E. Ruppert 1983. Comparative ultrastructure of the bryozoan funiculus: a blood vessel homologue. – Z. zool. Syst. Evolutionsforsch. **21**: 181-193.

Corrêa, D.D. 1948. A embriologia de *Bugula flabellata* (J.V. Thompson) (Bryozoa Ectoprocta). – Bolm Fac. Filos. Ciênc. Univ. S Paulo, Zool. **13**: 7-71,

d'Hondt, J.-L. 1983. Sur l'évolution des quatre macromères du pôle végétatif chez les embryons de Bryozoaires Eurystomes. – Cah. Biol. mar. **24**: 177-185, 1 pl.

Dyrynda, P.E.J. & P.E. King 1982. Sexual reproduction in *Epistomia bursaria* (Bryozoa: Cheilostomata), an endozooidal brooder without polypide recycling. – J. Zool. (Lond.) **198**: 337-352.

Dyrynda, P.E.J. & P.E. King 1983. Gametogenesis in placental and non-placental ovicellate cheilostome bryozoans. – J. Zool. (Lond.) **200**: 471-492.

Erdmann, W. 1935. Untersuchungen über die Lebensgeschichte der Auster. Nr. 5. Über die Entwicklung und die Anatomie der 'ansatzreifen' Larve von *Ostrea edulis* mit Bemerkungen über die Lebensgeschichte der Auster. – Wiss. Meeresunters., N.F., Helgoland **19**(6): 1-25, 8 pls.

Farmer, J.D. 1977. An adaptive model for the evolution of the ectoproct life cycle. – In R.M. Woollacott & R.L. Zimmer (eds): Biology of Bryozoans, pp 487-517. Academic Press, New York.

Franzén, Å. & T. Sensenbaugh 1983. Fine structure of the apical plate of the freshwater bryozoan *Plumatella fungosa* (Pallas) (Bryozoa: Phylactolaemata). – Zoomorphology **102**: 87-98.

Gerwerzhagen, A. 1913. Beiträge zur Kenntniss der Bryozoen. I. Das Nervensystem von *Cristatella mucedo* Cuv. – Z. wiss. Zool. **107**: 309-345, pls 12-14.

Gilmour, T.H.J. 1978. Ciliation and function of the food-collecting and waste-rejection organs of the lophophorates. – Can. J. Zool. **56**: 2142-2155.

Gordon, D.P. 1975. The resemblance of bryozoan gizzard teeth to 'annelid-like' setae. – Acta zool. (Stockh.) **56**: 283-289.

Gordon, D.P. 1977. The aging process in bryozoans. – In R.M. Woollacott & R.L. Zimmer (eds): Biology of Bryozoans, pp 335-376. Academic Press, New York.

Harmer, S.F. 1893. On the occurrence of embryonic fission in cyclostomatous Polyzoa. – Q. Jl microsc. Sci., N.S. **34**: 199-241, pls 22-24.

Hatschek, B. 1878. Studien über Entwicklungsgeschichte der Anneliden. – Arb. zool. Inst. Univ. Wien **1**: 277-404, pls 23-30.

Hatschek, B. 1891. Lehrbuch der Zoologie, 3. Lieferung (pp 305-432). – Gustav Fischer, Jena.

Hatschek, B. 1911. Das neue zoologische System. – Wilh. Engelmann, Leipzig.

Holborow, P.L., M.S. Laverack & V.C. Barber 1969. Cilia and other surface structures of the trochophore of *Harmothoë imbricata* (Polychaeta). – Z. Zellforsch. **98**: 246-261.

Hughes, R.L., Jr. & R.M. Woollacott 1979. Ultrastructure of potential photoreceptor organs in the larva of *Scrupocellaria bertholetti* (Bryozoa). – Zoomorphologie **91**: 225-234.

Hyman, L.H. 1959. Smaller Coelomate Groups. The Invertebrates, vol. 5. – McGraw-Hill, New York.

Jägersten, G. 1972. Evolution of the Metazoan Life Cycle. – Academic Press, London.

Jeuniaux, C. 1982. Composition chimique comparée des formations squelettiques chez les Lophophoriens et les Endoproctes. – Bull. Soc. zool. Fr. **107**: 233-249.

Kupelwieser, H. 1905. Untersuchungen über den feineren Bau und die Metamorphose des Cyphonautes. – Zoologica (Stuttg.) **47**: 1-50, 5 pls.

Lutaud, G. 1969. Le 'plexus' pariétal de Hiller et la coloration du système nerveux par le bleu de méthylène chez quelques Bryozoaires Chilostomes. – Z. Zellforsch. **99**: 302-314.

Lutaud, G, 1973. L'innervation du lophophore chez le Bryozoaire chilostome *Electra pilosa* (L.). – Z. Zellforsch. **140**: 217-234.

Lutaud, G, 1977. The bryozoan nervous system. – *In* R.M. Woollacott & R.L. Zimmer (eds): Biology of Bryozoans, pp 377-410. Academic Press, New York.

Lutaud, G. 1979. Étude ultrastructurale du 'plexus colonial' et recherche de connexions nerveuses interzoidiales chez le Bryozoaire chilostome *Electra pilosa* (Linné). – Cah. Biol. mar. **20**: 315-324, 4 pls.

Lutaud, G. 1982. Étude morphologique et ultrastructurale du funicule lacunaire chez le Bryozoaire Chilostome *Electra pilosa* (Linné). – Cah. Biol. mar. **23**: 71-81.

Lyke, E.B., C.G. Reed & R.M. Woollacott 1983. Origin of the cystid epidermis during metamorphosis of three species of gymnolaemate bryozoans. – Zoomorphology **102**: 99-110.

Malakhov, V.V. 1990. Description of the development of *Ascopodaria discreta* (Coloniales, Barentsiidae) and discussion of the Kamptozoa status in the animal kingdom. – Zool. Zh. **69**(10): 20-30. (In Russian, English summary. English translation available from: Library, Canadian Museum of Nature, P.O. Box 3443, Stn. D, Ottawa, Ontario, Canada K1P 6P4.)

Marcus, E. 1926a. Beobachtungen und Versuche an lebenden Meeresbryozoen. – Zool. Jb., Syst. **52**: 1-102, pls 1-2.

Marcus, E. 1926b. Beobachtungen und Versuche an lebenden Süsswasserbryozoen. – Zool. Jb., Syst. **52**: 279-350, pl. 6.

Marcus, E. 1934. Über *Lophopus crystallinus* (Pall.). – Zool. Jb., Anat. **58**: 501-606.

Marcus, E. 1938. Briozoarios marinhos brasileiros II. – Bolm Fac. Filos. Ciênc. Univ. S Paulo, Zool. **2**: 1-137, pls 1-29.

Marcus, E. 1939. Briozoários marinhos brasileiros III. – Bolm Fac. Filos. Ciênc. Univ. S Paulo, Zool. **3**: 111-299, pls 5-31.

Nielsen, C. 1970. On metamorphosis and ancestrula formation in cyclostomatous bryozoans. – Ophelia **7**: 217-256.

Nielsen, C. 1971. Entoproct life-cycles and the entoproct/ectoproct relationship. – Ophelia **9**: 209-341.

Nielsen, C. 1981. On morphology and reproduction of 'Hippodiplosia' insculpta and *Fenestrulina malusii* (Bryozoa, Cheilostomata). – Ophelia **20**: 91-125.

Nielsen, C. 1985. Animal phylogeny in the light of the trochaea theory. – Biol. J. Linn. Soc. **25**: 243-299.

Nielsen, C. 1987. Structure and function of metazoan ciliary bands and their phylogenetic significance. – Acta zool. (Stockh.) **68**: 205-262.

Nielsen, C. & K.J. Pedersen 1979. Cystid structure and protrusion of the polypide in *Crisia* (Bryozoa, Cyclostomata). – Acta zool. (Stockh.) **60**: 65-88.

Nitsche, H. 1869. Beiträge zur Kenntniss der Bryozoen. – Z. wiss. Zool. **20**: 1-36, pls 1-3.

Pace, R.M. 1906. On the early stages of the development of *Flustrella hispida* (Fabricius), and on the existence of a 'yolk nucleus' in the egg of this form. – Q. Jl microsc. Sci., N.S. **50**: 435-478, pls 22-25.

Patten, W. 1885. The embryology of *Patella*. – Arb. zool. Inst. Univ. Wien **6**: 149-174, pls 14-18.

Prouho, H. 1890. Recherches sur la larve de *Flustrella hispida* (Gray) structure et métamorphose. – Archs Zool. exp. gén., 2. sér. 8: 409-459, pls 22-24.

Prouho, H. 1892. Contribution à l'histoire des Bryozoaires. – Archs Zool. exp. gén., 2. ser., 10: 557-656, pls 23-30.

Reed, C.G. 1984. Larval attachment by eversion of the internal sac in the marine bryozoan *Bowerbankia gracilis* (Ctenostomata: Vesicularioidea): a muscle mediated morphogenetic movement. – Acta zool. (Stockh.) 65: 227-238.

Reed, C.G. 1988a. Organization of the nervous system and sensory organs in the larva of the marine bryozoan *Bowerbankia gracilis* (Ctenostomata: Vesiculariidae): functional significance of the apical disc and pyriform organ. – Acta zool. (Stockh.) 69: 177-194.

Reed, C.G. 1988b. The reproductive biology of the gymnolaemate bryozoan *Bowerbankia gracilis* (Ctenostomata: Vesiculariidae). – Ophelia 29: 1-23.

Reed, C.G. & R.A. Cloney 1982a. The larval morphology of the marine bryozoan *Bowerbankia gracilis* (Ctenostomata: Vesicularioidea). – Zoomorphology 100: 23-54.

Reed, C.G. & R.A. Cloney 1982b. The settlement and metamorphosis of the marine bryozoan *Bowerbankia gracilis* (Ctenostomata: Vesicularioidea). – Zoomorphology 100: 103-132.

Reed, C.G. & R.M. Woollacott 1982. Mechanisms of rapid morphogenetic movements in the metamorphosis of the bryozoan *Bugula neritina* (Cheilostomata, Cellularioidea): I. Attachment to the substratum. – J. Morph. 172: 335-348.

Reed, C.G. & R.M. Woollacott 1983. Mechanisms of rapid morphogenetic movements in the metamorphosis of the bryozoan *Bugula neritina* (Cheilostomata, Cellularioidea): II. The role of dynamic assemblages of microfilaments in the pallial epithelium. – J. Morphol. 177: 127-143.

Schäfer, P. 1991. Brutkammern der Stenolaemata (Bryozoa): Konstruktionsmorphologie und phylogenetische Bedeutung. – Cour. Forschungsinst. Senckenberg 136: 1-269.

Silén, L. 1945. The main features of the development of the ovum, embryo and ooecium in the ooeciferous Bryozoa Gymnolaemata. – Ark. Zool. 35A(17): 1-34.

Silén, L. 1966. On the fertilization problem in gymnolaematous Bryozoa. – Ophelia 3: 113-140.

Silén, L. 1972. Fertilization in Bryozoa. – Ophelia 10: 27-34.

Strathmann, R.R. 1973. Function of lateral cilia in suspension feeding of lophophorates (Brachiopoda, Phoronida, Ectoprocta). – Mar. Biol. (Berl.) 23: 129-136.

Strathmann, R.R. 1982. Cinefilms of particle capture by an induced local change of beat of lateral cilia of a bryozoan. – J. exp. mar. Biol. Ecol. 62: 225-236.

Strathmann, R.R. & L.R. McEdward 1986. Cyphonautes' ciliary sieve breaks a biological rule of inference. – Biol. Bull. Woods Hole 171: 754-760.

Stricker, S.A. 1987. Ultrastructure of the apical organ in a cyphonautes larva. – *In* J.R.P. Ross (ed.): Bryozoa: Present and Past, pp 261-268. Western Washington Univ., Bellingham.

Stricker, S.A. 1988. Metamorphosis of the marine bryozoan *Membranipora membranacea*: an ultrastructural study of rapid morphogenetic movements. – J. Morph. 196: 53-72.

Stricker, S.A. 1989. Settlement and metamorphosis of the marine bryozoan *Membranipora membranacea*. – Bull. mar. Sci. 45: 387-405.

Stricker, S.A., C.G. Reed & R.L. Zimmer 1988a. The cyphonautes larva of the marine bryozoan *Membranipora membranacea*. I. General morphology, body wall, and gut. – Can. J. Zool. 66: 368-383.

Stricker, S.A., C.G. Reed & R.L. Zimmer 1988b. The cyphonautes larva of the marine bryozoan *Membranipora membranacea*. II. Internal sac, musculature, and pyriform organ. – Can. J. Zool. 66: 384-398.

Ström, R. 1969. Sexual reproduction in a stoloniferous bryozoan, *Triticella koreni* (G.O. Sars). – Zool. Bidr. Upps. 38: 113-128, 4 pls.

Tavener-Smith, R. & A. Williams 1972. The secretion and structure of the skeleton of living and fossil Bryozoa. – Phil. Trans. R. Soc. B 264: 97-159, pls 6-30.

Taylor, P.D. 1981. Functional morphology and evolutionary significance of differing modes of tentacle eversions in marine bryozoans. – *In* G.P. Larwood & C. Nielsen (eds): Living and Fossil Bryozoa, pp 235-247. Olsen & Olsen, Fredensborg.

Taylor, P.D. 1985. Carboniferous and Permian species of the cyclostome bryozoan *Corynotrypa* Bassler, 1911 and their clonal propagation. – Bull. Br. Mus. nat. Hist., Geol. **38**: 359-372.

Taylor, P.D. 1990. Bioimmured ctenostomes from the Jurassic and the origin of the cheilostome Bryozoa. – Palaeontology **33**: 19-34.

Thompson, T.E. 1960. The development of *Neomenia carinata* Tullberg (Mollusca Aplacophora). – Proc. R. Soc. Lond. B **153**: 263-278.

Winston, J.E. 1978. Polypide morphology and feeding behavior in marine ectoprocts. – Bull. mar. Sci. **28**: 1-31.

Woollacott, R.M. & R.L. Zimmer 1971. Attachment and metamorphosis of the cheilo-ctenostome bryozoan *Bugula neritina* (Linné). – J. Morph. **134**: 351-382.

Woollacott, R.M. & R.L. Zimmer 1972a. Origin and fine structure of the brood chamber in *Bugula neritina* (Bryozoa). – Mar. Biol. (Berl.) **16**: 165-170.

Woollacott, R.M. & R.L. Zimmer 1972b. Fine structure of a potential photoreceptor organ in the larva of *Bugula neritina* (Bryozoa). – Z. Zellforsch. **123**: 458-469.

Woollacott, R.M. & R.L. Zimmer 1975. A simplified placenta-like system for the transport of extraembryonic nutrients during embryogenesis of *Bugula neritina* (Bryozoa). – J. Morph. **147**: 355-377.

Zimmer, R.L. 1973. Morphological and developmental affinities of the lophophorates. – *In* G.P. Larwood (ed.): Living and Fossil Bryozoa, pp 593-599. Academic Press, London.

Zimmer, R.L. & R.M. Woollacott 1977a. Structure and classification of gymnolaemate larvae. – *In* R.M. Woollacott & R.L. Zimmer (eds): Biology of Bryozoans, pp 57-89. Academic Press, New York.

Zimmer, R.L. & R.M. Woollacott 1977b. Metamorphosis, ancestrulae, and coloniality in bryozoan life cycles. – *In* R.M. Woollacott & R.L. Zimmer (eds): Biology of Bryozoans, pp 91-142. Academic Press, New York.

Zimmer & Woollacott 1989. Larval morphology of the bryozoan *Watersipora arcuata* (Cheilostomata: Ascophora). – J. Morph. **199**: 125-150.

PARENCHYMIA

Platyhelminthes and Nemertini are usually discussed together, because especially the platyhelminths appear to represent the most 'primitive' bilaterian type without a coelom, while the other bilaterians appear more specialized. Other theories regard the two phyla as secondarily simplified, because the acoelomate condition is regarded as secondary (see also Chapter 26).

It has been difficult to identify synapomorphies of the two phyla, but the ciliary bands of the Götte's larva (Platyhelminthes) and the pilidium (Nemertini) which have been studied in more detail show a number of similarities (Fig. 25.1). More detailed comparative studies are needed before more definitive statements can be made, but it appears that the general shape with a large circumferential band and a special band behind the mouth is identical, and no other metazoans show a similar structure. Compound cilia have only been observed in a pilidium larva, but compound cilia easily split up into the individual cilia at fixation, and those of the pilidium appear especially susceptible (Nielsen 1987), so it is possible that also the polyclad larvae actually have these structures. The function of the ciliary bands in particle collection has not been studied in sufficient detail in any of the groups, but the few observations show that the ciliary bands of both Götte's larva and pilidium beat towards the mouth except for an area behind the mouth which beats in the opposite direction (Lacalli 1982, Cantell 1969). The many different types of pilidium larvae make it difficult to recognize the ancestral larval type of the nemertines, but the above-mentioned similarities between one of the typical pilidium larvae and the Götte's larva of a polyclad turbellarian may indicate that this is close to the original type. According to the trochaea theory, these larvae should be modified trochophores, but it is difficult to recognize several of the trochophore characteristics. The downstream-collecting, preoral band of compound cilia develops, at least partially, on the $1a^2$-d^2-cells in pilidium larvae (Chapter 27) just as in trochophora larvae (Chapter 10), and it may represent a prototroch, and the ciliary band of the polyclad larvae may be homologous; further comparisons can only be guesses as long as both development, structure and function of the bands are so incompletely known.

Another characteristic shared by the two phyla is the lack of chitin (Jeuniaux 1982). The absence is most obvious in the platyhelminths, where the jaws, hooks and

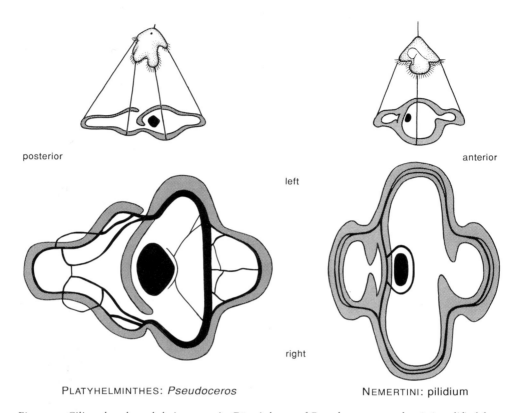

posterior

anterior

left

right

PLATYHELMINTHES: *Pseudoceros*

NEMERTINI: pilidium

Fig. 25.1. Ciliary bands and their nerves in Götte's larva of *Pseudoceros canadensis* (modified from Lacalli 1982) and in an unidentified pilidium larva from Friday Harbor (modified from Lacalli 1985). The larvae are seen from the underside with the lateral lobes spread out; the ciliary bands are heavily shaded and the mouth and nerves black.

copulatory structures are consolidated parts of the basement membranes or intracellular structures as opposed to the chitinous cuticular structures of for example annelids; also the egg shells are quinone tanned proteins without any chitin (Chapter 26). The nemertines lack all such structures, and the stylets of the hoplonemertine proboscis are intracellular structures (Chapter 27). The lack of a structure or process is of course a weak phylogenetic character, but the work of Jeuniaux (1982) shows that all other spiralians, except the sipunculans, have the ability to synthesize chitin, so this may be a synapomorphy after all.

The simple cuticle with only a glycocalyx between the microvilli is probably a symplesiomorphy (Rieger 1984).

There are no traces of coelomic cavities in the platyhelminths, but both the rhynchocoel and the blood vessels of the nemertines fit the usual definition of coeloms, i.e. fluid-filled spaces surrounded by epithelial mesoderm; both coeloms are formed by schizocoely, i.e. the hollowing out of a coelomic space in a solid mass of mesodermal cells (Chapter 27). However, the phylogenetically interesting question is whether these coelomic sacs are homologous with coelomic sacs in other spiralian phyla, and this is much more uncertain. The lateral blood vessels develop in much the same way as the coelomic spaces in a polychaete (Turbeville 1986), and this could speak for homology. The coelomic sacs of the leeches are fused and narrowed to canals functioning as blood vessels, but there is no trace of segmentation in the nemertines, neither in the adult anatomy nor during ontogeny. In the Teloblastica, the gonads are intimately associated with the coelom with the gametes passing through the coelom and the coelomoducts, but the nemertine gonads are separate and there are no 'coelomoducts'. The unpaired, dorsal rhynchocoel is even more difficult to homologize with other coeloms. So although a homology with other coeloms cannot be rejected outright, I prefer to interpret the coelomic spaces in the nemertines as apomorphies of the phylum.

The cleavage pattern shows that the parenchymians belong to the Spiralia, and this is further supported by the morphology of the apical organ, especially of the

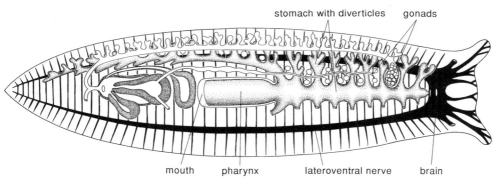

PLATYHELMINTHES: generalized turbellarian

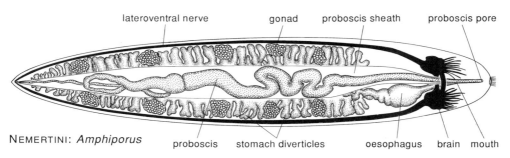

NEMERTINI: *Amphiporus*

Fig. 25.2. General body plans of a platyhelminth: a freshwater triclad (redrawn from Bresslau 1928-31) and a nemertine: *Amphiporus pulcher* (redrawn from Bürger 1895). The guts are shaded and the nervous systems black; the left half of the gut has been omitted in the drawing of the triclad to reveal the nervous system.

213

nemertines, and the presence of muscles between the apical organ and the region of the mouth.

The evolution of the parenchymians from the spiralian ancestor may be indicated by the morphology of the larval ciliary bands and the development of the central nervous systems. Both Götte's larva and the pilidium have extremely narrow postoral areas (Fig. 25.1), and the larvae could perhaps be interpreted as trochophores with strongly reduced hypospheres with no trace of the ventral nervous system and the anus. The adults show no trace of a ventral nervous system derived from fused blastopore lips; the lateral nerves are simple extensions from the brain (Fig. 25.2). There is no nervous concentration surrounding the oesophagus, because the ventral component of the central nervous system is absent, and the mouth may therefore occupy any position on the ventral side while the brain is always at the anterior end.

With this interpretation of the platyhelminths the following set of synapomorphies can be listed: larvae with strongly reduced hyposphere and completely reduced ventral nervous system and anus; adults only with apical nervous system consisting of a brain and various longitudinal extensions; lack of chitin and chitinase.

The position of the Parenchymia within the Spiralia is more difficult to ascertain. The lack of teloblasts indicates that they cannot be included in the Teloblastica, and there is no obvious indication of a sister-group relationship with the Bryozoa either (see Chapter 22). I have chosen to arrange the three groups Teloblastica, Parenchymia and Bryozoa in an unresolved trichotomy (Fig. 11.4).

References

Bresslau, E. 1928-1931. Turbellaria. – Handbuch der Zoologie, 2. Band, 1 Hälfte (2): 52-320. Walter de Gruyter, Berlin.

Bürger, O. 1895. Die Nemertinen des Golfes von Neapel. – Fauna Flora Golf. Neapel 22: 1-743, 31 pls.

Cantell, C.-E. 1969. Morphology, development, and biology of the pilidium larvae (Nemertini) from the Swedish west coast. – Zool. Bidr. Upps. 38: 61-112, 6 pls.

Jeuniaux, C. 1982. La chitine dans le règne animal. – Bull. Soc. zool. Fr. 107: 363-386.

Lacalli, T.C. 1982. The nervous system and ciliary band of Müller's larva. – Proc. R. Soc. Lond. B 217: 37-58, 8 pls.

Lacalli, T.C. 1985. The nervous system of a pilidium larva: evidence from electron microscopical reconstructions. – Can. J. Zool. 63: 1909-1916.

Nielsen, C. 1987. Structure and function of metazoan ciliary bands and their phylogenetic significance. – Acta zool. (Stockh.) 68: 205-262.

Rieger, R.M. 1984. Evolution of the cuticle in the lower Eumetazoa, – *In* J. Bereiter-Hahn, A.G. Maltoltsy & K.S. Richards (eds): Biology of the Integument, vol. 1, pp 389-399. Springer, Berlin.

Turbeville, J.M. 1986. An ultrastructural analysis of coelomogenesis in the hoplonemertine *Prosorhynchus americanus* and the polychaete *Magelona* sp. – J. Morph. 187: 51-60.

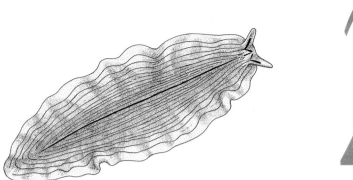

Phylum PLATYHELMINTHES

The flatworms comprise the mainly free-living, aquatic turbellarians and the parasitic, aquatic or terrestrial flukes and tapeworms; about 20 000 species have been described. The fossil records are dubious (Conway Morris 1985). The delimitation of the phylum is not much in question, the only uncertain members being *Xenoturbella* and the Gnathostomulida; the last-mentioned group is by many authors regarded as the sister group of the Platyhelminthes (Ax 1987, Ehlers 1985b), but it has here been included in the annelids, mainly on the basis of its chitinous cuticular structures (Chapter 17). *Xenoturbella bocki* is a very problematic organism; it is by some authors treated as a primitive turbellarian (Odening 1984), but this is founded only on the general shape and organization of the 'worm' – synapomorphies with the other platyhelminths have not been identified. A short discussion of this organism is given in Chapter 55.

The evolution of the platyhelminths has been studied by cladistic methods by a number of authors and there seems to be agreement about a phylogenetic tree resembling that presented by Ehlers (1985a, b) and Ax (1987) (see Fig. 26.1), although not all authors agree about the exact position of the Neodermata (see for example Rohde 1988). Many textbooks retain the economically important parasitic flukes (Digenea and Monogenea) and tapeworms (Cestoda) as classes parallel to Turbellaria, but this is mostly for practical reasons. Ehlers (1985b) included the two problematic groups Lecithoepitheliata and Prolecithophora in his cladogram at a level between Polycladida and Seriata, and Rieger *et al.* (1991) found some evidence to support this. The groups Catenulida, Acoelomorpha and Rhabditophora appear to be very distinct (Ehlers 1986, see below). The sister-group relationships between the three groups indicated in Fig. 26.1 was questioned by Smith, Tyler & Rieger (1986), but Rieger *et al.* (1991, p. 126) 'agree that the Platyhelminthes may indeed be a monophyletic group, (but) more evidence is needed ... to support such a conclusion and to trace the ancestry ... of the turbellarian clades.' For a discussion of the phylogenetic position of the Platyhelminthes it must therefore be sufficient to concentrate on the characters of the free-living 'lower' turbellarians, which probably exhibit the most ancestral characters of the phylum.

Chapter vignette: The polyclad *Prostheceraeus vittatus*. (Redrawn from Lang 1884.)

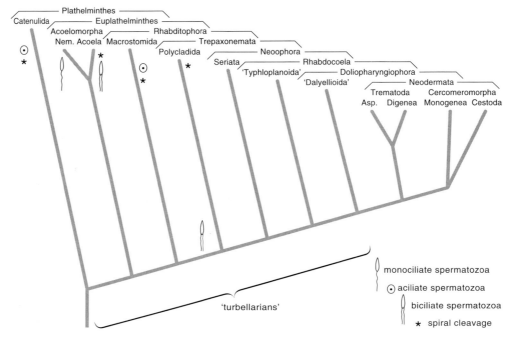

Fig. 26.1. The phylogeny of the Platyhelminthes. Nem. = Nemertodermata; Asp. = Aspidobothrii.
(Modified from Ehlers 1985a.)

The free-living flatworms are elongate, often dorso-ventrally flattened, mostly ciliated, with a ventral mouth leading to a gut without an anus. The parasitic flukes and tapeworms lack ciliated ectodermal cells except in early developmental stages.

The epidermis of the turbellarians is generally a single layer of multiciliate cells; a cuticle of collagenous fibrils like that observed in many other bilaterians is absent, but a more or less well-developed glycocalyx is present between microvilli in many groups (Rieger *et al.* 1991). There are apparently no true cuticular structures consisting of consolidated cuticle; all the jaw-like structures in the pharynx and the male copulatory stylets are specializations of the basement membrane (Rieger *et al.* 1991). This may be coupled with the complete absence of chitin in the platyhelminths (Jeuniaux 1982). Several types of mostly unicellular glands can be recognized, and the rhabdite glands have often been considered characteristic of turbellarians. There are, however, various types of rhabdites, and they may not be homologous. One special type, the lamellate rhabdite, is a unique characteristic of the group Rhabditophora (Ehlers 1985b, Rieger *et al.* 1991); the Neodermata lack these structures, but this is probably connected with their loss of the primary ectoderm. The structure and distribution of the other gland types are as yet so incompletely known, that their phylogenetic significance cannot be evaluated. The epidermis rests on a basement membrane except in acoels and nemertodermatids, and probably in some catenulids (Pedersen 1991). Some epidermal cells are 'insunk', i.e. the nucleus-bearing part of the cell extends below the layer of muscles; this is found in most of the supposedly primitive types which lack a well-developed basement membrane, but also in some of the groups with a normal basement membrane. The Neodermata have ciliated epithelial cells in the first larval stage, but these cells are cast off when the larva enters the first host and the later stages have an epidermis without nuclei outside the basement membrane (neodermis); all the hooks and other hard structures found in these parasitic forms are intraneodermal (Ehlers 1985b). Myoepithelial cells have not been observed.

Most of the turbellarians have a mouth which opens into a muscular, ectodermal pharynx; only a few species lack a mouth, and there is no pharynx in most of the acoelomorphs. The pharynx is either a simple tube surrounded by muscle cells or the epidermis is folded back around the muscles so that a more complicated, protrusible structure is formed; the uncomplicated type comprises several subtypes which are probably not homologous (Doe 1981, Rieger et al. 1991). The pharynx opens into the anterior end of the stomach in some forms, but a more posterior position of the mouth is characteristic of most groups.

The stomach is a simple tube of ciliated endodermal cells in catenulids and in members of several of the other groups. Lateral diverticula of many shapes characterize various groups, as reflected by names such as triclads and polyclads, and many of the more specialized groups lack the cilia. The gut is lacking completely in the cestodes and in the parasitic 'dalyellioid' family Fecampiidae (Rieger et al. 1991). An anus is normally absent, and the examples of one or several openings from gut diverticula to the outside must be seen as secondary specializations (Ehlers 1985b).

The internal cell mass of the acoels is not differentiated into an endodermal gut surrounded by mesoderm (Smith & Tyler 1985). Two zones can be recognized, a central mass involved in digestion, and a peripheral zone with muscles, nerves, gonads, and other cell types. The central zone consists of a central syncytium (at least in most species) surrounded by a layer of wrapping cells; the central syncytium is formed through fusion of wrapping cells and is in some species only formed after the ingestion of food and shed when the digestion is complete. Only *Paratomella* has a more permanent gut lumen, but the cells are not joined in an epithelium.

The nervous system comprises an apical, subepidermal brain, a number of longitudinal nerves with transverse commissures, and various nerve plexuses. Some of the peripheral nerve cords or nerves are basiepithelial and some subepidermal, and there is a bewildering variation in position and complexity even within the groups (Ehlers 1985b); Rieger et al. (1991, p. 34) conclude that 'the nervous system is now being viewed as showing a variable mixture of specialized and unspecialized, advanced and "primitive" components'. With this conclusion in mind it is understandable that the turbellarian nervous systems have been interpreted in many different ways (see below).

The mesoderm is a compact mass between ectoderm and endoderm and comprises muscle cells, neoblasts, apolar mesenchymal cells (Rieger 1986), and ample extracellular material in most types; only the acoels lack the extracellular matrix (Rieger et al. 1991). The structure is in many groups complicated by cell bodies which have sunk in from the ectoderm. The muscle cells are arranged in thin longitudinal and circular layers in the smallest forms, which glide on their cilia, while the larger forms have additional diagonal layers. The structure generally resembles smooth muscle, but various indications of striation have been reported in studies of the ultrastructure, and typical cross-striated muscle cells are found for example in the cercarian tail (Fried & Haseb 1991).

Hermaphroditism is the rule among the platyhelminths; the gonads are groups of cells without a special wall in catenulids and acoels (Ehlers 1985b), while the other flatworms have well-defined, sac-shaped gonads. The genital organs are quite complicated in most groups. Fertilization is always internal and there is a copulatory structure, which in many groups has hardened parts which are either intracellular or specializations of the basement membrane (Rieger *et al.* 1991). The copulation is a hypodermal impregnation or simple deposition of sperm or spermatophores on the partner in several groups, but sperm transfer through the female gonoducts is the rule in the more specialized forms.

Spermatozoa of the filiform type with one normal cilium with a $9 \times 2 + 2$ axoneme are found only in the nemertodermatids; the acoels have various types of biflagellate spermatozoa with two, one or no central filaments in the axonemes (Hendelberg 1986). The catenulids have very aberrant, aciliate spermatozoa with a pycnotic nucleus (Ehlers 1985b). The macrostomid sperm lacks cilia and the remaining platyhelminths, the group Trepaxonemata (Fig. 26.1) have highly characteristic filiform spermatozoa with two axonemes of a special $9 \times 2 + 1$ type and a unique intercentriolar body (Ehlers 1985b, Sopott-Ehlers 1989).

The female gonads are rather uncomplicated in catenulids, acoelomorphs, macrostomids, and polyclads (collectively called 'archoophorans'), which have endolecithal eggs. The neoophorans have yolk glands and the eggs are deposited in capsules together with a number of yolk cells. The fertilized eggs are released through rupture of the body wall in catenulids and acoelomorphs, but all the other groups have more or less complicated ducts (Ehlers 1985b, Rieger *et al.* 1991).

A normal spiral cleavage with quartets is found in the polyclads. Surface (1908) studied the acotylean *Hoploplana* and Anderson (1977) *Notoplana*, which have pelagic larvae, and Kato (1940) studied both cotylean and acotylean species comprising both species with direct development and species with Götte's larva and Müller's larva (see below). The cell-lineage appears to be quite similar in all species. The first two cleavages pass through the primary axis of the egg and the four resulting blastomers are of almost equal size, although the cells B and D are sometimes reported to be slightly larger than A and C. The third cleavage gives larger macromeres and smaller micromeres in most species. The 64-cell stage is highly characteristic with four quite small blastoporal 'macromeres' 4A-D and very large 4a-d-cells. These cells become covered by the cells of the micromere calotte in an epibolic gastrulation, but the 4A-D and 4a-c-cells disintegrate into yolk granules and the whole endoderm originates from descendants of 4d. According to Kato (1940) this cell divides horizontally, and each of these cells then gives rise to endoderm and mesoderm. Additional mesoderm (ectomesoderm) comes from the cells 2a-d. There is no sign of teloblastic growth or coelomic cavities in the mesoderm. The endodermal cells become arranged around a small gut, and a stomodaeum develops in front of the 'blastopore'. The brain develops from a small invagination of apical cells.

Spiral cleavage with quartets has been reported also from catenulids and macrostomids (Bogomolow, cited from Reisinger *et al.* 1974, Ax & Borkott 1969).

A modified spiral cleavage pattern is found in the acoels (Bresslau 1928-1931, Apelt 1969). It resembles a normal spiral cleavage with quartets, but lacking one of the two first meridional cleavages, so that duets are formed instead of quartets. The gastrulation and mesoderm formation resemble those of the polyclads, but the further development has not been followed.

A few experimental investigations on the development of polyclads and acoels show that the cleavages with quartets or duets resemble those of the 'higher' spiralians, but that the determination is less strong (Boyer 1986, 1987, 1989). The acoel *Childia* is able to form normal larvae when one blastomere of the 2-cell stage is deleted (Boyer 1971).

The most 'primitive' of the neoophorans, such as the lecithoepitheliate *Xenoprorhynchus* (Reisinger *et al.* 1974), have spiral cleavage with quartets, but the later development is influenced by the presence of the yolk cells. Eight micromeres, 2a-d and 3a-d, move to the periphery of the embryo and cover the yolk cells situated at the blastoporal pole, forming a very thin embryonic 'covering membrane'. This cell layer becomes resorbed when the yolk has been incorporated in the embryo. The 4D-cell gives rise to the endoderm and 4d divides laterally and gives rise to a pair of mesoblasts, which were stated to be mesoteloblasts; however, the illustrations show cells of equal size, and it is only stated that the cells divide – not that one large cell gives off a series of smaller cells, as the teloblasts in both mesoderm and ectoderm of for example annelids (Chapter 17). Members of the Proseriata (*Minona, Monocelis*; Reisinger *et al.* 1974) show increasing volumes of yolk cells and increasingly modified embryology, and the development of the parasitic groups is highly specialized (Ehlers 1985b).

The development is direct in most of the turbellarian groups, but planktonic, ciliated larvae are found in catenulids, polyclads and some of the parasitic groups. The only catenulid larva described is the 'Luther's larva' of the limnic *Rhyncoscolex* (Reisinger 1924); it resembles the adults, but has a stronger ciliation and a different statocyst; it should probably be regarded as a specialized juvenile rather than a true larva. Two types of polyclad larvae have been named, Götte's larva (Fig. 25.1), with four rather broad lateral lobes, and Müller's larva (Fig. 26.2), with eight usually more cylindrical lobes. Götte's larva has been found only in the Acotylea, but Müller's larva is known also from the Cotylea (Ruppert 1978). The larvae are completely ciliated with an apical tuft and a band of longer and more dense ciliation around the body following the sides of the lobes (Lacalli 1982, 1988); compound cilia have not been observed. The larvae swim with the ciliary band, which shows metachronal waves, and it appears that at least some of the larvae are planktotrophic (Ruppert 1978, Ballarin & Galleni 1987), but the method of particle collection has not been studied. There is no sign of an anus at any stage. The ciliary band is usually described as a ring, but Lacalli (1982) found that the ring is broken between the two lobes at the left side and that the anterior part of the band continues across the ventral side behind the mouth to the opposite side (Fig. 25.1). The band is two to eight cells wide and all the cells are multiciliate with the ciliary beat towards the mouth, except the median cell behind the mouth (the suboral rejectory cell), which has cilia beating away from the mouth (Ruppert 1978, Lacalli 1982, 1988). There is a system of intraepithelial nerves with monociliate sensory cells in or along the band. The larvae have an apical brain with anterior and posterior pairs of subepidermal nerves to the ciliated band, but there are apparently only few axons traversing the basement membrane to the basiepithelial nerve of the ciliary band (Lacalli 1982, 1983). The larvae have a pair of branched protonephridia (Ruppert 1978). The older larvae

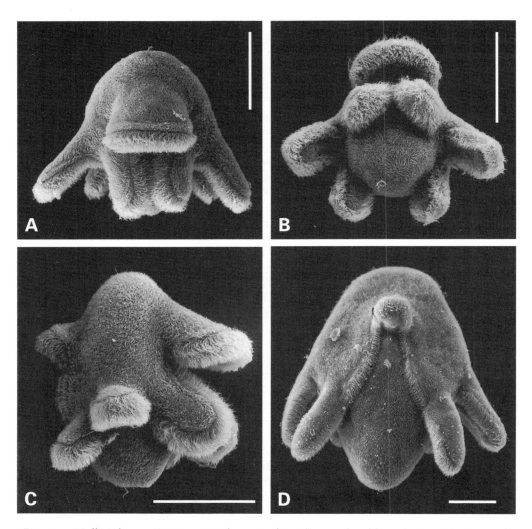

Fig. 26.2. Müller's larvae (SEM). – A-C, larvae in frontal, ventral and lateral views (plankton, off Bamfield Marine Station, Vancouver Island, Canada, August 1988). – D, larva in dorsal view (plankton, off Nassau, The Bahamas, September 1990). Scale bars: 100 µm.

gradually lose the lobes and the ciliated bands and transform into small adults (Lang 1884). The apical ganglion becomes incorporated in the adult brain, but the fate of the ciliary nerves is unknown. Dawydoff (1940) described a number of polyclad larvae resembling gigantic Müller's larvae with a general ciliation but without the prominent ciliary bands; this only shows that new investigations may reveal other types of larvae, which may add to our understanding of the development of the platyhelminths.

The parasitic forms have lecithotrophic, often ciliated larvae which shed the ciliated cells and the following life cycles comprise stages with asexual reproduction in some groups.

The three groups Catenulida, Acoelomorpha and Rhabditophora are very distinct: the catenulids have very peculiar aciliate spermatozoa with a pycnotic nucleus, a special, unpaired, biciliate protonephridium, and few cilia per cell; the acoelomorphs have a modified gut, a special type of epidermal cilia with a characteristic rootlet system, and no protonephridia; the rhabditophorans are characterized not only by the presence of rhabdites, but also by a special type of duo-gland adhesive organs, multi- (and mono-)ciliate epidermal receptors (the catenulids and acoelomorphs have only monociliate receptors), protonephridia with four to several hundred cilia (Rieger *et al.* 1991). It is difficult to be definite about the interrelationships of these three groups, and the interpretation is much dependent on the larger phylogenetic scheme into which the platyhelminths shall be fitted.

As pointed out by Rieger *et al.* (1991), it is difficult to identify apomorphies which characterize the platyhelminths, but there appears, nevertheless, to be general agreement about the monophyly of the group. The lack of an anus may be an apomorphy, but this depends on the the phylogenetic position assigned to the Platyhelminthes within the Bilateria. The ancestral platyhelminth probably combined the following characters: internal fertilization, filiform spermatozoa with one cilium, entolecithal eggs, spiral cleavage, bilateral symmetry, multiciliate cells forming both ectoderm and a sac-shaped gut, no anus, a pair of protonephridial excretory organs comprising more than one terminal cell with two or more cilia, compact mesoderm with only little mesenchyme and no trace of segmentation or coelomic sacs, and an apical brain (see also Ehlers 1985b, Ax 1987). The presence of a planktotrophic larva is suggested by the larvae of the polyclads, but some theories regard these larvae as secondarily evolved dispersal stages. Many of the characters are probably symplesiomorphies. Rhabdites have sometimes been regarded as homologous with cnidocysts, but the structures are quite dissimilar, and lamellate rhabdites are only found in the 'higher' turbellarians (Ehlers 1985b). Reisinger (1925, 1972) emphasized the supposedly primitive, 'orthogonal' type of nervous system observed for example in *Bothrioplana*, but as mentioned above, the nervous system shows so much variation that almost any interpretation can be supported by existing examples. Much more knowledge about both development, structure and function of nervous systems of many groups will be needed before a meaningful evaluation can be carried out.

The acoels have often been regarded as the most primitive type of bilaterians, i.e. neotenic planula larvae (Hyman 1951, Salvini-Plawen 1978), but the fact that many of the cnidarian larvae have a normal gut and that none of the planulae have a syncytial endoderm makes this theory very improbable. The alternative idea of re-garding the turbellarians with a gut as derived from planula larvae with a gut has to my knowledge not been prominent in the literature, but its consequences are the same as those of for example Ax's theory (discussed below).

Another well-known idea is connected with the enterocoel theory (for example Remane 1963) or its more recent form, the archicoelomate concept (Siewing 1976, 1980) which regards the bilaterian ancestor as a bilaterogastrea, a creeping organism with three pairs of gastral pouches which become coeloms; all acoelomate organisms are accordingly regarded as 'reduced'. It should be pointed out here that the platy-helminths show no signs of coelomic pouches or metanephridia, which are usually associated with the coeloms, and no signs of trimery. The interpretation of the platy-helminths as 'reduced' coelomates is based solely on speculations resulting from a phylogenetic theory.

Several authors, in modern times notably Ax (1987), have interpreted the Platy-helminthes (+ Gnathostomulida) as the sister group of the 'Eubilateria', the two groups together constituting the group Bilateria (the sister group of the Cnidaria), thus regarding the lack of an anus as a plesiomorphy. A consequence of this interpre-tation is that the spiral cleavage must be an ancestral bilaterian character (or that the cleavage type evolved twice), and I regard this as very improbable; the spiral cleavage with quartets is a very specialized pattern with rather well-defined fates for most of the resulting cells (see also Chapter 11). The cleavage of the deuterostomes shows no trace of a spiral pattern, and the blastomeres are capable of regulation in all stages of the development. Ax (1989) reached a similar conclusion and regarded the Spiralia as a monophyletic group with the Plathelminthomorpha, primarily without an anus, as the sister group of the Euspiralia, with an anus; the position of the aschelminths was not considered.

It should be clear that the phylogenetic position of this phylum cannot be de-duced directly by comparisons with other phyla in the hope of finding decisive synapomorphies defining sister-group relationships (see however the discussion of the relationships with the nemertines in Chapter 25). Understanding the position of the platyhelminths is intimately connected with the choice of a phylogenetic theory. The phylogenetic position of Platyhelminthes and Nemertini in the framework of this book is discussed in Chapter 25.

Interesting subjects for future research

1. Development of the nervous system in a number of types.
2. Cell-lineage studies of Götte's and Müller's larvae.

References

Anderson, D.T. 1977. The embryonic and larval development of the turbellarian *Notoplana australis* (Schmarda, 1859) (Polycladida: Leptoplanidae). – Aust. J. mar. Freshwat. Res. **28**: 303-310.

Apelt, G. 1969. Fortpflanzungsbiologie, Entwicklungszyklen und vergleichende Frühentwicklung acoeler Turbellarien. – Mar. Biol. (Berl.) **4**: 267-325.

Ax, P. 1987. The Phylogenetic System. The Systematization of Organisms on the Basis of their Phylogenesis. – John Wiley, Chichester.

Ax, P. 1989. Basic phylogenetic systematization of the Metazoa. – In B. Fernholm, K. Bremer & H. Jörnvall (eds): The Hierarchy of Life, pp 229-245. Excerpta Medica/Elsevier, Amsterdam.

Ax, P. & H. Borkott 1969. Organisation und Fortpflanzung von Macrostomum romanicum (Turbellaria, Macrostomida). – Zool. Anz., Suppl. 31: 344-347.

Ballarin, L. & L. Galleni 1987. Evidence for planctonic feeding in Götte's larva of Stylochus mediterraneus (Turbellaria - Polycladida). – Boll. Zool. 54: 83-85.

Boyer, B.C. 1971. Regulative development in a spiralian embryo as shown by cell deletion experiments on the acoel, Childia. – J. exp. Zool. 176: 97-105.

Boyer, B.C. 1986. Determinative development in the polyclad turbellarian, Hoploplana inquilina. – Int. J. Invert. Reprod. Dev. 9: 243-251.

Boyer, B.C. 1987. Development of in vitro fertilized embryos of the polyclad flatworm, Hoploplana inquilina, following blastomere separation and deletion. - Roux's Arch. dev. Biol. 196: 158-164.

Boyer, B.C. 1989. The role of the first quartet micromeres in the development of the polyclad Hoploplana inquilina. – Biol. Bull. Woods Hole 177: 338-343.

Bresslau, E. 1928-31. Turbellaria. – Handbuch der Zoologie, 2. Band, 1. Hälfte (2), pp 52-320. Walter de Gruyter, Berlin.

Conway Morris, S. 1985. Non-skeletized lower invertebrate fossils: a review. – In S. Conway Morris, J.D. George, R. Gibson & H.M. Platt (eds): The Origins and Relationships of Lower Invertebrates, pp 343-359. Oxford Univ. Press, Oxford.

Dawydoff, C. 1940. Les formes larvaires de polyclades et némertes du plancton indochinois. – Bull. biol. Fr. Belg. 74: 443-496.

Doe, D.A. 1981. Comparative ultrastructure of the pharynx simplex in Turbellaria. – Zoomorphology 97: 133-193.

Ehlers, U. 1985a. Phylogenetic relationships within the Platyhelminthes. – In S. Conway Morris, J.D. George, R. Gibson & H.M. Platt (eds): The Origins and Relationships of Lower Invertebrates, pp 143-158. Oxford Univ. Press, Oxford.

Ehlers, U. 1985b. Das phylogenetische System der Plathelminthes. – Gustav Fischer, Stuttgart.

Ehlers, U. 1986. Comments on a phylogenetic system of the Platyhelminthes. – Hydrobiologia 132: 1-12.

Fried, B. & M.A. Haseb 1991. Platyhelminthes: Aspidogastrea, Monogenea, and Digenea. – In F.W. Harrison (ed.): Microscopic Anatomy of Invertebrates, vol. 3, pp 141-209. Wiley-Liss, New York.

Hendelberg, J. 1986. The phylogenetic significance of sperm morphology in the Platyhelminthes. – Hydrobiologia 132: 53-58.

Hyman, L.H. 1951. Platyhelminthes and Rhynchocoela. The acoelomate Bilateria. The Invertebrates, vol. 5. – McGraw-Hill, New York.

Jeuniaux, C. 1982. La chitine dans le règne animal. – Bull. Soc. zool. Fr. 107: 363-386.

Kato, K. 1940. On the development of some Japanese polyclads. – Jap. J. Zool. 8: 537-573, pls 50-60.

Lacalli, T.C. 1982. The nervous system and ciliary band of Müller's larva. – Proc. R. Soc. Lond. B 217: 37-58, 8 pls.

Lacalli, T.C. 1983. The brain and central nervous system of Müller's larva. – Can. J. Zool. 61: 39-51.

Lacalli, T.C. 1988. The suboral complex in the Müller's larva of Pseudoceros canadensis (Platyhelminthes, Polycladida). – Can. J. Zool. 66: 1893-1895.

Lang, A. 1884. Die Polycladen (Seeplanarien) des Golfes von Neapel. – Fauna Flora Golf. Neapel 11: 1-688, 39 pls.

Odening, K. 1984. Stamm Plathelminthes, Plattwürmer. – In H.-E. Gruner (ed.): A. Kaestner's Lehrbuch der Speziellen Zoologie (4. ed.), 1. Band, 2. Teil, pp 341-442. Gustav Fischer, Stuttgart.

Pedersen, K.J. 1991. Invited review: Structure and composition of basement membranes and other basal matrix systems in selected invertebrates. – Acta zool. (Stockh.) 72: 181-201.

Reisinger; E. 1924. Die Gattung Rhynchoscolex. – Z. Morph. Ökol. Tiere 1: 1-37, pls 1-2.

Reisinger, E. 1925. Untersuchungen am Nervensystem der Bothrioplana semperi Braun. – Z. Morph. Ökol. Tiere 5: 119-149.

Reisinger, E. 1972. Die Evolution des Orthogons der Spiralier und das Archicoelomatenproblem. – Z. zool. Syst. Evolutionsforsch. 10: 1-43.

Reisinger, E., I. Cichocki, R. Erlach & T. Szyskowitz 1974. Ontogenetische Studien an Turbellarien: ein Beitrag zur Evolution der Dotterverarbeitung im ektolecitalen Ei. – Z. zool. Syst. Evolutionsforsch. 12: 161-195 & 241-278.

Remane, A. 1963. The enterocelic origin of the celom. – In E.C. Dougherty, Z.N. Brown, E.D. Hanson & W.D. Hartman (eds): The Lower Metazoa, pp 78-90. Univ. California Press, Berkeley.

Rieger, R.M. 1986. Über den Ursprung der Bilateria: die Bedeutung der Ultrastrukturforschung für ein neues Verstehen der Metazoenevolution. – Verh. dt. zool. Ges. 79: 31-50.

Rieger, R.M., S. Tyler, J.P.S. Smith III & G.E. Rieger 1991. Platyhelminthes: Turbellaria. – In F.W. Harrison (ed.): Microscopic Anatomy of Invertebrates, vol. 3, pp 7-140. Wiley-Liss, New York.

Rohde, K. 1988. Phylogenetic relationship of free-living and parasitic Platyhelminthes on the basis of ultrastructural evidence. – Fortschr. Zool. 36: 353-357.

Ruppert, E.E. 1978. A review of metamorphosis of turbellarian larvae. – In F.-S. Chia & M.E. Rice (eds): Settlement and Metamorphosis of Marine Invertebrate Larvae, pp 65-81. Elsevier, New York.

Salvini-Plawen, L.v. 1978. On the origin and evolution of the lower Metazoa. – Z. zool. Syst. Evolutionsforsch. 16: 40-88.

Siewing, R. 1976. Probleme und neuere Erkenntnisse in der Grossystematik der Wirbellosen. – Verh. dt. zool. Ges. 1976: 59-83.

Siewing, R. 1980. Das Archicoelomatenkonzept. – Zool. Jb., Anat. 103: 439-482.

Smith, J.P.S., III & S. Tyler 1985. The acoel turbellarians: kingpins of metazoan evolution or a specialized offshot? – In S. Conway Morris, J.D. George, R. Gibson & H.M. Platt (eds): The Origins and Relationships of Lower Invertebrates, pp 123-142. Oxford Univ. Press, Oxford.

Smith, J.P.S., III, S. Tyler & R.M. Rieger 1986. Is the Turbellaria polyphyletic? – Hydrobiologia 132: 13-21.

Sopott-Ehlers, B. 1989. On spermiogenesis in Invenusta aestus (Plathelminthes, Proseriata). An ultrastructural study with implications for plathelminth phylogeny. – Zoomorphology 109: 145-152.

Surface, F.M. 1908. The early development of a polyclad, Planocera inquilina Wh. – Proc. Acad. nat. Sci. Philad. 59: 514-559, pls 35-40.

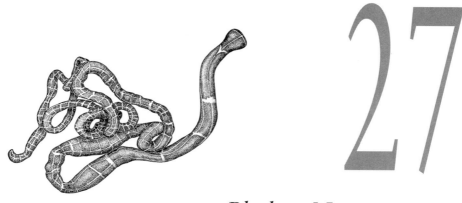

Phylum NEMERTINI

The ribbon worms are a phylum of about 900 described species; most species are benthic, marine, but several are pelagic; some groups have entered freshwater and a few are found in moist, terrestrial habitats (Gibson 1972). There is no reliable fossil record. Most nemertines are cylindrical to slightly flattened, but especially some of the pelagic forms are very flat. All species (except *Arhynchonemertes*, see below) can be recognized by the presence of an eversible proboscis, which is not a part of the alimentary canal, and which contains a coelomic cavity functioning as a hydrostatic skeleton at eversion. Two classes are recognized: Anopla (Palaeonemertini + Heteronemertini), with the mouth situated below or behind the cephalic ganglia and a separate, anterior proboscis pore, and Enopla (Hoplonemertini + Bdellonemertini), with the mouth in front of the ganglia, usually united with the proboscis pore; most of the enoplans have stylets on the tip of the everted proboscis (Gibson 1972).

The ectoderm (Turbeville 1991) is pseudostratified with multiciliate cells and several glands and sense organs. The ciliated cells have a border of branched microvilli, sometimes with a glycocalyx; there is never a cuticle of cross-arranged collagenous fibrils as that observed in most of the other spiralian phyla, and chitinous structures have not been reported either. Ectodermal cells may contain vacuoles with various types of hook-shaped or ovoid bodies which in some cases are calcified and in other cases non-calcified (Stricker & Cavey 1988). The ectoderm is resting on a basement membrane which is traversed by myofilament-containing processes from mesodermal muscles; myoepithelial cells have not been reported.

The gut is a straight ciliated epithelial tube extending from the anterior part of the ventral side to the posterior end. The foregut, which may be differentiated into a buccal cavity, an oesophagus and a stomach, is ectodermally derived like the rectum and the rhynchodaeum (see below). The midgut has various diverticula in almost all species; there are no muscles associated with the gut.

The proboscis (Gibson 1972, Turbeville 1991) is a long, tubular, eversible, muscular structure which is used for the capture of prey, defence, and in a few species for

---

Chapter vignette: *Tubulanus sexlineatus*. (Redrawn from Kozloff 1990.)

locomotion; when everted, it may be lost, but regeneration is apparently very rapid. It is an invagination of the ectoderm surrounded by a muscular coelomic sac which functions as a hydrostatic skeleton at eversion. The posterior end of the invagination is usually attached to the bottom of the coelomic sac by a retractor muscle. Hoplo-nemertines have one or more calcified stylets in the proboscis; the function of these structures appears to be the wounding of prey so that secretion from various poison glands in the epithelium of the proboscis can penetrate. The stylets are formed in in-tracellular vacuoles and brought into position when fully formed; they are replaced after use (Stricker & Cloney 1981). The proboscis has several layers of muscles, which are in fact continuations of the body wall musculature, and is capable of com-plex movements. Also the proboscis sheath has several muscle layers. *Arhyncho-nemertes* (Riser 1988) lacks any trace of a proboscis.

The brain comprises a pair of dorsal and a pair of ventral lobes connected by a dorsal and a ventral commissure, surrounding the rhynchodaeum and the anterior loop of the blood vessel system, but not the alimentary canal (Turbeville 1991). A pair of lateral, longitudinal nerve cords arise from the ventral lobes and extend to the posterior end where they are connected by a commissure dorsal to the anus; many dorsal and ventral transverse commissures are present in most forms. These nerves are situated either in the connective tissue layer below the ectoderm or between the muscle layers; a peripheral, basiepithelial plexus is also present.

Two or more layers of muscles with different orientation are situated below the basement membrane. The muscles of palaeonemertines are obliquely striated while those of heteronemertines and hoplonemertines are smooth. In some palaeonemer-tines some of the muscle cells make contact with the nerves through non-contractile extensions, but normal nerve contacts are found too (Turbeville 1991).

The blood vessel system consists of a pair of longitudinal vessels joined by anteri-or and posterior transverse vessels; additional longitudinal and transverse vessels are found in several groups. The vessels are lined with a continuous epithelium of meso-dermal cells joined with zonulae adherentes and sometimes having a rudimentary cil-ium; a discontinuous layer of muscle cells is found in the surrounding extracellular matrix (Turbeville & Ruppert 1985, Jespersen & Lützen 1988b). This is a unique sys-tem among the invertebrates, which as a rule have a haemal system, i.e. blood vessels surrounded by basement membrane and with no endothelium (Ruppert & Carle 1983). The origin of the vessels through hollowing out of narrow longitudinal bands of mesodermal cells (Turbeville 1986) and their epithelial character with cell junc-tions and cilia indicate that the vessels are coelomic cavities like the rhynchocoel.

The excretory system consists of a pair of branched canals with flame cells that are often in contact with the blood vessels (Bartolomaeus 1985, Jespersen 1987). Some of the terminal organs are quite complicated and may resemble glands or metanephridia, but the whole system is surrounded by a basement membrane and is clearly protonephridial (Jespersen & Lützen 1987, 1988a); there are no observations of openings between the terminal organs and the coelomic blood vessels.

The gonads are serially arranged sacs consisting of a mesodermal epithelium with germinal cells surrounded by a basement membrane; the gonoduct comprises an ectodermal invagination, but special copulatory or accessory glandular structures are generally lacking (Turbeville 1991). The gametes are usually spawned freely, but several species deposit the eggs in gelatinous masses and some are viviparous.

The primary axis of the egg is determined in the ovary, with the blastoporal pole at the attachment point to the ovarial epithelium (Wilson 1903). The eggs are spawned and fertilized in an early phase of meiosis, so the polar bodies can be followed at the apical pole inside the fertilization membrane (Friedrich 1979).

The cleavage is a spiral cleavage with quartets in all species studied so far. The four quadrants are of the same size, but the blastoporal 'macromeres' of the first cleavage are often smaller than the 'micromeres' (Iwata 1960, Friedrich 1979), and this is also the case with the 'macromeres' of the 5th and 6th cleavage for example in *Malacobdella* (Hammarsten 1918) and *Emplectonema* (Delsman 1915), so the terminology is only used to facilitate comparisons with other spiralians. Four quartets of micromeres are formed (Friedrich 1979), except in *Tubulanus* where Dawydoff (1928) observed only three. Cell-lineage studies are few, but Nusbaum & Oxner (1913) stated that *Lineus* forms endoderm from the 4a-c and the 4A-D-cells, and the mesoderm from the 4d-cell and cells of the invaginated endoderm. On the other hand, Hammarsten (1918) reported that *Malacobdella* forms endoderm from the 4A-D and 4a-d-cells while all the mesoderm originates from descendants of 2a-d-cells (the cells 2a^{1111}-d^{1111}). Dawydoff (1928) observed that the endoderm develops from 3A-C and the mesoderm from 3D and from four micromeres in *Tubulanus*. True teloblasts, i.e. large cells budding off smaller cells in one direction, have not been reported. Mesoderm formation is clearly one of the several details of nemertine ontogeny which require further studies. The blastula stage has a wide blastocoel and gastrulation is through invagination in many of the species with pilidium larvae, whereas many of the species with direct development have a stereoblastula and gastrulation by epiboly or ingression (Friedrich 1979). The narrow, apical cells become the apical organ which develops a tuft of long cilia, usually held together as one structure. Only some of the terrestrial nemertines lack apical organ and cilia completely (Hickman 1963).

Hörstadius (1937) stained the apical and the blastoporal cells of 8 and 16-cell stages and the distribution of the stain in the early pilidium larvae indicates that the cells of the ciliary band are descendants of the tiers 1a²-d² and 2a-d; the upper tier represents the primary trochoblasts in the usual spiral cleavage for example in annelids and molluscs (Table 11.1).

Experiments with isolated blastomeres of *Cerebratulus* have shown that apparently normal larvae develop from some of the isolated blastomeres of the 2-cell stage, and that only very few larvae develop from blastomeres of the 4-cell stage; the development of the juveniles from these larvae was not studied (Hörstadius 1937). These experiments indicate that the localization of developmental factors is relatively slow in the nemertines (Freeman 1978), as in the platyhelminths (Chapter 26), but in contrast to other spiralians, where isolated blastomeres cannot form complete larvae.

The following development is direct in palaeonemertines and enoplans, although with a pelagic stage in many species, whereas indirect development is found in the heteronemertines. The pelagic, planktotrophic pilidium larva is found in many genera, and also in *Hubrechtella*, which is usually placed in the palaeonemertines, but which could perhaps be a heteronemertine (Cantell 1969). Lecithotrophic development with larval types which probably represent modified planktotrophic larvae is known from *Micrura*, which has the pelagic Iwata's larva, and from *Lineus*, which has the non-pelagic Desor's and Schmidt's 'larvae', which develop inside the egg masses. In many other phyla it appears that the planktotrophic larvae exhibit the most ancestral developmental traits, but this is apparently not the case in the nemertines where the pilidium larvae show a development of the adult inside the larva which must be considered highly specialized. The direct developers will therefore be dealt with first.

The three palaeonemertines studied by Iwata (1960) had rather similar development. The blastula flattens so that the blastocoel disappears, and an invagination opposite the apical organ forms the archenteron; the outer part of the invagination is called stomodaeum in later stages. The ectodermal cells become ciliated and the juvenile breaks free from the fertilization membrane. The mesodermal cells become arranged in a continuous layer between ectoderm and gut. The nervous system develops from two large cells at the sides of the apical organ; these cells divide and invaginate, and the invaginations detach from the ectoderm and become compact, forming two ganglia that become the brain and differentiate into dorsal and ventral lobes with commissures; the longitudinal nerves develop as extensions from the ventral lobes along the sides of the gut below the epidermis. The apical organ is not in contact with the brain, and probably degenerates. The proboscis invagination and the anus were not formed at this stage, and their development has not been studied.

The hoplonemertines and bdellonemertines have been studied by a number of authors (see Friedrich 1979), but the descriptions show a good deal of variation between the species and are in some cases contradictory, and several important details of the development are still unclear. Gastrulation shows much variation, with invagination in *Prosorhochmus* (Salensky 1914), epiboly in *Oerstedia* (Iwata 1960), and ingression in *Malacobdella* (Hammarsten 1918) and *Geonemertes* (Hickman 1963). The archenteron loses contact with the ectoderm and becomes compact in most species, and a stomodaeum develops in front of the position of the closed blastopore. The compact embryos of the limnic *Prostoma* are reported to develop a new inner cavity and then to become syncytial (Reinhardt 1941); the development of this type should be reinvestigated. The stomodaeal mouth opening persists in *Drepanophorus* (Lebedinsky 1897), but it disappears and the oesophagus opens into the rhynchodaeum for example in *Tetrastemma* (Lebedinsky 1897) and *Prostoma* (Reinhardt 1941). An ectodermal invagination between the apical organ and the stomodaeum gives rise to the proboscis; it loses the connection to the ectoderm in many species and a special rhynchodaeum has been reported to develop from the degenerating apical organ. Hickman (1963) reported that embryos of *Geonemertes* become surrounded by a number of special, so-called ectoembryonic cells, which later become replaced by new, ciliated epidermal cells. The many different types of development of gut and proboscis (Friedrich 1979) give the impression of many lines of specialization, but the developmental type reported from the palaeonemertines appear to be the most ancestral.

The pilidium larvae develop inside the fertilization membrane and a long apical ciliary tuft and a wide circumblastoporal ciliary band appear shortly before a general ciliation makes the larvae begin to rotate in the capsules and finally to hatch (Wilson 1900). A blastoporal invagination forms a ciliated archenteron which bends posteriorly and differentiates into a funnel-shaped oesophagus (or stomodaeum) and a posterior, almost spherical stomach; an anal opening is not formed. The ciliation differentiates further, and the fully developed, feeding larvae have the circumblastoporal band specialized as compound cilia along the anterior and posterior ends and on a pair of lateral lobulate expansions (Nielsen 1987). There is also a postoral, transverse band of stronger ciliation (Lacalli & West 1985; Fig. 25.1). The apical tuft extends from a conspicuous apical organ which is a thickening of the epithelium; each cell may have one or several (up to 12) cilia, and the cilia of each cell are surrounded by a ring of long microvilli connected by a mucous structure (Cantell, Franzén & Sensenbaugh 1982). Two groups of small muscles extend from the apical organ to the anterolateral areas of the mouth (Wilson 1900). The ciliary bands have associated, basiepithelial nerves (Fig. 25.1), and a few nerve cells to the gut and towards the apical organ have been observed; no nerve cells have been observed in the apical organ itself (Lacalli & West 1985, Hay-Schmidt 1990).

The development of the adult nemertine inside the pilidium larva has only been studied in a few species, the only detailed description being that of Salensky (1914), which is based on the larvae called *Pilidium pyramidatum* and *P. gyrans*. Soon after completion of the larval gut, three pairs of ectodermal invaginations, called cephalic, cerebral and trunk sacs, develop from the underside of the larva close to the ciliary band. Slightly later, an unpaired dorsal sac develops from the dorsal epithelium above the ciliary band, possibly through delamination. Each sac becomes differentiated into a thick, internal cell plate, called an embryonic disc, and a very thin, exterior cell layer. The sacs expand and surround the gut, and the embryonic discs fuse to form the ectoderm of the juvenile while the outer layers form an amnionic membrane around the juvenile. The proboscis ectoderm develops as a narrow, ventral invagination of the ectoderm at the fusion line between the two cephalic discs. The ganglia develop from the ectoderm lateral to the proboscis invagination, i.e. completely isolated from the apical organ of the larva; they differentiate into dorsal and ventral lobes and the commissures around the rhynchodaeum develop; the lateral nerve cords develop as extensions from the cephalic ganglia. Mesodermal cells cover the ectoderm and endoderm, resembling peritoneal epithelia, and an anterior plus the paired longitudinal cavities fuse and become the blood vessel coelom. The origin of the rhynchocoel is more uncertain. A pair of small ectodermal invaginations in front of the blastopore give rise to the nephridia. The anus breaks through at a late stage. The juvenile finally breaks out of the larval body, which in several cases becomes ingested afterwards (Cantell 1966).

Micrura akkeshiensis (Iwata 1958) is a free spawner and the cleavage, gastrulation and development of the apical organ resemble those of the pilidium larvae; only the blastopore differs in being very narrow, and the lecithotrophic larva has no special ciliary bands. Five ectodermal invaginations develop, corresponding to the paired cephalic and trunk sacs and the unpaired dorsal sac of the pilidium, but cerebral sacs were not observed. The following development of the juvenile resembles that in the pilidium, but the main axis of the juvenile is rotated so that the anterior end faces the posterior pole of the larva (Fig. 27.1). The juvenile in the pilidium is rotated 90° from the primary axis, as usual in the gastroneuralians, so the orientation of the juvenile inside Iwata's larva is simply a further 90° rotation.

Indirect development with pilidium-like stages inside gelatinous egg masses are known from species of *Lineus*; two types have been distinguished: the Desor type, which is lecithotrophic, and the Schmidt type, which ingests other embryos from the same egg mass (Schmidt 1934, 1964). The development resembles that of Iwata's larva, but an apical organ is not developed and the ectoderm is apparently without cilia. The seven embryonic discs develop as in the pilidium, but it is stated that the proboscis originates from a separate anterior disc; this was also reported from pilidium larvae by some of the early authors but should be investigated further. The orientation of the juvenile relative to the larval ectoderm is unknown.

There are several more or less distinct types of pilidium larvae, many of which have been given names (Dawydoff 1940, Cantell 1969, see also Fig. 27.1). Only a few of these larvae can be related to their adult species (Friedrich 1979), and many larvae are known only from one developmental stage. *Pilidium incurvatum* (Fig. 27.1) has a separate posterior ring of cilia; the cilia along the edges of the anterior funnel were described as coarse, indicating compound cilia, and the photomicrographs of sections of the posterior ciliary band resemble those of bands with compound cilia (Cantell 1967, pl.2, fig. 1), but direct observations are lacking. The various larval types have inspired several phylogenetic speculations, and Jägersten's (1972) diagram of possible evolutionary connections between some of the larval types (see also Fig. 27.1) gives a picture of some of the types and of the orientation of the juvenile in relation to the axes of the larvae. A specialization from the identical orientation in the direct-developing species and *Pilidium recurvatum via* the more common type of pilidium with the juvenile axis perpendicular to the apical-blastoporal axis to the Iwata-larva with opposite orientation of the two axes may be indicated; however, it is not certain that the larval type with the most 'primitive' orientation of the juvenile is itself ancestral. Development with formation of the adult epithelium from embryonic discs can only be interpreted as highly specialized, so the life cycle involving a pilidium larva is definitely not ancestral. On the other hand, one could

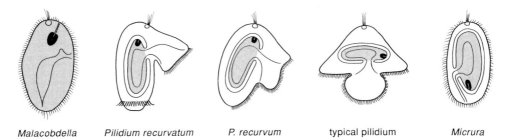

| Malacobdella | Pilidium recurvatum | P. recurvum | typical pilidium | Micrura |

Fig. 27.1. Larval types of nemertines and the orientation of the juvenile inside the larval body; *Malacobdella grossa* has direct development; the larva of *Micrura akkeshiensis* is often called Iwata's larva. The 'adult' structures are shaded and the primordia of the proboscis apparatus are black. (Redrawn from Hammarsten 1918 and Jägersten 1972.)

imagine that the ancestral nemertine had a trochophore-type larva and a creeping adult (as the ancestral protostome), and that specialization has taken two routes, one towards the direct development and the other towards the pilidium-type larvae, with the modified ciliary bands (see further in Chapter 25). This larval type then acquired the peculiar embryonic discs and lost the larval anus. This is admittedly highly speculative and it is premature to draw far-reaching conclusions as long as so little is known about whole life cycles of several interesting species.

The proboscis is a structure which has no counterpart in other spiralians and must be regarded as the most obvious apomorphy of the phylum. Its absence in *Arhynchonemertes* (Riser 1988, 1989) can be interpreted as a specialization, i.e. that the ancestor had a proboscis, or as a plesiomorphy, i.e. that the ancestral nemertine lacked a proboscis. This question seems impossible to answer at present, also because the systematic position of the species seems uncertain; investigations of its development may shed light on the problem. *Arhynchonemertes* has a simple, loop-shaped blood vessel system, and also the nervous system and musculature resemble those of the other nemertines, so its relationship with the nemertines can hardly be questioned.

Nemertini and Platyhelminthes are sometimes regarded as sister groups, constituting the Acoelomata or Parenchymia (Chapter 25), but the nemertines have also been regarded as the sister group of the other 'coelomate' spiralians (Brusca & Brusca 1990). A crucial question is whether the nemertine coeloms (blood vessel system and rhynchocoel) and the coeloms of for example annelids are homologous. I feel that the rhynchocoel cannot be homologized with any structure found in other spiralians; its position, structure and function are unique, and it is definitely not homologous with the pharyngeal structures in other phyla. The blood vessel system of the nemertines develops in a mode resembling that of the coeloms of some annelids (see above), but the system is not segmented and is neither associated with locomotion and somatic musculature nor with metanephridia or gonads as in the articulates, so I believe that the blood vessel coelom is a nemertine apomorphy and therefore not homologous with the coelom in other spiralians.

Interesting subjects for future research

1. Development of *Arhynchonemertes*.
2. Cell-lineage of pilidium larvae – origin of the embryonic discs.
3. Organogenesis at metamorphosis of the pilidium.

References

Bartholomaeus, T. 1985. Ultrastructure and development of the protonephridia of *Lineus viridis* (Nemertini). – Microfauna mar. 2: 61-83.
Brusca, R.C. & G.J. Brusca 1990. Invertebrates. – Sinauer Associates, Sunderland.
Cantell, C.-E. 1966. The devouring of the larval tissue during metamorphosis of pilidium larvae (Nemertini). – Ark. Zool., 2. ser., 18: 489-492, 1 pl.
Cantell, C.-E. 1967. Some developmental stages of the peculiar nemertean larva *pilidium recurvatum* Fewkes from the Gullmarfjord (Sweden). – Ark. Zool., 2. ser., 19: 143-147, 2 pls.
Cantell, C.-E. 1969. Morphology, development, and biology of the pilidium larvae (Nemertini) from the Swedish west coast. – Zool. Bidr. Upps. 38: 61-112, 6 pls.

Cantell, C.-E., Å. Franzén & T. Sensenbaugh 1982. Ultrastructure of multiciliated collar cells in the pilidium larva of *Lineus bilineatus* (Nemertini). – Zoomorphology 101: 1-15.

Dawydoff, C. 1928. Sur l'embryologie des protonémertes. – C. r. hebd. Séanc. Acad. Sci., Paris 186: 531-533.

Dawydoff, C. 1940. Les formes larvaires de polyclades et de némertes du plancton indochinois. – Bull. biol. Fr. Belg. 74: 443-496.

Delsman, H.C. 1915. Eifurchung und Gastrulation bei *Emplectonema gracile* Stimpson. – Tijdschr. ned. dierk. Vereen., 2. ser., 14: 68-114, pls 6-9.

Freeman, G. 1978. The role of asters in the localization of the factors that specify the apical tuft and the gut of the nemertine *Cerebratulus lacteus*. – J. exp. Zool. 206: 81-108.

Friedrich, H. 1979. Nemertini. – *In* F. Seidel (ed.): Morphogenese der Tiere, Deskriptive Morphogenese, 3. Lieferung, pp 1-136. Gustav Fischer, Jena.

Gibson, R. 1972. Nemerteans. – Hutchinson University Library, London.

Hammarsten, O.D. 1918. Beitrag zur Embryonalentwicklung der *Malacobdella grossa* (Müll.). – Arb. zootom. Inst. Univ. Stockh. 1: 1-96, 10 pls.

Hay-Schmidt, A. 1990. Catecholamine-containing, serotonin-like and neuropeptide FMRFamide-like immunoreactive cells and processes in the nervous system of the pilidium larva (Nemertini). – Zoomorphology 109: 321-244.

Hickman, V.V. 1963. The occurrence in Tasmania of the land nemertine, *Geonemertes australiensis* Dendy, with some account of its distribution, habits, variations and development. – Pap. Proc. R. Soc. Tasmania 97: 63-75, 2 pls.

Hörstadius, S. 1937. Experiments on determination in the early development of *Cerebratulus lacteus*. – Biol. Bull. Woods Hole 73: 317-342.

Iwata, F. 1958. On the development of the nemertean *Micrura akkeshiensis*. – Embryologia 4: 103-131.

Iwata, F. 1960. Studies on the comparative embryology of nemerteans with special reference to their interrelationships. – Publs Akkeshi mar. biol. Stat. 10: 1-51.

Jägersten, G. 1972. Evolution of the Metazoan Life Cycle. – Academic Press, London.

Jespersen, Å. 1987. Ultrastructure of the protonephridium in *Acteonemertes bathamae* Pantin (Rhynchocoela: Enopla: Hoplonemertini). – Acta zool. (Stockh.) 68: 115-125.

Jespersen, Å. & J. Lützen 1987. Ultrastructure of the nephridio-circulatory connections in *Tubulanus annulatus* (Nemertini, Anopla). – Zoomorphology 107: 181-189.

Jespersen, Å. & J. Lützen 1988a. The fine structure of the protonephridial system in the land nemertean *Pantinonemertes californiensis* (Rhynchocoela, Enopla, Hoplonemertini). – Zoomorphology 108: 69-75.

Jespersen, Å. & J. Lützen 1988b. Ultrastructure and morphological interpretation of the circulatory system of nemerteans (Phylum Rhynchocoela). – Vidensk. Meddr dansk naturh. Foren. 147: 47-66.

Kozloff, E.N. 1990. Invertebrates. – Saunders College Publishing, Philadelphia.

Lacalli, T.C. & J.E. West 1985. The nervous system of a pilidium larva: evidence from electron microscope reconstructions. – Can. J. Zool. 63: 1909-1916.

Lebedinsky, J. 1897. Beobachtungen über die Entwicklungsgeschichte der Nemertinen. – Arch. mikr. Anat. 49: 503-556, pls 21-23.

Nielsen, C. 1987. Structure and function of metazoan ciliary bands and their phylogenetic significance. – Acta zool. (Stockh.) 68: 205-262.

Nusbaum, J. & M. Oxner 1913. Die Embryonalentwicklung des *Lineus ruber* Müll. Ein Beitrag zur Entwicklungsgeschichte der Nemertinen. – Z. wiss. Zool. 107: 78-197, pls 1-8.

Reinhardt, H. 1941. Beiträge zur Entwicklungsgeschichte der einheimischen Süsswassernemertine *Prostoma graecense* (Böhmig). – Vierteljahrsschrift naturforsch. Ges. Zürich 86: 184-255, 4 pls.

Riser, N.W. 1988. *Arhynchonemertes axi* gen.n., sp.n. (Nemertini) – an insight into basic acoelomate bilaterial organology. – Fortschr. Zool. 36: 367-373.

Riser, N.W. 1989. Speciation and time – relationships of the nemertines to the acoelomate metazoan Bilateria. – Bull. mar. Sci. 45: 531-538.

Ruppert, E.E. & K.J. Carle 1983. Morphology of metazoan circulatory systems. – Zoomorphology 103: 193-208.

Salensky, W. 1912. Über die Morphogenese der Nemertinen. I. Entwicklungsgeschichte der Nemertine im Inneren des Pilidiums. – Mém. Acad. imp. Sci. St.-Petersb., 8. sér., Cl. phys.-math. 30(10): 1-74, 6 pls.

Salensky, W. 1914. Die Morphogenese der Nemertinen. 2. Über die Entwicklungsgeschichte des *Prosorhochmus viviparus*. – Mém. Acad. imp. Sci. St.-Petersb., 8. sér., Cl. phys.-math. 33(2): 1-39, 4 pls.

Schmidt, G.A. 1934. Ein zweiter Entwicklungstypus von *Lineus gessneriensis* O.F. Müll. (Nemertini). – Zool. Jb., Anat. 58: 607-660.

Schmidt, G.A. 1964. Embryonic development of littoral nemertines *Lineus desori* (mihi, species nova) and *Lineus ruber* (O.F. Mülleri, 1774, G.A. Schmidt, 1945) in connection with ecological relation changes of mature individuals when forming the new species *Lineus ruber*. – Zool. Polon. 14: 75-122.

Stricker, S.A. & M.J. Cavey 1988. Calcareous concretions and non-calcified hooks in the body wall of nemertean worms. – Acta zool. (Stockh.) 69: 39-46.

Stricker, S.A. & R.C. Cloney 1981. The stylet apparatus of the nemertean *Paranemertes peregrina*: its ultrastructure and role in prey capture. – Zoomorphology 97: 205-223.

Turbeville, J.M. 1986. An ultrastructural analysis of coelomogenesis in the hoplonemertine *Prosorhynchus americanus* and the polychaete *Magelona* sp. – J. Morph. 187: 51-60.

Turbeville, J.M. 1991. Nemertinea. – *In* F.W. Harrison (ed.): Microscopic Anatomy of Invertebrates, vol. 3, pp 285-328. Wiley-Liss, New York.

Turbeville, J.M. & E.E. Ruppert 1983. Epidermal muscles and peristaltic burrowing in *Carinoma tremaphoros* (Nemertini): correlatives of effective burrowing without segmentation. – Zoomorphology 103: 103-120

Turbeville, J.M. & E.E. Ruppert 1985. Comparative ultrastructure and the evolution of nemertines. – Am. Zool. 25: 53-71.

Wilson, C.B. 1900. The habits and early development of *Cerebratulus lacteus* (Verrill). – Q. Jl microsc. Sci., N.S. 43: 97-198, pls 9-11.

Wilson, E.B. 1903. Experiments on cleavage and localization in the nemertine-egg. – Arch. Entwicklungsmech. Org. 16: 411-460.

28

ASCHELMINTHES

A varying number of classes or phyla have for more than a century been united in a systematic group called Nemathelminthes, Aschelminthes or Pseudocoelomata; here I include the following phyla: Rotifera, Acanthocephala, Chaetognatha, Gastrotricha, Nematoda, Nematomorpha, Priapula, Kinorhyncha, and Loricifera. It has always been difficult to find good defining characters, and especially Acanthocephala and Chaetognatha have been excluded by many authors, while some other authors include Entoprocta. Hyman (1951) treated the groups she included as classes within the phylum Aschelminthes, and a similar arrangement can be found in several textbooks.

The cladistic method, which has formed the base for the present book, offers an alternative approach to the discussion of these difficult groups, which I have chosen to treat as phyla. The discussions on rotifers and acanthocephalans (Chapters 29-30) conclude that these two groups must be sister groups, and the discussions of the gastrotrichs, nematodes, nematomorphs, priapulids, kinorhynchs, and loriciferans (Chapters 33-40) conclude that these phyla constitute a monophyletic group, Cycloneuralia (Chapter 32). The question is therefore whether the three monophyletic groups Rotifera + Acanthocephala, Chaetognatha, and Cycloneuralia together form a monophyletic unit (Fig. 28.1).

The nervous systems of most aschelminth phyla are of the general protostomian (gastroneuralian) type (Fig. 10.2), but a larval apical organ with cilia, as seen in many spiralian larvae, is never developed. Some of the rotifers are planktotrophic ciliary feeders, and these forms have the typical protostomian ciliary system consisting of prototroch and metatroch of compound cilia functioning as downstream-collecting bands and an adoral zone of separate cilia. The ciliated epithelia are multiciliate, except in some of the gastrotrichs (Fig. 9.2). It can hardly be doubted that the aschelminths are protostomians.

The cleavage patterns are neither radial nor spiral; only the chaetognaths have an almost radial pattern. Classifications which include the aschelminths in the group Spiralia are not based on direct observations (Storch & Welsch 1991).

The mesoderm develops from cells around the margin of the blastopore in gastrotrichs and nematodes (Chapters 33, 35), but the embryology is poorly studied or

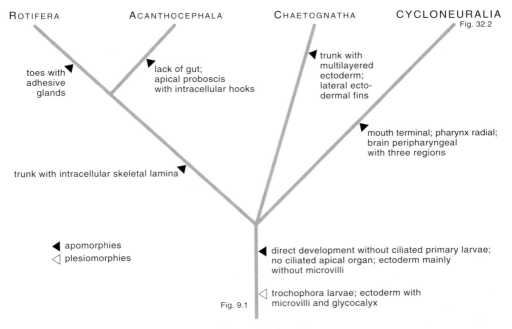

ROTIFERA   ACANTHOCEPHALA   CHAETOGNATHA   CYCLONEURALIA
Fig. 32.2

trunk with
multilayered
ectoderm;
lateral ecto-
dermal fins

toes with
adhesive
glands

lack of gut;
apical proboscis
with intracellular hooks

mouth terminal; pharynx radial;
brain peripharyngeal
with three regions

trunk with intracellular skeletal lamina

▸ apomorphies
◁ plesiomorphies

direct development without ciliated primary larvae;
no ciliated apical organ; ectoderm mainly
without microvilli

Fig. 9.1

◁ trochophora larvae; ectoderm with
microvilli and glycocalyx

Fig. 28.1. Phylogeny of the main groups of the Aschelminthes.

completely unknown in several phyla, so it is uncertain whether an apomorphy can be found here.

A very early morphological differentiation of the primordial germ cells, which are usually invaginated separately before gastrulation (rotifers, acanthocephalans, Chapters 29, 30) or become situated at the bottom of the archenteron (chaetognaths, nematodes, Chapters 31, 35), may turn out to be a synapomorphy of the aschelminths.

As mentioned above, the aschelminths are sometimes called pseudocoelomates, and this refers to the more or less spacious body cavity between the body wall and the gut, which is not surrounded by a peritoneum. The pseudocoel is regarded as a persisting primary body cavity, but the blastocoel disappears in early embryology in many of the groups and new spaces develop between different tissues. The pseudocoelomate body plan is contrasted to the coelomate plan, which has secondary body cavities, coeloms, surrounded by mesodermal epithelia. These definitions have played a strong role in phylogenetic discussions, but they have in my opinion been strongly overemphasized (Chapters 9, 12). Here, it should only be discussed whether any of the phyla here assigned to the Aschelminthes have body cavities which must be characterized as coelomic. Rotifers (Chapter 29) have a spacious body cavity and the body wall mesoderm is split up into separate muscles; there is nowhere a peritoneum-like covering of the body wall. Acanthocephalans (Chapter 30) have a body cavity and lack a gut; the body wall has a complete layer of mesodermal muscles, but there is nothing resembling a peritoneum. The mysterious ligament sacs resemble a basement membrane, but

235

their ontogeny is poorly described and they may be related to the gonads. A coelom in the usual meaning of the word is only found in the chaetognaths (Chapter 31), which have paired body cavities lined with mesodermal epithelium, and the coelomic sacs develop by a process resembling the enterocoely found in most deuterostomes. However, the homology of the two types of coelomic cavities is not supported by observations of specific similarities, and I have chosen to interpret the two systems as non-homologous (Chapter 31). Cycloneuralians generally lack a mesodermal covering of the gut, but some gastrotrichs (Chapter 33) have very voluminous longitudinal body muscles which fill the space around the gut, but there are no coelomic cavities. It must be concluded that only the chaetognaths have body cavities which can perhaps be characterized as coelomic, but these cavities are obviously not homologous with the coeloms found in spiralians or deuterostomes. The morphology of the body cavities does not show any character which can be regarded as an aschelminth apomorphy.

Some phylogenetic theories synonymize Bilateria and Coelomata because they regard the coelom as a synapomorphy of all bilaterians, and consequently the acoelomate and pseudocoelomate conditions as derived (Grobben 1910, Siewing 1985, Storch & Welsch 1991). These ideas are based exclusively on speculations; there are no actual observations which support the contention that aschelminths are derived from coelomate ancestors.

Monociliary cells and myoepithelial cells are sometimes mentioned as plesiomorphic characters within the metazoans and within the aschelminths (Rieger 1976, Ruppert 1982), but I believe that these characters are merely basal cellular potentials, which may be expressed when 'needed' in a tissue. The centrioles are present at every cell division, and a cilium may develop from one of the basal bodies in most cell types (nerve cells (in Cnidaria), sensory cells, ectodermal cells, peritoneal cells, endodermal cells). A monociliate epithelium is therefore not necessarily a plesiomorphic character. All cells have the ability to contract – at least at some stage during ontogeny – and myoepithelial cells have been reported from a number of different structures in both protostomian and deuterostomian phyla (ectoderm of tentacles and cirri of polychaetes, pharynges of ectoprocts and nematodes, tentacles of entoprocts, adhesive papillae of ascidian larvae, and in chaetognath mesoderm; Chapters 17, 23, 35, 24, 52, 31); they cannot be regarded as plesiomorphic characters either. Ciliated myoepithelial cells, which occur in the cnidarians and which should be the most 'plesiomorphic cell type', are found for example in the entoproct tentacle, in chaetognath mesoderm, and in mesenteria of the brachiopod *Lingula* (Chapters 23, 31, 45), and the scattered occurrence of this cell type in the bilaterians clearly demonstrates that it is not a plesiomorphy.

More than 40 years ago, Hyman (1951, p. 54) stated that 'the taxonomic disposition of the groups here under consideration ... has always been troublesome to zoologists' and the above discussion shows that the situation is not much changed. I have chosen to regard the Aschelminthes as the monophyletic sister group of the Spiralia (Chapter 11), but further studies are needed before reliable apomorphies can be identified.

Two names for this group are in current use: Nemathelminthes and Aschelminthes. Nemathelminthes was coined by Gegenbaur (1859) for a group comprising only Nematoda and Nematomorpha, with Chaetognatha as a closely related group. Aschelminthes was introduced by Grobben (1910) for a group comprising Rotifera, Gastrotricha, Kinorhyncha, Nematoda, Nematomorpha, and Acanthocephala, i.e. almost all the phyla included here. There are no nomenclatorial rules for the higher zoological taxa, and I have chosen to use Aschelminthes, which following Hyman (1951) has been in use in most textbooks and reviews.

References

Gegenbaur, C. 1859. Grundzüge der vergleichenden Anatomie, 2. ed. – Vilh. Engelmann, Leipzig.

Grobben, K. 1910. Claus-Grobben: Lehrbuch der Zoologie, 2. ed. – Elvert'sche Verlagsbuchhandlung, Marburg.

Hyman, L.H. 1951. Acanthocephala, Aschelminthes, and Entoprocta. The pseudocoelomate Bilateria. The Invertebrates vol. 5. – McGraw-Hill, New York.

Rieger, R.M. 1976. Monociliated epidermal cells in Gastrotricha: significance for concepts of early metazoan evolution. – Z. zool. Syst. Evolutionsforsch. 14: 198-226.

Ruppert, E.E. 1982. Comparative ultrastructure of the gastrotrich pharynx and the evolution of myoepithelial foreguts in Aschelminthes. – Zoomorphology 99: 181-220.

Siewing, R. 1985. Phylum Coelomata. – *In* R. Siewing (ed.): H. Wurmbach's Lehrbuch der Zoologie, Systematik, pp 203-217. Gustav Fischer, Stuttgart.

Storch, V. & U. Welsch 1991. Systematische Zoologie, 4. ed. – Gustav Fischer, Stuttgart.

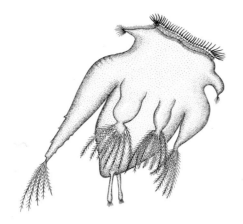

Phylum ROTIFERA

Rotifers are a group of usually less than millimetre-long, aquatic, mostly limnic organisms; about 1800 species have been described. Many types can be recognized by the ciliary 'wheel organ' or corona which has given the phylum its name, but the wheel organ is highly modified or completely absent in others. Three classes are recognized: Monogononta, Bdelloidea and Seisonidea. Some authors (for example Wallace & Colburn 1989) unite the two first-mentioned classes in the group Eurotatoria, while others (for example Beauchamp 1965) unite the two last-mentioned classes in the group Digononta; the three classes are actually very different.

Monogononts have a single ovary; their life cycles are complicated with parthenogenetic generations of females producing diploid eggs and sexual generations of females which produce haploid eggs; non-fertilized eggs develop into haploid males and fertilized eggs become resting eggs. The males are much smaller than the females and lack the gut in most species. Several types have a wheel organ which is a typical protostomian ciliary system with prototroch (called trochus), adoral ciliary zone and metatroch (called cingulum), but others are strongly modified.

Bdelloids are parthenogenetic without males. There is a pair of ovaries and the integument typically forms 16 slightly thickened rings which can telescope when the animals contract. Some forms have a ciliary system with the prototroch divided into a pair of trochal discs, an adoral ciliary zone and a metatroch, whereas others have a field of uniform cilia around the mouth.

Seisonidea comprises only the genus *Seison*, with two species which are ectoparasites on the crustacean *Nebalia*. Males and females are similar and their wheel organ is reduced to small bristles.

Many tissues of rotifers are syncytial, but the number of nuclei in most organs is nevertheless constant, and divisions do not occur after hatching; this implies that the powers of regeneration are almost absent (Martini 1912, Van Cleave 1922, Nachtwey 1925, Peters 1931).

Chapter vignette: *Hexarthra mira*. (Redrawn from Wesenberg-Lund 1952.)

The ectoderm of the ciliary bands consists of large cells with several nuclei and connected by various types of cell junctions, whereas the ectoderm of the main body region is a thin syncytium (Storch & Welsch 1969, Clément & Wurdak 1991). The ectoderm of the ciliary bands has the usual surface structure with microvilli and a layer of normal, extracellular cuticle between the tips of the microvilli (Clément 1977). Also some of the sensory organs have this type of cuticle (Clément & Wurdak 1991). The buccal epithelium appears to lack a cuticle and the cilia have modified, electron-dense tips. The pharyngeal epithelium has multiple layers of double membranes which also cover the cilia. The borderline between these two epithelia marks the origin of a flattened, funnel-shaped structure called the velum, which consists of a thick layer with a similar structure of parallel membranes (Clément et al. 1980). All ciliated epithelial cells are multiciliate. The body epithelium has a usually very thin extracellular cuticle probably consisting of glycoproteins, and an intracellular skeletal lamina (sometimes referred to as an intracellular cuticle) apposed to the inner side of the apical cell membrane. This intracellular lamina may be of different thickness in various parts of the body and in different species, homogeneous in *Asplanchna*, lamellate in *Notommata*, and with a honeycomb-like structure in *Brachionus*; it has characteristic pores with drop-shaped invaginations of the cell membrane, and the general structure is identical in monogononts and bdelloids (Storch & Welsch 1969, Storch 1984, Clément 1969) (*Seison* has not been investigated). The intracellular skeletal lamina appears to consist of intermediate filaments of the keratin-type and is a scleroprotein (Bender & Klenow 1988); chitin has not been found (Jeuniaux 1975). The mastax (see below) is a cuticular structure, which contains more than 50% chitin in *Brachionus* (Klusemann, Kleinow & Peters 1990). The various parts of the mastax were found to be thickened parts of a continuous membrane (Kleinow, Klusemann & Wratil 1990), and it appears that the whole structure is extracellular; the reports of intracellular mastax structures (see Clément & Wurdak 1991) are probably erroneous.

The wheel organ shows an enormous variation (Fig. 29.1). Some creeping types, as for example *Dicranophorus*, have a ventral, circumoral zone of single cilia used in creeping; predatory, planktonic forms, such as *Asplanchna*, have a preoral, almost complete ring of compound cilia used in swimming; planktotrophic forms which may be planktonic or sessile, such as *Hexarthra*, *Conochilus* and *Floscularia*, have an adoral zone of single cilia bordered by a preoral prototroch and a postoral metatroch, with the whole ciliary system surrounding the apical field; many other variations are found, and *Acyclus* and *Cupelopagis* lack the corona in the adult stage (Beauchamp 1965). Proto- and metatroch consist of compound cilia (Nielsen 1987, Clément & Wurdak 1991), and the whole complex is a downstream-collecting system (Strathmann, Jahn & Fonseca 1972).

Two sets of glands are associated with creeping: the retrocerebral complex and the pedal glands (Clément & Wurdak 1991). The first mentioned organ is present in most bdelloids, in creeping monogononts and possibly in *Seison* (Beauchamp 1909, Remane 1929-1933); it is situated on the dorsal side of the brain, but the glands may extend further posteriorly. In the most complicated form it consists of a retrocerebral sac, a pair of subcerebral glands and a paired duct opening on two small papillae in front of the brain. In bdelloids it may open on a small protrusible

239

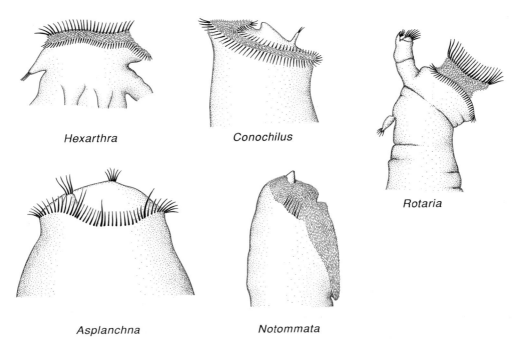

Fig. 29.1. Various types of ciliary bands in rotifers. The planktotrophic types have the trochophore type of ciliary bands (prototroch + adoral ciliary zone + metatroch). – The pelagic, solitary *Hexarthra mira* has the ciliary bands of a trochophore in the non-specialized shape; the pelagic, colonial *Conochilus unicornis* has similar ciliary bands, only with the lateral parts bent to the ventral side; the sessile *Rotaria magnicalcarata* has a prototroch which is divided into a pair of lateral, almost circular bands. – The pelagic, carnivorous *Asplanchna girodi* has only the prototroch. – The benthic, carnivorous *Notommata pseudocerebrus* mostly creeps on the extended adoral ciliary zone, but occasionally swims with a few prominent groups of compound cilia, which appear to be specialized parts of the prototroch. – Prototroch: long cilia; adoral ciliary zone: shaded; metatroch: small cilia; the arrows point at the mouth. (Redrawn from Beauchamp 1965 and Nielsen 1987.)

rostrum. The posterior end of the body is usually extended as a foot ending in a pair of toes with pedal glands. The secretions of the pedal glands are used in attachment and the secretions of both sets of glands are used in creeping (Clément & Wurdak 1991).

Lorenzen (1985) pointed out that some epithelial projections from the syncytia of the trochal discs in bdelloids (Zelinka 1886) may be homologous with the lemnisci of the acanthocephalans (see below); the ultrastructure and the function of these structures are unfortunately unknown.

Particles captured by the corona are transported through the ciliated buccal tube to the mastax, which is a muscular, ventral extension with a system of chitinous jaws (see above). The macrophagous species can protrude the jaws from the mouth and grasp algal filaments or prey. The movements of the hard parts (trophi) of the mastax are coordinated by the mastax ganglion, which receives input from ciliated sense organs at the bottom of its lumen and from the brain. Various types of trophi are characteristic of larger systematic groups and are correlated with feeding behaviour. A partly ciliated oesophagus leads to the stomach, which is syncytial and without cilia in bdelloids and cellular with cilia in monogononts (Clément & Wurdak 1991). There is a ciliated intestine opening into a short cloaca and a dorsal anus. A few genera lack the intestine, so only the protonephridia and the genital organs open into the cloaca.

The nervous system comprises a dorsal brain, a mastax ganglion (see below), a caudal ganglion ventral to the rectum, and a number of peripheral nerves with cells. A pair of lateroventral nerves connect the lateroposterior parts of the brain with the caudal ganglion. The brain comprises about 150 to 250 cells, the numbers being species-specific (Martini 1912, Nachtwey 1925, Peters 1931). Photoreceptors of a number of different types are found embedded in the brain of many species (Clément & Wurdak 1991). The caudal ganglion is usually associated with the foot and the cloaca, but separate ganglia for the two regions are found in some species (Remane 1929-1933); the caudal ganglion is present also in species which lack the foot. Many small sensory organs, usually with a tuft of cilia, are found in various parts of the body, especially in the anterior end, and some of them have a small basal ganglion. Each transverse muscle is innervated by one or two large nerve cells, which gives a superficial impression of segmentation (Zelinka 1888, Stossberg 1932). The low and constant numbers of cells have made it possible to elucidate many of the neuro-muscular pathways and to couple them with behavioural responses (Clément & Wurdak 1991).

Almost all the muscles are narrow bands with one nucleus. They attach to the body wall through an epithelial cell with hemidesmosomes and tonofibrils (Clément & Wurdak 1991). The two large retractor muscles of the corona are coupled to other muscles through gap junctions and send a cytoplasmic extension to the brain where synapses occur; other muscles are innervated by axons from the ganglia (Clément & Amsellem 1989).

There is practically no connective tissue, and the spacious body cavity functions as a hydrostatic skeleton in protrusion of the corona.

There is a paired protonephridial system with one to many flame cells; the structure of the filtering membrane shows important differences between monogononts and bdelloids (Clément & Wurdak 1991).

Female monogononts have an unpaired, sac-shaped germovitellarium, and the males have a single testis; both types of gonads open into the cloaca. The bdelloids have paired germovitellaria. The parthenogenetic eggs become surrounded by a chitinous shell secreted by the embryo (Jeuniaux 1975). Females of *Seison* have paired ovaria without vitellaria and the males have paired testes with a common sperm duct, where each sperm becomes enclosed in a small envelope (Remane 1929-1933).

Studies on monogonont development have centred on the pelagic genus *Asplanchna* (Jennings 1896, Tannreuther 1925, Nachtwey 1925, Lechner 1966), with additional observations on *Ploesoma* (Beauchamp 1956) and *Lecane* (Pray 1965). Lechner (1966) reinterpreted some of the earlier reports on the early development (Fig. 29.2 and Table 29.1) and Nachtwey (1925) described organogenesis. The cleavage is total and unequal and the 4-cell stage has three smaller A-C-blastomeres

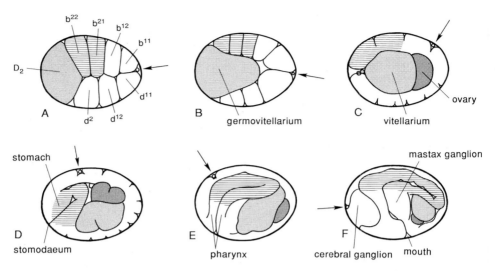

Fig. 29.2. Early development of *Asplanchna girodi*; median sections with the polar bodies (apical pole) indicated by a thick arrow. – A, 16-cell stage. – B, internalization of the D4-cell through an epibolic gastrulation. – C, the germovitellarium is completely internalized and divided into the primordial cells of the ovary and the vitellarium; a small blastopore is formed through further gastrulation movements. – D, gastrulation continues from the dorsal and lateral sides of the blastopore, forming the endodermal stomach, and the apical pole (indicated by the polar bodies) moves along the dorsal side. – E, further gastrulation movements from the whole area around the blastopore give rise to the inner part of the pharynx. – F, the pharynx is now fully internalized and the mastax ganglion differentiates from its ventral side; the brain has become differentiated from the ectoderm at the apical pole. (Modified from Lechner 1966.)

and a large D-blastomere; the polar bodies are situated at the apical pole. The cells B and D are in contact along the primary axis, separating the A and C-cells in the compact embryo; this is in contrast to the general configuration in spiralian embryos, so Costello & Henley's statement (1976, p. 287) 'Cleavage through the 4-cell stage is definitely spiral' must be rejected. The D-cell then divides unequally, and its large descendant (D_I) comes to occupy the blastoporal pole while the smaller descendant (d^I) forms an apical ring with the A-C-cells. The D_I-macromere cell gives off another small cell, and all the other blastomeres divide equally, with the spindles parallel to the primary axis. The embryo now consists of four rows of cells, with the large D_2-cell occupying the blastoporal pole. The smaller cells divide further and slide along the macromere, which becomes internalized as in an epibolic gastrulation; the movements continue as an invagination, forming an archenteron, where it appears that the stomach originates either from a-c-cells or exclusively from b-cells and the pharynx from all four quadrants. The D_2-cell gives off two abortive micromeres and the D_4-cell becomes the germovitellarium. The stronger gastrulation movements of the dorsal side (b-cells) moves the apical pole with the polar bodies towards the blastopore, so that the cells of the D-quadrant fi-

Table 29.1. Cell-lineage of the rotifer *Asplanchna girodi*. The original notation is given in parenthesis; the crosses indicate programmed cell death. (Modified from Lechner 1966).

$$
Z \begin{cases}
AB \begin{cases}
A(A_3) \ldots \text{lateral ectoderm, endoderm} \\
B(B_3) \begin{cases}
b^1(b4,2) \ldots \text{frontal ectoderm} \\
b_2(b4,1) \ldots \text{endoderm (stomach, stomach glands)}
\end{cases}
\end{cases} \\
CD \begin{cases}
C(C_3) \ldots \text{lateral ectoderm, endoderm} \\
D(D_3) \begin{cases}
d^1(d4,2) \ldots \text{dorsal and ventral ectoderm} \\
D_1(d4,1) \begin{cases}
d^2(d5,2) \ldots \text{ventral ectoderm, nephridia, bladder, uterus, cloaca} \\
D_2(d5,1) \begin{cases}
d^3(d6,2) \ldots + \\
D_3(d6,1) \begin{cases}
d^4(d7,2) \ldots + \\
D_4(d7,1) \begin{cases}
d^5(d8,1) \ldots \text{vitellarium} \\
D_5(d8,2) \begin{cases}
d^6(d9,3) \ldots + \\
D_6(d9,4) \text{ germ cells}
\end{cases}
\end{cases}
\end{cases}
\end{cases}
\end{cases}
\end{cases}
\end{cases}
$$

nally cover almost the whole dorsal and ventral sides. The small ectodermal cells of the apical region multiply and differentiate into the cerebral ganglion, which finally sinks in and becomes overgrown by the surrounding ectoderm. The mastax ganglion differentiates from the epithelium of the posterior (ventral) side of the pharynx, and the caudal ganglion differentiates from the ectoderm behind the blastopore/mouth. A small caudal appendix, perhaps with a pair of rudimentary toes (Car 1899) develops at an early stage but disappears in the adult *Asplanchna*. Protonephridia, bladder, oviduct and cloaca develop from the d^2-cell. The development of the ciliary bands is poorly described. The muscles of the body wall have been reported to differentiate from ectodermal cells (Nachtwey 1925), but this should be studied with modern methods.

Lechner (1966) showed that the eggs are highly determinate already before the polar bodies are given off, and that the powers of regulation are very limited.

Only Zelinka (1891) has studied the development of bdelloids. His report is difficult to follow in detail, but a 16-cell stage resembling that of *Asplanchna* was found; the D_2-cell was called endoderm, but the following development appears to resemble that of the monogononts.

The development of *Seison* has not been studied.

The interrelationships between the various groups of rotifers are still a matter of some controversy (Wallace & Colburn 1989), but the monophyly of the whole group is usually not questioned (see below). The ancestral rotifer probably possessed a mosaic of characters seen in living groups; the presence of morphologically almost similar males and females, as seen in *Seison*, must be regarded as primitive; paired gonads, as found in the bdelloids, is found in all the other aschelminth phyla and is therefore considered the ancestral character state; the complex of prototroch, adoral ciliary zone and metatroch is most typically represented in the monogononts, and if this ciliary system is homologous to that of many spiralian larvae it must be ancestral in the rotifers (see below). The intracellular skeletal lamina, the mastax, the retrocerebral organ, the toes with adhesive glands, and perhaps the origin of the germovitellarium from the D_2-cell (*Seison* has not been studied) are apomorphies which characterize the rotifers. The presence of toes in all major types may indicate that the ancestral rotifer was a benthic filter-feeder.

The acanthocephalans have a similar intracellular skeletal lamina and are, also for other reasons, by many authors regarded as the sister group of the rotifers (Chapter 30). However, Lorenzen (1985) pointed out that not only do the acanthocephalans share the intracellular skeletal lamina with the rotifers in general, but presumed homologues of the lemnisci and the proboscis of the acanthocephalans are found in bdelloid rotifers; he therefore regarded Acanthocephala as the sister group of Bdelloidea within the Rotifera. The bdelloid rostrum is a ciliated attachment organ used in locomotion when the wheel organ is retracted; its ultrastructure has not been described, and a similar organ is possibly found in notommatid monogononts (Remane 1929-1933). Both structure and function of the lemnisci of the rotifers are poorly known, but the lemnisci are associated with the dorsal proboscis in the acanthocephalans (Chapter 30). The homologies of the structures are therefore very uncertain, and since the embryology of the acanthocephalans indicates a circumpharyngeal origin of the brain (Chapter 30) in contrast to the apical origin of the rotifer brain, I have chosen to regard the two phyla as sister groups.

Hatschek (1878, 1891) stressed the similarities of the ciliary bands of rotifers and trochophora larvae of annelids and molluscs and proposed that the common ancestor of these groups had a larva of this type. He further believed that an even earlier common ancestor (trochozoon) had the morphology of the trochophore itself. The first part of his ideas is in complete accordance with the trochaea theory, and is further discussed in Chapters 2 and 10; the latter part of the ideas has been rejected by most subsequent authors. The alternative that the rotifers are neotenic, i.e. sexually mature trochophores, was proposed already by Lang (1888). The whole idea of ancestral trochophore-type ciliary bands in rotifers fell into disregard when Beauchamp (1907, 1909) published his comparative studies on the ciliary bands of several rotifers. He concluded that the types with the trochophore-type ciliation have evolved several times from an ancestral type with a circumoral ciliary field used in creeping, and that the rotiferan ciliation could be derived from the general ciliation of a flatworm via the ventral ciliation of the gastrotrichs. However, Jägersten (1972) hesitantly supported the old idea that the rotifers have the trochophore ciliation and that this is an 'original larval feature'.

I believe that the rotiferan wheel organ with proto- and metatroch of compound cilia, functioning as downstream-collecting bands, and bordering an adoral zone of single cilia is homologous with the similar bands of the spiralian trochophores. The various other types of wheel organs can be interpreted as adaptations to other feeding types. The trochophore is definitely a larval form (Chapters 2 and 10) and the rotifers must therefore be interpreted as neotenic – not as neotenic annelids, but as neotenic descendants of the protostomian ancestor, gastroneuron.

The planktotrophic rotifers must represent the ancestral type, which as mentioned above may have been temporarily or permanently attached; sessile forms have free juvenile stages, and changes between pelagic and sessile habits may have taken place several times; the macrophagous types, which may be pelagic or creeping, have reduced ciliary bands and must be regarded as specialized.

The trochophore-type morphology of some of the rotifers does not indicate closer relationships with for example annelids, because it is an ancestral character of all the protostomes, and the lack of any trace of spiral cleavage and of an apical organ and the very early differentiation of the primordial germ cell are characteristics found in many of the aschelminth phyla.

The acanthocephalans are regarded as the sister group of the rotifers (Chapter 30), but the interrelationships with other aschelminth groups are more difficult to ascertain (Fig. 28.1). The chaetognaths (Chapter 31) have chitinous jaws which may resemble a mastax, but their integument is very special and their embryology bears no resemblance to that of other aschelminths, except for the early determination of the germ cells. The nervous system with a well defined, dorsal cerebral ganglion derived from the apical region resembles that of the spiralians and is probably a plesiomorphic feature; it is very different from the circumpharyngeal brain of the cycloneuralians (Chapter 32). The mastax ganglion is intimately connected with the pharynx and embryologically derived from the pharynx epithelium, so it is probably not a ventral ganglion. A ventral nerve cord of the gastroneuralian type is absent.

Interesting subjects for future research

1. Ontogenetic origin of the body muscles.
2. Development, structure and function of rostrum and lemnisci.
3. Embryology and ultrastructure of *Seison*.

References

Beauchamp, P. de 1907. Morphologie et variations de l'apparail rotateur dans la série des Rotifères. – Archs Zool. exp. gén., 4. sér., 6: 1-29.

Beauchamp, P. de 1909. Recherches sur les Rotifères: les formations tégumentaires et l'appareil digestif. – Archs Zool. exp. gén., 4. sér., 10: 1-410, pls 1-9.

Beauchamp, P. de 1956. Le développement de *Ploesoma hudsoni* (Imhof) et l'origine des feuillets chez les Rotifères. – Bull. Soc. zool. Fr. 81: 374-383.

Beauchamp, P. de 1965. Classe des Rotifères. – Traité de Zoologie, vol. 4(3), pp 1225-1379. Masson, Paris.

Bender, K. & W. Klenow 1988. Chemical properties of the lorica and related parts from the integument of *Brachionus plicatilis*. – Comp. Biochem. Physiol. 89B: 483-487.

Car, L. 1899. Die embryonale Entwicklung von *Asplanchna brightwellii*. – Biol. Zbl. 19: 59-74.

Clément, P. 1969. Premières observations sur l'ultrastructure comparée des téguments de Rotifères. – Vie Milieu 20A: 461-482, 4 pls.

Clément, P. 1977. Ultrastructural research on Rotifera. – Ergebn. Limnol. 8: 270-297.

Clément, P. & J. Amsellem 1989. The skeletal muscles of rotifers and their innervation. – Hydrobiologia 186/187: 255-278.

Clément, P., J. Amsellem, A.-M. Cornillac, A. Luciani & C. Ricci 1980. An ultrastructural approach to feeding behaviour in *Philodina roseola* and *Brachionus calycifloreus* (Rotifera). I-III. – Hydrobiologia 73: 127-141.

Clément, P. & E. Wurdak 1991. Rotifera. – *In* F.W. Harrison (ed.): Microscopic Anatomy of Invertebrates, vol. 4, pp 219-297. Wiley-Liss, New York.

Costello, D.P. & C. Henley 1976. Spiralian development: a perspective. – Am. Zool. 16: 277-291.

Hatschek, B. 1878. Studien über Entwicklungsgeschichte der Anneliden. – Arb. zool. Inst. Univ. Wien 1: 277-404, pls 23-30.

Hatschek, B. 1891. Lehrbuch der Zoologie, 3. Lieferung (pp 305-432). – Gustav Fischer, Jena.

Jägersten, G. 1972. Evolution of the Metazoan Life Cycle. – Academic Press, London.

Jennings, H.S. 1896. The early development of *Asplanchna herrickii* de Guerne. A contribution to developmental mechanics. – Bull. Mus. comp. Zool. Harv. 30: 1-117, 10 pls.

Jeuniaux, C. 1975. Principes de systématique biochimique et application à quelques particuliers concernant les Aschelminthes, les Polychètes et les Tardigrades. – Cah. Biol. mar. 16: 597-612, 2 pls.

Kleinow, W., J. Klusemann & H. Wratil 1990. A gentle method for the preparation of hard parts (trophi) of the mastax of rotifers and scanning electron microscopy of the trophi of *Brachionus plicatilis* (Rotifera). – Zoomorphology 109: 329-336.

Klusemann, J., W. Kleinow & W. Peters 1990. The hard parts (trophi) of the rotifer mastax do contain chitin: evidence from studies on *Brachionus plicatilis*. – Histochemistry 94: 277-283.

Lang, A. 1888. Lehrbuch der vergleichenden Anatomie der wirbellosen Thiere, 1. Theil (pp 1-290). – Gustav Fischer, Jena.

Lechner, M. 1966. Untersuchungen zur Embryonalentwicklung des Rädertieres *Asplanchna girodi* de Guerne. – Roux Arch. Entwicklungsmech. Org. 157: 117-173.

Lorenzen, S. 1985. Phylogenetic aspects of pseudocoelomate evolution. – *In* S. Conway Morris, J.D. George, R. Gibson & H.M. Platt (eds): The Origins and Relationships of Lower Invertebrates, pp 210-223. Oxford Univ. Press, Oxford.

Martini, E. 1912. Studien über die Konstanz histologischer Elemente. III. *Hydatina senta*. – Z. wiss. Zool. 102: 425-645, pls 20-29.

Nachtwey, R. 1925. Untersuchungen über die Keimbahn, Organogenese und Anatomie von *Asplanchna priodonta* Gosse. – Z. wiss. Zool. 126: 239-492, pls 6-13.

Nielsen, C. 1987. Structure and function of metazoan ciliary bands and their phylogenetic significance. – Acta zool. (Stockh.) 68: 205-262.

Peters, F. 1931. Untersuchungen über Anatomie und Zellkonstanz von *Synchaeta* (*S. grimpei* Remane, *S. baltica* Ehrenb., *S. tavina* Hood und *S. triophthalma* Lauterborn). – Z. wiss. Zool. 139: 1-119.

Pray, F.A. 1965. Studies on the early development of the rotifer *Monostyla cornuta* Müller. – Trans. Am. microsc. Soc. 84: 210-216.

Remane, A. 1929-1933. Rotifera. – Bronn's Klassen und Ordnungen des Tierreichs, 4. Band, 2. Abt., 1. Buch, 1-4 Lief. (pp 1-576). Akademische Verlagsgesellschaft, Leipzig.

Storch, V. 1984. Minor pseudocoelomates. – *In* J. Bereiter-Hahn, A.G. Matoltsy & K.S. Richards (eds): Biology of the Integument, vol. 1, pp 242-268. Springer, Berlin.

Storch, V. & U. Welsch 1969. Über den Aufbau des Rotatorienintegumentes. – Z. Zellforsch. 95: 405-414.

Stossberg, K. 1932. Zur Morphologie der Rädertiergattungen *Euchlanis*, *Brachionus* und *Rhinoglaena*. – Z. wiss. Zool. 142: 313-424.

Strathmann, R.R., T.L. Jahn & J.R.C. Fonseca 1972. Suspension feeding by marine invertebrate larvae: clearance of particles by ciliated bands of a rotifer, pluteus, and trochophore. – Biol. Bull. Woods Hole 142: 505-519.

Tannreuther, G.W. 1920. The development of *Asplanchna ebbersbornii* (Rotifer). – J. Morph. 33: 389-437.

Van Cleave, H.J. 1922. A determination of the degree of constancy in the nuclei of certain organs in *Hydatina senta*. – Biol. Bull. Woods Hole 42: 85-98.

246

Wallace, R.L. & R.A. Colburn 1989. Phylogenetic relationships within phylum Rotifera: orders and genus *Notholca*. – Hydrobiologia 186/187: 311-318.

Wesenberg-Lund, C. 1952. De danske søers og dammes dyriske plankton. – Munksgaard, Copenhagen.

Zelinka, C. 1886. Studien über Räderthiere. I. Über die Symbiose und Anatomie von Rotatorien aus dem Genus *Callidina*. – Z. wiss. Zool. **44**: 396-506, pls 26-29.

Zelinka, C. 1888. Studien über Räderthiere. II. Der Raumparasitismus und die Anatomie von *Discopus synaptae* n.g., nov.sp. – Z. wiss. Zool. **47**: 141-246, pls 30-34.

Zelinka, C. 1891. Studien über Räderthiere. III. Zur Entwicklungsgeschichte der Räderthiere nebst Bemerkungen über ihre Anatomie und Biologie. – Z. wiss. Zool. **53**: 1-159, pls 1-6.

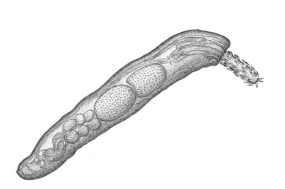

Phylum ACANTHOCEPHALA

Acanthocephalans are a phylum of about 900 aquatic or terrestrial, completely gut-less parasites, with the juveniles occurring in arthropods and the adults in the alimentary canal of vertebrates. The adults are from 2 mm up to almost 1 meter long. The eggs are fertilized in the female and develop into the acanthor stage before the eggs are shed and leave the host with the faeces. The eggs must become ingested by an intermediate host where the acanthor hatches in the intestine and enters the intestinal wall; here it develops into the acanthella stage and further into the cyst-acanth, which is the stage capable of infecting the final host. The intermediate host must be eaten by the final host (Schmidt 1985). Three orders are recognized, based mainly on differences in the morphology of the lacunar system (see below).

The body of the adult is cylindrical or slightly flattened and the anterior end has an inversible, cylindrical proboscis with recurved hooks, ideal for anchoring the parasite to the intestinal wall of the host; many species have rings of smaller or larger spines or hooks on the whole body. There is no trace of an alimentary canal at any stage, and it cannot be seen directly if the proboscis represents the anterior part of the body with a reduced, terminal mouth or a dorsal attachment organ, as indicated by Lorenzen (1985, see below). Also the dorsal-ventral orientation has been questioned (Haffner 1950).

The body wall consists of a syncytial ectoderm, a thick basement membrane, an outer layer of circular muscles and an inner layer of longitudinal muscles (Dunagan & Miller 1991). The ectoderm or tegument has very few gigantic nuclei with fixed positions; in a few families, the nuclei fragment during the later part of the development. The apical cell membrane shows numerous branched, tubular infoldings which penetrate an intracellular lamina (see below); these infoldings increase the surface area many times and are believed to facilitate nutrient uptake through the tegument (Graeber & Storch 1978, Whitfield 1984, Dunagan & Miller 1991). Apposed to the apical cell membrane is a thin, outer, electron-dense layer and a thicker, somewhat less electron-dense layer, which together are called the intrasyncytial lamina or intracellular cuticle (Storch & Welsch 1970, Whitfield 1984), and which will here be referred to as

Chapter vignette: *Acanthocephalus opsalichthydis*. (Redrawn from Yamaguti 1935.)

the intracellular lamina. The basal part of the epidermis is penetrated by a system of anastomosing canals, called the lacunar system, which apparently transports fluids through the body. Studies by Miller & Dunagan (1985a) show that the lacunar canals extend into conspicuous longitudinal canals on the inside of the muscle layers; the canals of the lacunar system are connected with canals inside both longitudinal and circular muscle cells. It is believed that the fluid in the lacunar system is moved by contractions of the body wall musculature. The ultrastructure of this system has not been worked out. The ultrastructure of the conspicuous hooks of larval and adult stages is not well known; Hutton & Oetinger (1980) described the development of the hooks of an acanthella larva, and although the description is not particularly clear it appears that the hooks develop as specialized cells originating below the basement membrane.

The proboscis has several associated sets of muscles which are involved in protrusion, eversion and retraction (Hammond 1966). The proboscis region can be protruded by the muscles of the body wall and retracted by the neck retractor muscles which surround the lemnisci and attach to the body wall. The inverted proboscis lies in a receptacle which has a single or double wall of muscles. The contraction of these muscles everts the proboscis with the receptacle fluid functioning as a hydrostatic skeleton, and a contraction of the neck retractors squeezes fluid from the lemnisci to the wall of the proboscis, which swells. A retractor muscle from the tip of the proboscis to the bottom of the receptacle inverts the proboscis, and the receptacle can be retracted further into the body by the contraction of the receptacle retractor which extends from the bottom of the receptacle to the ventral body wall.

The brain comprises a low, species-specific number of cells (review in Miller & Dunagan 1985b). Brain nerves have been traced to the muscles of the body wall and the proboscis, to paired genital ganglia, to a pair of sense organs at the base of the proboscis, and to a pair of sensory structures at the apical organ (Gee 1988). The apical organ is situated at the tip of the proboscis and comprises both nervous and glandular elements, but it appears that both structure and function are in need of further investigations based on a number of species before definite statements can be made (Miller & Dunagan 1985b, Dunagan & Miller 1991).

There is a spacious body cavity, which functions as a hydrostatic organ. It contains an enigmatic organ called the ligament sac(s), which develops in all types, but degenerates in some forms (Dunagan & Miller 1991). The ligament sac(s) and the gonads develop in the acanthella from a central mass of cells between the brain and the cloaca, and it is generally believed that a median string, called the ligament, represents endoderm. There is a single sac or a dorsal and a ventral sac which communicate anteriorly. The sacs are acellular, fibrillar structures which contain collagen (Haffner 1942). The posterior end of the (dorsal) ligament sac is connected with the uterine bell (see below).

Large excretory organs in the family Oligacanthorhynchidae (Dunagan & Miller 1986, 1991) have a pair of large, complicated, latero-dorsal protonephridia with short ducts opening into an unpaired bladder; each protonephridium is a syncytium with three nuclei situated centrally and many radiating flame bulbs with high numbers of cilia, which lack the central tubules. An unpaired, ciliated excretory canal opens into the urogenital canal.

The gonads are suspended by the ligament strand. The testes have ducts which open into a urogenital canal, which in turn opens on the tip of a small penis at the bottom of a bursa copulatrix. The primary ovaries develop on the dorsal side of the ligament strand but soon break up into fragments, often called ovarian balls, which float in the ligament sac or freely in the pseudocoel (Crompton 1985). The male injects the sperm into the uterus and the eggs become fertilized in the ovarial fragments and get surrounded by an oval, resistant, chitin-containing shell with a number of layers. The polar bodies are given off and cleavage begins before the egg detaches from the ovary; the polar bodies are situated at one pole of the ellipsoidal egg and mark the future anterior end. The uterine bell is supposed to function as a sorting apparatus which ensures that only eggs with a fully developed acanthor larva are shed (Miller & Dunagan 1985b).

The first two cleavages result in an embryo with one anterior, two median and one posterior cell; the blastomeres are usually of equal size, but the posterior cell is larger than the others in a few species. The embryo becomes syncytial at a stage of 4-36 cells according to the species. Meyer (1928, 1932-1933, 1936, 1938) followed the cell lineage (or rather the nuclear lineage) of *Macracanthorhynchus* (Fig. 30.1) and reported a modified spiral cleavage with a primary axis slightly oblique to the longitudinal axis of the egg; the A and C-cells of the 4-cell stage are in contact along the whole primary axis and the spindles of the following cleavages are almost parallel. A few smaller nuclei at the cleavage pole near the polar bodies migrate into the embryo, and these nuclei give rise to ligament and gonads; later stages show small nuclei migrating to the inside of the embryo from a ring-shaped area around the same region and giving rise to the ganglion. The area with inwandering nuclei was interpreted as the blastoporal area, and the nuclei at the opposite cleavage pole were observed to migrate to the anterior end of the egg where the proboscis develops. A ring of spines or hooks with associated myofibrils develops in the anterior end, and the acanthor larva is ready for hatching.

The eggs with fully developed acanthor larvae are shed and leave the host with the faeces. If the eggs become ingested by an intermediate host, the larva hatches and penetrates the wall of the midgut by means of the hooks and enters the host's haemocoel. The acanthor loses the hooks and their associated muscles degenerate, and the early acanthella stage is reached. The various organ systems differentiate from the groups of nuclei seen in the acanthor stage (Hamann 1891, Meyer 1932-1933, 1938), but the details of organogenesis have not been worked out. Most tissues remain syncytial, but the nervous system and the muscles become cellular. The lemnisci develop as a pair of long, syncytial protrusions from the ectoderm around the proboscis invagination (Hamann 1891); in *Macracanthorhynchus*, Meyer (1938) observed a ring of 12 very large nuclei which slowly migrated into the early, cytoplasmic protrusions. The proboscis apparatus is at first enclosed by the syncytial ectoderm, but an opening is formed, and the larva is now in the cystacanth stage, which has almost the adult morphology, and is ready for infection of the final host.

The systematic position of the acanthocephalans has been a matter of discussion for a long time, but almost all newer papers place them in or near the aschelminths. Conway Morris & Crompton (1982) compared the Burgess Shale priapulan *An-*

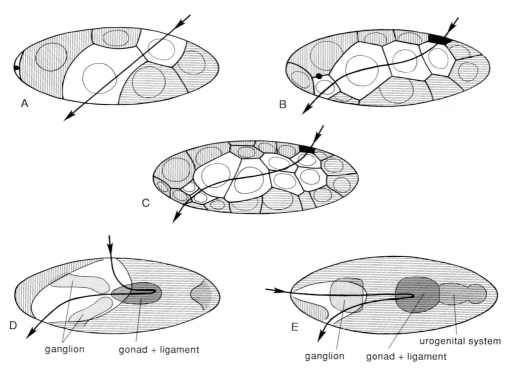

Fig. 30.1. Embryology of *Macracanthorhynchus hirudinaceus*; embryos seen from the left side. The first three stages are cellular and the two last syncytial. – A, 8-cell stage. – B, 17-cell stage. – C, 34-cell stage. – D, early stage of the internalization of the condensed nuclei of the inner organs and of the movement of the apical pole. – E, stage with fully organized primordia of the inner organs and with the apical pole at the anteroir end. – The a-quadrant is white, the b-quadrant horizontally hatched, the c-quadrant black, and the d-quadrant vertically hatched; optical sections of the inner primordia are shown by shading. (Modified from Meyer 1928, 1936.)

calagon with the living acanthocephalans and found so many similarities that they considered priapulans and acanthocephalans as sister groups. The overall resemblance between the two groups is considerable, but the intracellular nature of both the 'cuticle' and the spines on the proboscis in acanthocephalans is in strong contrast to the true cuticular structure of these organs in priapulans and demonstrates that the resemblance is completely superficial.

Several recent papers emphasize the importance of the syncytial ectoderm with an intracellular lamina found also in rotifers (Storch & Welsch 1970, Storch 1979), and Lorenzen (1985) regarded the acanthocephalans as a group of parasitic rotifers closely related to the bdelloids. I find that both the ectoderm and the characteristic embryology with the movement of the apical cleavage pole from a position near the posterior end of the embryo to the anterior end are characters which demonstrate that acanthocephalans and rotifers are a monophyletic unit. The proposed homology of the bdelloid retrocerebral organ and the acanthocephalan proboscis is not supported by the embryological origin of the organs in the two groups, and the function of the lemnisci cannot be the same in the two groups, because they are associated with the trochal discs in the bdelloids and with the proboscis in the acanthocephalans. The two phyla must be regarded as sister groups.

The position of the acanthocephalans within the aschelminthes is indicated by their lack of spiral cleavage, lack of a larval apical organ, and early differentiation of the germinative cells (nuclei). The brain is possibly a modified circumpharyngeal ganglion like that of the cycloneuralians (Chapter 32), but I regard this as a convergence. The typical protostomian (gastroneuralian) nervous system with a midventral longitudinal pair of nerves is not present either, but a highly modified nervous system is not unexpected in strongly specialized parasites.

The embryology is very unusual. It appears that the polar bodies are situated closer to the blastoporal pole than to the apical pole in the early embryos, but several other aschelminths, for example nematodes (Chapter 35), have polar bodies which move away from the apical pole. The report of a modified spiral cleavage in *Macracanthorhynchus* must be rejected; the cleavage pattern shows no spiral traits, and the origin of the mesoderm is not documented. If Meyer's (1938) interpretation of the developmental processes is correct, the inward migration of condensed nuclei could be interpreted as a modified gastrulation and the ganglion as a circumpharyngeal nerve ring which has become compact; the ligament could be the reduced endoderm and the proboscis with the terminal 'apical' sensory organ an apical structure. There is thus a considerable similarity with the embryology of the rotifers (Chapter 29, Fig. 29.1), which have an early internalization of the primordial cell of the germovitellarium followed by an invagination which gives rise to the gut; the migration of the apical pole along the dorsal side to the anterior end of the adult is also observed (Fig. 29.1). The only important difference seems to be that the rotiferan brain is derived from the apical pole, whereas the acanthocephalan brain appears to be a circumpharyngeal structure. This interpretation of the embryology implies that the proboscis is an apical structure and therefore not homologous with the introverts of the introvert phyla (Chapter 34). It further implies that the brain is ventral, which is in accordance with the orientation accepted by most authors. It is furthermore clear that acanthocephalans and rotifers must be sister groups, and that the acanthocephalans therefore cannot be 'parasitic rotifers'.

Interesting subjects for future research

1. Cell (or nucleus) lineages of representatives of major groups.
2. Development and ultrastructure of the hooks of larvae and adults.
3. Development and ultrastructure of the ligament sac.

References

Conway Morris, S. & D.W.T. Crompton 1982. The origins and evolution of the Acanthocephala. – Biol. Rev. 57: 85-115.
Crompton, D.W.T. 1985. Reproduction. – *In* D.W.T. Crompton & B.B. Nickol (eds): Biology of the Acanthocephala, pp 213-271. Cambridge Univ. Press, Cambridge.

Dunagan, T.T. & D.M. Miller 1986. A review of protonephridial excretory systems in Acantho-cephala. – J. Parasit. **72**: 621-632.

Dunagan, T.T. & D.M. Miller 1991. Acanthocephala. – *In* F.W. Harrison (ed.): Microscopic Ana-tomy of Invertebrates, vol. 4, pp 299-332. Wiley-Liss, New York.

Gee, R.J. 1988. A morphological study of the nervous system of the praesoma of *Octospinifer mali-centus* (Acanthocephala: Noechinorhynchidae). – J. Morph. **196**: 23-31.

Graeber, K. & V. Storch 1978. Elektronenmikroskopische und morphometrische Untersuchungen am Integument der Acanthocephala (Aschelminthes). – Z. Parasitenkunde **57**: 121-135.

Haffner, K.v. 1942. Untersuchungen über das Urogenitalsystem der Acanthocephalen. I.-III. – Z. Morph. Ökol. Tiere **38**: 251-333.

Haffner, K.v. 1950. Organisation und systematische Stellung der Acanthocephalen. – Zool. Anz. **145**(Suppl): 243-274.

Hamann, O. 1891. Monographie der Acanthocephalen. – Jena. Z. Naturw. **25**: 113-231, pls 5-14.

Hammond, R.A. 1966. The proboscis mechanism of *Acanthocephalus ranae*. – J. exp. Biol. **45**: 203-213.

Hutton, T.L. & D.F. Oetinger 1980. Morphogenesis of the proboscis hooks of an archiacantho-cephalan, *Moniliformis* (Bremser 1911) Travassos 1915. – J. Parasit. **66**: 965-972.

Lorenzen, S. 1985. Phylogenetic aspects of pseudocoelomate evolution. – *In* S. Conway Morris, J.D. George, R. Gibson & H.M. Platt (eds): The Origins and Relationships of Lower Invertebrates, pp 210-223. Oxford Univ. Press, Oxford.

Meyer, A. 1928. Die Furchung nebst Eibildung, Reifung und Befruchtung des *Gigantorhynchus gigas*. – Zool. Jb., Anat. **50**: 117-218, pls 1-5.

Meyer, A. 1932-1933. Acanthocephala. – Bronn's Klassen und Ordnungen des Tierreichs, 4. Band, 2. Abt., 2. Buch. Akademische Verlagsgesellschaft, Leipzig.

Meyer, A. 1936. Die plasmodiale Entwicklung und Formbildung des Riesenkratzers (*Macracan-thorhynchus hirudinaceus*). I. Teil. – Zool. Jb., Anat. **62**: 111-172, pls 2-4.

Meyer, A. 1938. Die plasmodiale Entwicklung und Formbildung des Riesenkratzers (*Macracan-thorhynchus hirudinaceus* (Pallas)). III. Teil. – Zool. Jb., Anat. **64**: 131-197, pl. 8.

Miller, D.M. & T.T. Dunagan 1985a. New aspects of acanthocephalan lacunar system as revealed in anatomical modelling by corrosion cast method. – Proc. helminth. Soc. Wash. **52**: 221-226.

Miller, D.M. & T.T. Dunagan 1985b. Functional morphology. – *In* D.W.T. Crompton & B.B. Nickol (eds): Biology of the Acanthocephala, pp 73-123. Cambridge Univ. Press, Cambridge.

Schmidt, G.D. 1985. Development and life cycles. – *In* D.W.T. Crompton & B.B. Nickol (eds): Bi-ology of the Acanthocephala, pp 273-305. Cambridge Univ. Press, Cambridge.

Storch, V. 1979. Contributions of comparative ultrastructural research to problems of invertebrate evolution. – Am. Zool. **19**: 637-645.

Storch, V. & U. Welsch 1970. Über den Aufbau resorbierender Epithelien darmloser Endoparasiten. – Zool. Anz., Suppl. **33**: 617-621.

Whitfield, P.J. 1984. Acanthocephala. – *In* J. Bereiter-Hahn, A.G. Matoltsy & K.S. Richards (eds): Biology of the Integument, pp 234-241. Springer, Berlin.

Yamaguti, S. 1935. Studies on the helminth fauna of Japan. Part 8. Acanthocephala, I. – Jap. J. Zool. **6**: 247-278.

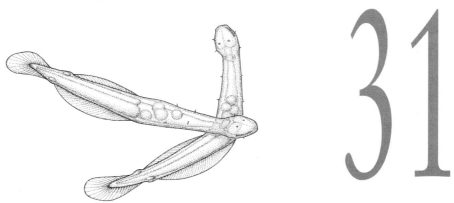

Phylum CHAETOGNATHA

Arrowworms are a small phylum of marine, mainly holopelagic 'worms'; some species, mainly of the genera *Spadella* and *Paraspadella*, are benthic, but it appears that many deep-water forms remain to be described. A recent estimate of the total number of species, known and unknown, is in the order of 200 (Bieri 1991). The Burgess Shale fossil *Amiskwia* has usually been interpreted as a chaetognath, but this is rejected by Bieri (1991), so the oldest reliable fossil appears to be the Carboniferous *Paucijaculum* (Schram 1973).

The body is cylindrical with a rounded head and a tapering tail; there are one or two pairs of lateral fins and a large, horizontal tail fin. The body is usually quite transparent and the fins are completely hyaline. The head has a ventral mouth surrounded by various chitinous teeth and paired lateral groups of large spines (hooks) used in grasping prey organisms, mainly copepods. The prey is apparently poisoned by a tetrodotoxin produced by bacteria somewhere in the head of the chaetognaths, but their exact location is unknown (Thuesen 1991). An ectodermal fold, the hood, can enwrap the head almost completely giving it a streamlined shape, and it can be retracted rapidly to expose the teeth and spines; ventrally it originates at the posterior border of the head, laterally the attachment curves forward, and the dorsal attachment forms an inverted V which reaches almost to the anterior tip of the head (Burfield 1927).

The ectoderm is monolayered with a cuticle at the inner side of the hood and on the anterior and ventral sides of the head, whereas the remaining parts of the body have a multilayered ectoderm. The cuticle with teeth and spines forms a continuous structure (Ahnelt 1984). The teeth and spines are complicated structures formed by several cells; they have a high content of crystalline α-chitin with very little protein and are impregnated with zinc and silicon (Bone, Ryan & Pulsford 1983). The multilayered epithelium consists of an outer layer of polygonal cells covering two or more layers of interdigitating cells with abundant bundles of tonofilaments (Welsch & Storch 1983). The ectoderm rests on a basement membrane which comprises a

Chapter vignette: Two courting *Spadella cephaloptera*. (Redrawn from G. Thorson's Christmas card 1966.)

network of crossing bundles of collagen filaments (Duvert & Salat 1990b). The fins are extensions of the basement membrane covered by the multilayered epithelium and stiffened by elongate fin ray cells with a paracrystalline body of filaments with aligned substructures (Duvert & Salat 1990a).

Two ectodermal organs of unknown function are located anteriorly on the dorsal side: the corona and the retrocerebral organ. The corona is an oval band of monociliate cells surrounding a glandular epithelium (Kuhl 1938, Malakhov & Frid 1982); the ciliary cells appear to be sensory and innervated by a pair of coronal nerves, but the function of the organ is totally unknown. The retrocerebral organ consists of a pair of sacs located at the posterior side of the cerebral ganglion with fine ducts opening in a common anterior pore. The sacs consist of large cells with numerous intertwined microvilli with a core filament (Bone & Goto 1991).

The mouth opens into a pharynx, or oesophagus, which leads to a tubular intestine, which has a pair of anterior diverticula in some species, and further to a short rectum; absorptive cells of the intestine and all rectal cells are multiciliate (Parry 1944).

The nervous system comprises a number of ganglia, notably the cerebral ganglion, a pair of vestibular ganglia sending nerves to the muscles of the spines, and a large ventral ganglion; paired nerves connect the cerebral ganglion with the vestibular ganglia and the ventral ganglion (Goto & Yoshida 1987). The ventral ganglion is a longitudinally elongate, rectangular structure situated at the median part of the ventral side of the body. It has 12 pairs of lateral nerves and a pair of posterior nerves, which pass beyond the anus. Mapping of individual large cells and fibres in the ventral ganglion has begun (Bone & Pulsford 1984, Goto & Yoshida 1987, Bone & Goto 1991). The reported absence of the ventral ganglion in *Bathybelos* (Bieri & Thuesen 1990) requires confirmation. The nervous structures are all situated outside the basement membrane or sunken down but still surrounded by the basement membrane (Salvini-Plawen 1988), and the innervation of muscles is through the basement membrane, in some cases via thin-walled pits in the membrane (Duvert & Barets 1983). Sensory structures comprise paired eyes, each with numerous ciliary sensory cells and one pigment cell, and vibration-sensitive ciliary fence organs with monociliate sensory cells (Bone & Goto 1991, Reisinger 1970).

There are a number of body cavities surrounded by a mesodermal epithelium and thus falling within the usual definition of coeloms (Duvert & Salat 1979, Welsch & Storch 1982). The head contains one narrow cavity and there is a pair of lateral cavities in the body; the lateral cavities are divided by a transverse septum at the level of the anus into an anterior part containing the ovaries and a posterior part containing the testes. The body musculature consists of four areas of longitudinal 'primary' muscles, dorsally and ventrally separated by the basement membrane of the mesenteries and narrow bands of 'secondary' muscles and laterally by narrow bands of 'secondary' muscles (Duvert 1989). The primary muscles are cross-striated and consist of two types of muscle fibres (Duvert & Salat 1980). The secondary muscles likewise consist of cross-striated fibres, but the ultrastructure is unique, with two alternating types of sarcomeres (Duvert 1991). The body musculature and the gut are covered by a peritoneum, which forms a dorsal and a ventral mesentery. Some of the epithelial cells associated with the secondary muscles contain myofilaments and carry single cilia, and also the peritoneal cells of the gut are myoepithelial (Duvert 1989, Welsch & Storch 1982). The posterior lateral cavities are subdivided by incomplete longitudinal mesenteries, which may be partially ciliated (Alvariño 1983).

There are no haemal or nephridial systems.

Both testes and ovaries are elongate bodies covered by the peritoneum and attached to the lateral body wall by a mesentery (Kuhl 1938). Clusters of spermatogonia break off from the testes and circulate in the posterior cavities during differentiation. Mature spermatozoa pass into a pair of seminal vesicles and are passed in a loosely organized spermatophore to the body surface of the partner at a pseudocopulation; the spermatozoa then migrate to the female gonopore (Ghirardelli 1968, Pearre 1991). An oviduct, consisting of an inner, syncytial layer and an outer, cellular layer, lies laterally in the ovary and opens on a small dorsolateral papilla at the level of the anus; the inner part of the oviduct functions as a seminal receptacle (Shinn 1992). The ripe eggs are surrounded by a thick membrane with a small micropyle occupied by two accessory cells which form a canal for the penetration of the sperm from the oviduct (Stevens 1903). The fertilized eggs enter the oviduct and are shed freely in the water or attached in small packets near the gonopore or to objects on the bottom (Kapp 1991).

The development of *Sagitta* and *Spadella* has been studied by a number of authors (Hertwig 1880, Doncaster 1902, Burfield 1927, John 1933); no important differences have been reported between the genera and species, but many stages are still in need of further studies.

After fertilization and formation of the polar bodies, a small, round body, the 'germ cell determinant', develops in the cytoplasm near the blastoporal pole, and this body remains undivided during the first five cleavages, situated in one of the cells at the blastoporal pole (Elpatiewsky 1910; Fig. 31.1). Buchner (1910) believed that this organelle was a degenerate accessory cell, but this was refuted by Stevens (1910). At the sixth cleavage the body divides and the material becomes distributed to two daughter blastomeres, which can also be recognized by their large nuclei; these cells are the primordial germ cells. The cleavage is total and equal, and the early development is rather easy to follow because of the transparency of the embryos. The 4-cell stage shows a spiral cleavage configuration with two opposite blastomeres in wide contact at the apical pole and the other pair in contact at the blastoporal pole (Hertwig 1880, Elpatiewsky 1910). A blastula with a narrow blastocoel develops; one side flattens at a stage of about 64 cells, and a typical invagination gastrula is formed. The two primordial germ cells are situated at the bottom of the archenteron, but they soon detach from its wall, move into the archenteron and divide once; one cell eventually ends up in the posterior mesodermal compartments of each side. The anterolateral parts of the archenteron wall form a pair of folds which grow towards the blastopore carrying the germ cells at the tips. The anterior part of the archenteron thus becomes divided into a median gut compartment and lateral mesodermal sacs. The blastopore closes and a new opening breaks through from an anterior, stomodaeal invagination to the gut. The anterior parts of the mesodermal sacs become pinched off and become the head mesoderm. The embryo now elongates and curves

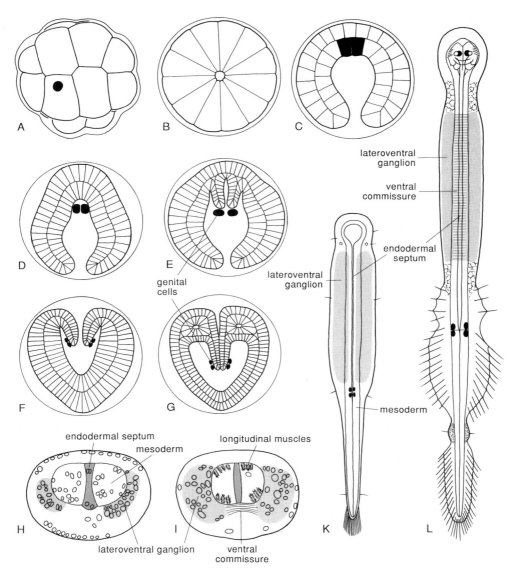

Fig. 31.1. Development of *Sagitta* sp. – A, 16-cell stage seen from the blastoporal pole; the germ cell determinant is indicated in black. – B, blastula. – C, gastrula; the germ cells (black) are still situated in the endoderm. – D, late gastrula; the germ cells are situated in the archenteron. – E, early stage of mesoderm formation. – F, development of the mouth. – G, development of the anterior pair of coelomic pouches. – H, transverse section of a stage like that shown in Fig. G, showing the lateroventral ganglionic cells (grey). – I, transverse section of a stage like that in Fig. L, showing the ventral ganglion. – K, newly hatched juvenile. – L, juvenile four days after hatching, with brain and ventral ganglion. (A and C redrawn from Elpatiewsky 1910; B and D-G redrawn from Burfield 1927; H-L redrawn from Doncaster 1902.)

257

inside the egg membrane, and all cavities inside the embryo disappear. The gut becomes a flat, median column bordered by the lateral mesodermal masses, which meet along the midline at the posterior end of the embryo. The brain ganglion develops from a thickening of the dorsal ectoderm of the head, and the ventral ganglion develops as a pair of lateral ectodermal thickenings, which fuse ventrally some time after the hatching.

The newly hatched juvenile (Doncaster 1902) is completely compact, but cavities develop as slits both in the gut and the mesoderm; the two mesodermal cavities in the head fuse, and the lateral mesodermal sacs become quite thin-walled with two primordial germ cells at each side of the median mesenterium. The mesoderm with the germ cells develops a fold which finally divides each lateral cavity into an anterior and a posterior compartment, each with one of the germ cells at the lateral wall; the anterior germ cells become incorporated into the ovaries and the posterior ones into the testes. The anus breaks through at the level of the septa between the lateral cavities.

The monophyletic character of the Chaetognatha has not been questioned, and most authors regard the phylum as one of the most isolated groups among the metazoans. Both the multilayered epithelium, the structure of the fins, and the morphology of the ventral ganglion can be mentioned as apomorphies characterizing the phylum. Its phylogenetic position has consequently been a matter of much discussion, and attempts have been made to relate them to almost every other metazoan phylum, both protostomian and deuterostomian (Kuhl 1938, Ghirardelli 1968, Bone, Kapp & Pierrot-Bults 1991).

The embryology and especially the differentiation of the archenteron resembles the enterocoely of the deuterostomes (Chapter 43) and particularly that of brachiopods (Chapter 45), and this is the reason why so many authors have placed the chaetognaths in the Deuterostomia. However, a more specific position or sister group has never been pointed out, and most other characters point to protostomian relationships. The lateral body cavities of the chaetognaths are definitely coeloms, but their homology with coeloms of other groups is unproven. The coelomic sacs are formed through enterocoely as in the deuterostomes, but the cavities cannot be homologized with the proto-, meso- and metacoelomic sacs characteristic of the deuterostomes; there are no tentacles on a mesosome, and the division of the posterior coeloms into anterior cavities with ovaries and posterior cavities with testes bears no resemblance to the metacoels of the deuterostomes. The coelomic cavities of the chaetognaths must be regarded as an apomorphy of the phylum.

The multilayered epithelium of the body has been interpreted as a deuterostome character, but within the deuterostomians such epithelia are only known from the vertebrates, and to my knowledge nobody has proposed that chaetognaths and vertebrates should be sister groups. A multilayered epithelium of the body has also been described in the polychaete *Travisia* (Chapter 17), so this character appears unreliable.

The nervous system is decidedly protostomian (gastroneuralian) both in the adults and in the embryos. The development of a longitudinal, median ganglion by the fusion of lateral, intraepithelial ganglia is one of the key characters of the protostomians (Chapter 10). The chaetognath brain has not been related to an apical pole, but it is characteristic of aschelminths that they lack larvae and a ciliated apical organ (Chapter 28). The cleavage pattern is almost radial, which is unusual in the aschelminths, but the early differentiation of the germ cells is seen in many other aschelminths. A more specific relationship could perhaps be indicated in the structure of the chitinous cuticle of the head with the teeth and spines, which resembles the chitinous cuticular membrane with the mastax apparatus of the rotifers (Chapter 29). Both structures have a very high content of chitin. I find it premature to identify rotifers (+ acanthocephalans) as the sister group of the chaetognaths, but it could be interesting to make some comparative studies. Until this has been done, I prefer to place the chaetognaths in an unresolved trichotomy as shown in Fig. 28.1.

Interesting subjects for future research

1. Embryology of a number of species – origin of the mouth, origin and differentiation of the mesoderm
2. Comparative studies of rotifers and chaetognaths.

References

Ahnelt, P. 1984. Chaetognatha. – In J. Bereiter-Hahn, A.G. Matoltsy & K.S. Richards: Biology of the Integument, vol. 1, pp 746-755. Springer, Berlin.

Alvariño, A. 1983. Chaetognatha. – In K.G. Adiyodi & R.G. Adiyodi (eds): Reproductive Biology of Invertebrates, vol. 2, pp 531-544. John Wiley & Sons, Chichester.

Bieri, R. 1991. Systematics of the Chaetognatha. – In Q. Bone, H. Kapp & A.C. Pierrot-Bults (eds): The Biology of Chaetognaths, pp 122-136. Oxford Univ. Press, Oxford.

Bieri, R. & E.V. Thuesen 1990. The strange worm Bathybelos. – Am. Scient. 78: 542-549.

Bone, Q. & T. Goto 1991. The nervous system. – In Q. Bone, K.P. Ryan & A.C. Pierrot-Bults (eds): The Biology of Chaetognaths, pp 18-31. Oxford Univ. Press, Oxford.

Bone, Q., H. Kapp & A.C. Pierrot-Bults 1991. Introduction and relationships of the group. – In Q. Bone, H. Kapp & A.C. Pierrot-Bults (eds): The Biology of Chaetognaths, pp 1-4. Oxford Univ. Press, Oxford.

Bone, Q. & A. Pulsford 1984. The sense organs and ventral ganglion of Sagitta (Chaetognatha). – Acta zool. (Stockh.) 65: 209-220.

Bone, Q., K.P. Ryan & A.L. Pulsford 1983. The structure and composition of the teeth and grasping spines of chaetognaths. – J. mar. biol. Ass. U.K. 63: 929-939.

Buchner, P. 1910. Keimbahn und Ovogenese von Sagitta. – Zool. Anz. 35: 433-443.

Burfield, S.T. 1927. L.M.B.C. Memoir 28: Sagitta. – Proc. Trans. Lpool biol. Soc. 41(Appendix 2): 1-101, 12 pls.

Doncaster, L. 1902. On the development of Sagitta; with notes on the anatomy of the adult. – Q. Jl microsc. Sci., N.S. 46: 351-395, pls 19-21.

Duvert, M. 1989. Etude de la structure et de la fonction de la musculature locomotrice d'un invertébré. Apport de la biologie cellulaire a l'histoire naturelle des Chaetognathes. – Cuad. Invest. Biol. 15: 1-130.

Duvert, M. 1991. A very singular muscle: the secondary muscle of chaetognaths. – Phil. Trans R. Soc. B 332: 245-260.

Duvert, M. & A.L. Barets 1983. Ultrastructural studies of neuromuscular junctions in visceral and skeletal muscles of the chaetognath Sagitta setosa. – Cell Tissue Res. 233: 657-669.

Duvert, M. & C. Salat 1979. Fine structure of muscle and other components of the trunk of Sagitta setosa (Chaetognatha). – Tissue Cell 11: 217-230.

Duvert, M. & C. Salat 1980. The primary body-wall musculature in the arrow-worm Sagitta setosa (Chaetognatha): an ultrastructural study. – Tissue Cell 12: 723-738.

Duvert, M. & C. Salat 1990a. Ultrastructural studies on the fins of chaetognaths. – Tissue Cell 22: 853-863.

Duvert, M. & C. Salat 1990b. Ultrastructural and cytochemical studies on the connective tissue of chaetognaths. – Tissue Cell 22: 865-878.

Elpatiewsky, W. 1910. Die Urgeschlechtszellenbildung bei Sagitta. – Anat. Anz. 35: 226-239.

Ghirardelli, E. 1968. Some aspects of the biology of the chaetognaths. – Adv. mar. Biol. 6: 271-375.

Goto, T. & M. Yoshida 1987. Nervous system in Chaetognatha. – In M.A. Ali (ed.): Nervous Systems in Invertebrates (NATO ASI, Ser. A 141), pp 461-481. Plenum Press, New York.

Hertwig, O. 1880. Die Chaetognathen. Eine Monographie. – Jena. Z. Naturw. 14: 196-311, pls 9-14.

John, C.C. 1933. Habits, structure, and development of Spadella cephaloptera. – Q. Jl microsc. Sci., N.S. 75: 625-696, pls 34-38.

Kapp, H. 1991. Morphology and anatomy. – In Q. Bone, H. Kapp & A.C. Pierrot-Bults (eds): The Biology of Chaetognaths, pp 5-17. Oxford Univ. Press, Oxford.

Kuhl, W. 1938. Chaetognatha. – Bronn's Klassen und Ordnungen des Tierreichs, 4. Band, 4. Abt., 2. Buch, 1. Teil. Akademische Verlagsgesellschaft, Leipzig.

Malakhov, V.V. & M.G. Frid 1982. The structure of ciliary loop and retrocerebral organ of Sagitta glacialis (Chaetognatha). – Dokl. Akad. Nauk. USSR 227: 763-765 (In Russian).

Parry, D.A. 1944. Structure and function of the gut in Spadella cephaloptera and Sagitta setosa. – J. mar. biol. Ass. U.K. 26: 16-36.

Pearre, S. Jr. 1991. Growth and reproduction. – In Q. Bone, H. Kapp & A.C. Pierrot-Bults (eds): The Biology of Chaetognaths, pp 61-75. Oxford Univ. Press, Oxford.

Reisinger, E. 1970. Zur Problematik der Evolution der Coelomaten. – Z. zool. Syst. Evolutionsforsch. 8: 81-109.

Salvini-Plawen, L.v. 1988. The epineural (vs. gastroneural) cerebral complex of Chaetognatha. – Z. zool. Syst. Evolutionsforsch. 26: 425-429.

Schram, F.R. 1973. Pseudocoelomates and a nemertine from the Illinois Pennsylvanian. – J. Palaeont. 47: 985-989.

Shinn, G.L. 1992. Ultrastructure of somatic tissues in the ovaries of a chaetognath (Ferosagitta hispida) – J. Morph. 211: 221-241.

Stevens, N.M. 1903. On the ovogenesis and spermatogenesis of Sagitta bipunctata. – Zool. Jb., Anat. 18: 227-240, pls 20-21.

Stevens, N.M. 1910. Further studies on reproduction in Sagitta. – J. Morph. 21: 279-319.

Thuesen, E.V. 1991. The tetrodotoxin venom of chaetognaths. – In Q. Bone, H. Kapp & A.C. Pierrot-Bults (eds): The Biology of Chaetognaths, pp 55-60. Oxford Univ. Press, Oxford.

Welsch, U. & V. Storch 1982. Fine structure of the coelomic epithelium of Sagitta elegans (Chaetognatha). – Zoomorphology 100: 217-222.

Welsch, U. & V. Storch 1983. Fine structural and enzyme histochemical observations on the epidermis and the sensory cells of Sagitta elegans (Chaetognatha). – Zool. Anz. 210: 34-43.

32

CYCLONEURALIA

A number of aschelminth phyla have an anterior mouth, a cylindrical pharynx and a collar-shaped, peripharyngeal brain (Fig. 32.1), and these characters are here interpreted as apomorphies of a monophyletic group. It could perhaps be tempting to use the old name Nemathelminthes for this group, but it has been used to cover quite varying phyla over the years, and it could be confusing to give different definitions to the groups Aschelminthes and Nemathelminthes, which are often used as synonyms. I have therefore chosen to give the group a new name, which relates to the shape of the brain.

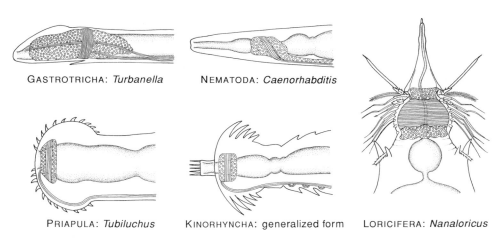

GASTROTRICHA: *Turbanella*   NEMATODA: *Caenorhabditis*

PRIAPULA: *Tubiluchus*   KINORHYNCHA: generalized form   LORICIFERA: *Nanaloricus*

Fig. 32.1. Brain structure in cycloneuralian phyla; left views except *Nanaloricus*, which is in dorsal view. The collar-shaped brains have an anterior and a posterior zone with perikarya (small circles) and a mid-zone with neuropil (fine lines). – Gastrotricha: *Turbanella cornuta* (based on Teuchert 1977). – Nematoda: *Caenorhabditis elegans* (based on White *et al.* 1986). – Priapula: *Tubiluchus philippinensis* (based on Calloway 1975 and Rehkämper *et al.* 1989). – Kinorhyncha: generalized form (based on Hennig 1984 and Kristensen & Higgins 1991). – Loricifera: *Nanaloricus mysticus* (based on Kristensen 1991). – The nematomorph brain is not well described, and a reliable drawing cannot be made.

A triradiate pharynx with radiating musculature is found in several cycloneuralian groups. It consists of myoepithelial cells in nematodes, gastrotrichs and loriciferans, but of mesodermal cells covered by an epithelium in the homalorhagid kinorhynchs, as in a number of other protostomian groups (Ruppert 1982). A myoepithelial, triradiate pharynx is also found in tardigrades (Chapter 21) and in ectoprocts, where each cell has a central vacuole (Chapter 24). The triradiate structure is obviously a successful solution to a functional problem, namely sucking in food, and it has very probably evolved several times. The triradiate shape of most myoepithelial pharynx types is probably derived independently from circular types within each phylum; this is indicated for example in gastrotrichs where juvenile *Lepidodasys* have a circular pharynx, whereas the adult pharynx has the inverted Y-shape characteristic of the macrodasyoids (Ruppert 1982). The triradiate shape is the simplest shape of a subdivided ring which can function as a sucking organ; a pharynx with only two bands of muscles (and no accessory structures) will at contraction become oval in cross-section without opening the lumen.

The myoepithelial cell is almost the only type of muscle cell in the cnidarians (Chapter 8), and ectodermal epithelia of this cell type has therefore been regarded as

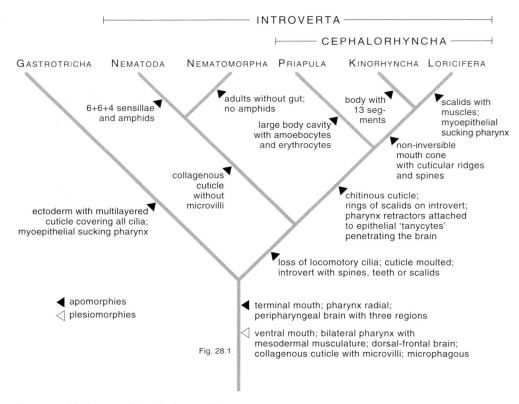

Fig. 32.2. Phylogeny of the Cycloneuralia.

a plesiomorphic character when it is found in bilaterian groups. Also the monociliate cell has been regarded as plesiomorphic when found in bilaterians, but this has been doubted (Nielsen 1987). These characters are discussed in Chapter 28, and it is concluded that both myoepithelial cells and monociliate cells (and monociliate myoepithelial cells) occur scattered over so many bilaterian groups that they cannot a priori be regarded as plesiomorphies.

The anterior mouth and the cylindrical pharynx differ distinctly from the ventral mouth and bilateral pharynx (often with paired jaws) of the other aschelminth phyla. Protostomian ontogeny indicates that the ventral mouth and the bilateral pharynx are ancestral to the group, and the cycloneuralian mouth/pharynx must therefore be regarded as apomorphic.

The collar-shaped brain with the perikarya in anterior and posterior rings separated by a zone of neuropil is a unique structure, which is apparently linked with the complete absence of an apical organ in all stages. The brain develops around the mouth-pharynx as the brain plus circumoesophageal nerves in rotifers and chaetognaths as well as in the spiralians, and the cycloneuralian organization of the central nervous system appears to be a reliable apomorphy.

The proposed phylogeny of the Cycloneuralia is given in Fig. 32.2.

References

Calloway, C.B. 1975. Morphology of the introvert and associated structures of the priapulid *Tubiluchus corallicola* from Bermuda. – Mar. Biol. (Berl.) **31**: 161-174.
Hennig, W. 1984. Taschenbuch der Zoologie, Band 2, Wirbellose I. – Gustav Fischer, Jena.
Kristensen, R.M. 1991. Loricifera. – *In* F.W. Harrison (ed.): Microscopic Anatomy of Invertebrates, vol. 4, pp 351-375. Wiley-Liss, New York.
Kristensen, R.M. & R.P. Higgins 1991. Kinorhyncha. – *In* F.W. Harrison (ed.): Microscopic Anatomy of Invertebrates, vol. 4, pp 377-404. Wiley-Liss, New York.
Nielsen, C. 1987. Structure and function of metazoan ciliary bands and their phylogenetic significance. – Acta zool. (Stockh.) **69**: 205-262.
Rehkämper, G., V. Storch, G. Alberti & U. Welsch 1989. On the fine structure of the nervous system of *Tubiluchus philippinensis* (Tubiluchidae, Priapulida). – Acta zool. (Stockh.) **70**: 111-120.
Ruppert, E.E. 1982. Comparative ultrastructure of the gastrotrich pharynx and the evolution of myoepithelial foreguts in Aschelminthes. – Zoomorphology **99**: 181-220.
Teuchert, G. 1977. The ultrastructure of the marine gastrotrich *Turbanella cornuta* Remane (Macrodasyoidea) and its functional and phylogenetic importance. – Zoomorphologie **88**: 189-246.
White, J.G., E. Southgate, J.N. Thomson & S. Brenner 1986. The structure of the nervous system of the nematode *Caenorhabditis elegans*. – Phil. Trans. R. Soc. B **314**: 1-340.

Phylum GASTROTRICHA

Gastrotrichs are a small phylum of small to microscopic, aquatic animals; about 430 species representing two orders have been described. The recent review by Ruppert (1991) is an invaluable source for information about anatomy and ultrastructure. The Macrodasyoida are marine and characterized by a pharynx which is inverted Y-shaped in cross-section with a dorsal and a pair of lateral furrows; the lateral furrows open to the exterior through a lateral pore in all genera but *Lepidodasys*. The Chaetonotoida are marine or limnic and have a pharynx which is Y-shaped with a ventral and a pair of lateral furrows and lacks pharyngeal pores. *Neodasys* is usually classified with the chaetonotoids, but it is intermediate between the two orders in several characters.

The elongate body has a flattened ventral side with a ciliated ventral sole. The mouth is anterior, and an almost cylindrical gut consisting of a myoepithelial sucking pharynx and an intestine of cells with microvilli leads to the ventral anus near the posterior end. The two different shapes of the pharynx naturally lead to speculations about the shape of the ancestral pharynx. Ruppert (1982) reported that the juvenile gut of the macrodasyoid *Lepidodasys* is circular and that both the anterior and the posterior ends of the pharynx of other genera may be circular, quadriradiate or multiradiate and concluded that the circular shape must be ancestral.

The ectoderm including the pharynx epithelium is a monolayer of unciliated, monociliate or multiciliate cells (Rieger 1976, Rieger & Rieger 1977, Ruppert 1991). The whole surface including the cilia is covered by a lamellate exocuticle, whereas an inner, granular or fibrillar endocuticle, which may be thrown into complicated scales, hooks or spines, covers the epithelial surface but not the cilia (Ruppert 1991). The lamellae of the exocuticle resemble cell membranes, but variations occur. The cuticular structures contain an extension of the epithelium in *Xenodasys*, whereas the other structures are hollow or solid. The cuticle consists of proteinaceous compounds without any trace of chitin (Jeuniaux 1975).

The myoepithelial pharynx has a cuticle which in some species forms teeth, hooks or more complicated, scraper-like structures. Some of the cells bear one

Chapter vignette: *Turbanella cornuta*. (Redrawn from Remane 1926.)

kinocilium, and some cells have a few microvilli which penetrate the cuticle (Ruppert 1982, 1991). The tubular midgut consists of a single layer of microvillous cells surrounded by mesodermal muscles; cilia are absent except in *Xenodasys* (Rieger *et al.* 1974). The chaetonotoids, except *Neodasys*, have a short, cuticle-lined rectum, but the other groups have a simple pore between the midgut and the epidermis.

The nervous system (Teuchert 1977b) comprises a circumpharyngeal brain with an anterior concentration of perikarya, a median ring of neuropil and a posterior concentration of perikarya. Two pairs of nerves extend from the median ring to the anterior end and two pairs of nerves extend posteriorly and unite, forming a pair of ventrolateral nerves/nerve cords. Gastrotrichs have poorly developed basement membranes, so the relative position of the nervous system has not been ascertained. The muscle cells are innervated by synapses between short processes of the muscle cells and the longitudinal nerves. There are several types of sensory organs: chemoreceptors, mechanoreceptors and photoreceptors, which all consist of modified monociliate cells (Ruppert 1991).

The body wall has an outer layer of circular muscles and an inner layer of longitudinal muscles, which do not form continuous sheets (Ruppert 1991). The muscles are usually striated, but smooth muscles are found in *Lepidodasys*. Macrodasyoids have a row of large so-called Y-cells on each side of the gut. These cells have large vacuoles and are believed to function as a sort of hydrostatic skeleton. Teuchert (1977a) and Teuchert & Lappe (1980) pointed out that the Y-cells and the gonads are surrounded by muscle cells, i.e. mesoderm, and that spaces between muscles and organs could be interpreted as coeloms, but Ruppert (1991) stressed that there is no open body cavity; the only open spaces are the small lacunae around the terminal organs of the protonephridia. There is no circulatory system.

One to several pairs of protonephridia are located laterally (Neuhaus 1987, Ruppert 1991). Each protonephridium consists of one or a few monociliate terminal cells, a monociliate duct cell and a sometimes monociliate pore cell situated in the epithelium.

The gonads are sac-shaped and sometimes surrounded by muscle cells, but the detailed structure has not been completely cleared up (Ruppert 1991). Most species have complicated accessory reproductive structures associated with copulation, and it appears that all species have internal fertilization (Hummon & Hummon 1989); some are viviparous.

The embryology of a number of species representing both orders has been studied; the macrodasyoid *Turbanella* is best known (Teuchert 1968) and will be described first (Figs 33.1, 33.2). Fertilization takes place in the ovary but the polar bodies are not given off until after spawning. The egg is ovoid and the polar bodies usually become situated at the blunt end, which has been called the animal pole, but a few eggs with the polar bodies at the more pointed end have been observed too. The movements of the cells during the early developmental stages make it difficult to define the axes of the embryo and the lack of an apical sense organ makes the use of the term apical misleading (in the following description the blunt end of the egg is called the anterior pole). The first cleavage is equatorial, and the second cleavage is parallel to the longitudinal egg axis with the two divisions being perpendicular to each other.

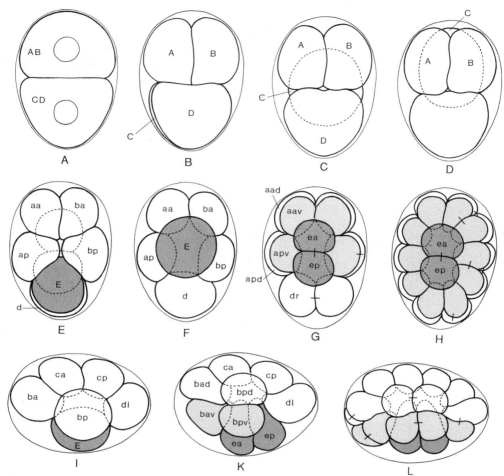

Fig. 33.1. Cleavage of *Turbanella cornuta*. A-H, embryos seen from the blastoporal/ventral side; I-L, embryos seen from the left side. – A, 2-cell stage. – B, 4-cell stage just after the cleavage. – C-D, 4-cell stages showing the movement of the C-cell to the anterior pole. – E, 8-cell stage with the E-cell at the posterior pole. – F and I, 8-cell stage where the E-cell has moved to the blastoporal side. – G and K, 14-cell stage. – H and L, 30-cell stage; the mesodermal cells form a ring around the two endoderm cells. – Endodermal cells are dark grey and mesodermal cells light grey. (Redrawn from Teuchert 1968.)

One of the posterior cells then moves towards the anterior pole of the egg and the resulting embryo consists of three cells at the anterior end and one at the posterior end. The embryo is now bilateral and the four cells can be named and related to the orientation of the adult: the two descendants from the anterior cell (A and B) are situated anteroventrally to the right and left, respectively; the third cell at the anterior pole (C) is anterodorsal; and the fourth cell (D) is posterior.

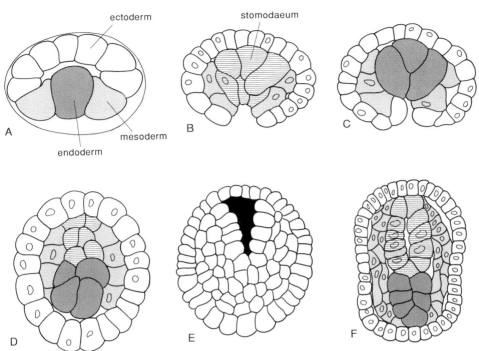

Fig. 33.2. Early development of *Turbanella cornuta*. The ectoderm is white, the mesoderm light grey , the endoderm dark grey, and the stomodaeum cross-hatched. – A, cross-section of a 30-cell stage showing the infolding of the endoderm. – B-C, cross-sections of a stage with an elongate blastopore; B is in the anterior region with the stomodaeum and C in the posterior region with the endoderm. – D, horizontal section of a similar stage. – E, embryo in ventral view showing the elongate stomodaeal invagination (black); the endoderm is fully covered. – F, horizontal section of a similar stage. – Ectodermal cells white, mesodermal cells light grey, endodermal cells dark grey, stomodaeum cross-hatched. (Redrawn from Teuchert 1968).

At the fourth cleavage the A-, B- and C-cells divide almost parallel to the longitudinal egg axis, while D divides perpendicular to the axis. The ventral descendant of the D-cell slides to a midventral position, so that the embryo now consists of two dorsal cells and a ventral cell, separated by a ring of five cells. During the following cleavages the dorsal cells form two longitudinal rows of four cells each and the ventral cell a longitudinal row of two cells, while the ring of cells through two cleavages develops into a double-ring each with ten cells. The dorsal cell rows and the dorsal-most ring of cells become ectoderm, the cells of the ventral-most ring become ectoderm plus mesoderm, while the two ventral cells become endoderm. The fates of the cells are summarized in Table 33.1. The precursors of the mesodermal cells appear to form a ring around the coming blastoporal invagination, but it must be noted that the fate of the single cells has not been followed further on, and since the mesoderm after gastrulation forms a pair of lateral bands, it is not certain that the mesoderm surrounds the blastopore completely.

Table 33.1. Cell lineage of a 30-cell stage of *Turbanella cornuta* (see Figs 33.1, 33.2; based on Teuchert 1968). The first letter in each combination denotes the descent from the first four cells A–D; the following letters indicate: a – anterior, p – posterior, d – dorsal, v – ventral.

```
                                            aada
                                    aad  {        } 'secondary' ectoderm
                                            aadp
                             aa {
                                            aava
                                    aav  {        } mesectoderm
                        A {                  aavp

                                            apda
                                    apd  {        } 'secondary' ectoderm
                             ap {            apdp

                                            apva
                                    apv  {        } mesectoderm
                   AB {                      apvp

                        B – as A

                                            cala
                                    cal  {
                                            calp
                             ca {
                                            cara
                                    car  {
        Z {                                 carp
                        C {                              } 'primary' ectoderm
                                            cpla
                                    cpl  {
                                            cplp
                             cp {
                                            pra
                                    cpr  {
                   CD {                      cprp

                                            drd    'secondary' ectoderm
                                    dr  {
                                            drv    mesectoderm
                             d {
                                            dld    'secondary' ectoderm
                                    dl  {
                        D {                  dlv    mesectoderm

                                    ea
                             E {                     } endoderm
                                    ep
```

The 30-cell embryo has a small blastocoel and during the following cell divisions the endodermal cells and the cells of the stomodaeum (pharynx) invaginate as a longitudinal furrow. The posterior parts of the blastopore lips fuse and the endodermal cells form a compact mass of cells at the end of the stomodaeum with mesodermal cells on both sides. The blastocoel becomes obliterated. The brain develops from ectodermal cells in the pharyngeal region. Genital cells are presumed to originate from mesodermal cells.

The embryo now becomes curved ventrally, and during the following development the endoderm becomes arranged as the gut and an anus opens. All other organ systems develop during the later part of the embryonic period, so that the hatching worm is a miniature adult. No traces of larval organs have been described and the embryos hatch as small juveniles which resemble the adults in all major features.

There are two studies on the embryology of chaetonotoids, viz. of *Neogossea* by Beauchamp (1929) and of *Lepidodermella* by Sacks (1955); the two authors followed the cell-lineage to a stage of about 64 cells and the further development in optical sections. The two descriptions agree in all major points, but differ from that of *Turbanella* in several significant details: the primary endoderm cell should come from a descendant of the A-cell and thus be located at the anterior part of the embryo; the C- and D-cells should form the right and left halves of the posterior part of the embryo (compare with Fig. 33.1: A); there should be a large proctodaeum formed from the C- and D-cells and the genital cells should originate from a pair of cells just lateral to the stomodaeal opening. Teuchert (1968) pointed out that a partial correspondence between the descriptions could be obtained if the antero-posterior orientation of the two older descriptions was reversed. However, it appears that both Beauchamp and Sacks had rather good markers of the main axis of the embryos all along the development, whereas Teuchert (1968, pp 379-380) pointed out that the 30-cell stage has no markers which distinguish the anterior from the posterior pole, so it is also possible that Teuchert had reversed the antero-posterior axis. The differences regarding the formation of the gut are more difficult to explain, and the whole development must be studied in both orders before a meaningful discussion can be carried out.

Be that as it may, the cleavage pattern has no resemblance with the spiral pattern, but it is also very difficult to compare cell-lineages of the two gastrotrich types with that of the nematodes.

The cuticle with a lamellar exocuticle surrounding also the cilia and a granular or fibrous endocuticle is a unique structure, and the gastrotrichs are undoubtedly a monophyletic group.

Two characters have had a prominent position in the discussion of the phylogenetic position of the gastrotrichs: the myoepithelial pharynx and the mono- versus multiciliate cells of the epithelia.

Ruppert (1982) considered the myoepithelial foreguts of the gastrotrichs, nematodes, tardigrades and ectoproct bryozoans as homologous and as a symplesiomorphy because the myoepithelial cell type is characteristic of the cnidarians. He even went so far as to indicate a sister-group relationship between nematodes and chaetonotoids on the basis of pharynx structure. As also indicated in Chapter 28, a cell with a cilium and contractile structures is of course primitive, and is even characteristic of the sister group of all the metazoans, viz. the choanoflagellates. During ontogeny, all cells have centrioles, which are used in the cell division and which can differentiate as ciliary basal bodies, and most blastomeres are contractile, but both of these characters disappear in most of the fully differentiated cell types. This shows that all cells have the potential to organize organelles such as cilia and myofilaments, and this potential may become realized in tissues where they have a function. I interpret the suction pharynx of the just-mentioned groups as convergent adaptations to this special way of food intake.

The ciliation of the epithelia of a number of species was studied by Rieger (1976), who found that a number of macrodasyoid genera and *Neodasys* have monociliate epithelia, whereas the other chaetonotoids and other macrodasyoid genera have multiciliate epithelia; the type of ciliation appears to be uniform within genera but not within families. The monociliate and some of the multiciliate cells have an accessory centriole at the base of each cilium. Rieger (1976, p. 214) interpreted the monociliate condition as primitive within the gastrotrichs, but admitted that it could be the result of a reduction from the multiciliate condition. It should be clear that this question can only be resolved by comparisons with other phyla, and I have come to the conclusion (Chapter 9) that the multiciliate condition is an apomorphy of the Protostomia, and that the presence of monociliate epithelia in a few protostomian taxa, such as some gastrotrichs and annelids (*Owenia* and the gnathostomulids; Chapter 17) must represent reversals to the monociliate condition.

The terminal mouth and the cylindrical pharynx surrounded by the collar-shaped brain with a median ring of neuropil characterize the gastrotrichs as cyclo-neuralians. The same characters together with the special cuticle, which is not moulted, and the lack of an introvert characterize them as the sister group of the introvertans (Chapter 34). General aschelminth characters are more difficult to point out (Chapter 28), but the initially asymmetrical cleavage, the origin of the mesoderm from cells surrounding the blastopore and the lack of an apical organ are all in accordance with the definition of the group applied here.

Interesting subjects for future research

1. Embryology and cell lineages of both chaetonotoids and macrodasyoids.
2. Development of the pharynx in species of both orders.

References

Beauchamp, P. de 1929. Le développement des Gastrotriches. – Bull. Soc. zool. Fr. **54**: 549-558.
Hummon, W.D. & M.R. Hummon 1989. Gastrotricha. – In K.G. Adiyodi & R.G. Adiyodi (eds): Reproductive Biology of Invertebrates, vol. 4A, pp 201-206. John Wiley & Sons, Chichester.
Jeuniaux, C. 1975. Principes de systématique biochimique et application à quelques problèmes particuliers concernant les Aschelminthes, les Polychètes et les Tardigrades. – Cah. Biol. mar. **16**: 597-612.
Neuhaus, B. 1987. Ultrastructure of the protonephridia in *Dactylopodalia baltica* and *Mesodasys laticaudatus* (Macrodasyoida): implications for the ground pattern of the Gastrotricha. – Microfauna mar. **3**: 419-438.
Remane, A. 1926. Morphologie und Verwandtschaftsbeziehungen der aberranten Gastrotrichen I. – Z. Morph. Ökol. Tiere **5**: 625-754.
Rieger, R.M. 1976. Monociliated epidermal cells in Gastrotricha: significance for concepts of metazoan evolution. – Z. zool. Syst. Evolutionsforsch. **14**: 198-226.
Rieger, G.E. & R.M. Rieger 1977. Comparative fine structure study of the gastrotrich cuticle and aspects of cuticle evolution within the Aschelminthes. – Z. zool. Syst. Evolutionsforsch. **15**: 81-124.
Rieger, R.M., E. Ruppert, G.E. Rieger & C. Schöpfer-Sterrer 1974. On the fine structure of gastrotrichs with description of *Chordodasys antennatus* sp.n. – Zool. Scr. **3**: 219-237.
Ruppert, E.E. 1982. Comparative ultrastructure of the gastrotrich pharynx and the evolution of myoepithelial foreguts in Aschelminthes. – Zoomorphology **99**: 181-220.
Ruppert, E.E. 1991. Gastrotricha. – In F.W. Harrison (ed.): Microscopic Anatomy of Invertebrates, vol. 4, pp 41-109. Wiley-Liss, New York.
Sacks, M. 1955. Observations on the embryology of an aquatic gastrotrich, *Lepidodermella squamata* (Dujardin, 1841). – J. Morph. **96**: 473-495.

Teuchert, G. 1968. Zur Fortpflanzung und Entwicklung der Macrodasyoidea (Gastrotricha). – Z. Morph. Tiere **63**: 343-418.

Teuchert, G. 1977a. Leibeshöhlenverhältnisse von dem marinen Gastrotrich *Turbanella cornuta* Remane (Ordnung Macrodasyoidea) und eine phylogenetische Bewertung. – Zool. Jb., Anat. **97**: 586-596.

Teuchert, G. 1977b. The ultrastructure of the marine gastrotrich *Turbanella cornuta* Remane (Macrodasyoidea) and its functional and phylogenetic significance. – Zoomorphologie **88**: 189-246.

Teuchert, G. & A. Lappe 1980. Zum sogennanten 'Pseudocoel' der Nemathelminthes. – Ein Vergleich der Leibeshöhlen von mehreren Gastrotrichen. – Zool. Jb., Anat. **103**: 424-438.

34

INTROVERTA

The phyla Nematoda, Nematomorpha, Priapula, Kinorhyncha, and Loricifera share a number of characters, which I interpret as apomorphies, and I have chosen to regard them as a monophyletic taxon and to name it Introverta after the inversible anterior end (Fig. 34.1). Other apomorphies include the absence of locomotory cilia and the presence of a cuticle which is moulted.

The introvert is the anteriormost part of the body, which can be invaginated. This definition appears quite straightforward, but difficulties arise when the position of the mouth opening shall be defined to delimit the introvert. The buccal cavity and the pharynx with teeth can be everted, for example in priapulans, and it is difficult to find an anatomical criterion for the position of the mouth. I have chosen to emphasize the functional differences between the introvert and the buccal cavity plus pharynx: the introvert is used for penetration of a substratum, whereas the teeth in the buccal cavity and the pharynx are used for grasping and ingesting prey; this means that the spines/scalids of the introvert point away from the midgut when the introvert is invaginated, whereas the spines/scalids/jaws of buccal cavity and pharynx point towards the midgut. This is of no help when the mouth cone of kinorhynchs and loriciferans is discussed, but can give a definition of the introvert in the nematodes.

If the above definition is followed, an introvert is well developed in larval nematomorphs, priapulans, kinorhynchs, and loriciferans; nematodes generally lack an introvert, but *Kinonchulus* (Riemann 1972; Fig. 34.1) has an unmistakable introvert with six double rows of cuticular spines of varying length.

All the five phyla have compact cuticles which are moulted, but the chemical composition is not the same in all groups. Nematoda and Nematomorpha have a thick layer of collagenous fibres in the inner layer of the cuticle, whereas the Cephalorhyncha (Priapula + Kinorhyncha + Loricifera) have cuticles with chitin. It is possible that the cuticle of the ancestor of the Introverta was a stiff, proteinaceous cuticle which was moulted, and that the nematode line reinforced the cuticle with collagen whereas the cephalorhynch line impregnated it with chitin. The last assumption could be supported by the observations on the priapulan cuticle by Carlisle

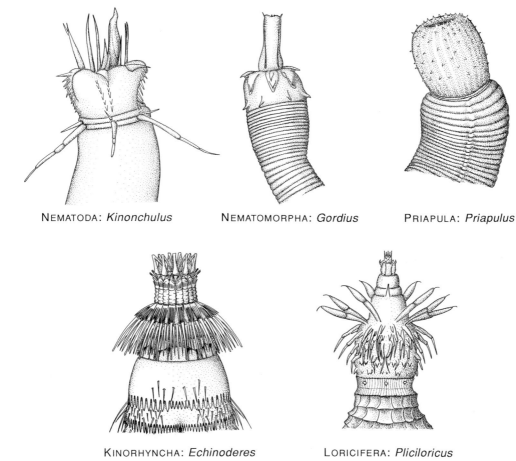

NEMATODA: *Kinonchulus*   NEMATOMORPHA: *Gordius*   PRIAPULA: *Priapulus*

KINORHYNCHA: *Echinoderes*   LORICIFERA: *Pliciloricus*

Fig. 34.1. Anterior regions of introvertans. – Nematoda: *Kinonchulus sattleri* (redrawn from Rie-mann 1972). – Nematomorpha: larva of *Gordius aquaticus* (redrawn from Dorier 1930). – Priapu-la: *Priapulus horridus* (redrawn from Theel 1911). – Kinorhyncha: *Echinoderes aquilonius* (see the chapter vignette to Chapter 39). – Loricifera: larva of *Pliciloricus gracilis* (redrawn from Higgins & Kristensen 1986).

(1959), who could not detect chitin in the cuticle of newly moulted *Priapulus*, where-as a clear reaction was obtained in the exuviae. The moulting of the cuticle must be controlled by hormones of some sort, but the present knowledge of such compounds in introvertans is limited to nematodes, and the phylogenetic interpretation of the available information is most uncertain. The ectodermal cells appear to be completely without microvilli, and this is probably connected with the moulting.

The lack of locomotory cilia is complete in the five phyla, but this character has apparently evolved independently, for example in arthropods, chaetognaths and acanthocephalans.

The triradiate pharynx with radiating musculature is found in several groups. It consists of myoepithelial cells in gastrotrichs, nematodes and loriciferans, but of mesodermal cells covered by an epithelium in the homalorhagid kinorhynchs. A myoepithelial, triradiate pharynx is also found in ectoprocts, where each cell has a central vacuole (Chapter 24) and in tardigrades (Chapter 21). The triradiate structure is obviously a successful solution to a functional problem, namely sucking in food, and it has very probably evolved several times; it must therefore be treated with suspicion in phylogenetic discussions (Chapter 32).

References

Carlisle, D.B. 1959. On the exuvia of *Priapulus caudatus* Lamarck. – Ark. Zool., 2. Ser. **12**: 79-81.

Dorier, A. 1930. Recherches biologiques et systématiques sur les Gordiacés. – Annls Univ. Grenoble, N.S., Sci.-Med. **7**: 1-183, 3 pls.

Higgins, R.P. & R.M. Kristensen 1986. New Loricifera from southeastern United States coastal waters. – Smithson. Contr. Zool. **438**: 1-70.

Riemann, F. 1972. *Kinonchulus sattleri* n.g. n.sp. (Enoplida, Tripyloidea), an aberrant freeliving nematode from the lower Amazoans. – Veröff. Inst. Meeresforsch. Bremerh. **13**: 317-326.

Theel, H. 1911. Priapulids and sipunculids dredged by the Swedish Antarctic expedition 1901–1903 and the phenomenon of bipolarity. – K. svenska Vetenskapsakad. Handl. **47**(1): 1–36, 5 pls.

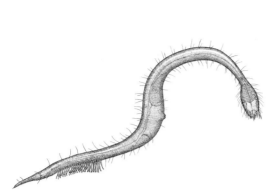

Phylum Nematoda

The nematodes are one of the most successful phyla today; about 20 000 species have been described, but it is believed that the number of living species should be counted in millions. Nematodes occur in almost all habitats, both aquatic and terrestrial, and many parasitize plants or animals and are of great economic importance.

In spite of the variations in habitats, all nematodes are of a remarkably similar body plan: almost all species are cylindrical with tapering ends and with a thick elastic cuticle, which may be smooth, often with a microrelief; a few types have ring- or scale-shaped thickenings, and some others have various types of spines or bristles. The cuticle maintains a turgor in the body cavity, which functions as a hydrostatic skeleton with bands of longitudinal body-wall muscles as antagonists; ring muscles are absent in the body wall and locomotory cilia are completely absent. This very specialized body plan has obviously made the invasion of very different niches possible without strong modifications. Classification has not reached a stable stage, but two classes are generally recognized. Lorenzen (1981) has proposed a cladistic classification of all the free-living groups and contributed several new characters of phylogenetic importance.

A most characteristic feature of nematodes is that the numbers of cells/nuclei in most organs appear to be constant within species (Nigon 1965). The small species *Caenorhabditis elegans* (length about 1.3 mm, diameter about 80 µm), which has been the subject of a long series of detailed studies, has a total number of 959 somatic nuclei in the newly hatched hermaphrodite and 1031 in the male (Kenyon 1985); the intestine comprises 34 cells and the nervous system 302 neurons in the adult hermaphrodite (Sulston *et al.* 1983). Much larger species, for example *Ascaris lumbricoides*, have about the same arrangement and number of nuclei in the nervous system (about 250; see Stretton *et al.* 1978) and the numbers in most other organ systems are also constant, but only few organ systems have been investigated (Nigon 1965). Some tissues, especially the epidermis and the endoderm, but also some of the muscles, consist of large syncytia with a more indefinite number of nuclei. The development is strongly determined and the whole cell lineage of *Caenorhabditis* and the

Chapter vignette: *Draconema cephalatum*. (Redrawn from Steiner 1919.)

cells of its adults have been mapped (see below); regenerative powers are very small, since cell division in the somatic tissue is completed at hatching.

The body wall consists of a cuticle, a layer of ectodermal cells (often called hypodermis), a basement membrane, and a layer of longitudinal body-wall muscle cells (Wright 1991). The cuticle also lines the buccal cavity, the pharynx and the rectum. The outermost layer of the cuticle is thin epicuticle, which is a double membrane resembling the cell membrane, but both its thickness and its freeze-fracture pattern differ from those of a normal cell membrane (Wright 1991). It is the first layer to be secreted during ontogeny and also during synthesis of the new cuticle at moulting. The main layer of the cuticle comprises several zones which consist mainly of collagen; chitin is lacking (Jeuniaux 1975). Its median zone shows wide variation and contains highly organized structures in many species. Its inner zone contains more or less complex layers of fibrils with different orientations. A few parasitic genera have only a partial cuticle or no cuticle at all (Wright 1991).

The ectodermal cells are arranged in longitudinal rows with the nuclei concentrated in a pair of lateral thickenings or cords; an additional dorsal and ventral cord are found in the larger species, which may also have nuclei between the cords. The cords divide the somatic musculature into longitudinal bands. The muscle cells are connected to the cuticle via hemidesmosomes and tonofibrils in the ectodermal cells. Each muscle cell has an axon-like extension which reaches to the motor neurons located at the longitudinal nerves (see below). There are oblique muscles in the posterior region of the male, and a few muscular cells are associated with the intestine and the rectum.

The terminal mouth is surrounded by various labial organs and sensilla (see below) and opens into a buccal cavity often with cuticular structures such as thorns, hooks, jaws, or spines (Lorenzen 1981, 1988). Many genera of the order Dorylaimida have a grooved tooth which is used to penetrate food organisms; this tooth is secreted by one of the ventrolateral myoepithelial cells of the pharynx (Grootaert & Coomans 1980); the groove is almost closed in some genera and the tooth is a long hypodermic needle with the base surrounding the opening to the pharynx completely in others (Coomans & de Coninck 1963). The remarkable genus *Kinonchulus* has a conspicuous introvert with six longitudinal double-rows of short and long spines in front of the normal rings of sensilla (see below); there is a large grooved or hollow tooth in contact with the dorsal side of the pharynx (Riemann 1972; see Fig. 34.1).

The long pharynx (sometimes called oesophagus) is cylindrical, often with one or two swellings and has a lumen which is Y-shaped in cross section. The one layered epithelium consists of myoepithelial cells which form a dorsal and two lateral thickenings, while epithelial cells with conspicuous tonofilaments attaching to hemidesmosomes line the bottoms of the grooves (Albertson & Thomson 1976). All the cells secrete a cuticle which may form various masticatory structures. Kenyon (1985, p. 27) has proposed that the ancestral pharynx had only the lateral muscle bands and that the dorsal band is a later specialization. The nematode pharynx is definitely a suction organ, and it appears that a pharynx with only a pair of lateral muscle bands will become oval in cross section at contraction without opening the lumen. A pharynx with only two bands of muscles is therefore not a likely ancestral character.

The intestine is a straight tube of unciliated cells with a microvillous border. There are generally no muscles around the intestine, except the four posterior cells mentioned below, and the food particles are apparently pressed through the intestine from the pharynx (Wright 1991).

The nervous system is sometimes characterized as orthogonal (Beklemischew 1960, Reisinger 1972), but this appears to be quite misleading. The nervous system of *Caenorhabditis* (Fig. 35.1) has been mapped in every detail (Albertson & Thomson 1976, White *et al.* 1986) and other well-studied species appear to have very similar systems, even with cells being recognizable between species. It is clear that there is a collar-shaped brain around the pharynx, a ventral longitudinal nerve cord and a concentration of cells surrounding the rectum; the few nerve cells located outside this typical protostomian (gastroneuralian) central nervous system are sensory cells. The brain consists of an anterior and a posterior ring with concentrations of perikarya and a middle zone of neuropil. The nerves along the lateral and dorsal ectodermal thickenings, which have been described from all nematodes investigated during more than a century, consist of bundles of axons from the sensory cells or from motor cell bodies in the brain or the ventral nervous system; none of the axons extend directly from the brain along these nerves, which are thus of a character quite distinct from that of the ventral cord. The whole nervous system (with the exception of six cell bodies) is situated between the epidermis and its basement membrane, and the neuromuscular junctions with the mesodermal muscles connect the two types of cells across the basement membrane (White *et al.* 1986). This is possible because the muscle cells have the axon-like protrusions mentioned above. The ventral cord contains a row of 'ganglia' containing the cell bodies of the motor neurons of the body-wall muscles; their connections to the lateral and dorsal longitudinal nerves are via lateral commissures, which form species-constant patterns. In *Ascaris* there are five repeating units of neurons each with 11 cells; each unit has six right-side and one left-side commissure (Johnson & Stretton 1980). Similar structures are known from other genera. The units have sometimes been described as segments, but it must be emphasized that the repeating units are not formed from teloblasts as are the segments of the articulates (see below).

Various types of sensory organs have been described (Ward *et al.* 1975, Wright 1980). Some sensory cells are internal, such as receptors just below the cuticle of the pharynx and the stretch receptors with a modified cilium in the lateral epithelial cords of enoplids (Hope & Gardiner 1982). Other sense organs respond to exterior

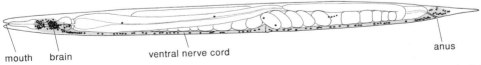

mouth brain ventral nerve cord anus

Fig. 35.1. Central nervous system of *Caenorhabditis elegans*. The nuclei of the nerve cells in the left half of the animal are indicated as black dots. The central nervous system (anterior ganglion, dorsal ganglion, lateral ganglion, ventral ganglion, retro-vesicular ganglion, ventral cord, pre-anal ganglion, dorso-rectal ganglion, and lumbar ganglion) is shaded. The few nuclei outside the CNS are exclusively of sensory cells. (Based on White *et al.* 1986.)

stimuli, such as chemoreceptors, mechanoreceptors and eyespots. Sensilla are chemoreceptors (or perhaps sometimes mechanoreceptors) which typically consist of a nerve cell with one or a few short terminal, modified cilia (often called dendritic processes), a sheath cell and a socket cell, but there may be more than one nerve cell in each sensillum (Ward *et al.* 1975, Coomans 1979, Wright 1980). The distal extensions of the socket cell and the sheath cell surround a pit or another modified area of the cuticle into which the short cilia protrude. Many of these sensory organs are situated on the flat surface, for example on the lips, while others have the shape of spines, for example the spicules of the male copulatory apparatus. There are normally three rings of sensilla around the mouth with four sensilla in the posterior ring and six in each of the anterior rings. A pair of large, in the free-living species usually very conspicuous, sensilla, the amphids, is found laterally in the head or neck region. Each amphid contains 4-13 sensory cells, which in some species have many cilia. The amphidial pore or canal is large and varying in shape, funnel-shaped, circular or spirally coiled. Most amphids have an additional unit lying deeply in the canal and consisting of an extended, distal part of the sheath cell with strongly folded or microvillous cell membranes around the canal; some of the receptor cells may have a microvillous zone in the same area.

The body cavity is lined by ectodermal cells of the epithelial cords, mesodermal muscles and endodermal intestine and is therefore a blastocoel, although it forms at a late stage of the development (see below).

The excretory organ consists of 1-4 cells with considerable differences between the families (Nigon 1965). Many of the marine forms have one large glandular cell (called ventral gland or renette cell), but the system may be complicated by the presence of one or a pair of longitudinal tubular extensions. *Caenorhabditis* represents the most complicated type with four cells: an H-shaped excretory cell, a binucleate, A-shaped secretory cell, a duct cell, and a pore cell (Nelson, Albert & Riddle 1983). The large excretory cell apparently secretes a fluid into its long, narrow lumen; the secretory cell contains many membrane-bounded secretion granules which are supposedly given off by exocytosis. Pulsation of the excretory system has been observed in several species, and the pulsations are correlated with the osmolarity of the surrounding fluid, so osmoregulation is one function of the system (Atkinson & Onwuliri 1981). It has been suggested that the excretory system secretes a 'moulting fluid' through the excretory pore into the space between the old and the new cuticle and that this fluid is necessary for the moulting. However, later observations indicate that the loosening of the old cuticle begins at the mouth rather than at the excretory pore, and laser ablation experiments with nuclei of the excretory cells have not prevented moulting (Singh & Sulston 1978). The role of the excretory system in moulting must therefore be characterized as uncertain. Ecdysteroids have been found in many nematodes, but 'the hormonal role of these substances (in nematodes) and their biological functions, however, are still obscure. They seem to be involved in growth regulation, embryogenesis, vitellogenesis, and molting' (Franke & Käuser 1989, p. 302). Their occurrence and postulated/proven functions in protozoans and metazoans indicates that only future investigations can unravel their phylogenetic importance (Käuser 1989, Barker, Chitwood & Rees 1990).

The gonads are sac-shaped, consisting of germinal cells surrounded by a basement membrane; they open through a ventral pore. Fertilization is internal, and the males have a complicated copulatory apparatus. The sperm is aflagellate and amoeboid, and its final differentiation usually takes place within the female (Wright 1991). The fertilized egg becomes surrounded by an egg shell, which contains chitin (Jeuniaux 1975).

The embryology of *Parascaris equorum* was studied by a number of authors around the turn of the century (Strassen 1896, 1906, Boveri 1899, Müller 1903), but the recent studies of the two small nematodes *Caenorhabditis elegans* and *Panagrellus redivivus* have set a new standard for the study of cell lineages (Sulston *et al.* 1983, Sternberg & Horvitz 1982). The two last-mentioned species belong to two rhabditoid families; *Panagrellus* is about twice as long as *Caenorhabditis* and has a slightly higher number of cells, but the cell lineages of the two species are remarkably similar (Sternberg & Horvitz 1982). As far as can be seen, the embryology of the much larger *Parascaris* does not deviate much from the pattern of the two small species, but a detailed comparison has not been made.

The fertilized egg is usually somewhat elongate, often with a slightly flattened or even concave side; the polar bodies are in some species formed at or near one end of the egg, but in other species at the flat/concave side (Strassen 1959). The following development shows that one end of the egg becomes the anterior end of the embryo and the flat/concave side becomes the blastoporal/ventral side, but there is no indication of an apical organ. It is clear, however, that the position of the polar bodies is not fixed at an apical pole, and it appears that the polar bodies may slide along the embryo during the early cleavages.

The first cleavage is often almost equal, but the two blastomeres can be recognized by differences in yolk content. The second cleavage shows considerable variation, both with respect to timing and directions of the spindles. The 4-cell stage may be a line of cells, a T-shaped figure with the A- and B-cells as the horizontal part and the P_2- and EMS-cells (see below and Table 35.1) as the handle, tetrahedral, or rhomboidal. However, the rhomboidal shape is usually reached after a while through movements of the cells. The complete cell lineage has been worked out for a few species, and the blastomeres have traditionally been given letters as seen in Table 35.1 and Fig. 35.2. The rhomboidal 4-cell stage has the bilateral symmetry of the adult, with the four blastomeres representing roughly the four parts of the embryo: A – anterior, B – dorsal, P_2 – posterior, EMS – ventral (Fig. 35.2). The A- and B-cells divide into right and left blastomeres, whereas the two other cells divide into anterior and posterior blastomeres: EMS into the anterior MS and the posterior E, which becomes the endoderm, and P_2 into the ventral P_3 and the dorsal C. The MS-cell divides into a right and a left blastomere, which in turn divide into S- and M-cells, where the S-cells give rise to mesodermal muscle cells, coelomocytes, somatic gonad cells, and glia cells, whereas the S-cells become the ectodermal pharynx, with muscle cells, gland cells and motor neurons, and the mid-ventral body muscles (Kenyon 1985). A coeloblastula is formed and gastrulation by invagination moves the E-cells to the interior of the embryo with an elongate blastopore surrounded by S-, M- and D-cells; the P_4-cell (the primordial germ cell) lies at the posterior edge of the blastopore, but

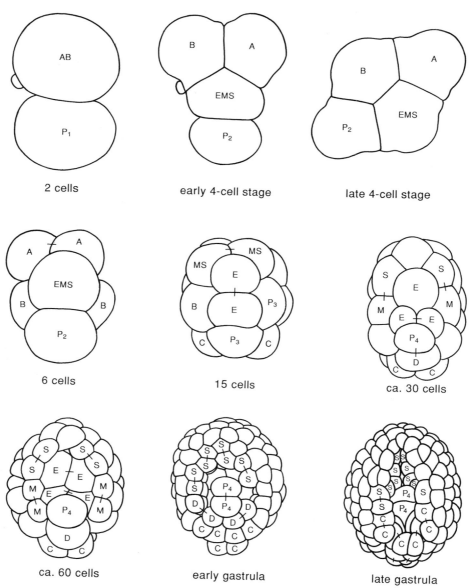

Fig. 35.2. Early development of the nematode *Parascaris megalocephala*. The 2- and 4-cell stages are seen from the right side, the following stages from the ventral side. Daughter cells of the last cleavage are united by a short line. The letters on the blastomeres are explained in Table 35.1. The A- and B-cells are not labelled in the later stages. The large endoderm cells (E) become partially covered by the primordial germ cells (P4) in the early gastrula stage, and completely covered in the late gastrula. In the late gastrula and blastopore-closure stages, the C-descendants which become ectoderm are not indicated. Stages from 2-cells to early gastrula redrawn from Boveri (1899); late gastrula redrawn from Müller (1903).

Table 35.1. The cell lineage of *Caenorhabditis elegans* (based on Sulston *et al.* 1983).

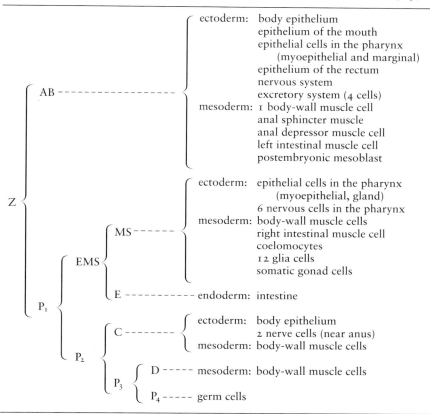

soon disappears into the blastocoel, which then becomes obliterated. The cells which will give rise to mesoderm lie in a horseshoe around the blastopore (Fig. 35.3). The lateral blastopore lips become pressed together and fuse (Malakhov 1986), leaving a large funnel-shaped stomodaeum, while the posterior opening, which corresponds to the anus, soon constricts completely, and a small proctodaeum is formed at a later stage. When the mesodermal cells have disappeared and the blastoporal lips fused, the surface of the embryo is covered by descendants of the AB- and C-cells with the last-mentioned cells forming a ventral stripe on the posterior part of the embryo (Fig. 35.3); the stomodeal invagination, which develops into the pharynx, is composed of a wider variety of cells. The embryo has now started to curve so that part of the dorsal side can be seen in ventral view. On both sides of the blastopore lie extensive areas with cells which will give rise to neurons in the ventral part of the central nervous system; when these cells migrate into the embryo, the most lateral cells, which form a row of six on each side (Fig. 35.3), finally meet and become arranged in a single, midventral line. These cells each give off an epidermal cell, and the ventral

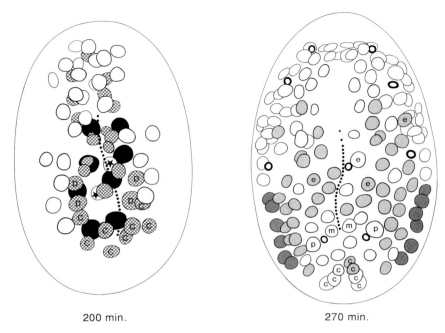

200 min.   270 min.

Fig. 35.3. Ventral aspects of *Caenorhabditis elegans* embryos; only the position of the nuclei is indicated and only of nuclei on the surface and in the blastoporal cleft. (Based on Sulston *et al.* 1983.)

Blastopore closure (200 minutes). The nuclei of the eight endodermal cells are black; the two germ-line nuclei are marked with asterisks; the cross-hatched nuclei indicate the main mesodermal precursor cells – the cells originating from C and D are marked with the respective letters and the cells without letters are descendants of MS. The position of the blastoporal cleft is indicated by the dotted line.

Late gastrulation (270 minutes). The mesodermal cells are covered ventrally by cells which are the precursors of the ventral nervous system. All nuclei are of AB-cells except the small posterior wedge of cells marked c. The nuclei of cells which will give rise to cells in the ventral nervous system (including ventral ganglion, retrovesicular ganglion, ventral cord, pre-anal ganglion, and lumbar ganglion) are shaded. The lightly shaded cells will invaginate as the last part of the gastrulation bringing the darkly shaded lateral lines of cells in contact midventrally, where they become arranged in one line; these cells then each divide into a nerve cell and an epithelial cell. The unshaded cell marked e becomes the H-shaped excretory cell; the uppermost shaded cell marked e divides to form the excretory duct cell and some ganglionic cells; the lowermost shaded cells marked e divide to form the two excretory duct cells and some ganglionic cells. The two cells marked m are situated at the proctodaeum; they are precursors of the following mesodermal muscle cells: the AB body-wall muscle cell, the anal sphincter, the anal depressor, and the left intestinal muscle. The two cells marked p become the socket cells of the posterior sensilla called phasmids. The small thick rings indicate the results of programmed cell death.

part of the central nervous system thus becomes internalized. The cells of each row are descendants of the cells ABp(r/l)ap, but the cell lineage (see Table 35.2) shows that these cells are not derived from a stem cell like the neuroblasts of the articulates (see for example Chapter 14) and the two cells give rise to several other types of cells.

Table 35.2. Cell-lineage of the right and left row of lateral cells which form the ventral cord (see Fig. 35.3). 1/2 indicates that the position in the midventral row of cells after complete blastopore closure can be either number 1 or 2. (Based on Sulston et al. 1983)

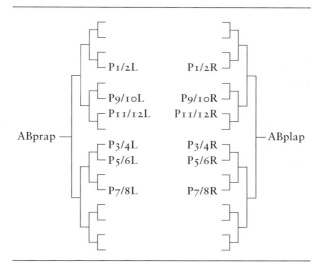

The two AB-cells which give rise to mesodermal muscles (right: the AB-derived longitudinal body muscle cell and the anal sphincter; left: anal depressor and intestinal muscle L) lie symmetrically at the sides of the posteriormost part of the blastopore which becomes the proctodaeum. The ring ganglion originates exclusively from AB-cells. The precursors of the four cells of the excretory organ are located in the field of nervous precursor cells (Fig. 35.3).

The cleavage is highly determined, and although the experiments with cell ablations using laser beams are not fully conclusive because the cells are not removed, it appears that if a cell is ablated during development the corresponding cells/organs are generally lacking in the later stages (Sulston et al. 1983). The cell lineage (Table 35.1) shows a rather complicated pattern with some of the early blastomeres giving rise to many different cell types. Mixed embryological origin is shown for example by the longitudinal body-wall muscles, which are composed of descendants from the cells MS, D, C, and AB; the epithelium comprises large syncytial areas, and the dorsal syncytium is formed by the fusion of cells from AB and C. There are examples of cells where the two daughter cells of the last cleavage become a neuron and a cell which fuses with other epidermal cells.

The development is invariably through four moults, some of which take place inside the egg in some species. The developmental stages are rather similar, but the more prominent spines or hooks are not fully developed until the final stage.

The nematodes are clearly a monophyletic group. The ground plan with a thick collagenous cuticle, which is moulted, and longitudinal muscle bands around a pseudocoelomic hydrostatic skeleton is not seen in any other phylum except in the nematomorphs (Chapter 36), and the three separate rings of sensilla, with 6 sensilla on the lips and 6+4 sensilla on the body, seem to be a very reliable apomorphy (Lorenzen 1981, 1985). Also the pair of large chemosensilla, the amphids, which are apparently always present in the cephalic region, may be considered a nematode apomorphy. The relationship with the nematomorphs is discussed in Chapter 36.

The cleavage of the nematodes has been compared to a spiral cleavage with only one quartet, but although the spiralian development has not been described in comparably fine detail, I find it impossible to find any specific similarity between the two cleavage patterns (compare Tables 11.1 and 35.1). The ultrastructure of the sensilla resembles that of some arthropod sense organs, but this 'probably reflects evolution of convergent solutions to the problem of how to get a sensory neuron into or through an exoskeleton' (Ward *et al.* 1975, p. 335).

The origin of the tube-shaped gut through fusion of the lateral blastopore lips, and the development of the ventral nerve cord along the fusion line unequivocally define the nematodes as protostomians (gastroneuralians). The aschelminth (or pseudocoelomate) phyla have often caused problems in phylogenetic discussions (they are simply omitted from the cladogram presented by Brusca & Brusca 1990, see Fig. 1.2), but the information on the development of *Caenorhabditis* should have removed any remaining doubt about the protostomian character of the nematodes. Development is direct without any trace of a trochophore stage, but similar cases of disappearance of larval characters are known from many other groups, where they may be found even within genera, so this does not in any way detract from the interpretation of nematodes as protostomians.

The asymmetrical cleavage pattern without resemblance to the spiralian pattern, the origin of the mesoderm from a zone along the blastopore, and the direct development with lack of an apical organ are aschelminth characters. The terminal mouth and the cylindrical pharynx surrounded by the collar-shaped ganglion show that the nematodes are cycloneuralians and clearly identify their position on the phylogenetic tree (Fig. 32.2).

Interesting subjects for future research

1. Ultrastructure of *Kinonchulus*.
2. Physiology of the moulting.

References

Albertson, D.G. & J.N. Thomson 1976. The pharynx of *Caenorhabditis elegans*. – Phil. Trans. R. Soc. B **275**: 299-325.

Atkinson, H.J. & C.O.E. Onwuliri 1981. *Nippostrongylus brasiliensis* and *Haemonchus contortus*: Function of the excretory ampulla of the third-stage larva. – Expl Parasit. **52**: 191-198.

Baker, G.C., D.J. Chitwood & H.H. Rees 1990. Ecdysteroids in helminths and annelids. – Invert. Reprod. Dev. **18**: 1-11.

Beklemischew, W.N. 1960. Die Grundlagen der vergleichenden Anatomie der Wirbellosen, 2 vols. – VEB Deutscher Verlag der Wissenschaften, Berlin.

Boveri, T. 1899. Die Entwickelung von *Ascaris megalocephala* mit besonderer Rücksicht auf die Kernverhältnisse. – Festschrift zum siebzehnten Geburtstag von Carl von Kupffer, pp 383-430, pls 40-45. Gustav Fischer, Jena.

Brusca, R.C. & G.J. Brusca 1990. Invertebrates. – Sinauer Associates, Sunderland.

Coomans, A. 1979. The anterior sensilla of nematodes. – Revue Nématol. **2**: 259-283.

Coomans, A. & L. de Coninck 1963. Observations on spear-formation in *Xiphinema*. – Nematologica **9**: 85-96.

Franke, S. & G. Käuser 1989. Occurrence and hormonal role of ecdysteroids in non-arthropods. – *In* J. Koolman (ed.): Ecdysone, pp 296-307. Georg Thieme, Stuttgart.

Grootaert, P. & A. Coomans 1980. The formation of the anterior feeding apparatus in *Dorylaimus*. – Nematologica **26**: 406-431.

Hope, W.D. 1988. Ultrastructure of the feeding apparatus of *Rhabdodemania minima* Chitwood, 1936 (Enoplida: Rhabdodemaniidae). – J. Nematol. **20**: 8-140.

Hope, W.D. & S.L. Gardiner 1982. Fine structure of a proprioceptor in the body wall of the marine nematode *Deontostoma californicum* Steiner et Albin, 1933 (Enoplida: Leptosomatidae). – Cell Tissue Res. **225**: 1-10.

Jeuniaux, C. 1975. Principes de systématique biochimique et application à quelques problèmes particuliers concernant les Aschelminthes, les Polychètes et les Tardigrades. – Cah. Biol. mar. **16**: 597-612.

Johnson, C.D. & A.O.W. Stretton 1980. Neural control of locomotion in *Ascaris*: anatomy, electrophysiology, and biochemistry. – *In* B. Zuckerman (ed.): Nematodes as Biological Models, pp 159-195. Academic Press, New York.

Käuser, G. 1989. On the evolution of ecdysteroid hormones. – *In* J. Koolman (ed.): Ecdysone, pp 327-336. Georg Thieme, Stuttgart.

Kenyon, C. 1985. Cell lineage and the control of *Caenorhabditis elegans* development. – Phil. Trans. R. Soc. B **312**: 21-38.

Lorenzen, S. 1981. Entwurf eines phylogenetischen Systems der freilebenden Nematoden. – Veröff. Inst. Meeresforsch. Bremerh., Suppl. **7**: 1-472.

Lorenzen, S. 1985. Phylogenetic aspects of pseudocoelomate evolution. – *In* S. Conway Morris, J.D. George, R. Gibson & H.M. Platt (eds): The Origins and Relationships of Lower Invertebrates, pp 210-223. Oxford Univ. Press, Oxford.

Malakhov, V.V. 1986. Nematodes. Anatomy, Development, Systematics, and Phylogeny. – Nauka, Moskva. (In Russian.)

Müller, H. 1903. Beitrag zur Embryonalentwicklung der *Ascaris megalocephala*. – Zoologica (Stuttg.) **17**(41): 1-30.

Nelson, F.K., P.S. Albert & D.L. Riddle 1983. Fine structure of the *Caenorhabditis elegans* secretory-excretory system. – J. Ultrastruct. Res. **82**: 156-171.

Nigon, V. 1965. Développement et reproduction des Nematodes. – Traité de Zoologie, vol **4**, pp 218-386. Masson, Paris.

Reisinger, E. 1972. Die Evolution des Orthogons der Spiralier und das Archicölomatenproblem. – Z. zool. Syst. Evolutionsforsch. **10**: 1-43.

Riemann, F. 1972. *Kinonchulus sattleri* n.g. n.sp. (Enoplida, Tripyloidea), an aberrant freeliving nematode from the lower Amazoanas. – Veröff. Inst. Meeresforsch. Bremerh. **13**: 317-326.

Singh, R.N. & J.E. Sulston 1981. Some observations on moulting in *Caenorhabditis elegans*. – Nematologica **24**: 63-71.

Steiner, G. 1919. Untersuchungen über den allgemeinen Bauplan des Nematodenkörpers. – Zool. Jb., Morph. **43**: 1-96, pls 1-3.

Sternberg, P.W. & H.R. Horvitz 1982. Postembryonal nongonadal cell lineages of the nematode *Panagrellus redivivus*: description and comparison with those of *Caenorhabditis elegans*. – Dev. Biol. **93**: 181-205.

Strassen, O. zur 1896. Embryonalentwicklung der *Ascaris megalocephala*. – Arch. Entwicklungsmech. Org. **3**: 27-105, 131-190, pls 5-9.

Strassen, O. zur 1906. Die Geschichte der T-Riesen von *Ascaris megalocephala* als Grundlage zu einer Entwicklungsmechanik dieser Species. – Zoologica (Stuttg.) **17**(40): 1-342, 5 pls.

Strassen, O. zur 1959. Neue Beiträge zur Entwicklungsgeschichte der Nematoden. – Zoologica (Stuttg.) 38(107): 1-142.

Stretton, A.O.V., R.M. Fishpool, E. Southgate, J.E. Donmoyer, J.P. Walrond, J.E.R. Moses & I.S. Kass 1978. Structure and physiological activity of the motoneurons of the nematode Ascaris. – Proc. natn. Acad. Sci. USA 75: 3493-3497.

Sulston, J.E., E. Schierenberg, J.G. White & J.N. Thomson 1983. The embryonic cell lineage of the nematode Caenorhabditis elegans. – Dev. Biol. 100: 64-119.

Ward, S., N. Thompson, J.C. White & S. Brenner 1975. Electron microscopical reconstruction of the anterior sensory anatomy of the nematode Caenorhabditis elegans. – J. comp. Neurol. 160: 313-338.

White, J.G., E. Southgate, J.N. Thomson & S. Brenner 1986. The structure of the nervous system of the nematode Caenorhabditis elegans. – Phil. Trans. R. Soc. B 314: 1-340.

Wright, K.A. 1980. Nematode sense organs. – In B.M. Zuckerman (ed.): Nematodes as Biological Models, vol. 2, pp 237-295. Academic Press, New York.

Wright, K.A. 1991. Nematoda. – In F.W. Harrison (ed.): Microscopic Anatomy of Invertebrates, vol. 4, pp 111-195. Wiley-Liss, New York.

Phylum NEMATOMORPHA

Hair worms are a small phylum of nematode-like parasites which spend the larval stages in the body cavities of arthropods, while the adult sexual stage is free, non-feeding, but can nevertheless live for several months. About 325 species have been described, representing two orders: the marine Nectonematoidea, with the only genus *Nectonema* with four species, and the limnic-terrestrial Gordioidea. The *Nectonema* larvae parasitize decapod crustaceans and the adult stage is pelagic with dorsal and ventral double rows of swimming bristles. The gordioid larvae are found in insects and the adults crawl in or cling to the vegetation.

The body is an extremely slender cylinder, in some species more than a meter long. The anterior end is rounded without any appendages in gordioids, with a small mouth opening at the ventral side in some species, whereas other species lack a mouth; the pharynx or oesophagus is a solid strand of cuticle in most species. *Nectonema* has a minute buccal cavity with a pair of teeth leading to a narrow, cuticularized pharynx/oesophagus, which consists of only one cell (Bresciani 1991). The intestine is a narrow tube of monolayered epithelial cells with microvilli (Eakin & Brandenburger 1974, Skaling & MacKinnon 1988); it opens into a cuticle-lined cloaca (Bresciani 1991). It is clear that nutrient uptake does not take place through the digestive system, but the intestine is involved in storage of substances taken up through the cuticle (Bresciani 1991).

The body wall consists of a monolayered ectoderm, covered by a cuticle with several layers, and a layer of longitudinal muscle cells (Bresciani 1991). The thinner outer layer of the cuticle comprises an osmiophilic and a non-osmiophilic layer, and the thick inner layer consists of crossed layers of thick collagenous fibres. The natatory bristles of *Nectonema* consist of parallel fibrils and are loosely attached to the outer layer of the body cuticle (Bresciani 1991). The ectoderm is thin except for a dorsal longitudinal thickening or cord; *Nectonema* has both a dorsal and a ventral cord. The layer of longitudinal muscles is interrupted at the cords.

The gordioid nervous system (Montgomery 1903, Eakin & Brandenburger 1974) comprises a brain surrounding the pharynx and having a large, dorsal organ

---

Chapter vignette: *Gordius aquaticus* just emerged from its host. (Redrawn from Bresciani 1991.)

which has been interpreted as an eye, a ventral nerve cord and an anal ganglion. The whole nervous system is basiepithelial, but the ventral cord is situated in the mesenchyme surrounded by a mid-ventral extension of the ectodermal basement membrane. In *Nectonema*, the brain surrounds the pharynx and contains four giant nerve cells with axons to the ventral cord (Feyel 1936). The ultrastructure and the innervation of the longitudinal muscles have not been described.

The muscles are very unusual, with a peripheral layer of very thick paramyosin filaments with a unique organization and thin (?actin) filaments. The muscle cells are completely surrounded by extracellular matrix and are anchored to the cuticle via hemidesmosomes and tonofibrils in the ectoderm. Extensions from the muscle cells to the nerves like those of nematode muscle cells have not been described. The thick, fibrous cuticle and the longitudinal muscles work together around the hydrostatic skeleton formed by the pseudocoel.

The gordioids have the primary body cavity almost completely filled with a mesenchymatous tissue; narrow pseudocoelomic canals surround the intestine and the gonads (Eakin & Brandenburger 1974). *Nectonema* has a more spacious pseudocoel which is divided into a small cephalic chamber and a long body cavity by a septum (Feyel 1936, Bresciani 1991). Haemal systems and excretory organs are absent.

Paired, sac-shaped gonads are situated in the pseudocoel, surrounded by the parenchyma, and open into the cloaca.

The gordioid eggs are fertilized in the uterus and deposited in strings. Cleavage is total and usually equal, and there is much variation in the arrangement of the blastomeres (Inoue 1958). A coeloblastula is formed, and in *Gordius* it soon becomes filled with a compact mass of cells (Mühldorf 1914, Malakhov & Spiridonov 1984). A thick-walled, shallow, anterior invagination becomes the introvert, and a thinner, posterior invagination becomes the gland (see below) and the intestine (Montgomery 1904, Inoue 1958). The cuticular lining of the anterior invagination indicates that it represents the stomodaeum. The full-grown larva (Zapotosky 1974, 1975) has an anterior, preseptal region with the introvert and a posterior, postseptal region; both regions have an annulated epidermis with a thick cuticle. The two body regions are separated by the septum, which is a diaphragm of six mesodermal cells surrounded by a thickened basement membrane only pierced by the pharynx (oesophagus) cell. The introvert carries three rings each with six cuticular spines and a cylindrical proboscis, which can be withdrawn but not inverted. The proboscis has a smooth cuticle with three longitudinal cuticular rods, which fuse anteriorly around the narrow mouth opening. Protrusion of the introvert and proboscis appears to be caused by contraction of parietal muscles from the septum to the body wall just behind the introvert; retractor muscles from the septum insert on the introvert in front of the anterior ring of spines and inside the proboscis. One epithelial cell containing a contorted cuticular canal extends from the mouth opening to the septum, where the canal continues through two postseptal cells to a gland consisting of eight large cells; the function of gland + duct has not been demonstrated. The postseptal region contains the gland, an intestine with a short rectum, six longitudinal muscles, and a number of undifferentiated cells. The intestine contains a number of granules of unknown function. Mühldorf (1914) and Malakhov &

Spiridonov (1984) observed a ventral double-row of ectodermal nuclei which they supposed represented the ventral nervous cord.

The larva hatches and must penetrate the body wall or gut of a host by means of the introvert and proboscis in order to develop further; this may happen directly or, perhaps more normally, after a period of encystment where the cysts become ingested by the host (Dorier 1930). The following development takes place in the body cavity of the arthropod. The introvert disappears and the preseptal region diminishes while the postseptal region grows enormously, so that the preseptal region of the adult is represented only by the anterior, hemispherical calotte. The sexually mature specimens break out of the host, usually through the anal region (Dorier 1930).

The embryology of *Nectonema* has not been described. Huus (1931) observed copulation and subsequent spawning of eggs, but the eggs did not develop. The youngest larva observed by Huus (1931) was already inside the host; it was cylindrical with a short introvert with two rings of six hooks each and a pair of anterior spines. The drawings show a gut without a lumen in the anterior end, but a somewhat older stage, still with hooks on the introvert, showed the thin, curved cuticle-lined oesophagus observed in the adults. The larvae moult several times in the host, and the adult stage with the natatory bristles has been found inside the last exuvium (Leslie, Campbell & Daborn 1981).

The nematomorphs are probably a monophyletic group, but it is difficult to list reliable apomorphies (Lorenzen 1985). They share several apomorphies with the nematodes, such as the cuticle with layers of crossing collagenous fibrils and lack of chitin, the body wall with only longitudinal muscles, and the ectodermal longitudinal cords, and it appears unquestionable that the two groups together constitute a monophyletic unit. The nematomorphs have been interpreted as specialized nematodes, perhaps derived from the parasitic mermithoids, which have a similar life cycle with juveniles parasitizing arthropods and free-living adults which do not feed, and with a thin pharynx without connection to the intestine. However, the mermithoids have the usual nematode sensillae and amphids, ovaria with an anterior and a posterior branch extending from the median genital opening and no trace of an introvert. The characteristic axonal extensions from the muscle cells to the nerve cords found in nematodes have not been described in nematomorphs, but studies of both structure and development of the nematomorphs may reveal new information of phylogenetic importance. With the present state of knowledge, I can only agree with Lorenzen (1985), who concluded that the similarities between mermithoid nematodes and nematomorphs are due to convergence, and that both nematodes and nematomorphs have evolved from an ancestor with an introvert. The nematodes have then specialized their feeding mechanism by the evolution of a myoepithelial sucking pharynx, which has secondarily been lost in some of the endoparasites; the nematomorphs have retained the introvert, and almost lost the gut as another line of specialization to parasitism.

Interesting subjects for future research

1. Development of *Nectonema*.
2. Innervation of the longitudinal body muscles.
3. Structure of the brain and the putative eye.

References

Bresciani, J. 1991. Nematomorpha. – In F.W. Harrison (ed.): Microscopic Anatomy of Inverte-
brates, vol. 4, pp 197-218. Wiley-Liss, New York.
Dorier, A. 1930. Recherches biologiques et systématiques sur les Gordiacés. – Annls Univ. Gren-
oble, N.S., Sci.-Med. 7: 1-183, 3 pls.
Eakin, R.E. & J.L. Brandenburger 1974. Ultrastructural features of a gordian worm (Nematomor-
pha). – J. Ultrastruct. Res. 46: 351-374.
Feyel, T. 1936. Recherches histologiques sur Nectonema agile Verr. Étude de la forme parasite. –
Archs Anat. microsc. 32: 195-234.
Huus, J. 1931. Über die Begattung bei Nectonema munidae Br. und über den Fund der Larve von
dieser Art. – Zool. Anz. 97: 33-37.
Inoue, I. 1958. Studies on the life history of Chordodes japonensis, a species of Gordiacea. I. Devel-
opment and structure of the larva. – Jap. J. Zool. 12: 203-218.
Leslie, H.A., A. Campbell & G.A. Daborn 1981. Nectonema (Nematomorpha: Nectonematoidea)
a parasite of decapod Crustacea in the Bay of Fundy. – Can. J. Zool. 59: 1193-1196.
Lorenzen, S. 1985. Phylogenetic aspects of pseudocoelomate evolution. – In S. Conway Morris, J.D.
George, R. Gibson & H.M. Platt (eds): The Origins and Relationships of Lower Invertebrates,
pp 210-223. Oxford Univ. Press, Oxford.
Malakhov, V.V. & S.E. Spiridonov 1984. The embryogenesis of Gordius sp. from Turkmenia, with
special reference to the position of the Nematomorpha in the animal kingdom. – Zool. Zh. 63:
1285-1296 (In Russian, English summary).
Montgomery, T.H. 1903. The adult organisation of Paragordius varius (Leidy). – Zool. Jb., Anat.
18: 387-474, pls 37-43.
Montgomery, T.H. 1904. The development and structure of the larva of Paragordius. – Proc. Acad.
nat. Sci. Philad. 56: 738-755, pls 49-50.
Mühldorf, A. 1914. Beiträge zur Entwicklungsgeschichte und zu den phylogenetischen Beziehungen
der Gordiuslarve. – Z. wiss. Zool. 111: 1-75, pls 1-3.
Skaling, B. & B.M. MacKinnon 1988. The absorptive surfaces of Nectonema sp. (Nematomorpha,
Nectonematoidea) from Pandalus montagui: histology, ultrastructure, and absorptive capa-
bilities of the body wall and intestine. – Can. J. Zool. 66: 289-295.
Zapotosky, J.E. 1974. Fine structure of the larval stage of Paragordius varius (Leidy, 1851) (Gor-
dioidea: Paragordiidae). I. The preseptum. – Proc. helminth. Soc. Wash. 41: 209-221.
Zapotosky, J.E. 1975. Fine structure of the larval stage of Paragordius varius (Leidy, 1851) (Gordioidea: Paragordiidae). II. The postseptum. – Proc. helminth. Soc. Wash. 42: 103-111.

37

CEPHALORHYNCHA

Close relationships between priapulans and kinorhynchs have been suggested by several authors. Lang (1953) united the two groups together with the acanthocephalans in his Rhynchaschelminthes or Rhynchohelminthes. Malakhov (1980) introduced the name Cephalorhyncha for a group consisting of priapulans, kinorhynchs and nematomorphs; the newly discovered group Loricifera was added later (Adrianov, Malakhov & Yushin 1990). I believe that Priapula, Kinorhyncha and Loricifera form a monophyletic unit, and instead of creating a new name, I have chosen to use Malakhov's name, but to exclude the Nematomorpha.

The three phyla share a number of characters which can best be interpreted as apomorphies (Figs. 32.2 and 37.1): 1) chitinous cuticle (not demonstrated in the loriciferans); 2) rings of scalids on introvert; 3) two rings of introvert retractors attaching to the mouth region in the zone of the collar shaped brain; one ring runs inside the nerve ring and the other outside the ring.

The cuticle of nematodes and nematomorphs, which together are regarded as the sister group of the cephalorhynchs, have a collagenous cuticle without any trace of chitin. This is not a result of a lack of the enzymes which are necessary for the synthesis of chitin, as demonstrated by the presence of chitin in nematode egg shells, but the two components are apparently mutually exclusive in a cuticle; the same distribution of chitin and collagen in the cuticles is observed in the sister groups annelids and arthropods (Chapter 16). Chitinous cuticles are not known in any of the other aschelminth phyla, although the mastax of rotifers and the jaws of chaetognaths consist mainly of chitin (Chapters 29 and 31).

Characteristic chitinous spines or scales, called scalids, form pentagonal or hexagonal patterns around the introvert; they contain cilia from one or more monociliate sensory cells and are not observed in any other aschelminth group. Their shape is highly variable, as indicated by the names of the different types, spinoscalids, trichoscalids, clavoscalids etc, but their essential structure is identical. The nematodes and nematomorphs have cuticular sense organs called sensilla, which consist of three cell types (Chapter 35); sensilla may protrude as papillae or spines, but their special structure with three cell types does not resemble that of the scalids.

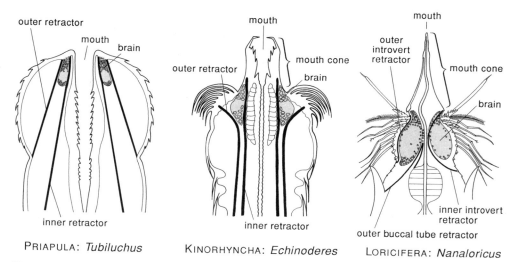

PRIAPULA: *Tubiluchus*   KINORHYNCHA: *Echinoderes*   LORICIFERA: *Nanaloricus*

Fig. 37.1. Longitudinal sections of introverts with brain and retractor muscles of the three cephalo-rhynch phyla. – Priapula: *Tubiluchus* sp. (interpretation based on Calloway 1975 and Rehkämper et al. 1989). – Kinorhyncha: *Echinoderes capitata* (based on Nebelsick 1993). – Loricifera: *Nanaloricus mysticus* (based on Kristensen & Higgins 1991 and personal communications from Dr R.M. Kristensen, Univ. Copenhagen).

The sensory setae on the trunk of loriciferans and kinorhynchs have a similar structure, but the chemical composition of the cuticle is of course different.

The retractor muscles of the introvert show a highly characteristic pattern. There are two rings of muscles which both attach to the ectoderm around the mouth or at the base of the mouth cone; the inner ring of muscles runs along the inner side of the brain ring, in some cases penetrating its anterior part, and the outer ring runs along the outside of the brain (Fig. 37.1). Anterior muscle attachments penetrating the brain with elongate, tonofibril-containing ectodermal cells (sometimes called tanycytes) have been observed in all three phyla (Kristensen & Higgins 1991), but their position has only been documented in a few cases.

The sister-group relationships of the three phyla have been discussed by Kristensen (1983) and Higgins & Kristensen (1986), who emphasize the presence of a non-inversible mouth cone and detailed similarities between several types of scalids in loriciferans and kinorhynchs, and regard these two phyla as sister groups. The loriciferan pharynx has a myoepithelium whereas kinorhynchs and priapulans have exclusively mesodermal musculature. The myoepithelial pharynx has usually been regarded as a symplesiomorphy within the aschelminth groups, but I regard it as probable that this structure has evolved independently several times (Chapter 32), so this character should not make the proposed sister-group relationships unreasonable.

References

Adrianov, A.V., V.V. Malakhov & V.V. Yushin 1990. Loricifera – a new taxon of marine inverte-brates. – Soviet J. mar. Biol. **15**: 136-138.

Calloway, C.B. 1975. Morphology of the introvert and associated structures of the priapulid *Tubilu-chus corallicola* from Bermuda. – Mar. Biol. (Berl.) **31**: 161–174.

Higgins, R.P. & R.M. Kristensen 1986. New Loricifera from southeastern United States coastal waters. – Smithson. Contr. Zool. **483**: 1-70.

Kristensen, R.M. 1983. Loricifera, a new phylum with Aschelminthes characters from the meioben-thos. – Z. zool. Syst. Evolutionsforsch. **21**: 163-180.

Kristensen, R.M. & R.P. Higgins 1991. Kinorhyncha. – *In* F.W. Harrison (ed.): Microscopic Ana-tomy of Invertebrates, vol. 4, pp 377-404. Wiley-Liss, New York.

Lang, K. 1953. Die Entwicklung des Eies von *Priapulus caudatus* Lam. und die systematische Stel-lung der Priapuliden. – Ark. Zool., 2. ser., **5**: 321-348.

Malakhov, V.V. 1980. Cephalorhyncha, a new type of animal kingdom uniting Priapulida, Kino-rhyncha, Gordiacea, and a system of Aschelminthes worms. – Zool. Zh. **59**: 485-499 (In Rus-sian, English summary).

Nebelsick, M. 1993. Introvert, mouth cone, and nervous system of *Echinoderes capitatus* (Kino-rhyncha, Cyclorhagida) and implications for the phylogenetic position of the Kinorhyncha. – Zoomorphology **113**: 211-232.

Rehkämper, G., V. Storch, G. Alberti & U. Welsch 1989. On the fine structure of the nervous system of *Tubiluchus philippinensis* (Tubiluchidae, Priapulida). – Acta zool. (Stockh.) **70**: 111-120.

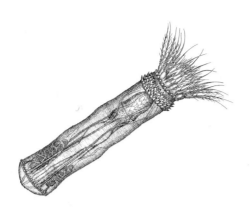

38

Phylum PRIAPULA

Priapulans are a small phylum comprising a few well-known, marine, macrobenthic genera, such as *Priapulus* and *Halicryptus*, and three meiobenthic genera, *Tubiluchus*, *Meiopriapulus* and *Maccabeus*, which have been described within the last decades. A total of about 17 living species are now recognized; they are placed in three families, sometimes with *Maccabeus* placed in a separate order (Salvini-Plawen 1974, Por 1983, van der Land & Nørrevang 1985). A number of genera from the Middle Cambrian bear striking resemblance to some of the living genera (Conway Morris 1977, Conway Morris & Robison 1986).

The body consists of a cylindrical trunk and a large introvert; some genera have one or two caudal appendages. *Maccabeus* (see the chapter vignette) has a double circle of hollow setose tentacles and a ring of setose spines around the mouth (Por & Bromley 1974). The introvert carries many rings of different types of scalids, mostly with a monociliate sensory cell (Storch, Higgins & Morse 1989b, Storch 1991); the scalids point posteriorly when the introvert is everted. The introvert can be everted by the contraction of the trunk muscles with the spacious, fluid-filled body cavity functioning as a hydrostatic skeleton. Retraction of the introvert is by contraction of two sets of muscles, an outer ring extending from the mouth region to the posterior limit of the introvert, and an inner ring from the mouth region to the mid-region of the trunk (see below). Burrowing through the substratum is by eversion and contraction of the introvert combined with peristaltic movements of the whole body (Hammond 1970). The caudal appendages are extensions of the body wall with musculature and body cavity; they are believed to have respiratory functions (Fänge & Mattisson 1961, Storch 1991).

The ectoderm is monolayered and covered by a chitinous cuticle (Carlisle 1959, an observation based only on chitosan tests), with an electron-dense outer and a homogeneous, electron-lucent inner zone in most genera, but the inner zone has a layer of crossing fibres of unknown chemical composition in *Meiopriapulus* (Storch, Higgins & Morse 1989b).

Chapter vignette: *Maccabeus tentaculatus*. (Redrawn from Por 1972.)

The mouth opening is usually surrounded by a narrow field of buccal papillae with many sensory cells (Storch, Higgins & Rumohr 1990) and opens into a pharynx with circles of cuticular teeth, which lack the cilia found in the scalids; the teeth point towards the midgut when the pharynx is not everted. The macrobenthic species have a large muscular pharynx and a very short oesophagus with a sphincter muscle just in front of the midgut. *Meiopriapulus* and *Tubiluchus* (Morse 1981, Storch & Alberti 1985, Storch, Higgins & Morse 1989a, Storch 1991) have been described as having a mouth cone and lacking pharyngeal teeth, but since the epithelium with the teeth can be inverted it is rather the pharynx which can be everted; a non-inversible mouth cone like that of kinorhynchs and loriciferans is not present. *Meiopriapulus* and *Tubiluchus* have the pharynx followed by a long oesophagus with longitudinal folds and a muscular, gizzard-like swelling with an anterior ring of small cuticular projections and a posterior ring of long, comb-like plates. The pharynx has both longitudinal, circular and oblique muscles. The larger species are carnivores, which grasp annelids and other prey organisms and ingest them through the action of the pharyngeal teeth (van der Land 1970). *Tubiluchus* and *Meiopriapulus* scrape bacteria and other small organisms from sediment particles (Higgins & Storch 1991, Storch, Higgins & Morse 1989a), whereas *Maccabeus* apparently swallows larger prey caught by the pharynx, which is then retracted by the unique pharynx retractors attached to the posterior end of the body (Por & Bromley 1974, see the chapter vignette). It appears that none of the species use the pharynx as a suction pump.

The midgut is a long tube of cells with microvilli; it is surrounded by longitudinal and circular muscles (Candia Carnevali & Ferraguti 1979). The rectum has a folded cuticle.

The central nervous system consists of a peribuccal nerve ring, a ventral cord and a caudal ganglion ventral to the anus; all these structures are intraepithelial. The collar-shaped brain consists of an anterior and a posterior ring of perikarya separated by a zone of neuropil. The sense organs of the introvert are innervated from the brain (van der Land & Nørrevang 1985, Rehkämper *et al.* 1989).

There is a rather spacious body cavity filled with a fluid with erythrocytes and amoebocytes (Mattisson & Fänge 1973). There is no peritoneal cell layer, and the isolated cells observed at the inner side of the body wall are amoebocytes (McLean 1984).

The body wall of trunk and introvert has an outer layer of ring muscles and an inner layer of longitudinal muscles. There are two rings of introvert retractor muscles (Fig. 37.1). The shorter outer retractors extend from the body wall along the anterior side of the brain to the base of the introvert, and the eight longer inner retractors extend from the body wall at the posterior side of the brain to the middle region of the trunk (Candia Carnevali & Ferraguti 1979; Storch, Higgins & Rumohr 1990). The inner retractors of *Tubiluchus* penetrate the inner side of the brain (Calloway 1975), and Rehkämper et al. (1989) reported the presence of tanycytes in this region. The pharynx is surrounded by longitudinal, oblique and circular muscles (Candia Carnevali & Ferraguti 1979; Storch, Higgins & Morse 1989a). *Maccabeus* has two rings of muscles between the oesophagus and the body wall, but only one ring of introvert retractors (Por & Bromley 1974, Por 1983).

Priapulus has large, branched protonephridia with monociliate terminal cells, which make sinuous weirs between each other (Kümmel 1964). *Tubiluchus* has biciliate terminal cells (Alberti & Storch 1986). The protonephridia and the sac-shaped gonads have common ducts with a pair of openings lateral to the anus in *Priapulus* and *Tubiluchus*, but with openings in the rectum in *Meiopriapulus* (van der Land & Nørrevang 1885, Storch, Higgins & Morse 1989a). Candia Carnevali & Ferraguti (1979) reported myoepithelial cells in the urogenital organs of *Halicryptus*, particularly at the mesenteria.

Priapulus is a free spawner (Lang 1953), whereas *Tubiluchus* has internal fertilization (Alberti & Storch 1988). The first cleavages of *Priapulus* and *Halicryptus* are apparently radial; a coeloblastula is formed, and gastrulation resembles polar ingression (Zhinkin 1949, Zhinkin & Korsakova 1953, Lang 1953). The small species have not been investigated, and the stages between gastrula and the feeding larva are unknown. All species but *Meiopriapulus* develop through a characteristic larval stage with a ring of longitudinal chitinous shields on the trunk (Kirsteuer 1976, Higgins & Storch 1991; Fig. 38.1). All species go through a series of moults, both in the larval stage with the lorica and in the adult stage. The larvae move by eversion and retraction of the introvert (Hammond 1970).

The Burgess Shale fossils show that the living priapulans are the few surviving representatives of a much larger group. Their close relationships with kinorhynchs and loriciferans are indicated by the similarities in the cuticle with scalids, the structure of the introvert with the two rings of retractor muscles, and the structure of the brain with tanycytes. The spacious body cavity with erythrocytes and amoebocytes appears to be a reliable apomorphy. Kinorhynchs and loriciferans have a well-defined, non-inversible mouth cone which is not found in priapulans. The priapulans are therefore regarded as the sister group of these two phyla.

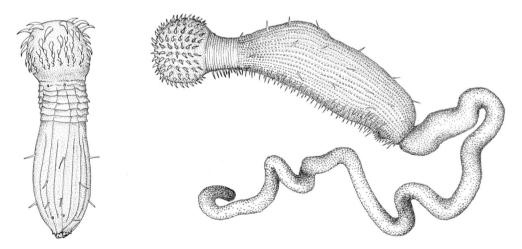

*Fig. 38.1. Larva and adult male of *Tubiluchus corallicola*; only the male has the cuticular spines on the ventral side of the abdomen. (Redrawn from van der Land 1970, 1975 and Calloway 1975.)*

Interesting subjects for future research

1. Embryology and early development of several species.
2. Secretion and chemical composition of the cuticle of *Meiopriapulus* and a species with a compact cuticle.

References

Alberti, G. & V. Storch 1986. Zur Ultrastruktur der Protonephridien von *Tubiluchus philippinensis* (Tubiluchidae, Priapulida). – Zool. Anz. **217**: 259-271.

Alberti, G. & V. Storch 1988. Internal fertilization in a meiobenthic priapulid worm: *Tubiluchus philippinensis* (Tubiluchidae, Priapulida). – Protoplasma **143**: 193-196.

Calloway, C.B. 1975. Morphology of the introvert and associated structures of the priapulid *Tubiluchus corallicola* from Bermuda. – Mar. Biol. (Berl.) **31**: 161–174.

Candia Carnevali, M.D. & M. Ferraguti 1979. Structure and ultrastructure of muscles in the priapulid *Halicryptus spinulosus*: functional and phylogenetic significance. – J. mar. biol. Ass. U.K. **59**: 737-744.

Carlisle, D.B. 1959. On the exuvia of *Priapulus caudatus* Lamarck. – Ark. Zool., 2. ser., **12**: 79-81.

Conway Morris, S. 1977. Fossil priapulid worms. – Spec. Pap. Palaeont. **20**: 1-95, 30 pls.

Conway Morris, S. & R.A. Robison 1986. Middle Cambrian priapulids and other soft-body fossils from Utah and Spain. – Paleont. Contr. Univ. Kansas **117**: 1-22.

Fänge, R. & A. Mattisson 1961. Function of the caudal appendage of *Priapulus caudatus*. – Nature **190**: 1216-1217.

Hammond, R.A. 1970. The burrowing of *Priapulus caudatus*. – J. Zool. (Lond.) **162**: 469-480.

Higgins, R.P. & V. Storch 1989. Ultrastructural observations on the larva of *Tubiluchus corallicola* (Priapulida). – Helgoländer wiss. Meeresunters. **43**: 1-11.

Higgins, R.P. & V. Storch 1991. Evidence for direct development in *Meiopriapulus fijiensis* (Priapulida). – Trans. Am. microsc. Soc. **110**: 37-46.

Kirsteuer, E. 1976. Notes on adult morphology and larval development of *Tubiluchus corallicola* (Priapulida), based on in vivo and scanning electron microscopic examinations of specimens from Bermuda. – Zool. Scr. **5**: 239-255.

Kümmel, G. 1964. Die Feinstruktur der Terminalzellen (Cyrtocyten) an den Protonephridien der Priapuliaden. – Z. Zellforsch. **62**: 468-484.

Lang, K. 1953. Die Entwicklung des Eies von *Priapulus caudatus* Lam. und die systematische Stellung der Priapuliden. – Ark. Zool., 2. Ser. **5**: 321-348.

Mattisson, A. & R. Fänge 1973. Ultrastructure of erythrocytes and leucocytes of *Priapulus caudatus* (De Lamarck) (Priapulida). – J. Morph. **140**: 367-379.

McLean, N. 1984. Amoebocytes in the lining of the body cavity and mesenteries of *Priapulus caudatus* (Priapulida). – Acta zool. (Stockh.) **65**: 75-78.

Morse, M.P. 1981. *Meiopriapulus fijiensis* n.gen., n.sp.: an interstitial priapulid from coarse sand in Fiji. – Trans. Am. microsc. Soc. **100**: 239-252.

Por, F.D. 1972. Priapulida from deep bottoms near Cyprus. – Israel J. Zool. **21**: 525-528.

Por, F.D. 1983. Class Seticoronaria and phylogeny of the phylum Priapulida. – Zool. Scr. **12**: 267-272.

Por, F.D. & H.J. Bromley 1974. Morphology and anatomy of *Maccabeus tentaculatus* (Priapulida: Seticoronaria). – J. Zool. (Lond.) **173**: 173-197.

Rehkämper, G., V. Storch, G. Alberti & U. Welsch 1989. On the fine structure of the nervous system of *Tubiluchus philippinensis* (Tubiluchidae, Priapulida). – Acta zool. (Stockh.) **70**: 111-120.

Salvini-Plawen, L.v. 1973. Zur Morphologie und Systematik der Priapulida: *Chaetostephanus praeposteriens*, der Vertreter einer neuen Ordnung Seticoronaria. – Z. zool. Syst. Evolutionsforsch. **12**: 31-54.

Storch, V. 1991. Priapulida. – *In* F.W. Harrison (ed.): Microscopic Anatomy of Invertebrates, vol. 4, pp 333-350. Wiley-Liss, New York.

Storch, V. & G. Alberti 1985. Zur Ultrastruktur des Darmtraktes von *Tubiluchus philippinensis* (Tubiluchidae, Priapulida). – Zool. Anz. **214**: 262-272.

Storch, V., R.P. Higgins & M.P. Morse 1989a. Internal anatomy of *Meiopriapulus fijiensis* (Priapulida). – Trans. Am. microsc. Soc. **108**: 245-261.

Storch, V., R.P. Higgins & M.P. Morse 1989b. Ultrastructure of the body wall of *Meiopriapulus fijiensis* (Priapulida). – Trans. Am. microsc. Soc. **108**: 319-331.

Storch, V., R.P. Higgins & H. Rumohr 1990. Ultrastructure of introvert and pharynx of *Halicryptus spinulosus* (Priapulida). – J. Morph. **206**: 163-171.

van der Land, J. 1970. Systematics, zoogeography, and ecology of the Priapulida. – Zool. Verh. (Leiden) **112**: 1-118.

van der Land, J. 1975. Priapulida. – *In* A.C. Giese & J.S. Pearse (eds): Reproduction of Marine Invertebrates, vol. 2, pp 55-65. Academic Press, New York.

van der Land, J. & A. Nørrevang 1985. Affinities and intraphyletic relationships of the Priapulida. – *In* S. Conway Morris, J.D. George, R. Gibson & H.M. Platt (eds): The Origins and Relationships of Lower Invertebrates, pp 261-273. Oxford Univ. Press, Oxford.

Zhinkin, L. 1949. Early stages in the development of *Priapulus caudatus*. – Dokl. Akad. Nauk SSSR **65**: 409-412 (In Russian).

Zhinkin, L. & G. Korsakova 1953. Early stages in the development of *Halicryptus spinulosus*. – Dokl. Akad. Nauk SSSR **88**: 571-573 (In Russian).

Phylum KINORHYNCHA

Kinorhynchs or mud dragons are a small phylum comprising about 150 benthic, marine species, which are almost all less than one millimetre long. Two orders are generally recognized, but the group is actually very homogeneous.

The body consists of 13 segments: the introvert, the neck and 11 trunk segments, each with a dorsal and one or a pair of ventral cuticular plates. The introvert can be retracted into the anterior part of the trunk; there is a closing structure consisting of a ring of small plates at the neck in the Cyclorhagida, and of 2-4 dorsal and 2-4 ventral platelets in the Homalorhagida.

The anterior end of the fully extended introvert carries a mouth cone with a ring of oral styles surrounding the mouth; the mouth cone is retracted, but not inverted when the introvert retracts. The median part of the introvert carries several rings of scalids with locomotory and sensory functions. The sockets of the scalids contain a number of monociliate sensory cells with the cilia extending to the tip of the scalids (Moritz & Storch 1972b, Brown 1989, Kristensen & Higgins 1991). The introvert is everted by contraction of the body musculature with the body cavity functioning as a hydrostatic skeleton. Retraction is by contraction of two sets of muscles between the introvert and a number of trunk segments (see below).

The mouth leads to a short oral cavity inside the oral cone and further to a muscular pharynx, which is circular or slightly nine-lobed in cross-section in the Cyclorhagida and tri-radiate in cross-section with a ventral and a pair of lateral muscular swellings in the Homalorhagida (Kristensen & Higgins 1991). The ectoderm of the oral cavity bears several rings of scalids of different types (Brown 1989). Strong squeezing may rupture some of the pharyngeal muscles and force an eversion of the pharyngeal wall with the scalids (Brown 1989). The muscular pharynx bulb has a layer of ectoderm with a thin cuticle, and the mesodermal musculature comprises both radial and circular muscle cells (Kristensen & Higgins 1991). The feeding biology is poorly studied, but many of the cyclorhagids swallow diatoms whereas

Chapter vignette: *Echinoderes aquilonius*. Disko, Greenland, July 1988 (drawn from a SEM preparation courtesy of Dr R.M. Kristensen, Univ. Copenhagen, Denmark.)

most of the homalorhagids ingest detritus by opening the mouth and sucking the material with the pharyngeal bulb (Zelinka 1928).

The midgut is a tube of large endodermal cells with a thick layer of microvilli surrounded by a thin layer of longitudinal and circular, mesodermal muscle cells. There is a cuticle-lined rectum.

The monolayered ectoderm is generally covered by a compact, chitinous cuticle without microvilli (Jeuniaux 1975, Kristensen & Higgins 1991), but microvilli have been observed in the cuticle of the pharyngeal crown surrounding the anterior rim of the pharynx bulb (Moritz & Storch 1972a). The plates of the trunk have large, ventral apodemes for the attachment of the longitudinal muscles between the segments. Various spines with associated glands occur on the cuticular plates, especially in the posterior end.

The central nervous system comprises a collar-shaped brain surrounding the anterior part of the pharynx at the base of the mouth cone. Three regions can be recognized; an anterior region with ten clusters of perikarya, a middle region consisting mainly of neuropil, and a posterior ring with eight or ten clusters of perikarya. The anterior ganglia innervate the mouth cone, which has a small nerve ring with ten small ganglia, and the scalids on the introvert. There is a paired, midventral, intraepithelial cord with ganglionic swellings and commissures in each segment, and ending in a ventral anal ganglion; the ventral nerve cord appears to be exclusively motor. Three pairs of longitudinal nerves with segmental ganglia are situated laterally and dorsally (Kristensen & Higgins 1991). The different types of scalids all have cilia extending to the tip and are believed to be either chemoreceptors or mechanoreceptors, while at the same time functioning in locomotion and perhaps food manipulation.

The muscles of the body wall are cross-striated and attach to the cuticle by way of ectodermal cells with tonofilaments and hemidesmosomes (Kristensen & Higgins 1991). The longitudinal muscles between the segmental cuticular plates and the dorso-ventral muscles are segmentally arranged. There is a narrow cavity with amoebocytes between the muscles of the body wall and the gut, and this space must be interpreted as a primary body cavity. A ring of 16 outer retractor muscles extend from the epithelium of the introvert and along the outer side of the ring–shaped brain, whereas a ring of 12 inner retractor muscles extend from the base of the mouth cone and along the inner side of the brain; both sets of muscles attach to the epithelium of a number of the anterior segments (Nebelsick 1993, Kristensen & Higgins 1991; see also Fig. 37.1). The presence of tanycytes in the brain was reported by Kristensen & Higgins (1991), but without indication of their precise position.

There is a pair of protonephridia opening in the 11th segment. Each protonephridium comprises a number of biciliate cells and a microvillous weir (Neuhaus 1988, Kristensen & Hay-Schmidt 1989, Kristensen & Higgins 1991).

A pair of sac-shaped gonads opens in a posteroventral pore. Spermatophores have been observed in some genera, and the presence of seminal receptacles in all species investigated indicates internal fertilization (Kristensen & Higgins 1991). This fits well with the aberrant type of sperm with a large, sausage-shaped head and a very short flagellum (Nyholm & Nyholm 1982).

The fertilized eggs of *Echinoderes* are deposited singly in small, muddy tubes; the cleavage could not be followed, but eggs containing almost fully developed juveniles and the hatching of small, 11-segmented juveniles were observed by Kozloff (1972). The post-embryonal development of representatives of both orders was described by Higgins (1974), who observed a series of 5-6 moultings, which gradually transformed the juvenile into an adult. Earlier reports of juvenile stages with fewer segments are now regarded as misunderstandings (Nyholm 1947).

The 'segmented' musculature and nervous system corresponding to the 11 rings of cuticular plates of the trunk clearly demonstrate the monophyletic character of the group. The 'segmentation' resembles that found in the arthropods, but there is nothing to indicate that the segments arise from a posterior teloblastic zone; a similar 'pseudosegmentation' is seen in the longitudinal muscles of bdelloid rotifers (Chapter 29). The whole structure of the radially symmetrical introvert, of the pharynx with teeth or scalids, and of the collar-shaped brain surrounding the pharynx excludes any possibility of closer relationships between the kinorhynchs and arthropods. The large introvert with different types of scalids, the chitinous cuticle, the two rings of introvert retractor muscles, and the presence of tanycytes are characters shared with priapulans and loriciferans, demonstrating the monophyly of the Cephalorhyncha. The presence of a non-inversible mouth cone with cuticular ridges and spines indicates the sister-group relationship with the loriciferans (Chapter 40).

Interesting subjects for future research

1. Embryology.
2. Secretion of the cuticle at moulting.

References

Brown, R. 1989. Morphology and ultrastructure of the sensory appendages of a kinorhynch introvert. – Zool. Scr. **18**: 471-482.

Higgins, R.P. 1974. Kinorhyncha. – *In* A.C. Giese & J.S. Pearse (eds): Reproduction of Marine Invertebrates, vol. 1, pp 507-518. Academic Press, New York.

Jeuniaux, C. 1975. Principes de systématique biochimique et application à quelques problèmes particuliers concernant les Aschelminthes, les Polychètes et les Tardigrades. – Cah. Biol. mar. **16**: 597-612.

Kozloff, E.N. 1972. Some aspects of development in *Echinoderes* (Kinorhyncha). – Trans. Am. microsc. Soc. **91**: 119-130.

Kristensen, R.M. & A. Hay-Schmidt 1989. The protonephridia of the arctic kinorhynch *Echinoderes aquilonius* (Cyclorhagida, Echinoderidae). – Acta zool. (Stockh.) **70**: 13-27.

Kristensen, R.M. & R.P. Higgins 1991. Kinorhyncha. – *In* F.W. Harrison (ed.): Microscopic Anatomy of Invertebrates, vol. 4, pp 377-404. Wiley-Liss, New York.

Moritz, K. & V. Storch 1972a. Zur Feinstruktur des Integumentes von *Trachydemus giganteus* Zelinka (Kinorhyncha). – Z. Morph. Tiere **71**: 189-202.

Moritz, K. & V. Storch 1972b. Über den ultrastrukturellen Bau der Skaliden von *Trachydemus giganteus* (Kinorhyncha). – Mar. Biol. (Berl.) **16**: 81-89.

Nebelsick, M. 1993. Introvert, mouth cone, and nervous system of *Echinoderes capitatus* (Kinorhyncha, Cyclorhagida) and implications for the phylogenetic position of the Kinorhyncha. – Zoomorphology **113**: 211-232.

Neuhaus, B. 1988. Ultrastructure of the protonephridia in *Pycnophyes kielensis* (Kinorhyncha, Homalorhagida). – Zoomorphology 108: 245-253.

Nyholm, K.-G. 1947. Contributions to the knowledge of the postembryonic development in Echinoderida Cyclorhagae. – Zool. Bidr. Upps. 25: 423-428.

Nyholm, K.-G. & P.-O. Nyholm 1982. Spermatozoa and spermatogenesis in Homalorhaga Kinorhyncha. – J. Ultrastruct. Res. 78: 1-12.

Zelinka, K. 1928. Monographie der Echinodera. – Wilh. Engelmann, Leipzig.

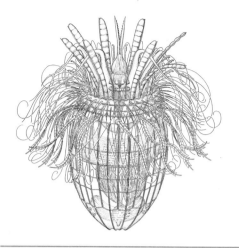

Phylum LORICIFERA

The newly discovered phylum Loricifera is now known to comprise more than 100 species, but only few have been formally described. The microscopic animals occur in sediments of all types, from coarse, shallow-water sediments to fine, deep-sea mud. All the primary information is found in the four papers listed at the end of this chapter, and I have therefore omitted specific references.

The body consists of a trunk with an exoskeleton called the lorica, a neck region, and an introvert which can be invaginated into the anterior part of the trunk (Fig. 40.1).

The introvert has a complicated mouth cone with stiffening stylets or ridges; it is retracted but not inverted when the introvert invaginates. The introvert proper bears nine circles of scalids of different types; these scalids contain several monociliate sensory cells; the spinoscalids have intrinsic muscles. The trichoscalids on the neck contain one cilium, arise from a socket in one or two small cuticular plates, and can be moved by a pair of small muscles at the socket. The trichoscalids of *Pliciloricus* are so thin that they may act as locomotory cilia (see the chapter vignette). Other types are flattened, clavate, shaped like a pea-pod, spiny or hairy, branched, or articulated. The neck may have rows of cuticular plates with additional scalids.

The trunk is covered by the lorica which consists of 6-30 longitudinal cuticular plates; the plates are rather stiff in *Nanaloricus*, but those of *Pliciloricus* and *Rugiloricus* are very thin and may form longitudinal folds. The posterior end has a number of smaller plates surrounding the anus.

In *Nanaloricus*, the tip of the mouth cone continues in a hexagonal, telescoping mouth tube with six rows of small cuticular teeth, which in turn continues in a long, flexible, heavily cuticularized and therefore non eversible buccal canal. The buccal canal passes through the collar-shaped brain and ends in a triradiate, myoepithelial pharynx bulb, followed by a short oesophagus. The midgut consists of large cells with microvilli, and the rectum has a thin cuticle. *Rugiloricus* has a less complicated mouth cone without an eversible mouth tube; the mid-ventral pair of anterior scalids

Chapter vignette: *Pliciloricus enigmaticus*. (Modified from Higgins & Kristensen 1986; courtesy of Dr R.M. Kristensen.)

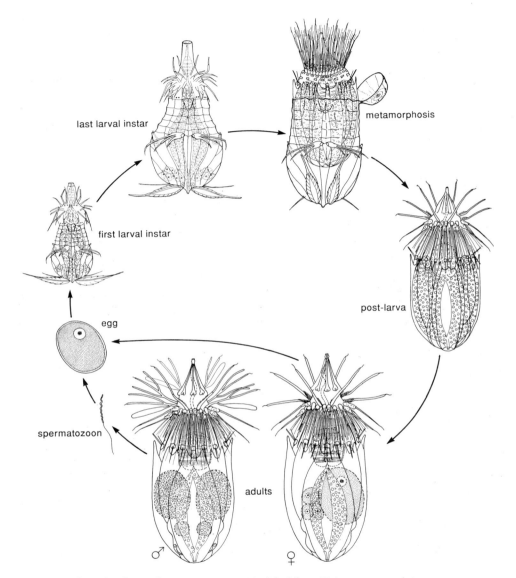

Fig. 40.1. Life cycle of *Nanaloricus mysticus*. (Modified from Kristensen 1991b.)

are more or less fused and have the shape of serrated stylets. The pharynx bulb is situated in the mouth cone; it is hexaradiate with weak radiating muscles and large gland cells.

The ectoderm is monolayered and covered by a compact, supposedly chitinous cuticle.

The central nervous system is intraepithelial and comprises a large brain in the anterior part of the introvert surrounding the gut, and a number of longitudinal nerves. The brain is collar-shaped with anterior and posterior concentrations of ganglionic cell bodies and a median zone of neuropil; it is insunk from the epithelium but surrounded by an extension of the basement membrane, as shown by the ring-shaped connection between the two structures. The cells in the forebrain are concentrated in a ring of eight ganglia which innervate the scalids and the mouth cone; the cells in the hindbrain are arranged in a ring of ten ganglia connected with ten longitudinal nerve cords. The ventral pair of nerve cords is larger than the others and passes through a subpharyngeal ganglion and a a number of double ganglia to a paired caudal ganglion. The brain is surrounded by circular muscles, and special epidermal cells with long tonofilament bundles (tanycytes) attach to the circular muscles and to the inner introvert retractors.

The muscular system consists of a high number of small muscles, many comprising only one or two myomeres, but not all muscles have been described. A varying number of outer introvert retractors extend from the base of the mouth cone to the body wall at the transition between introvert and neck. Five inner introvert retractors extend from the base of the mouth cone to the anterior part of the neck, penetrating the anterior part of the collar–shaped brain. The adult *Nanaloricus* has a system of five inner and six outer muscles situated in and behind the brain with long tendons to the mouth region; these muscles can retract the mouth tube (Fig. 37.1).

Small ring muscles are situated along some of the rings of scalids and between the plates of the lorica. *Nanaloricus*, with only six plates in the lorica, has more individualized muscles between the plates.

The sexes are separate, and there are obvious differences between the sexes, for example in the scalids (Fig. 40.1). There is a pair of combined gonads and excretory organs. The terminal cells of the protonephridia are monociliate. The ducts and their openings have not been described in detail.

The embryology has not been described, but the first postembryonic stage is the so-called Higgins larva (Fig. 40.1), which has the general anatomy of the adult, but with fewer scalids. The most characteristic structure of the Higgins larva is a pair of large, pointed moveable spines, situated in the posterior end; these appendages are wide and flipper-like and can be used in swimming in the larvae of *Nanaloricus*, whereas they are more narrow and have adhesive glands in other genera. There is a varying number of moults, and a post-larva without the posterior appendages is found in *Nanaloricus*. The sexual dimorphism is seen only in the adult stage. The exuvia comprise all the cuticularized regions and show the armature of the pharynx/oesophagus especially clearly.

The larval stages of all the described genera lack the eversible buccal tube found in *Nanaloricus*, and have various rings of teeth around the mouth opening and in the buccal tube. Their pharynx bulb is small and situated in the mouth cone as in the adults of *Pliciloricus* and *Rugiloricus*. This may indicate that the long, eversible buccal tube and the large pharynx bulb are apomorphies of *Nanaloricus*. The larvae have shorter plates in the lorica, and additional plates around the posterior and anterior ends. The larvae have groups of locomotory spines on the ventral side along the anterior border of the lorica.

The feeding biology is unknown, but the very narrow mouth, which is either surrounded by spines or situated at the tip of an eversible mouth tube with small teeth, and the pharyngeal bulb, which seems ideal for sucking, suggest that at least the adults pierce other animals and suck their body fluids. The larvae have larger mouth zones, but probably also feed by sucking.

The Loricifera are definitely a monophyletic group. The life cycle with the Higgins larva, the mouth cone with the very narrow mouth, the myoepithelial pharynx, and the scalids with intrinsic muscles separate the group from priapulans and kinorhynchs, which must be regarded as the sister groups. The interrelationships of the three phyla are discussed in Chapter 37.

Interesting subjects for future research

1. Embryology.
2. Feeding biology of larvae and adults.
3. Chemical composition of the cuticle.

References

Higgins, R.P. & R.M. Kristensen 1986. New Loricifera from southeastern United States coastal waters. – Smithson. Contr. Zool. **438**: 1-70.

Kristensen, R.M. 1983. Loricifera, a new phylum with Aschelminthes characters from the meiobenthos. – Z. zool. Syst. Evolutionsforsch. **21**: 163-180.

Kristensen, R.M. 1991a. Loricifera. – *In* F.W. Harrison (ed.): Microscopic Anatomy of Invertebrates, vol. 4, pp 351-375. Wiley-Liss, New York.

Kristensen, R.M. 1991b. Loricifera – A general biological and phylogenetic overview. – Verh. dt. zool. Ges. **84**: 231-246.

41

PROTORNAEOZOA

Ctenophores and deuterostomes are here treated as sister groups because of their similar modes of mesoderm formation (see also Chapter 9 and Fig. 9.1). The ctenophores (Chapter 42) have important plesiomorphic characters shared with the cnidarians, notably the gastrula-type organization (although with mesoderm), but they also show a series of apomorphies, such as the comb plates, the complicated apical organ and possibly the colloblasts, which characterize them as a monophyletic group. Cnidarian and ctenophore eggs share a characteristic first cleavage which begins at the blastoporal pole; the significance of this is unknown. The deuterostomes (Chapter 43) share a whole suite of apomorphies, including the bilateral symmetry with separate mouth and anus, the mesoderm with three pairs of coelomic sacs, and the dorsal central nervous system.

The ctenophores have a mesoderm consisting of a mesogloea with various cell types including muscle cells. The mesoderm develops from micromeres budded off from the oral side of the macromeres (Fig. 41.1 and Table 42.1).

The deuterostomes show mesoderm formation of two types. Primary mesenchyme is formed from the blastoporal micromeres in echinoderms (Fig. 41.1), while coelomic sacs are formed through various types of enterocoely in most phyla (Fig. 43.2). The primary mesenchyme, which has not been reported from the other deuterostome phyla, resembles the mesoderm of the ctenophores and the two tissues are probably homologous. The coelomic sacs formed through enterocoely in the deuterostomes may represent a new type of mesoderm.

The mesoderm of the protostomes develops from the blastopore edge (Chapter 10) and is probably not homologous with any of the mesoderm types of the protornaeozoans.

The similarity in the development of the mesoderm of the ctenophores and the primary mesenchyme of the echinoderms indicates that the Ctenophora and Deuterostomia are sister groups, and they have been united in the group Protornaeozoa, a name proposed together with the trochaea theory (Nielsen 1985). Some of the parts of this theory around the origin of the protornaeozoans and deuterostomes may not be valid (see Chapter 43) and the name therefore not descriptive, but this does not invalidate the phylogeny, which is based on a cladistic argumentation.

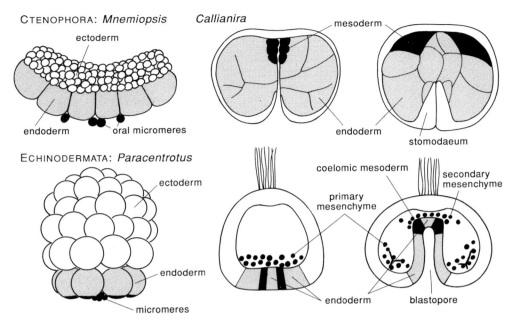

Fig. 41.1. Mesoderm formation in ctenophores and echinoderms. – Ctenophores: 128-cell stage (the 4E-cells have not divided yet) of *Mnemiopsis leidyi* (based on Freemann & Reynolds 1973), early and late gastrula of *Callianira bialata* (redrawn from Metschnikoff 1885). – Echinoderms: 64-cell stage, blastula with micromere ingression and gastrula of *Paracentrotus lividus* (based on Hörstadius 1939).

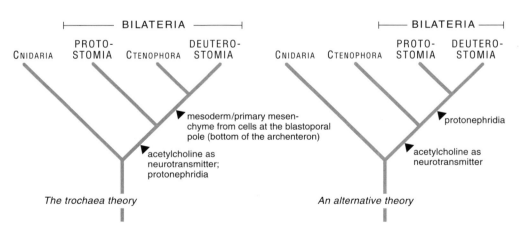

Fig. 41.2. Alternative phylogenies of the basal eumetazoan groups. The trochaea theory interprets the mesoderm of ctenophores and deuterostomes as homologous, whereas the alternative theory regards the 'mesoderm' of the ctenophores as non-homologous with the mesoderm of the deuterostomes.

Other apomorphies of the Protornaeozoa are difficult to point out. Multiciliarity has probably developed independently in ctenophores and cyrtotretes (Chapter 43, see also Fig. 9.2).

It should be stressed that the phylogeny outlined here is based on the assumption that the mesoderm of the ctenophores is homologous with the primary mesoderm of the echinoderms. If new investigations on ctenophore embryology make this unlikely, it will be necessary to reevaluate the available characters, and a different phylogenetic tree is likely to emerge. It could perhaps place the ctenophores as the sister group of the bilaterians, and the distribution of the apomorphies would then be as seen in Fig. 41.2.

References

Freemann, G. & G.T. Reynolds 1973. The development of bioluminescence in the ctenophore *Mnemiopsis leidyi.* – Dev. Biol. 31: 61-100.

Hörstadius, S. 1939. The mechanisms of sea urchin development studied by operative methods. – Biol. Rev. 14: 132-179.

Metschnikoff, E. 1885. Vergleichend-embryologische studien. – Z. wiss. Zool. 42: 648-673, pls 24-26.

Nielsen, C. 1985. Animal phylogeny in the light of the trochaea theory. – Biol. J. Linn. Soc. 25: 243-299.

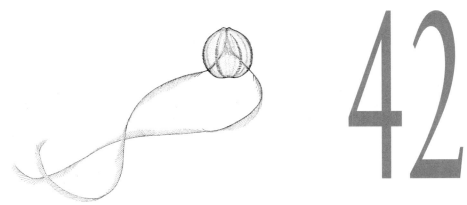

Phylum CTENOPHORA

Ctenophores or comb jellies are a small phylum of about 80 living species of marine organisms. Most species are holopelagic, transparent and sometimes very fragile so that they can hardly be collected with any of the more conventional methods. The few creeping, benthic species are more compact, and the sessile *Tjalfiella* is almost leathery. Surprisingly, two fossil species have been described from the Lower Devonian Hunsrück Slate (Stanley & Stürmer 1987). A recent review of the phylum recognizes 7 orders (Harbison & Madin 1982), but the many new types which are currently being discovered, for example during blue-water diving, indicate that our knowledge of the phylum is still quite incomplete. A considerable morphological variation is found within the phylum, but it is nevertheless very well delimited and its phylogenetic position in the animal kingdom is rather isolated.

The ctenopores are in principle built as a gastrula, with the blastopore remaining as the mouth-anus and the archenteron becoming the sac-shaped adult gut; only the presence of a mesodermal (mesogloeal) layer between ectoderm and endoderm indicates a higher level of organization. The apical-blastoporal axis is retained throughout life as the main axis. A pair of tentacles and the main axis define the tentacular plane, and the perpendicular plane is called oral (or sagittal) because the mouth and stomodaeum are flattened in this plane; a few types lack tentacles in all stages, but the two planes can be identified on the basis of the shape of the gut and the shape of the apical pole. The two planes both divide the ctenophores into symmetrical halves, and this type of symmetry is called biradial.

The body is spherical in the more 'primitive' forms, such as many members of the order Cydippida, but various parts of the body may be expanded into folds or lappets or the whole body may be band-shaped with an extreme flattening in the tentacular plane; the benthic forms are creeping or attached with the oral side (Harbison & Madin 1982). The tentacles are cylindrical, can usually be retracted into tentacle sheaths and have specialized side branches, tentillae, in many species. The adhesive colloblasts are specialized ectodermal cells of the tentacles (see below). Eight

Chapter vignette: *Pleurobrachia pileus*. (Based on Brusca & Brusca 1990.)

meridional rows of comb plates, which are very large compound cilia, are the main locomotory organs in most pelagic forms and can be recognized in developmental stages also of the benthic species, which lack the comb plates in the adult phase. Some of the pelagic forms have short, apically located comb rows, and the expanded oral lobes or the tentacles appear to be more important in locomotion. The benthic forms creep by means of cilia on the expanded pharyngeal or oral epithelium.

The ectoderm (Hernandez-Nicaise 1991) is monolayered in the early developmental stages, but the adults have both an external epithelium, with ciliated cells, glandular cells and 'supporting' cells, and nerve and muscle cells (parietal muscles) which are not in connection with the surface. The ectoderm is underlain by a conspicuous basement membrane.

The ciliated cells are multiciliate (except in some sense organs) and several specialized types can be recognized. The comb plates consist of high numbers of aligned cilia from several cells and show an orthoplectic beat pattern; they show a unique structure with compartmenting lamellae between the lateral doublets (nos 3 and 8) of the axoneme and the cell membrane. The comb plates are used in swimming, which is normally with the apical pole in front, i.e. the effective stroke is towards the oral pole, but their beat can be reversed locally or on the whole animal so that oriented swimming, for example associated with feeding, is possible (Tamm & Moss 1985). The cells at the base of the apical organ and the polar fields and ciliated furrows, which extend from the apical organ, have separate cilia.

The apical organ (Tamm 1982) is a statocyst with four compound cilia, balancers, carrying a compound statolith. The balancer cells are monociliate and the individual otoliths are formed as specialized cells from a region adjacent to the balancer cells. The whole structure is enclosed in a dome-shaped cap consisting of cilia from cells at the periphery of the organ. The organ protrudes from the apical pole in most species, but it is situated in an invagination in *Coeloplana* (Abbott 1907).

The unique macrocilia with several hundred axonemes are only found at the mouth of the beroids (Tamm & Tamm 1988a,b).

The tentacles have the very characteristic colloblasts (Fig. 42.1) which are formed continuously from undifferentiated ectodermal cells of the basement growth zone of the tentacles (Hernandez-Nicaise 1991). The full-grown colloblasts have a very characteristic structure with a spirally coiled thread around the stalk and a head with numerous small peripheral granules with a mucous substance which is released by contact with a prey (Franc 1978). They bear no resemblance to the nematocysts of the cnidarians, which are intracellular organelles (compare with Fig. 8.1).

The tentacles of *Haeckelia* lack colloblasts but contain nematocysts (cleptocnidia), which originate from ingested medusae (Carré & Carré 1980, Mills & Miller 1984). They are enclosed in a vacuole in an innervated cell which lacks the cnidocil and the other structures associated with the nematocyst when it is in the normal position in a cnidarian nematocyte (Carré & Carré 1989). The tentacles also have many cells of a type called pseudocolloblasts (Carré & Carré 1989), which are used in prey capture before the cleptocnidia have become functional, but their homology with the colloblasts seems uncertain.

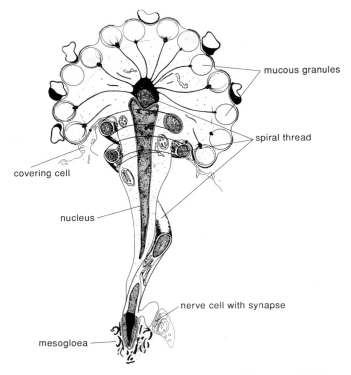

covering cell

mucous granules

spiral thread

nucleus

nerve cell with synapse

mesogloea

Fig. 42.1. Diagrammatic longitudinal section of a resting colloblast. (Modified from Franc 1978.)

The nervous system consists of a nerve net at the base of the ectoderm, with conspicuous concentrations along the comb rows, and at the base of the endoderm; nerve nets occur also in the mesoderm, especially concentrated in the tentacles (Hernandez-Nicaise 1991). The synapses have a unique structure with a 'presynaptic triad' and a thickened postsynaptic membrane. Acetylcholinesterase has been demonstrated in a variety of synapses, and Anctil (1985) observed that luminescence was elicited by acetylcholine in *Mnemiopsis*. There is a large concentration of nerve cells in the epithelium below the apical organ, and this ganglion serves many of the functions of a brain.

The ribbon-shaped, smooth parietal muscles (Hernandez-Nicaise 1991) are found at the base of the ectoderm of the body and the pharynx, but are apparently lacking in the tentacles.

The endodermal gut or gastrovascular system is a complicated system of branched canals with eight major, meridional canals along the comb-plate rows. The pharynx should probably be interpreted as a stomodaeum, as indicated by its origin from the apical micromeres (see below) and by its innervation and parietal musculature (Hernandez-Nicaise 1991). A narrow apical extension of the gut reaches to the underside of the apical organ where it gives off a pair of Y-shaped canals in the oral plane. One branch on each side ends in a small ampulla while the other forms a small pore to the outside; these pores may function as anal openings (Main 1928), but the undigested remains of the prey are usually egested through the mouth.

The walls of peripheral parts of the gastrovascular system show 'ciliated rosettes' (Franc 1972) consisting of a double ring of endodermal cells surrounding a pore, which can be constricted by the ring of cells in plane with the gut wall. This ring of cells has a conical tuft of cilia which beat towards the gut lumen. The other ring of cells protrudes into the mesogloea and has long cilia which beat towards the mesogloea. Experiments indicate that the rosettes can transport water between gut and mesogloea.

The mesoderm or mesogloea (Hernandez-Nicaise 1991) is a hyaline, gelatinous extracellular matrix with muscle cells, nerve cells and mesenchyme cells; epithelial cell layers are not formed. The matrix contains a meshwork of fibrils, which are banded like collagen in certain areas, especially in the tentillae, but it is now assumed that a type of collagen is a major component of the whole meshwork. The matrix appears to be secreted mainly by the ectodermal cells, but also some of the smooth muscle cells of the mesoderm secrete collagen. The muscle cells in the body are very large, branched and smooth. The smooth, longitudinal muscle cells of the tentacles are arranged around a core of matrix with nerves (Fig. 42.2). The tentillae have more complex musculature, and striated muscles have been found in *Euplokamis* (Mackie, Mills & Singla 1988). The mesenchymal cells are of two types, but their functions are unknown.

The gonads differentiate from the endoderm of the eight meridional gastrovascular canals. The developing oocyte is connected with three clusters of nurse cells through intercellular bridges, and the polar bodies are given off in a constant relation to these bridges. The gametes are shed through pores in the epidermis above the go-

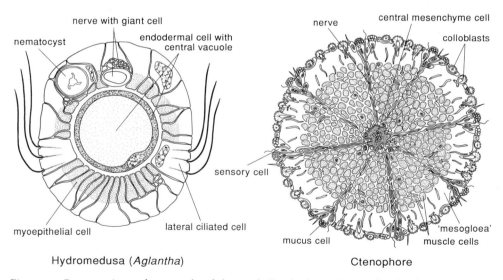

Hydromedusa (*Aglantha*) Ctenophore

Fig. 42.2. Cross-sections of a tentacle of the trachyline hydromedusa *Aglantha digitale* (redrawn from Mackie, Nielsen & Singla 1989) and a cydippid ctenophore (based on Hernandez-Nicaise 1973).

313

nads (Pianka 1974). Fertilization takes place at spawning, and it is believed that the first cleavage furrow begins to form at the point of entry of the sperm, which thus determines the position of the primary axis of the embryo. This axis is not fixed during oogenesis, and the fact that the polar bodies are often given off at the blastoporal pole is probably because the area with the polar bodies is the first part of the egg which comes in contact with the exterior at spawning and where fertilization is therefore most likely to take place (Freemann 1977).

The development is highly determined (Freemann & Reynolds 1973, Pianka 1974). The first cleavage is in the oral plane and the second is in the tentacular plane. The first cleavage begins at the blastoporal pole, and the cleavage furrow cuts up towards the apical pole so that a peculiar stage with the two blastomeres connected by an apical 'handle' is formed just before the blastomeres finally separate; similar shapes are seen in the following two cleavages (Freemann 1977). The third cleavage is in the oral plane and the 8-cell stage accordingly consists of two rows of four cells; the four median cells are called M-cells and the four external E-cells; their fates are illustrated in Table 42.1. Ectoderm develops from micromeres given off at the apical

Table 42.1. The cell-lineage of ctenophores based on Reverberi & Ortolani (1963), Farfaglio (1963), Ortolani (1964, 1989), and Freemann & Reynolds (1973) (modified from Nielsen 1985). Capital letters: macromeres; lower case letters: apical micromeres; boldface italic letters: oral micromeres; the superscripts p and d denote proximal (median) and distal (external) blastomeres, respectively.

EM
├─ E
│ ├─ 1e – ectoderm: cells of the comb rows, cells of the ciliary dome of the apical organ
│ └─ 1E
│   ├─ 2e – ectoderm: cells between the comb rows, tentacle bulbs, tentacles
│   └─ 2E
│   ├─ 3e – ectoderm: mouth and pharynx
│   └─ 3E
│   ├─ 3Ep { $3E^{p1}$; $3e^{p2}$ }
│   └─ 3Ed { $3E^{d1}$; $3e^{d2}$ }   } mesoderm and endoderm
│
└─ M
   ├─ 1m – ectoderm: cells of the balancer cilia of the apical organ
   └─ 1M
      ├─ 2m – ectoderm: mouth and pharynx
      └─ 2M
         ├─ 2Mp { $2M^{p1}$; $2m^{p2}$ }
         └─ 2Md { $2M^{d1}$; $2m^{d2}$ }   } mesoderm and endoderm (photocytes from $2M^d$)

side of the macromeres and mesoderm from micromeres given off from the aboral side; the macromeres themselves appear to become endoderm. Some of the earlier authors (for example Siewing 1977) interpreted the oral micromeres as a special type of ectoderm, but this can only be regarded as a rather far-fetched attempt to make the ctenophores diploblastic. The young embryo is compact with a shallow apical concavity and a rather thick ring of apical micromeres (Fig. 41.1). An epibolic gastrulation begins when the apical micromeres spread orally over the macromeres and ends with an invagination of the blastoporal (oral) side of the macromeres carrying the oral micromeres to the bottom of the archenteron (Fig. 41.1). The first stages of the tentacles are formed just after gastrulation as ectodermal thickenings which invaginate to form the tentacle sheaths, and the tentacles sprout from the bottom of these invaginations. Metschnikoff (1885) reported similar epithelial thickenings in *Beroe*, which lacks tentacles, but this should be reinvestigated. The studies summarized in Table 42.1 indicate the cell lineage of most organs; only the origin of the ectodermal (parietal) muscles and of the muscles of the tentacles remains unknown. However, there is not complete agreement about the origin and differentiation of the endodermal and mesodermal elements, and cell-lineage studies using modern staining techniques are needed to clarify a number of important details.

The monophyly of the Ctenophora can hardly be questioned. The comb plates and the apical organ are unique, and further apomorphies can be pointed out, for example the very unusual oogenesis (see above) and the naked extracellular bundles of tubulin structures resembling ciliary axonemes found in grooves along the smooth muscle cells (Tamm & Tamm 1991).

The ancestral ctenophore probably resembled a cydippid, and it may have had tentacles with colloblasts. The lack of tentacles in the beroids has been regarded as a plesiomorphy, but since tentacles are lacking in representatives of the order Lobata (Harbison & Madin 1982) it may just as well be an apomorphy of the beroids. The opposite view (Harbison 1985) implies that only the comb plates and the apical organ are synapomorphies of the ctenophores. Both theories are compatible with the phylogenetic position of the ctenophores advocated here.

The ctenophores have traditionally been treated together with the cnidarians, because of the superficial similarity between the gelatinous, pelagic forms. Other authors have viewed the ctenophores as related to turbellarians and derived the ctenophores from the flatworms (Hadzi 1953). The latter view is based on overall similarities in body-plans but not supported by synapomorphies and is only of historical interest.

Most textbooks unite ctenophores and cnidarians in the supraphyletic group Coelenterata or Radiata, often with the ctenophores interpreted as derived from holopelagic trachyline medusae, and sometimes with *Haeckelia* and/or the enigmatic *Hydroctena* as 'missing links' (e.g. Möhn 1984, Brusca & Brusca 1990). However, it is well documented that *Haeckelia* is a cydippid ctenophore which lacks colloblasts and which acquires nematocysts (cleptocnidia) from ingested trachyline medusae (see above). *Hydroctena* (Dawydoff 1903) is more problematic; it has only been found once, and the material is apparently lost. Dawydoff (1953) subsequently reported drawings of some sections made from another of his three specimens, but his report merely throws some doubt about his first observations on the aboral sense organ. The absence of colloblasts and the presence of nematocysts may be interpreted as a specialization like that found in *Haeckelia*, the invaginated apical organ resembles that of the ctenophore *Coeloplana*, and the tentacles are of a structure completely different from that of the trachylines: *Hydroctena* has tentacles with a core of longitudinal mesodermal muscles, whereas the trachylines have non-retractile tentacles with epitheliomuscular ectodermal cells surrounding a row of endodermal cells with a large vacuole (Fig. 42.2). I believe that *Hydroctena* could best be interpreted as an aberrant ctenophore and it should definitely not be used in phylogenetic discussions (Nielsen 1987).

A sister-group relationship between cnidarians and ctenophores is only indicated by the unusual type of the first embryonic cleavages. The nematocysts of the cnidarians are completely different from the colloblasts of the ctenophores (compare Figs 8.1 and 42.1). The cnidarians have remained at the gastrula stage with an acellular mesogloea, while the ctenophores have a 'mesogloea' with mesodermal cells derived from the blastoporal pole of the blastula like the primary mesenchyme cells of the echinoderm embryos (Chapter 48; see Fig. 41.1). The last-mentioned character links the ctenophores to the deuterostomes, and these two groups are regarded as sister groups. It is evident that this conclusion will have to be reconsidered if new studies on cell-lineage lead to a changed interpretation of the origin of the endo-mesoderm (Chapter 41).

Interesting subjects for future research

1. Origin of the parietal and tentacular muscles.
2. Cell-lineage studies with modern staining methods.

References

Abbott, J.F. 1907. The morphology of *Coeloplana*. – Zool. Jb., Anat. **24**: 41-70, pls 8-10.
Anctil, M. 1985. Cholinergic and monoaminergic mechanisms associated with control of bioluminescence in the ctenophore *Mnemiopsis leidyi*. – J. exp. Biol. **119**: 225-238.
Brusca, R.C. & G.J. Brusca 1990. Invertebrates. – Sinauer Associates, Sunderland.
Carré, C. & D. Carré 1980. Les cnidocystes du cténophore *Euchlora rubra* (Kölliker 1853). – Cah. Biol. mar. **21**: 221-226, 4 pls.
Carré, D. & C. Carré 1989. Acquisition de cnidocystes et différenciation de pseudocolloblastes chez les larves et les adultes de deux cténophores du genre *Haeckelia* Carus, 1863. – Can. J. Zool. **67**: 2169-2179.
Dawydoff, C. 1903. *Hydroctena salenskii* (Étude morphologique sur un nouveau coelentére pélagique). – Mém. Acad. imp. Nat. St. Pétersb., 8. Sér., **14**(9): 1-17, 1 pl.
Dawydoff, C. 1953. Contribution à nos connaissances de l'*Hydroctena*. – C. r. hebd. Séanc. Acad. Sci., Paris **237**: 1301-1302.
Farfaglio, G. 1963. Experiments on the formation of the ciliated plates in ctenophores. – Acta Embryol. Morph. exp. **6**: 191-203.
Franc, J.-M. 1972. Activités des rosettes ciliés et leurs supports ultrastructuraux chez les Cténaires. – Z. Zellforsch. **130**: 527-544.
Franc, J.-M. 1978. Organization and function of ctenophore colloblasts: an ultrastructural study. – Biol. Bull. Woods Hole **155**: 527-541.

Freemann, G. 1977. The establishment of the oral-blastoporal axis in the ctenophore embryo. – J. Embryol. exp. Morph. **42**: 237-260.

Freemann, G. & G.T. Reynolds 1973. The development of bioluminescence in the ctenophore *Mnemiopsis leidyi.* – Dev. Biol. **31**: 61-100.

Hadzi, J. 1953. An attempt to reconstruct the system of animal classification. – Syst. Zool. **2**: 145-154.

Harbison, G.R. 1985. On the classification and evolution of the Ctenophora. – *In* S. Conway Morris, J.D. George, R. Gibson & H.M. Platt (eds): The Origins and Relationships of Lower Invertebrates, pp 78-100. Oxford Univ. Press, Oxford.

Harbison, G.R. & L.P. Madin 1982. Ctenophora. – *In* S.P. Parker (ed.): Synopsis and Classification of Living Organisms, vol. 1, pp 707-715, pls 68-69. McGraw-Hill, New York.

Hernandez-Nicaise, M.-L. 1973. Le système nerveux des Cténaires. I. Structure et ultrastructure des réseaux épithéliaux. – Z. Zellforsch. **143**: 117-133.

Hernandez-Nicaise, M.-L. 1991. Ctenophora. – *In* F.W. Harrison (ed.): Microscopic Anatomy of Invertebrates, vol. 2, pp 359-418. Wiley-Liss, New York.

Mackie, G.O., C.E. Mills & C.L. Singla 1988. Structure and function of the prehensile tentilla of *Euplokamis* (Ctenophora, Cydippida). – Zoomorphology **107**: 319-337.

Mackie, G.O., C. Nielsen & C.L. Singla 1989. The tentacle cilia of *Aglantha digitale* (Hydrozoa: Trachylina) and their control. – Acta zool. (Stockh.) **70**: 133-141.

Main, R.J. 1928. Observations on the feeding mechanism of a ctenophore, *Mnemiopsis leidyi.* – Biol. Bull. Woods Hole **55**: 69-78.

Metschnikoff, E. 1885. Vergleichend-embryologische Studien. – Z. wiss. Zool. **42**: 648-673, pls 24-26.

Mills, C.E. & R.L. Miller 1984. Ingestion of a medusa (*Aegina citrea*) by the nematocyst-containing ctenophore *Haeckelia rubra* (formerly *Euchlora rubra*): phylogenetic implications. – Mar. Biol. (Berl.) **78**: 215-221.

Möhn, E. 1984. System und Phylogenie der Lebewesen, Band 1: Physikalische, chemische und biologische Evolution. Prokaryonta, Eukaryonta (bis Ctenophora). – E. Schweizerbart'sche Verlagsbuchhandlung, Stuttgart.

Nielsen, C. 1985. Animal phylogeny in the light of the trochaea theory. – Biol. J. Linn. Soc. **25**: 243-299.

Nielsen, C. 1987. *Haeckelia* (= *Euchlora*) and *Hydroctena* and the phylogenetic interrelationships of the Cnidaria and Ctenophora. – Z. syst. Zool. Evolutionsforsch. **25**: 9-12.

Ortolani, G. 1964. Origine dell'organo apicale e di derivati mesodermici nello sviluppo embrionale de ctenophori. – Acta Embryol. Morph. exp. **7**: 191-200.

Ortolani, G. 1989. The ctenophores: a review. – Acta Embryol. Morph. exp., N.S. **10**: 13-31.

Pianka, H.D. 1974. Ctenophora. – *In* A.C. Giese & J.S. Pearse (eds): Reproduction of Marine Invertebrates, vol. 1, pp 201-265. Academic Press, New York.

Reverberi, G. & G. Ortolani 1963. On the origin of the ciliated plates and of the mesoderm in the ctenophores. – Acta Embryol. Morph. exp. **6**: 175-190.

Siewing, R. 1977. Mesoderm bei Ctenophoren. – Z. zool. Syst. Evolutionsforsch. **15**: 1-8.

Stanley, G.D. & W. Stürmer 1987. A new fossil ctenophore discovered by X-rays. – Nature **328**: 61-63.

Tamm, S.L. 1982. Ctenophora. – *In* G.A.B. Shelton (ed.): Electrical Conduction and Behaviour in 'Simple' Invertebrates, pp 266-358. Oxford Univ. Press, Oxford.

Tamm, S.L. & A.G. Moss 1985. Unilateral ciliary reversal and motor responses during prey capture by the ctenophore *Pleurobrachia.* – J. exp. Biol. **114**: 443-461.

Tamm, S. & S.L. Tamm 1988a. Development of macrociliary cells in *Beroë.* I. Actin bundles and centriole migration. – J. Cell Sci. **89**: 67-80.

Tamm, S. & S.L. Tamm 1988b. Development of macrociliary cells in *Beroë.* II. Formation of macrocilia. – J. Cell Sci. **89**: 81-95.

Tamm, S. & S.L. Tamm 1991. Extracellular ciliary axonemes associated with the surface of smooth muscle cells of ctenophores. – J. Cell Sci. **94**: 713-724.

43

DEUTEROSTOMIA (= NOTONEURALIA)

Deuterostomia is a large, very well-defined group of bilateral animals; many of their apomorphic characters have been mentioned in Chapter 9, but they will be discussed in more detail here.

As indicated by the name, the deuterostomes should have a blastopore which becomes the anus in the adult, while the adult mouth should be a new opening from the bottom of the archenteron. This type of development can actually be followed in a number of species in most of the phyla, but there is a good deal of variation and, as already stressed in the discussion of the protostomes (Chapter 10), I regard the blasto- ˙ poral fate as an unreliable character because it shows considerable intraphyletic variation, both in protostomes and deuterostomes.

Phoronids and brachiopods show the most deviating developmental types. The brachiopods show considerable variation (Chapter 45). *Crania* has a normal embolic gastrula where the blastopore is situated at the posterior end of the larva until it closes; the adult mouth breaks through at the anterior end. This corresponds well with the deuterostome type. The articulate brachiopods have a different pattern where the blastopore constricts from behind and the adult mouth is said to break through near the anterior end of the constriction, but a comparison with settling stages in *Crania* indicates that the body of the articulate larva is strongly curved so that the ventral side must be very short, which makes it difficult to identify the relative positions of blastopore and mouth. Phoronid gastrulation is usually a quite schematic invagination, and the blastopore is said to become elongate and to close from behind in some species, whereas it remains open and becomes the adult mouth in others (Chapter 44). This resembles the protostome pattern, but so many other characters indicate that the phoronids are deuterostomes, that I regard their blastopore fate as a deviation from the normal deuterostome pattern.

Echinoderms exhibit the least variation: the blastopore becomes the larval anus in almost all species where the archenteron is formed through invagination, and the larval mouth is formed as a new opening from the archenteron. Species with large yolk-rich eggs have modified development, but the two types of development can be found even within the same genus (Chapter 48) and corresponding areas of the eggs give rise to identical adult structures in the two types. Later stages are complicated in several of the major groups; a new adult mouth is formed at metamorphosis in asteroids (Fig. 48.3) and echinoids, and the adult ophiuroids have no anus. There are no reports of a development where the blastopore becomes the mouth.

The enteropneusts show limited variation too, and there are no reports of blastopore fates which deviate from the general deuterostome pattern (Chapter 50).

The chordates have no primary larvae, and their development is complicated through early development of the dorsal nerve cord, which encloses the blastopore at the posterior end, forming the neurenteric canal. The anus appears to be a new opening, and this appears to be the case for the mouth too (Chapter 51). These modifications of the developmental stages must be related to the lecithotrophic development or viviparity in urochordates and vertebrates, and to a very early differentiation of the adult feeding structures in amphioxus (Chapter 53).

It must be concluded that most of the phyla which are here regarded as the 'more primitive' deuterostomes have a blastopore fate which corresponds well with the predictions of the trochaea theory, and that the few deviations from this pattern can be seen as variations within a major group which for other reasons must be considered as monophyletic.

The cleavage of the deuterostomes is radial in all the cases where a total cleavage is found. There are of course many species with aberrant cleavage patterns related to large amount of yolk or placental nourishment, just as in the protostomes, but all reports of spiral-like cleavage patterns in deuterostomes have been shown to be erroneous.

The first cleavage is median in most groups, i.e. the first cleavage separates the right and left halves of the embryo. In phoronids, markings of the first two blastomeres have shown that there is a considerable variation in the orientation of the first cleavage, but that the first cleavage is transverse in the majority of the cases (Chapter 44). There are no published observations on the brachiopods, but Dr Gary Freeman (Univ. of Texas at Austin) has marked 2-cell stages of *Terebratalia* and found that the first cleavage is median. The pterobranchs have not been studied in this respect. The echinoderms show some variation, with oblique cleavage planes dividing the zygote into oral, right, aboral, and left parts in some species, while a more 'normal' right-left first division is found in other species (Chapter 48). Studies of early development of both enteropneusts, urochordates and cephalochordates (Chapters 50, 52, 53) have demonstrated that the first cleavage is median. The vertebrates show much variation in connection with yolk-rich eggs or placentally nourished embryos, but a first median cleavage is well-documented, for example in anurans (Chapter 54).

The first two cleavages divide the ctenophore embryo into four symmetrical segments (Chapter 42) in accordance with the biradial symmetry of the adults, and it is possible that the first cleavage corresponds to the median cleavage of the deuterostomes.

The general rule among deuterostomes (and ctenophores), is thus that the first cleavage is median, and this is in contrast to the cleavage patterns of both the spiralians, which have the first two cleavages dividing the embryo into four cells, which are roughly anterior, right, posterior, and left (Chapter 11), and the aschelminths, which generally have an asymmetrical cleavage (Chapter 28).

The degree of determination varies somewhat between the deuterostome phyla, but each cell of the 2-cell stage is usually competent of developing into a complete embryo. This is also the case for cells of the 4-cell stage in some species, but it appears that the degree of determination has evolved more or less independently in the various phyla so that this character contains only little phylogenetic information.

The cleavage leads to the formation of blastulae and gastrulae in all the forms with small eggs developing into free larvae, and also in many species with yolk-rich eggs.

An apical organ with sensory cells with long cilia at the position of the polar bodies is found in almost all the deuterostome phyla. Only larvae of urochordates and vertebrates, which lack cilia, have no apical organ. The apical organs are slightly thickened epithelial areas with basiepithelial neuropil and single neuronal connections to other areas. This is in contrast to the spiralian apical organs, which are generally very conspicuous, almost onion-shaped and with nerves, often accompanied by muscle cells, to the prototroch or other larval organs (the aschelminths lack apical organs).

Another important deuterostome character is that the small apical organ in no case becomes incorporated into the adult brain as in the spiralians (Chapter 11). In phoronids, the apical organ is situated at the part of the hood of the actinotroch larva which becomes shed at metamorphosis (Fig. 44.2). The brachiopod larvae either have no apical organ, as for example the larva of *Crania*, or the apical organ appears not to become integrated into the adult nervous system (Chapter 45). The larva of the pterobranch *Rhabdopleura* has a small apical sense organ, which probably disappears at metamorphosis (Chapter 47). Echinoderm larvae have very small apical centres which disappear at an early stage, and the whole apical part of the body is shed at metamorphosis in several species; the adult echinoderms lack a brain in the normal sense of the word (Chapter 48). The tornaria larvae of enteropneusts usually have a conspicuous apical ciliary tuft and some species have a pair of eye spots lateral to the apical organ. The protocoel forms an extension which attaches to the basal side of the apical organ and this connection does in most cases become a compact string with muscle cells, but apparently never with nerve cells. At metamorphosis, the apical organ degenerates (Chapter 50). The small apical tuft of the early amphioxus embryo disappears after a short period, and there is apparently no connection with the adult nervous system (Chapter 53).

The central nervous system (CNS) of the adults is a dorsal area of specialized ectoderm in the region of the mesosome (see below). The brain is obviously not derived from the apical organ, and the longitudinal nerve cord is dorsal and in no way derived from the laterally compressed blastopore lips; this is a very conspicuous difference from the protostomes (Chapter 10, Figs 10.1 and 10.2). The phoronid CNS is an oval area occupying almost the whole dorsal side between mouth and anus (Fig. 44.1). The brachiopods have very inconspicuous nervous centres (as could be expected from their sessile habits), and it is not possible to make definitive statements about the existence of a 'brain' (Chapter 45). The pterobranchs have a CNS which resembles that of the phoronids (Chapter 47). The echinoderms lack a brain (Chapter 48).

The CNS of enteropneusts is not well understood; a short, mesosomal neural tube, which is in open connection with the ectoderm at both ends, at least in some species, is apparently not homologous with the neural tube of the chordates, although it is situated in the same region (Chapter 50). The development of the central nervous system of the chordates through the fusion of the dorsal neural folds and the complete internalization of the neural plate is described in all the text-books (Chapter 51).

The archimeric regionation is perhaps the most obvious characteristic of the deuterostomes. All the non-chordate phyla show a division of the body into three regions, prosome, mesosome and metasome, which cannot always be recognized externally, but which each has well-defined coelomic compartments, protocoel, mesocoel and metacoel, respectively.

The prosome is an externally well-defined region only in pterobranchs and enteropneusts; it is small in phoronids and completely integrated in the body in the remaining phyla.

The protocoel is unpaired in phoronids and brachiopods and it is clearly recognizable only in the larval stages. It is present as a more or less open cavity in the hood of the actinotroch larva of phoronids, and it is retained in the adult in the shape of a more or less well-defined cavity in the mesoderm of the preoral lip (Chapter 44). In the brachiopods, it is well-developed in the larva of *Crania*, where it gives rise to the ventral muscles which contract the body at metamorphosis; it has not been observed in the embryos of the other brachiopods, and it has not with certainty been identified in any of the adults (Chapter 45).

In the non-chordate neorenalians (pterobranchs, echinoderms and enteropneusts), the protocoel becomes integrated into an excretory organ, the axial complex, typically comprising a heart, a glomerulus and a nephridial capsule (Fig. 46.1). The heart is a median blood vessel surrounded by a muscular heart sac which is a special compartment of the protocoel; it pumps blood through a narrower, somewhat convoluted glomerular vessel which has a peritoneal wall consisting of podocytes. This is obviously the place of formation of the primary urine, which fills the nephridial capsule formed by the main compartment of the protocoel. The coelomoduct of the protocoel functions as a nephridial canal.

The chordates lack the axial complex, but the protocoel can be recognized during ontogeny and in strongly modified shapes also in the adults of cephalochordates and vertebrates (Table 51.1). Only the urochordates show no coelomic sacs, and it is not possible to identify the three body regions at any stage (Chapter 52).

The mesosome is a well-defined body region in phoronids, brachiopods and pterobranchs, where it carries tentacles, and in enteropneusts, where it forms the collar. In the remaining phyla it cannot be recognized from the exterior.

The mesocoel (Fig. 43.1) is paired, surrounds the pharynx/oesophagus, and sends canals into the tentacles in phoronids, brachiopods and pterobranchs, and its left pouch is specialized as the hydrocoel with radial canals with extensions into the podia in the echinoderms. It is a pair of simple coelomic sacs in the enteropneusts. It has not been identified with certainty in the chordates (Chapter 51).

The metasome is the main part of the body in all the phyla, but it cannot be distinguished from the mesosome in the chordates.

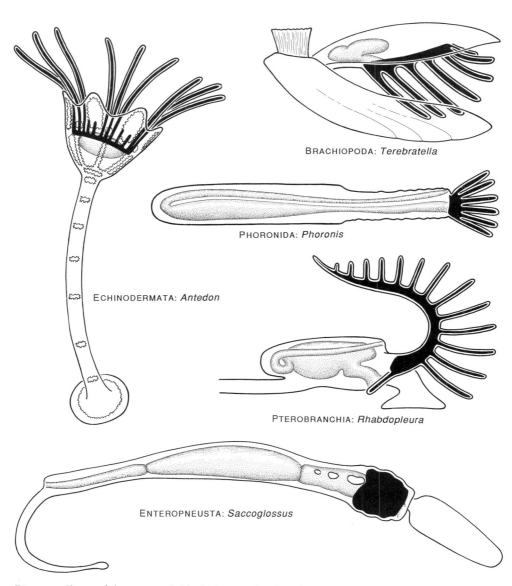

BRACHIOPODA: *Terebratella*

PHORONIDA: *Phoronis*

ECHINODERMATA: *Antedon*

PTEROBRANCHIA: *Rhabdopleura*

ENTEROPNEUSTA: *Saccoglossus*

Fig. 43.1. Shape of the mesocoel (black) in non-chordate deuterostome phyla. – Phoronida: *Phoronis ovalis* (based on Marcus 1949). – Brachiopoda: *Terebratella inconspicua* (based on Atkins 1961). – Echinodermata: pentacrinoid-stage of *Antedon rosaceus* (based on Thomson 1865). – Pterobranchia: *Rhabdopleura* sp. (based on Dawydoff 1948). – Enteropneusta: juvenile *Saccoglossus horsti* (based on Burdon-Jones 1952).

The metacoel is paired in all the phyla, but has different functions in the 'lower' deuterostomes and in the neorenalians. In phoronids and brachiopods, each meso-coel has a gonad and a large metanephridium with a large, ciliated funnel, which also selects ripe gametes and carries them to the nephridiopore/gonopore. In the non-chordate neorenalians, the excretory function has been taken over by the axial com-plex (see above), and the gametes are shed through separate ducts from the gonads; the metacoels have no metanephridia.

The origin of the mesoderm and the type of coelom formation have been given high rank in earlier phylogenies. A hypothetical ancestor in the shape of a creeping bilateral organism (often called dipleurula, but see below) with three pairs of coelomic sacs, of which the two first pairs were connected to each other on each side through a common coelomoduct, has had a central position in many discussions of the phylogeny of the deuterostomes. This organism resembles the benthic stage of notoneuron, except that the coelom was not considered in my earlier definition (Chapter 2).

A view of the variation in the way of forming mesoderm and coelom, for example within the enteropneusts (Fig. 43.2), demonstrates that there is an enormous intra-phyletic variation, indicating that these characters are unreliable in general. The mesoderm is given off from the archenteron as coelomic pouches, enterocoely, in a number of species, but in the shape of solid outgrowths or diffuse delamination or ingression in other species where the cavities accordingly arise through schizocoely. The mesoderm formation observed in many echinoids and asteroids, viz. a pocket pinched off from the apical end of the archenteron which then gives off mesocoel + metacoel on each side, is often regarded as typical of the deuterostomes, but there is much variation both between and within the phyla. There is a series of intermediate forms between the echinoid type and a type where all the mesoderm comes from the area of the archenteron closest to the blastopore, as for example in the brachiopod *Crania*. The protocoel is pinched off from the apical end of the archenteron in most groups, but it is difficult to make any generalizations about the formation of meso- and metacoel. However, it should be noted that the mesoderm is always given off from the archenteron, and not from the rim of the blastopore as in most protostomes (Chapter 9). The trochaea theory regards the formation of the five coelomic com-partments through division of an anterior extension of the archenteron as the primi-tive state (Chapter 2), but a phylogenetic tree based on a modified theory which re-gards the formation of five isolated coelomic pouches as the primitive stage will be identical to those in Figs 9.1 and 43.7.

Coelomoducts have various functions in the phyla. The functional significance of the coelomoducts as metanephridia and gonoducts in phoronids and brachiopods has been mentioned above, as have the coelomoducts from the protocoel which function as the nephridial capsule in the axial complex of the neorenalians (except chordates). The origin of the coelomoducts is not clear in all cases, and their func-tion is sometimes unknown. It appears that the pore is formed when an extension from the coelomic sac contacts the ectoderm, which may sometimes form a small in-vagination, resembling a stomodaeum or proctodaeum (for example in larvae of enteropneusts and echinoderms, see Ruppert & Balser 1986). The sea-cucumbers

Labidoplax and *Stichopus* are exceptional in that the protocoel is formed from the bottom of the archenteron, which curves dorsally, gets contact with the ectoderm and forms a coelomopore before the archenteron makes the usual ventral turn to form the mouth (Chapter 48). This special type has given rise to the part of the trochaea theory which describes the evolution of the mouth, coelomopore and the two primary gill openings as four canals from the bottom of the archenteron to the exterior. It should be stressed that this is a highly speculative attempt at explaining the origins of the new openings from the archenteron, viz. mouth, gill pores and hydropore as well as the obviously closely connected problem of the origin of the coelomic sacs. The trochaea theory is definitely open for alternative explanations on this point, which would not in any way affect its main points or the resulting phylogenetic trees.

The shape, structure and function of the ciliary bands used in feeding and sometimes also in locomotion of deuterostome larvae, and in feeding of some of the adults too, are very uniform among the phyla (Fig. 43.3) and very different from the ciliary bands found in the protostomes (Fig. 10.4). The ancestral deuterostome larva in all probability had a circumoral band of separate cilia on monociliate cells functioning as an upstream-collecting system. Ciliary bands of the above-mentioned structure and function are characteristic of larvae of phoronids, brachiopods, echinoderms, and enteropneusts (Fig. 43.4), i.e. all the deuterostome phyla where planktotrophic larvae are known. Adults of phoronids, brachiopods and pterobranchs have the same type of ciliary band on their tentacles. Ciliary bands of this type have never been encountered among cnidarians or protostomes, and must be considered as one of the principal deuterostome apomorphies.

The structure of the perianal ring of compound cilia of actinotrochs and tornaria larvae differs from that of trochophores, but the two are not of similar structure. The trochophore telotroch is a circle of compound cilia on multiciliate cells (Fig. 43.5). In the actinotrochs it consists of cilia from monociliate cells (Nielsen 1987) which in the young larvae form a ring of almost cylindrical compound cilia (Fig. 43.5) whereas the much wider bands of the adult larvae indicate a structure of a wide belt of cells with compound cilia consisting of a few orthoplectic rows with a high

Fig. 43.2. Development of the coelomic cavities in selected deuterostomes; the diagrams show some features which do not occur simultaneously in the embryos: the protocoel is usually pinched off before the meso- and metacoels develop, and the blastopore is usually closed at an earlier stage; the blastopore is in most cases not in the plane of the diagram. – ectoderm: white; mesoderm: grey; endoderm: black. – coe1, protocoel; coe2, mesocoel; coe3, metacoel; coe4, the small fourth coelomic compartment of Crania; l, left. – Phoronida: Phoronis vancouverensis (based on Zimmer 1964). – Brachiopoda: Crania anomala (based on Nielsen 1991) and Terebratalia transversa (based on Long 1964). – Pterobranchia: Rhabdopleura normani (based on Lester 1988). – Echinodermata: Strongylocentrotus lividus (based on Ubisch 1913), Solaster endeca (based on Gemmill 1912), Asterina gibbosa (based on Ludwig 1882), and Oxychomatus japonicus (based on Holland 1991). – Enteropneusta: Saccoglossus pusillus (based on Davis 1908), tornaria larva 1 (based on Dawydoff 1944, 1948), Saccoglossus kowalevskii (based on Bateson 1884), tornaria larva 2 (based on Dawydoff 1944, 1948), and Balanoglossus clavigerus (based on Stiasny 1914). – Cephalochordata: Branchiostoma lanceolatum (based on Hatschek 1881 and Conklin 1932).

number of cilia, as in many trochophores, but the structure of the 'adult' compound cilia has not been studied. In the tornaria larvae the ring consists of cells with large cylindrical compound cilia, in some species arranged in a hexagonal pattern (Spengel 1893; Fig. 43.5). It appears that the perianal rings develop after the perioral neo-

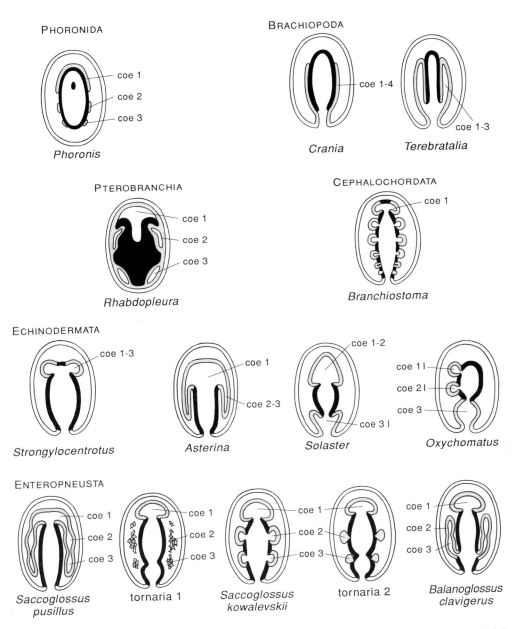

PHORONIDA

coe 1
coe 2
coe 3

Phoronis

BRACHIOPODA

coe 1-4

coe 1-3

Crania   Terebratalia

PTEROBRANCHIA

coe 1
coe 2
coe 3

Rhabdopleura

CEPHALOCHORDATA

coe 1

Branchiostoma

ECHINODERMATA

coe 1-3

Strongylocentrotus

coe 1

coe 2-3

Asterina

coe 1-2

coe 3

Solaster

coe 1
coe 2
coe 3

Oxychomatus

ENTEROPNEUSTA

coe 1
coe 2
coe 3

Saccoglossus
pusillus

coe 1
coe 2
coe 3

tornaria 1

coe 1
coe 2
coe 3

Saccoglossus
kowalevskii

coe 1
coe 2
coe 3

tornaria 2

coe 1
coe 2
coe 3

Balanoglossus
clavigerus

troch in both types of larvae, and this may indicate that they are phylogenetically younger structures. The phylogenetic implications of these facts are discussed below.

The complete lack of multiciliate cells in phoronids, brachiopods, pterobranchs, and echinoderms is interpreted as a plesiomorphic character (Fig. 9.2).

A number of the characters normally considered typical of deuterostomes thus show much intraphyletic variation and must therefore be regarded with caution in phylogenetic discussions.

On the other hand there is a set of characters which appear to be unique to the deuterostomes:

1. The apical organ of the larva disappears at metamorphosis and the central nervous system of the adult develops from a dorsal area behind the apical pole. It may remain as an intraepithelial (basiepithelial) zone of the dorsal side or it may invaginate to form a tube. It is situated in the mesosomal region in the more 'primitive', non-chordate phyla, but extends along the metasome all the way to the anus in the chordates. This is in strong contrast to the apical-ventral CNS of the protostomes (Chapter 10, Fig. 10.2).
2. The larvae and adults of the 'lower' phyla have primary filter-feeding larvae, tornaria or dipleurula larvae (see below), with ciliary bands of the upstream-collecting type consisting of separate cilia on monociliate cells. This is fundamentally different from the trochophora larvae with downstream-collecting bands of compound cilia on multiciliate cells found in the protostomes (Chapter 10).
3. The body is archimeric, consisting of three regions, prosome, mesosome and metasome, each with its characteristic coelomic cavities, protocoel, mesocoel and metacoel, respectively. A similar regionation is unknown among protostomes.

The larval characters are not found in the chordates, which lack primary larvae, but other characteristics link these three phyla firmly to the deuterostomes (Chapter 51).

To me, there is no doubt that the Deuterostomia, as defined here, is a monophyletic group. There is almost unanimous agreement about this in the recent literature, but only if the phoronids and brachiopods are excluded from the group (which then becomes synonymous with Neorenalia). Careful comparisons of the characters of the nervous system, including the fate of the apical organ of the larva, the archimeric regionation with the three pairs of characteristic coelomic pouches, and the structure and function of the ciliary feeding structures show however that the two

---

Fig. 43.3. SEM of locomotory and particle-collecting ciliary bands (neotrochs) in various deuterostomes. – A, lateral ciliary band on a tentacle of an unidentified phoronid larva (plankton, off Phuket Marine Biological Center, Thailand, March 1982). – B, lateral ciliary band on a tentacle of a young, shelled larva of the brachiopod *Discinisca* sp. (plankton, Phangnga Bay, Thailand, March 1982). – C, lateral ciliary band on a pinnule of the pterobranch *Cephalodiscus gracilis* Harmer (Bluebird Ridge region, Bermuda, June 1981; SEM preparation courtesy of Dr. P.N. Dilly, Univ. London, UK.) – D, ciliary band on an arm of an auricularia larva of the sea star *Astropecten irregularis* (plankton, off Kristineberg Marine Biological Station, Sweden, October 1984). – E, ciliary band on an arm of an ophiopluteus larva of an unidentified brittle star (plankton, off Friday Harbor Laboratories, WA, USA, August 1981). – F, ciliary band of an unidentified enteropneust larva (plankton, off Nassau, The Bahamas, October 1990). – Scale bar: 5 μm.

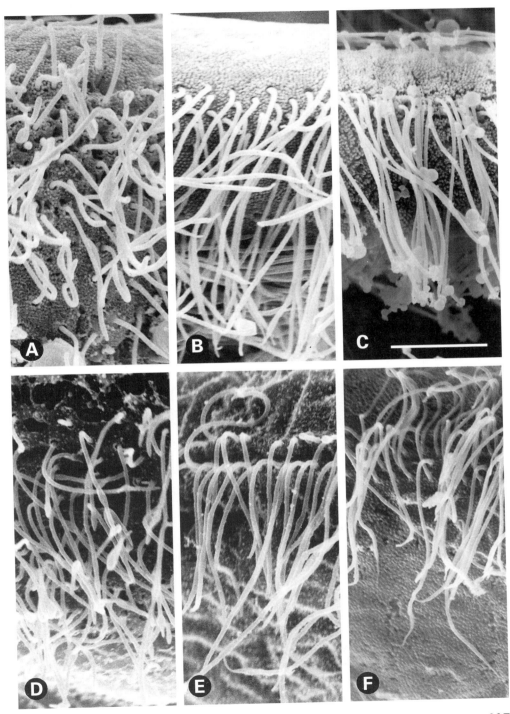

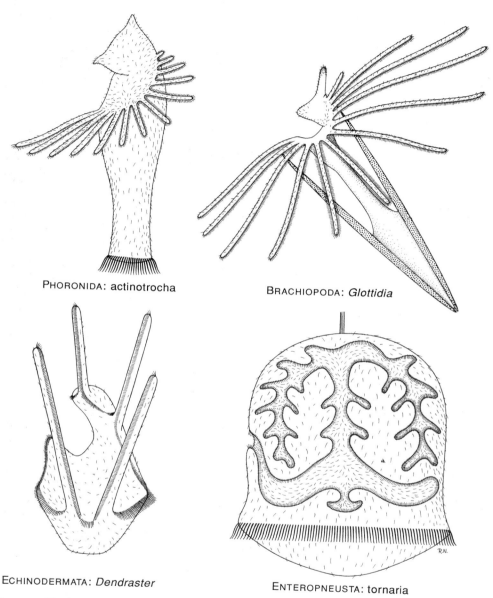

PHORONIDA: actinotrocha

BRACHIOPODA: *Glottidia*

ECHINODERMATA: *Dendraster*

ENTEROPNEUSTA: tornaria

Fig. 43.4. Planktotrophic deuterostome larvae showing the ciliary band (neotroch) considered ancestral in the Deuterostomia. – Phoronida: an unidentified actinotrocha larva. – Brachiopoda: shelled larva of *Glottidia* sp. – Echinodermata: auricularia larva of *Dendraster ecxentricus*. – Enteropneusta: an unidentified tornaria larva. (Modified from Nielsen 1985.)

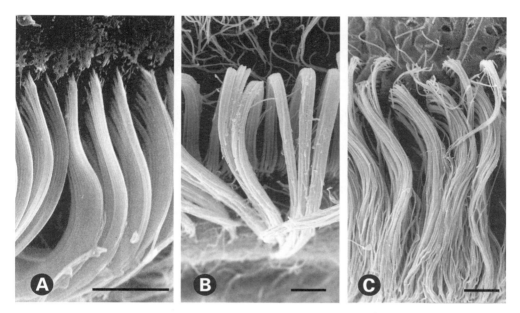

Fig. 43.5. Structures of perianal ciliary bands of protostomes (A) and deuterostomes (B-C). – A, telotroch of a trochophora larva of the annelid *Polygordius* sp. (plankton, off San Salvador Island, The Bahamas, October 1990). – B, perianal ring of an actinotrocha larva of the phoronid *Phoronis psammophila*. – C, perianal ring of a large unidentified tornaria larva of an enteropneust. (See Nielsen 1987.)

phyla share such fundamental apomorphies with the neorenalians that the monophyletic nature of the Deuterostomia as defined here can hardly be questioned.

Two phyla are often included in the Deuterostomia, but are excluded here: chaetognaths and ectoproct bryozoans. The chaetognaths (Chapter 31) have an enterocoelic mode of coelom formation, but as mentioned above this character shows a high degree of intraphyletic variation, especially in the Protostomia, and is therefore of low phylogenetic value. On the other hand, their nervous system has the typical protostomian ontogeny and morphology, and although their more precise phylogenetic position is uncertain, they must definitely be referred to that group. The ectoproct bryozoans, which are usually united with phoronids and brachiopods, bear only superficial resemblance to these two groups (Chapter 24) and are here considered protostomes.

According to the trochaea theory, the ancestor of the deuterostomes, notoneuron, was a pelagobenthic form with a tornaria-like larva and a creeping, deposit-feeding adult (Chapter 2). The above considerations must result in some modifications of this hypothetical ancestor.

The larva of the deuterostome ancestor could have been a tornaria with only monociliate cells and with a perioral system of upstream-collecting single cilia and a perianal ring of compound cilia, as that described in the trochaea theory, but judging from the available information about the deuterostome larvae it is equally probable that it lacked the perianal ring. A larval type only with a perioral ring is well-known in all discussions of larval types of echinoderms, where it is usually called dipleurula (see for example Nichols 1962, Fell 1967), a name introduced by Semon (1888) for the bilaterally symmetrical larval stage of the ancestral echinoderm.

Unfortunately, several authors around the turn of the century (review in Holland 1988), most prominently Bather (1900), used the same name for a hypothetical benthic ancestor of the echinoderms, an organism characterized by the presence of three pairs of coelomic sacs, and this usage of the term has continued too (see for example Hyman 1955, Gruner 1980). Ubaghs (1967) used the term for both the larva and the adult. I propose to remove this ambiguity by reserving the term dipleurula for the ancestral larval stage with a perioral, upstream-collecting ciliary ring but without a perianal ring of compound cilia (Fig. 43.6) and to call the adult ancestor (described below) notoneuron.

The structure of the adult, non-chordate deuterostomes indicates that the adult deuterostome ancestor was benthic with ciliated tentacles and three pairs of coelomic sacs. This differs from the notoneuron described earlier, which was envisaged as a deposit feeder without tentacles (Nielsen & Nørrevang 1985, Nielsen 1985; Chapter 2), but the revised phylogeny of the deuterostomes (see below and Fig. 43.8) has clearly placed the phyla with ciliated tentacles at the bottom of the tree. As mentioned above, it appears confusing to call this hypothetical ancestor dipleurula, and it also differs from Bather's concept in having a new, dorsal central nervous system instead of the apical organ. I have chosen to modify the concept of notoneuron to fit the new conclusions.

The deuterostome ancestor can then be described as a pelago-benthic notoneuron with a larva which was either a tornaria or a dipleurula (Figs 43.6, 43.7). The perioral ciliary band of the larva was taken over by the adult and extended on mesosomal tentacles on the dorsal side lateral to the new central nervous system.

The relationships of the deuterostomes to the other metazoans is still a matter of discussion. The most commonly held opinion is that all the bilateral animals form a monophyletic group, and that the coelomic phyla evolved from one 'archecoelomate'. These theories must derive deuterostomes from protostomes, or vice versa. The several important differences between these two main groups are however so fundamental, that it is very unlikely that one should be derived from the other. My conclusion (Nielsen & Nørrevang 1985, Nielsen 1985, 1987; Fig. 9.1) has been that the two groups have evolved separately from a radially symmetrical, non-coelomate ancestor.

One of the more speculative points of the trochaea theory is the derivation of the deuterostomes. Originally, I believed that the ring of compound cilia surrounding the mouth of the trochaea, the archaeotroch, was formed from multiciliate cells because practically all the bands of compound cilia in protostomes and in the enteropneust larva are formed from multiciliate cells. It was therefore a surprise to find that the archaeotroch of the phoronid larva was formed from monociliate cells, and this fact made me change my mind about the character of the archaeotroch of trochaea and of the origin of multiciliarity (Nielsen 1987). I now believe that the deuterostomes are originally monociliate, a plesiomorphy shared with the cnidarians, and that multiciliarity evolved in the cyrtotretes (enteropneusts + chordates). An indication of this is perhaps seen in the occurrence in enteropneusts of multiciliate cells where each cilium has an accessory centriole (Chapter 50). Together with the increased emphasis now placed on the axial complex, this new information has resulted in a modified

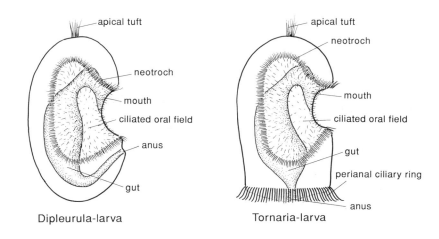

Dipleurula-larva Tornaria-larva

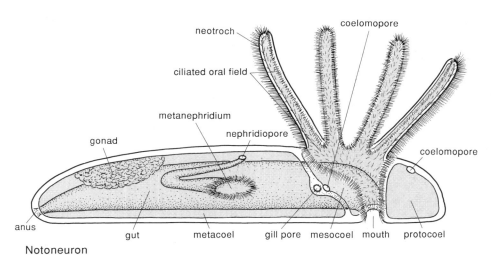

Notoneuron

Fig. 43.6. Larval and adult stages of notoneuron, the ancestral deuterostome; two alternative larval forms are shown. The adult is drawn in accordance with the modified interpretation of deuterostome phylogeny; it has mesosomal tentacles carrying loops of the neotroch and metasomal metanephridia, which additionally function as gonoducts. Note the apical brain of the larvae and the mesosomal nervous centre of the adult. A gill pore is indicated in accordance with the trochaea theory, but its existence in this ancestral type is uncertain.

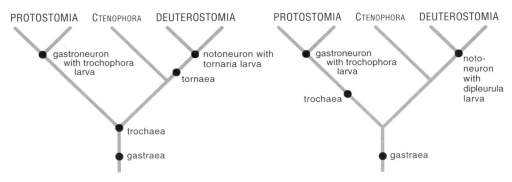

Fig. 43.7. Alternative phylogenies of the Bilateria (see also Fig. 41.2). The left tree shows the phylogeny proposed by the trochaea-theory and the right tree shows the modification which interprets the perianal bands of compound cilia in phoronid and enteropneust larvae as independently evolved apomorphies. Note that the branching patterns are identical.

phylogenetic tree (Fig. 43.8). The deuterostomes are derived either from a trochaea with an archaeotroch formed on monociliate cells, or from a dipleurula. The group called Brachiata (Pterobranchia + Phoronida + Brachiopoda) in the earlier papers (Nielsen 1985, 1987) is now considered paraphyletic and the name has therefore disappeared from the scheme.

The protostome-deuterostome dichotomy is accepted in most modern textbooks, although usually with the 'lophophorates' dangling between the two main groups, but there are a number of authors who with various arguments derive the deuterostomes (in the narrower sense, i.e. neorenalians) from some point within the protostomes.

Løvtrup (1975) discussed the 'classical' differences between protostomes and deuterostomes and concluded that several of the characters are rather ambiguous; he did not mention the different structure and position of the nervous systems, and the differences between the two larval types, called trochophore and pluteus, were circumvented by postulating that the lecithotrophic echinoderm larvae resemble mollusc larvae (sic). The main part of the paper is a comparative enumeration of characters shared by two of the three groups Mollusca, Echinodermata and Vertebrata. Only three characters are found to be shared by molluscs and echinoderms, and three by echinoderms and vertebrates, whereas 51 characters are shared by vertebrates and molluscs. The choice of characters can of course be criticized, and other authors may come up with quite different numbers, but my main objections to the three lists of characters are: 1) that some of the characters are actually shared by all three phyla (the vertebrate skeleton contains calcium carbonate), 2) that some of the statements are directly wrong (the molluscs do not have a horny ectodermal skeleton), 3) that absence of a structure or compound (as for example chitin) is a very weak character, and 4) that the presence of for example pericardium, liver tissue, epithelial glands, kidney, optic sense organs, and brain in two groups is not a sign of relationship if these structures are not homologous in the two groups. In my view, the last objection is so strong that the whole argument must be rejected.

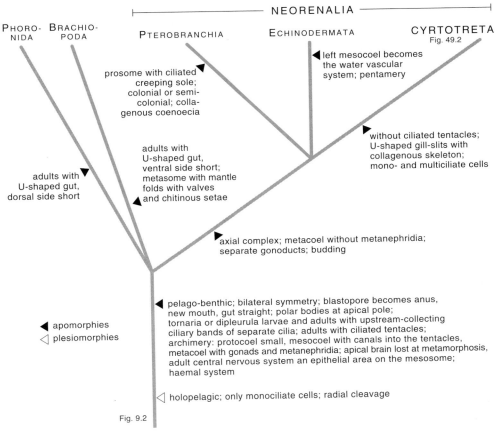

Fig. 43.8. Phylogeny of the deuterostomes. The trochaea theory implies that notoneuron had a pair of gill openings, and that these were then lost in the brachiopods (and *Rhabdopleura*); if this is not accepted, the gill openings must be regarded as an apomorphy of the neorenalians; the character is not included in the tree.

Salvini-Plawen (1982) regarded the Oligomera (Tentaculata + Deuterostomia) as the sister group of the Sipuncula, with this compound as the sister group of a group consisting of Echiura, Myzostomida and Articulata. Although I will not pretend to understand all the details of the argument, I believe that a central part is a list of spiralian versus deuterostomian characters of the phoronids (op. cit., table I). The deuterostomian characters can be accepted with few reservations, whereas the characters purported to indicate spiralian relationships 1) are known to show much variation both within and between related phyla, viz. the blastopore fate and the method of coelom formation, 2) are probably symplesiomorphies, viz. the protonephridia and the apical ganglion of the larvae, 3) may be misinterpreted, viz. the cilia along the rim of the hood in the actinotroch which probably have nothing to

333

do with a prototroch (Chapter 44), 4) are definitely misinterpreted, viz. the identification of the ciliary band of the tentacles as a metatroch, and 5) are completely enigmatic, viz. the protocoel, a character which is also found in the list of deuterostome characters. The phoronid metacoel is homologized with the sipunculan body cavity, which is a sort of starting point; but the coelomic canals of the labial tentacles of the sipunculans are believed to come from the protocoel (a mesocoel is not mentioned), whereas the phoronid tentacles come from the mesosome. So, although I don't understand all the arguments, I am inclined to reject the idea altogether, because none of the characters in the apparently central table can stand scrutiny.

A less detailed theory based mainly on biochemical and molecular data was presented by Bergström (1986). It suggested a phylogenetic tree with the Chordata + Hemichordata + Echinodermata as the sister group of the Mollusca and further to the Crustacea; only the morphological evidence will be considered here. The evolution from a 'pre-molluscan stock' to an ancestral hemichordate should have involved the following transitions: 'enterocoel, pharyngeal slits and tentacles form as a result of a new mode of life; creeping sole and ventral nerve cords disappear; ciliated bands of larva strongly modified; loss of spiral cleavage and other spiralian characters' (op. cit., fig. 5). The difference between protostome larvae with downstream-collecting bands of compound cilia and deuterostome larvae with upstream-collecting band of separate cilia is mentioned but dismissed with the sentence 'This is a difference, but what does it mean and how did it come about?' (op. cit., p. 198). Using argumentation of this type, just about any phylogenetic tree can be defended.

I have not seen any convincing theory which derives the deuterostomes from any of the coelomate protosomes, and there have to my knowledge not been any recent attempts to derive them from platyhelminths or aschelminths. The trochaea theory on the other hand interprets the deuterostomes (together with the ctenophores) as the sister group of the protostomes. The common ancestor of deuterostomes and ctenophores, trochaea, should have been a gastraea with a ring of compound cilia around the blastopore, but without any coelomic structures (Chapter 2). The modified phylogeny discussed above places trochaea on the protostomian line, but this only means that the common bilaterian ancestor was a gastraea. The theory implies that mesoderm and coelom in protostomes and deuterostomes are not homologous and that the only part of the central nervous system which is homologous in the two groups is the larval apical organ. It further emphasizes the fundamental differences between the ciliary feeding structures of the trochophora- and tornaria/pluteus/dipleurula-type larvae. The original version of the theory further homologized the perianal ring of compound cilia in actinotrochs and tornaria larvae with all the bands of compound cilia in the trochophores, but this point now appears somewhat questionable.

It must be concluded that the ancestral deuterostome was a pelago-benthic organism with a dipleurula (or possibly a tornaria) larva with the upstream-collecting ciliary bands seen in both larvae and adults of several deuterostome phyla and a benthic adult with a lophophore bearing the ciliary bands (Fig. 43.6). The adult had the archimeric coelomic compartments, viz. paired or unpaired protocoel and paired mesocoel and metacoel with coelomoducts. The coelomoducts of the metacoels functioned both as metanephridia and gonoducts.

The phylogeny of the deuterostomes has been discussed extensively in all modern textbooks and in a number of review papers (see for example Schaeffer 1987), but there is no agreement about the details of the shape of the tree. One reason for this may be that the discussions have been muddled by the habit of keeping pterobranchs and enteropneusts together in the group Hemichordata, which according to my interpretation is polyphyletic (Chapter 47).

Jefferies' interpretation of some early echinoderm-like fossils as ancestors of most of the deuterostomes, the so-called calcichordate theory, is discussed in Chapter 48.

In constructing a phylogenetic tree of the deuterostomes, I have put special emphasis on the development, morphology and functions of the coelomic compartments, and this has resulted in the trees in Figs 43.8 and 49.2 (see also Fig. 51.2). The details in the branching pattern will be discussed in the following chapters.

References

Atkins, D. 1961. A note on the growth stages and structure of the adult lophophore of the brachiopod *Terebratella (Waltonia) inconspicua* (G.B. Sowerby). – Proc. zool. Soc. Lond. **136**: 255-271.

Bateson, W. 1884. The early stages of the development of *Balanoglossus* (sp. incert.). – Q. Jl microsc. Sci., N.S. **24**: 208-236, pls 18-21.

Bather, F.A. 1900. The Echinoderma. – *In* E. Ray Lankester: A Treatise on Zoology, vol. 3. Adam & Charles Black, London.

Bergström, J. 1986. Metazoan evolution – a new model. – Zool. Scr. **15**: 189-200.

Burdon-Jones, C. 1952. Development and biology of the larva of *Saccoglossus horsti* (Enteropneusta). – Phil. Trans. R. Soc. B **236**: 553-590.

Conklin, E.G. 1932. The embryology of *Amphioxus*. – J. Morph. **54**: 69-118.

Davis, B.M. 1908. The early life-history of *Dolichoglossus pusillus* Ritter. – Univ. Calif. Publs Zool. **4**: 187-226, pls 4-8.

Dawydoff, C. 1944. Formation des cavités coelomiques chez les tornaria du plancton indochinois. – C. r. hebd. Séanc. Acad. Sci., Paris **218**: 427-429.

Dawydoff, C. 1948. Classe des Entéropneustes + Classe des Ptérobranches. – Traité de Zoologie, vol. 11, pp 369-489. Masson, Paris.

Fell, H.B. 1967. Echinoderm ontogeny. – *In* R.C. Moore (ed.): Treatise of Invertebrate Paleontology, part S, vol. 1, pp 60-85. Geological Society of America, Lawrence, Kansas.

Gemmill, J.G. 1912. The development of the starfish *Solaster endeca* Forbes. – Trans. zool. Soc. Lond. **20**: 1-71, pls 1-5.

Gruner, H.-E. 1980. Einführung. – *In* H.-E. Gruner (ed.): A. Kaestner's Lehrbuch der speziellen Zoologie (4th ed.), 1. Band, 1. Teil, pp 15-156. Gustav Fischer, Stuttgart.

Hatschek, B. 1884. Studien über Entwicklung des *Amphioxus*. – Arb. zool. Inst. Univ. Wien **4**: 1-88, pls 1-9.

Holland, N.D. 1988. The meaning of developmental asymmetry for echinoderm evolution: a new interpretation. – *In* C.R.C. Paul & A.B. Smith (eds): Echinoderm Phylogeny and Evolutionary Biology, pp 13-25. Oxford Univ. Press, Oxford.

Holland, N.D. 1991. Echinodermata: Crinoidea. – *In* A.C. Giese, J.S. Pearse & V.B. Pearse (eds): Reproduction of Marine Invertebrates, vol. 6, pp 247-299. Boxwood Press, Pacific Grove, CA.

Hyman, L.H. 1955. Echinodermata. – The Invertebrates, vol. 4. McGraw-Hill, New York.

Lester, S.M. 1988. Ultrastructure of adult gonads and development and structure of the larva of *Rhabdopleura normani* (Hemichordata, Pterobranchia). – Acta zool. (Stockh.) **69**: 95-109.

Long, J.A. 1964. The embryology of three species representing three superfamilies of articulate brachiopods. – Ph.D. thesis, Univ. Washington.

Løvtrup, S. 1975. Validity of the Protostomia-Deuterostomia theory. – Syst. Zool. **24**: 96-108.

Ludwig, H. 1882. Entwicklungsgeschichte der *Asterina gibbosa* Forbes. – Z. wiss. Zool. **37**: 1-98, pls 1-8.

Marcus, E. du B.-R. 1949. *Phoronis ovalis* from Brazil. – Bolm Fac. Filos. Ciênc. Univ. S Paulo, Zool. **14**: 157-166, 3 pls.

Nichols, D. 1962. Echinoderms. – Hutchinson Univ. Library, London.

Nielsen, C. 1985. Animal phylogeny in the light of the trochaea theory. – Biol. J. Linn. Soc. **25**: 243-299.

Nielsen, C. 1987. Structure and function of metazoan ciliary bands and their phylogenetic significance. – Acta zool. (Stockh.) **68**: 205-262.

Nielsen, C. 1991. The development of the brachiopod *Crania (Neocrania) anomala* (O.F. Müller) and its phylogenetic significance. – Acta zool. (Stockh.) **71**: 7-28.

Nielsen, C. & A. Nørrevang 1985. The trochaea theory: an example of life cycle phylogeny. – *In* S. Conway Morris, J.D. George, R. Gibson & H.M. Platts (eds): The Origin and Relationships of Lower Invertebrate Groups, pp 28-41. Oxford Univ. Press, Oxford.

Ruppert, E.E. & E.J. Balser 1986. Nephridia in the larvae of hemichordates and echinoderms. – Biol. Bull. Woods Hole **171**: 188-196.

Salvini-Plawen, L.v. 1982. A paedomorphic origin of the oligomerous animals. – Zool. Scr. **11**: 77-81.

Schaeffer, B. 1987. Deuterostome monophyly and phylogeny. – Evol. Biol. **21**: 179-235.

Semon, R. 1888. Die Entwicklung der *Synapta digitata* und die Stammesgeschichte der Echinodermen. – Jena. Z. Naturw. **22**: 1-135, pls 1-6.

Spengel, J.W. 1893. Die Enteropneusten des Golfes von Neapel. – Fauna Flora Golf. Neapel **18**: 1-758, pls 1-37.

Stiasny, G. 1914. Studien über die Entwicklung des *Balanoglossus clavigerus* Delle Chiaje. II. Darstellung der weiteren Entwicklung bis zur Metamorphose. – Mitt. zool. Stn Neapel **22**: 255-290, pls 6-9.

Thomson, W. 1865. On the embryogeny of *Antedon rosaceus*, Linck (*Comatula rosacea* of Lamarck). – Phil. Trans. R. Soc. **155**: 513-544, pls 23-27.

Ubaghs, G. 1967. General characters of echinoderms. – *In* Treatise of Invertebrate Paleontology, part S, vol. 1, 3-60. Geol. Soc. Am., Lawrence, Kansas.

Ubisch, L.v. 1913. Die Entwicklung von *Strongylocentrotus lividis* (*Echinus microtuberculatus, Arbacia pustulosa*). – Z. wiss. Zool. **106**: 409-448, pls 5-7.

Zimmer, R.L. 1964. Reproductive Biology and Development of Phoronida. – Ph.D. thesis, Univ. Washington.

Phylum PHORONIDA

Phoronids are one of the smallest animal phyla with only about 12 recognized species referred to two genera, *Phoronis* and *Phoronopsis*; there is no reliable fossil record. All species are marine and benthic, building chitinous tubes covered by mud or sand or boring into calcareous material. The general life form is solitary, but several species occur in smaller or larger masses, and *P. ovalis* may even form lateral buds from the body so that small, temporary colonies arise (Marcus 1949). Regenerative powers are considerable: an autotomized tentacle crown is easily regenerated and autotomized tentacle crowns of *P. ovalis* appear to regenerate completely (Silén 1955, Marsden 1957).

The adult phoronid has a cylindrical body with a lophophore bearing cylindrical, ciliated tentacles around the mouth. The smallest species, *P. ovalis*, has an almost circular tentacle crown; the somewhat larger species have lophophores in the shape of a simple horseshoe with a double row of tentacles; and the largest species have spirally coiled lophophore arms (Abele, Gilmour & Gilchrist 1983). The gut is U-shaped with the anus situated near the mouth, and the ontogeny shows that the short area between mouth and anus is the dorsal side. The epithelia of all three germ layers are monolayered and their ciliated cells are all monociliate (Nielsen 1987, Bartolomaeus 1989). There are no chitinous structures in the epidermis but the tubes contain chitin (Jeuniaux 1982).

Both larval and adult phoronids are obviously archimerics.

The small prosome, the epistome, is a round or crescentic flap on the dorsal side of the mouth and along the lophophore in the species with many tentacles. Its coelomic cavity, the protocoel, which is partly obliterated by strands of muscles in some species (Siewing 1973), is not in communication with the mesocoel in the juveniles, but a communication may develop later also in anterior ends formed by regeneration (Zimmer 1978).

The lophophore bears from ten to several hundred tentacles, which have characteristic ciliary bands. Most conspicuous is the lateral band which is an upstream-collecting system of single cilia which strain particles from the water current they create.

Chapter vignette: *Phoronis hippocrepia* (Redrawn from Emig 1982.)

The strained particles are transported toward the mouth along the frontal ciliary band and the general ciliation of the oral field. A row of laterofrontal sensory cells is found along the sides of the frontal band (Nielsen 1987). The lophophore has a more spacious coelomic cavity, the mesocoel, which surrounds the oesophagus and sends a small coelomic channel into each tentacle.

The elongate body has a large coelom, the metacoel, which is separated from the mesocoel by a conspicuous transverse septum, often called a diaphragm, at the base of the lophophore. The metasomal peritoneum forms a number of mesenteries which suspend the gut (Cori 1890).

The nervous system is intraepithelial (basiepithelial) with a strong concentration in the short dorsal side between mouth and anus, overlying the dorsal side of the mesocoel (Silén 1954a; Fig. 44.1).

There is a haemal system in the shape of well-defined vessels formed between the basement membranes of peritoneum and endoderm or between apposing peritoneal layers. Small blind vessels in the tentacles are formed by folds of the frontal side of the mesocoelic lining (Pardos *et al.* 1991). Horseshoe-shaped vessels run along the lophophore and two or three longitudinal vessels run along the gut to the 'posterior' end of the body where a system of lacunae surround the gut (Emig 1979). Podocytes have been observed in many of the blood vessels of the metasomal region of *P. muelleri* (Storch & Herrmann 1978), and they are probably the site of the formation of

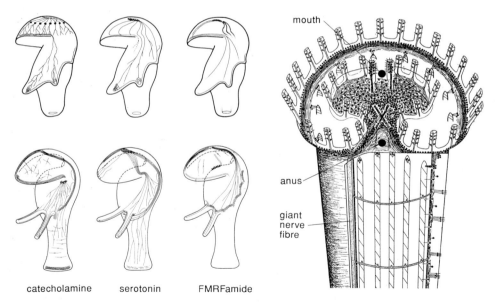

catecholamine   serotonin   FMRFamide

Fig. 44.1. Larval and adult nervous systems of phoronids. The actinotrochs are seen from the right with the staining methods for the three types of nerve cells indicated below; the 3-tentacle stage is of *Phoronis vancouverensis* (modified from Hay-Schmidt 1990b) and the 10-tentacle stage is of *P. muelleri* (modified from Hay-Schmidt 1990a). The adult is seen from the dorsal side with the distal parts of the tentacles and parts of the dorsal body wall removed (modified from Silén 1954a).

primary urine. A pair of large metanephridia with large funnels formed by ciliated epithelio-muscular cells are situated in the metacoel (Bartolomaeus 1989). These nephridia drain the metasome and function as gonoducts. The gonads are formed from the peritoneum at the stomach part of the gut. The sperm becomes enclosed in elaborate spermatophores which are shed and float in the water (Zimmer 1967). The spermatophores become caught by the tentacles of another specimen (or even engulfed); the mass of sperm becomes amoeboid and lyses the body wall, entering the metacoel (and the mesosomal-metasomal septum if caught by a tentacle) (Zimmer 1991). This type of internal fertilization has been observed in a number of species, and the reports of external fertilization are not based on direct observations and should be checked (Emig 1977). The polar bodies are given off after the eggs have been shed (Zimmer 1964); they have been reported to move back into the egg and become resorbed in some species (Herrmann 1986)

Development has been studied in a number of species. Silén (1954b) showed that *P. ovalis*, with the largest eggs, has direct development, whereas other species with comparatively smaller eggs develop through the well-known actinotrocha stage. Some species shed the eggs free in the water but others retain them in the lophophore until a stage of about four tentacles. Zimmer (1964) showed that isolated blastomeres of 2- and 4-cell stages are able to develop into actinotrochs, at least in some cases. The two first cleavages take place along the primary axis of the egg, as shown by the polar bodies (Freemann 1991). Cleavage is total and usually equal, and although traces of a spiral pattern have been reported by a few authors it is now agreed that the cleavage is radial (Zimmer 1964, Emig 1977). Experiments with vital staining of blastomeres have shown that the orientation of the first cleavage shows considerable variation, but the first cleavage is transverse in about 70% of the embryos (Freemann 1991).

A coeloblastula is formed, and gastrulation is by invagination. The blastopore constricts from the posterior side and closes completely in some species but it remains open as the larval mouth for example in *P. vancouverensis* (Zimmer 1980). At a later stage the anus breaks through at the posterior pole of the larva, possibly in the posteriormost end of the constricted blastopore.

The origin and differentiation of the mesoderm have been interpreted variously by the earlier authors (usefully tabulated by Zimmer 1964 and Emig 1974), but the recent descriptions by Zimmer (1964: several species, 1980: *P. vancouverensis*), Emig (1974: *P. psammophila*) and Herrmann (1986: *P. muelleri*) are well documented and in general agreement and the most characteristic types of development are described below. In *P. muelleri*, mesodermal cells proliferate into the blastocoel from the whole archenteron and become arranged as a more or less continuous peritoneum surrounding the blastocoel; this large coelomic cavity becomes divided into protocoel, mesocoel and metacoel. In *P. vancouverensis*, the mesoderm is proliferated from the anterior part of the archenteron and becomes arranged as a thin hemispherical protocoel around the anterolateral parts of the blastopore; more laterally situated cells of similar origin appear to give rise to the mesocoel, but they remain as a pair of compact lateral bands until a late larval stage; the origin of the metacoel is more unclear, but it may originate similar to the mesocoel. The coelomic cavities

339

arise through schizocoely or through the arrangement of mesodermal cells as linings of parts of the blastocoel; there is no sign of enterocoely. The three coelomic units protocoel, mesocoel and metacoel are clearly present in all species in which fully developed larvae have been studied, although the protocoel may have lost parts of its lining or become compressed almost completely (Zimmer 1978).

The first pair of tentacle buds develop on either side of the ventral midline between mouth and anus, and additional tentacles develop laterally along curved lines almost meeting dorsally behind the apical organ. The blastulae are uniformly ciliated, but when the first tentacles develop, a band of longer and more closely set cilia can be recognized along the lateral faces of the tentacles. This band becomes the upstream-collecting ciliary band of the tentacles (Nielsen 1987), which is the feeding organ and the sole locomotory organ in the early stages. Later, a perianal ring of large compound cilia, likewise formed from monociliate cells (Nielsen 1987), develops and becomes the main locomotory organ of the larva. The larval tentacles of *P. vancouverensis* contain narrow extensions of the lophophoral mesocoel, and it appears that the adult tentacles develop directly from the larval ones (Zimmer 1964). In other species, such as *P. psammophila*, small abfrontal knobs at the bases of the larval tentacles become the adult tentacles with the coelomic canals while the distal parts of the larval tentacles, which lack the coelomic canals, are shed after metamorphosis (Herrmann 1979). Large larvae of *P. muelleri* develop an extra row of tentacles just behind the larval tentacles (Silén 1954b), and the larval tentacles are shed together with their common bases (Herrmann 1980).

An apical ciliary tuft develops at the apical pole already at the blastula stage and can be followed to the metamorphosis. A complicated nervous system develops gradually in the larvae (Hay-Schmidt 1989, 1990a, 1990b, Lacalli 1990), with an additional sensory organ with cilia frontal to the apical organ in old larvae of for example *P. muelleri* and *P. architecta* (Zimmer 1978); this organ is protruded when the larva is testing the substrate, ready for settling (Silén 1954b). Early larvae (of *P. vancouverensis*, see Hay-Schmidt 1990b) still without a perianal ciliary ring develop a wide ring of catecholamine-containing cell bodies around a kidney-shaped concentration of serotonin-containing cell bodies and scattered FMRFamide-containing cells; all these cells send neurons to the tentacles (Fig. 44.1). Older larvae (of *P. muelleri*; Hay-Schmidt 1990a) have a ring of serotonin-containing cell bodies surrounding a kidney-shaped concentration of FMRFamide-containing cells at the apical pole and develop more complicated innervation of the tentacles, a ring mainly consisting of catecholamine-containing nerve processes at the base of the cells of the perianal ciliary ring, and additional concentrations of cell bodies at the oral field. A connection between the apical ganglion and the perianal nerve has not been observed, and the nerve is separated from the ciliated cells by a basement membrane; contacts between the nerves and the ciliated cells have not been observed (Hay-Schmidt 1989).

A long, tubular invagination of the ventral body wall, the metasomal sac, develops at some stage, and in larvae which are about to be ready for metamorphosis it occupies much of the space around the gut (see Fig. 44.2).

A pair of protonephridia each with several solenocytes develop from a median ectodermal invagination just anterior to the anus in the early larvae; the common

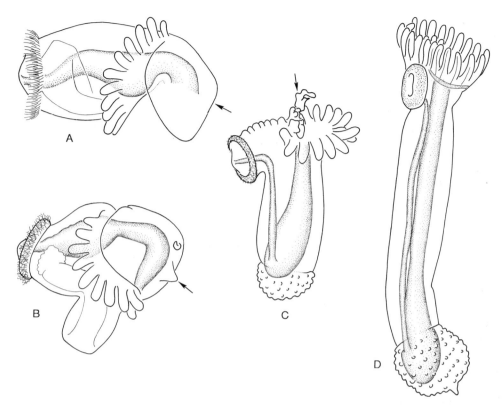

Fig. 44.2. Metamorphosis of *Phoronopsis harmeri*; the position of the apical organ is indicated by an arrow. – A, a full-grown actinotrocha larva. – B, beginning of metamorphosis; the metasomal sac is halfway everted. – C, eversion of the metasomal sac completed; the preoral hood is disintegrating and the dorsal side is strongly shortened. – D, metamorphosis completed; the preoral hood has been cast off and the perianal ciliary ring will soon be discarded too. (Redrawn from Zimmer 1964, 1991.)

part of the nephridial duct soon disappears and the nephridiopores of older larvae are situated near the coelomopores of the metacoels. In the advanced larvae the solenocytes form clusters in the tentacle region where they drain the blastocoel (Hay-Schmidt 1987, Bartolomaeus 1989).

The metamorphosis is rapid and dramatic (Fig. 44.2). The metasomal sac everts, pulling the gut into a U-shape, and contractions in the larval body bring the mouth and anus close to each other; this establishes a new main body axis perpendicular to the larval main axis. The major part of the hood with the apical organ (and the accessory sensory organ when present) is cast off together with the larval tentacles in species where these do not become the adult tentacles; the large ciliary ring around the anus is either resorbed or cast off. These larval organs are often ingested by the juvenile (Herrmann 1979).

The juvenile has the general shape of the adult, and the larval meso- and meta-coelic cavities are taken over almost without modifications. The main part of the protocoel is lost with the hood, but the proximal (posterior) part can be followed through to the epistome in several species (Zimmer 1978).

The dorsal central nervous concentration develops in the ectoderm without any connection with the apical region. There is no indication of nerve cells in the corresponding position in the larval stages described so far, but they may be found in actinotrochs ready for metamorphosis.

The larval nephridia undergo a major reorganization at metamorphosis (Bartolomaeus 1989). The solenocytes and the terminal ducts break off and become phagocytosed by other duct cells and the duct now ends blindly. At a later stage areas of the mesocoelic epithelium differentiate into a pair of ciliated funnels which gain connection with the ducts to form metanephridia.

The phylogenetic position of the Phoronida has been the subject of many discussions. They are often united with Brachiopoda and Bryozoa (= Ectoprocta) to a supraphyletic taxon called Tentaculata in the older literature and Lophophorata in the newer literature (following the suggestion of Hyman 1959). Most authors following this approach have concluded that the lophophorates must be placed near the split between protostomes and deuterostomes because they have characters indicating relationships with both groups (see for example Gruner 1980, Barnes 1986, Brusca & Brusca 1990). In my opinion, this approach is counterproductive, and I will here discuss the phylogenetic relationships of the three phyla independently. It should be clear that it is futile to discuss the position of a group which may consist of very distantly related taxa, i.e. a polyphyletic group.

The phoronid cleavage pattern is now considered radial, as characteristic of the deuterostomes. The anterior part of the blastopore persists as the adult mouth while the anus is formed anew, which is a protostome character, but this character shows a high degree of variation within several phyla of both protostomes (Chapter 10) and deuterostomes (Chapter 43) and is therefore of low phylogenetic value.

The mesoderm is formed from the archenteron through ingression of single cells, and the coelomic cavities are not formed through the characteristic enterocoelous mode seen for example in echinoderms and many enteropneusts. However, there is no sign of either 4d-mesoderm or ectomesoderm as found in most spiralians and similarities to the aschelminth mode of mesoderm formation are difficult to envisage. The morphology of the three coelomic regions, protocoel, mesocoel and metacoel, is on the other hand highly characteristic of the deuterostomes (Fig. 43.1). The protocoel is well-defined in some of the actinotrochs and persists as a small, somewhat uncharacteristic structure in the adults, but its character as a separate coelomic cavity cannot be questioned. As in the brachiopods (Chapter 45), the protocoel is not connected with the haemal system to form an axial complex (Chapter 46). The mesocoel surrounds the mouth and sends canals into the tentacles as in brachiopods, pterobranchs and echinoderms (the water vascular system) (Fig. 43.1). The metacoel is the large body cavity with metanephridia and gonads as in the brachiopods. This architecture of the coelom with three regions with special structure and function, called archimery (or less precisely oligomery), is characteristic of the deuterostomes, and nothing similar has been reported from protostomes. This strongly suggests that the phoronids are deuterostomes.

The nervous system of the larva comprises an apical, and sometimes an additional preapical, centre which becomes cast off at metamorphosis, and the adult central nervous system is a dorsal, intraepithelial concentration, which occupies the very short area between mouth and anus. There are no ventral nerve cords. This is characteristic of the deuterostomes and in contrast to the protostomes, which have a brain derived from or incorporating the apical ganglion, and a paired or secondarily unpaired ventral nerve cord.

Both the actinotrocha larva and the adult phoronid have tentacles with upstream-collecting ciliary bands, which consist of single cilia on monociliate cells; all these three characters of the ciliary bands are characteristic of the dipleurula- or tornaria-type larva considered ancestral of the deuterostomes. On the contrary the trochophora larvae of the protostomes have downstream-collecting ciliary bands consisting of compound cilia on multiciliate cells.

An alternative interpretation of the actinotroch was given by Jägersten (1972) and Farmer (1977), who regarded the cyphonautes larva of some of the ectoproct bryozoans as an ancestral form from which the actinotroch was derived; Lacalli (1990) read the evolution the opposite way. These theories interpret the cilia along the edge of the actinotroch hood as homologous with the prototroch or corona of the cyphonautes, and further with the prototroch of the trochophore. There are, however, very considerable differences between these structures. The ciliated bands of the actinotroch tentacles are monociliate with single cilia beating away from the mouth, whereas both the corona/prototroch of the ectoprocts and the prototroch of the trochophores are multiciliate and beat towards the mouth (or away from the apical organ in species without a mouth); only the prototroch is particle-collecting, being a downstream-collecting band with compound cilia. The particle-collecting ciliated ridge of the cyphonautes larva is of a unique structure and function (Chapter 24). With so completely different structures and functions there seems to be no way of changing one type of ciliary band to the other. Lacalli (1990) favoured the idea that the cilia along the edge of the actinotroch hood represent the prototroch, but these cilia are not engaged in particle collecting and their role in locomotion is unclear. He further demonstrated that the conspicuous nerve cells which run along the edge are not in contact with the ciliated cells and concluded that they may be responsible for the rather complicated movements of the hood. Thus, there is no direct observation which supports this idea.

The fate of the blastopore appears to be the only character which points to a relationship with the protostomes, and this character is known to show considerable intraphyletic variation and is therefore of limited phylogenetic importance.

It can therefore be concluded that a number of characters of structure and function of both larval and adult phoronids are typical deuterostomian (as opposed to protostomian): 1) The early development shows a regular radial cleavage. 2) The body is divided into three regions, prosome, mesosome and metasome, each with a coelomic compartment, protocoel, mesocoel and metacoel; the mesosome carries ciliated tentacles and the metasome contains the large body cavity with gonads and metanephridia. 3) Both larvae and adults have upstream-collecting ciliary bands formed by single cilia on monociliate cells. 4) The central nervous system of the adults is a dorsal, intraepithelial concentration in the mesosomal area without any connection with the larval apical organ, which is shed at metamorphosis.

Both phoronids and brachiopods lack the rather complicated axial complex, a nephridium consisting of the small protocoel and a specialized haemal system with podocytes, which characterizes the 'higher' deuterostomes (Neorenalia; Chapter 46) and there is no sign that they have ever possessed one. The two phyla are usually considered closely related, and a sister-group relationship could therefore be considered. However, recent investigations on brachiopod embryology have shown that their ventral side is the short one, and this makes it more reasonable to interpret the two phyla as evolved independently from an ancestor with a straight gut (further discussion in Chapter 45).

Interesting subjects for future research

1. Origin of mesocoel and metacoel.
2. Development and function of the nervous system of one species through all stages from the youngest actinotroch to the adult.

References

Abele, L.G., T. Gilmour & S. Gilchrist 1983. Size and shape in the phylum Phoronida. – J. Zool. (Lond.) **200**: 317-323.

Barnes, R.D. 1986. Invertebrate Zoology, 5th ed. – Saunders College Publishing, Philadelphia.

Bartolomaeus, T. 1989. Ultrastructure and relationship between protonephridia and metanephridia in *Phoronis muelleri* (Phoronida). – Zoomorphology **109**: 113-122.

Brusca, R.C. & G.J. Brusca 1990. Invertebrates. – Sinauer Associates, Sunderland.

Cori, C.J. 1890. Untersuchungen über die Anatomie und Histologie der Gattung *Phoronis*. – Z. wiss. Zool. **51**: 480-568, pls 22-38.

Emig, C.C. 1974. Observations et discussions sur le développement embryonnaire des Phoronida. – Z. Morph. Tiere **77**: 317-335.

Emig, C.C. 1977. Embryology of Phoronida. – Am. Zool. **17**: 21-37.

Emig, C.C. 1979. British and other phoronids. – Synopses Br. Fauna, N.S. **13**: 1-57.

Emig, C.C. 1982. Phoronida. – *In* S.P. Parker (ed.): Synopsis and Classification of Living Organisms, vol. 2, p. 741, pls 126-127. McGraw-Hill, New York.

Farmer, J.D. 1977. An adaptive model for the evolution of the ectoproct life cycle. – *In* R.M. Woollacott & R.L. Zimmer (eds): Biology of Bryozoans, pp 487-517. Academic Press, New York.

Freemann, G. 1991. The bases for and timing of regional specification during larval development of *Phoronis*. – Dev. Biol. **147**: 157-173.

Gruner, H.-E. 1980. Einführung. – *In* H.-E. Gruner (ed.): A. Kaestner's Lehrbuch der speziellen Zoologie, 4th ed., 1. Band, 1. Teil, pp 15-156. Gustav Fischer, Stuttgart.

Hay-Schmidt, A. 1987. The ultrastructure of the protonephridium of the actinotroch larva (Phoronida). – Acta zool. (Stockh.) **68**: 35-47.

Hay-Schmidt, A. 1989. The nervous system of the actinotroch larva of *Phoronis muelleri* (Phoronida). – Zoomorphology **108**: 333-351.

Hay-Schmidt, A. 1990a. Distribution of catecholamine-containing, serotonin-like and neuropeptide FMRFamide-like immunoreactive neurons and processes in the nervous system of the actinotroch larva of *Phoronis muelleri* (Phoronida). – Cell Tissue Res. **259**: 105-118.

Hay-Schmidt, A. 1990b. Catecholamine-containing, serotonin-like, and FMRFamide-like immuno-reactive neurons and processes in the nervous system of the early actinotroch larva of *Phoronis vancouverensis* (Phoronida): distribution and development. – Can. J. Zool. **68**: 1525-1536.

Herrmann, K. 1979. Larvalentwicklung und Metamorphose von *Phoronis psammophila* (Phoronida, Tentaculata). – Helgoländer wiss. Meeresunters. **32**: 550-581.

Herrmann, K. 1980. Die archimere Gliederung bei *Phoronis mülleri* (Tentaculata). – Zool. Jb., Anat. **103**: 234-249.

Herrmann, K. 1986. Die Ontogenese von *Phoronis mülleri* (Tentaculata) unter besonderer Berück-sichtigung der Mesodermdifferenzierung und Phylogenese des Coeloms. – Zool. Jb., Anat. **114**: 441-463.

Hyman, L.H. 1959. The lophophorate phyla – Phylum Phoronida. – The Invertebrates, vol. 5, pp 228-274. McGraw-Hill, New York.

Jägersten, G. 1972. Evolution of the Metazoan Life Cycle. – Academic Press, London.

Jeuniaux, C. 1982. Composition chimique comparée des formations squelettiques chez les Lopho-phoriens et les Endoproctes. – Bull. Soc. zool. Fr. **107**: 233-249.

Lacalli, T.C. 1990. Structure and organization of the nervous system in the actinotroch larva of *Phoronis vancouverensis*. – Phil. Trans. R. Soc. **327**: 655-685.

Marcus, E. du B.-R. 1949. *Phoronis ovalis* from Brazil. – Bolm Fac. Filos. Ciênc. Univ. S Paulo, Zool. **14**: 157-171.

Marsden, J.R. 1957. Regeneration in *Phoronis vancouverensis*. – J. Morph. **101**: 307-323.

Nielsen, C. 1987. Structure and function of metazoan ciliary bands and their phylogenetic signifi-cance. – Acta zool. (Stockh.) **68**: 205-262.

Pardos, F., C. Roldán, J. Benito & C.C. Emig 1991. Fine structure of the tentacles of *Phoronis australis* Haswell (Phoronida, Lophophorata). – Acta zool. (Stockh.) **72**: 81-90.

Siewing, R. 1973. Morphologische Untersuchungen zum Archicoelomatenproblem. 1. Die Körper-gliederung bei *Phoronis ijimai* Oka (Phoronidea). – Z. Morph. Tiere. **74**: 17-36.

Silén, L. 1954a. On the nervous system of *Phoronis*. – Ark. Zool., 2. ser., **6**(1): 1-40.

Silén, L. 1954b. Developmental biology of Phoronidea of the Gullmar Fiord area (West coast of Sweden). – Acta zool. (Stockh.) **35**: 215-257.

Silén, L. 1955. Autotomized tentacle crowns as propagative bodies in *Phoronis*. – Acta zool. (Stockh.) **36**: 159-165.

Storch, V. & K. Herrmann 1978. Podocytes in the blood vessel linings of *Phoronis muelleri* (Phoronida, Tentaculata). – Cell Tissue Res. **190**: 553-556.

Zimmer, R.L. 1964. Reproductive Biology and Development of Phoronida. – Ph. D. thesis, Univ. Washington.

Zimmer, R.L. 1967. The morphology and function of accessory reproductive glands in the lopho-phores of *Phoronis vancouverensis* and *Phoronopsis harmeri*. – J. Morph. **121**: 159-178.

Zimmer, R.L. 1978. The comparative structure of the preoral hood coelom in Phoronida and the fate of this cavity during and after metamorphosis. – *In* F.-S. Chia & M. E. Rice (eds): Settle-ment and Metamorphosis of Marine Invertebrate Larvae, pp 23-40. Elsevier, New York.

Zimmer, R.L. 1980. Mesoderm proliferation and formation of the protocoel and metacoel in early embryos of *Phoronis vancouverensis* (Phoronida). – Zool. Jb., Anat. **103**: 219-233.

Zimmer, R.L. 1991. Phoronida. – *In* A.C. Giese, J.S. Pearse & V.B. Pearse (eds): Reproduction of Marine Invertebrates, vol. 6, pp 1-45. Boxwood Press, Pacific Grove, CA.

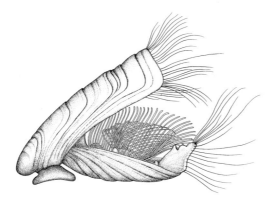

Phylum BRACHIOPODA

Brachiopods or lamp-shells are a highly characteristic group of benthic, marine organisms. The two shells, usually called dorsal and ventral (but see discussion below), make both living and extinct brachiopods immediately recognizable. About 300 living and over 12 000 fossil species have been recognized, with the fossil record going back to the Lower Cambrian. With such an extensive fossil record, which comprises many extinct major groups, the phylogeny is largely built on the fossils (see for example Williams & Rowell 1965). The living forms are usually arranged in two main groups, Inarticulata and Articulata, but this classification has been challenged (Nielsen 1991), and four main groups of living brachiopods, Lingulidae, Discinidae, Craniidae, and Articulata will be recognized here.

The anatomy of brachiopods is still in much need of study. The three 'inarticulate' groups are particularly poorly known, the only detailed studies being those of Blochmann (1892-1900) on *Lingula*, *Discinisca* and *Crania (Neocrania)*; all information on these groups go back to these two papers if nothing else is stated.

The body of the adult brachiopods has an upper and a lower mantle fold which secrete calcareous or chitinophosphatic valves. The mantle folds with the valves enclose the lophophore and the main body with gut, gonads and excretory organs. One valve, often called the dorsal valve, has the gut and the lophophore attached to it and will here be called the brachial valve, while the other valve which carries the stalk, or pedicle, or is cemented to the substratum will be called the pedicle valve. The lingulids have chitinophosphatic shells and are anchored in the bottom of a burrow by a stalk which protrudes between the posterior edges of the valves. The discinids also have chitinophosphatic shells, but they are attached to a hard substratum by a short stalk which protrudes through a slit in the pedicle valve. The craniids have calcareous valves and are cemented to a hard substratum by the pedicle valve. The articulates, which are by far the most diverse of the living groups, have calcareous shells and are attached by a stalk which protrudes through a hole in the pedicle valve; the umbo of the pedicle valve with the hole for the stalk is usually curved towards the

---

Chapter vignette: The articulate *Pumilus antiquatus*. (Redrawn from Atkins 1958.)

brachial valve so that the brachial valve is against the substratum (see the chapter vignette).

In the articulates the gut is incomplete with a blind-ending intestine at the ventral side of the stomach; the inarticulate groups have a complete gut, with the anus situated in the mantle cavity on the right side in *Lingula* and *Discinisca*, but in the posterior midline in *Crania*.

All epithelia are monolayered and the many types of ciliated cells are all monociliate, both in the ectoderm (Reed & Cloney 1977, Nielsen 1987) and in the endoderm (Storch & Welsch 1975). A ciliary stub has been observed on some of the peritoneal cells (Reed & Cloney 1977). Compound cilia have not been reported.

The lophophore has the shape of an almost closed horseshoe in the newly metamorphosed specimens; it is situated just behind the ventral side of the mouth and with the arms extending dorsally, almost meeting in the midline some distance anterior to the mouth. In the larger species it becomes coiled and wound into various complicated shapes. The cylindrical tentacles are arranged in a single row in the juveniles, but as additional tentacles become added at the tips of the two lophophore arms they become arranged alternatingly in two parallel rows, frontal (inner, adlabial) tentacles closest to the mouth/food groove and abfrontal (outer, ablabial) tentacles (Atkins 1961). A narrow upper lip (brachial lip, epistome) borders the anterior side of the mouth and follows the base of the tentacle row so that a furrow, the brachial groove, is formed between the tentacle bases and the lip as an extension of the lateral corners of the mouth.

The frontal and abfrontal tentacles show the same ciliary bands but with somewhat different positions on the two types of tentacle (Reed & Cloney 1977). The frontal surfaces have a narrow longitudinal band of cilia which beat towards the base of the tentacles where the bands unite with the ciliation of the food groove leading to the mouth. A row of laterofrontal cilia have been observed at the sides of the frontal band (Gilmour 1981, Nielsen 1987). A lateral ciliary band is found on each side of the tentacle; this band is situated on thickened epithelial ridges on the laterofrontal sides of the outer tentacles and on the lateroabfrontal sides of the inner tentacles so that the bands of adjacent tentacles bridge the gaps between the tentacles (Rudwick 1970). Other cells are mainly unciliated, but *Lingula* has cilia on most tentacle cells.

The lateral and frontal ciliary bands of the tentacles function as an upstream-collecting system in which the lateral cilia create the water current from which particles become deflected to the frontal side of the tentacle where the frontal cilia transport the particles towards the mouth (Strathmann 1973, Nielsen 1987).

The peritoneum of the tentacles and of the lophophoral arms comprises myoepithelial cells with smooth or striated myofilaments which control the various flexions of the tentacles and the more restricted movements of the arms, whereas the thick basement membrane with collagenous fibrils is mainly responsible for the straight, relaxed posture (Reed & Cloney 1977). The arms of the lophophore are supported by a quite complicated connective tissue consisting of a hyaline matrix with scattered cells (Hoverd 1985). A calcareous skeleton consisting of more or less fused spicules is secreted by cells of the connective tissue of the lophophores, and sometimes also of the mantle folds, of several terebratulid articulates (Schumann 1973); the histo-logy/cytology of these structures appears to be undescribed. The spicules are secreted as single crystals in special scleroblasts, and the spicules may later on fuse to form a stiff endoskeleton supporting the lophophore base and short, almost tubular 'joints' in the tentacles.

The structure of lophophore and tentacles is very similar to that of the phoro-nids, and the embryology of *Crania* (see below) demonstrates that the coelomic canal along the tentacle bases (the small brachial canal) with channels into the tentacles is the mesocoel. The other large coelomic cavity in the lophophore, the large brachial channel, has been interpreted as the protocoel (see for example Pross 1980), but its ontogenetic development is unknown and the apparent absence of a protocoel in the embryos of lingulids and articulates makes this interpretation entirely conjectural.

The main body cavity, the metacoel, is spacious and sends extensions into the mantle folds and into the stalk of lingulids and discinids. It is completely separated from the mesocoel in *Crania*, but the septum may be incomplete in other forms. *Cra-nia* has a small fourth coelomic cavity around the rectum.

The mantle folds secrete two valves. Their outer layer, periostracum, is mainly proteinaceous and contains β-chitin in lingulids and discinids; the reports of chitin in the articulate periostracum has been questioned and chitin appears to be absent also in *Crania* (Jeuniaux 1982). The periostracum is secreted by a narrow band of cells at the inner side of the mantle edge. The mineralized shell material is secreted by the outer surface of the mantle epithelium; it is calcium carbonate in articulates and crani-ids and fluoroapatite in lingulids and discinids (Lowenstam & Weiner 1989). Craniids and certain articulates have characteristic extensions of the mantle epithelium extend-ing into channels in the calcified shells. The extensions (unfortunately called caeca) are branched and do not reach the periostracum in craniids. The stouter extensions of the articulates have a distal 'brush border' of microvilli which each extend through a narrow canal in the shell to the periostracum in the young stage; in the later stages the microvilli retract and additional layers of periostracum separate the epithelial exten-sion from the outer periostracum (Williams 1973). The function of these structures is unknown. The mantle edges carry chitinous chaetae (setae), each formed by one ecto-dermal chaetoblast (Storch & Welsch 1972, Gustus & Cloney 1972).

The nervous systems are not well known; the only detailed study of the articulates is that of Bemmelen (1883), mainly based on *Gryphus*. The overall pattern of the ner-vous systems appears similar in the four groups, but one difference must be noted. In *Gryphus*, *Lingula* and *Discinisca*, the largest nervous concentration is a suboesophageal ganglion which sends nerves to the mantle folds, the adductor muscles and the stalk; in *Crania* the only nervous element in this position is a narrow transverse commissure con-necting a pair of elongate ganglia lateral to the anterior occlusor muscles. The far less conspicuous supraoesophageal ganglion is transversely elongate and continues laterally into the main nerve of the lophophore. Small nerves connect the two systems.

A haemal system surrounded by a basement membrane is found in all brach-iopods but is poorly known. A contractile vessel is found in the dorsal mesentery of articulates, and *Crania* is said to have several such vessels, whereas the haemal sys-tem should be poorly developed in *Discinisca*. At least some of the peripheral circulation appears to be through larger haemal spaces. Each tentacle has a small vessel in the shape of a fold of the frontal side of the peritoneum with an inner basement membrane and an outer layer of peritoneal cells with myofibrils (Storch & Welsch 1976, Reed & Cloney 1977).

A pair of large ciliated metanephridia (two pairs in certain articulates), which also function as gonoducts, is found in the metacoel. The gonads are formed from the peritoneum and extend into the mantle canals in most species. The ripe gametes are shed through the metanephridia.

The sperm is shed directly into the water and fertilization appears to be external. The eggs are spawned free in *Lingula* and *Crania*, but are retained in the lophophore in most articulates (Long & Stricker 1991).

Most stages of development have been studied in a number of articulates and in *Lingula* and *Crania*, while the embryology of discinids is unknown. The development of *Crania* (Nielsen 1991, see Fig. 45.1) appears to be the easiest to interpret and it will therefore be described first. The cleavage is total and radial and the following stages are a coeloblastula and an invagination gastrula. The anterodorsal part of the archenteron wall becomes the endodermal gut, while the posteroventral part

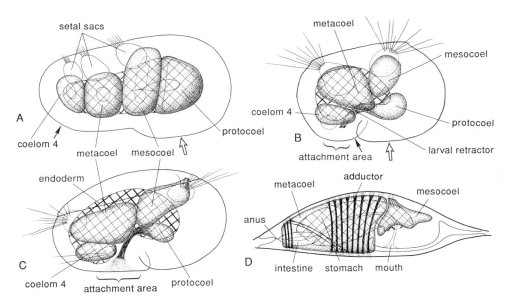

Fig. 45.1. Four stages of the development of *Crania (Neocrania) anomala*. – A, larva with short chaetae and fully formed coelomic sacs; the longitudinal muscles from the protocoel are not developed yet. – B, contracted, full-grown larva; the contracted muscles from the protocoel are conspicuous. – C, young bottom-stage; the first stage of the brachial valve is indicated above the two remaining bundles of chaetae; the attachment of the mesocoel to the valve can be seen. – D, juvenile with mouth, anus, the two shells with adductor muscles, and the short extensions of the mesocoel to the first three pairs of tentacles. The black arrows point at the position of the closed blastopore and the white arrows of the position of the adult mouth. (Modified from Nielsen 1991.)

becomes the mesoderm, which subsequently slides forwards laterally between ectoderm and endoderm. The mesoderm of each side divides into four parts, and each
part folds up so that four coelomic sacs are finally found on each side of the larva.
The blastopore remains as a posteroventral opening from the posteriormost part of
the archenteron, which becomes the fourth pair of mesodermal sacs when the blastopore finally closes. The full-grown larva has a pair of ectodermal thickenings with
bundles of chaetae at the dorsal side of each of the three posterior coelomic sacs (Fig.
45.2); the chaetae appear identical to those of the adults.

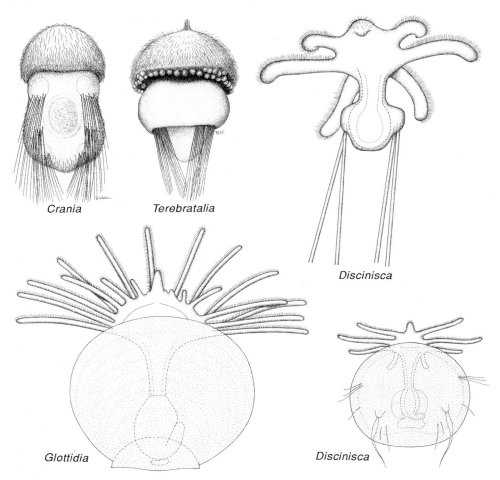

Fig. 45.2. Larval types of the four main types of brachiopods. – Craniacea: full-grown larvae of
Crania (Neocrania) anomala. – Articulata: full-grown larva of *Terebratalia transversa.* – Disci-
nacea: young and full grown-larvae of *Discinisca* sp. (not to scale, the distal parts of the larval
chaetae have been omitted). – Lingulacea: full-grown larva of *Glottidia* sp. (Modified from Nielsen
1991.)

At settling, the larva curls up through the contraction of a pair of muscles from the first pair of coelomic sacs to the posterior end of the larva just behind the area where the blastopore closed. A secretion from epithelial cells at the same area attaches the larva to the substratum, and the posterior pair of chaetae is usually shed. The brachial valve becomes secreted from a special area of the dorsal ectoderm in the region of the second and third pair of coelomic sacs; this area expands strongly at the periphery, and the whole organism soon becomes covered by the valve. The periphery of the attachment area also expands and a thin pedicle valve becomes secreted; the pedicle valve is thus secreted by the posterior end of the larva behind the area of the closed blastopore and can therefore not be described as ventral. The first pair of coelomic sacs (protocoel) apparently disappears, while the second (mesocoel) and third pairs (metacoel) develop into the lophophore coelom and the large body cavity, respectively. The fourth pair of coelomic sacs has not been followed through the metamorphosis. The adult mouth breaks through at the anteroventral side of the metamorphosed larva; the anus apparently develops from a proctodaeal invagination of the posterior ectoderm between the valves, i.e. behind the attachment area, and accordingly quite some distance from the closed blastopore.

The early development of the articulates follows the same general pattern, but the differentiation of the archenteron into gut and coelom is somewhat different in the articulates (see for example Conklin 1902, Percival 1944 and Long & Stricker 1991). The rather wide archenteron becomes divided by a U-shaped fold of its dorsal lining so that an antero-median gut becomes separated from a posterior and lateral coelomic cavity still connected to the exterior through the blastopore. This is apparently a more 'primitive' way of forming the coelom, but the process in *Crania* is actually not very different if one imagines the folding of the basement membrane of the archenteron when the mesoderm slides along the lateral sides of the endoderm. The coelomic cavity becomes divided into an anterior and a posterior pair of coelomic sacs and the blastopore closes. A ring-shaped thickening containing extensions of the posterior coelomic sacs and two pairs of chaetal bundles develop around the equator of the larva behind the closing blastopore. The full-grown larva is lecithotrophic and resembles the *Crania* larva, except for the annular thickening and the position of the two pairs of chaetal bundles (see for example Stricker & Reed 1985, Hoverd 1985, and Fig. 45.2). At settling, the ring-shaped fold and the chaetal bundles fold anteriorly and the larva attaches with the posterior pole, where the stalk develops; the reflexed folds begin to secrete the two valves (Stricker & Reed 1985). The tentacles develop from an area near the anterior part of the closed blastopore.

The early development of *Lingula* resembles that of *Crania* but the development of the mesoderm is difficult to make out; the free-swimming larvae have a pair of shells and a horseshoe of ciliated tentacles used both in swimming and feeding (Yatsu 1902; Fig. 45.2). The youngest known larvae of *Discinisca* lack shells but have tentacles like those of the *Lingula* larva and four groups of very long chaetae, but the shells develop at a later stage and the long chaetae are shed (Chuang 1977; Fig. 45.2).

Long (1964) separated the blastomeres of 2-cell stages of *Hemithyris* and observed that several of the isolated blastomeres developed into small but normal blastulae; a few blastulae developed further into apparently normal gastrulae with mesoderm. Unpublished experiments with marking of one blastomere of 2-cell stages by Dr Gary Freeman (University of Texas at Austin) have shown that the first cleavage is median.

The cleavage pattern resembles that of most other deuterostomes with small eggs, and the formation of the mesoderm from part of the archenteron through a more or less modified enterocoely is one of the important deuterostome characteristics. The fate of the blastopore is somewhat difficult to make out in the articulates and *Lingula*, where it appears to close from behind, and the adult mouth subsequently develops from the area of its anterior end. However, the blastopore of *Crania* is definitely situated in the posterior end of the embryo while the adult mouth breaks through at the anterior end. This is the blastoporal fate characteristic of the deuterostomes, and the embryology of the other brachiopods has been interpreted as modified through a precocious curving of the embryo bringing the blastoporal and oral areas so close to each other that a distinction becomes difficult (Nielsen 1991).

The number of the coelomic sacs appears confusing at first, but the differentiation of the second pair of sacs in *Crania* demonstrates that this sac represents the mesocoel. This interpretation is cemented by the many detailed similarities between tentacle structure and function in brachiopods, phoronids and pterobranchs: the ciliary bands of single cilia on monociliate cells functioning as upstream-collecting systems, the laterofrontal sensory cells, the myoepithelial peritoneal cells, and the frontal blood vessel. Also the shape and general position of the lophophore with the ciliary band passing just behind the mouth are identical in the three groups. The existence of a protocoel in adult brachiopods is uncertain, but the metacoel, which forms the large body cavity with gonads and metanephridia, is of identical structure in brachiopods and phoronids.

An apical organ is clearly seen in articulate larvae but is apparently lacking in *Crania* and has not been identified with certainty in *Lingula* and *Discinisca*. There is no indication that an apical organ becomes incorporated in the adult nervous system; the dorsal components of the nervous system are poorly developed, but this deviation from the usual deuterostome pattern with a dorsal nervous centre is probably associated with the sessile habits.

The valves have a structure resembling that of mollusc valves, but the general structure of all calcified exoskeletons is rather similar (Lowenstam & Weiner 1989), so this may just show that a calcareous exoskeleton in an organic matrix can easily be secreted by any ectodermal epithelium. The endoskeleton of the terebratulids has no counterparts in protostomes or cnidarians, but has a stereomic structure resembling that of the endoskeleton of the echinoderms.

The chaetae appear identical to those described from annelids (Chapter 17) and some ectoprocts have a gizzard with denticles of a very similar ultrastructure formed from one cell with microvilli (Chapter 24). This has been interpreted as a synapomorphy of annelids and brachiopods, but the two phyla are so different in most morphological and embryological respects that a sister-group relationship must be considered highly unlikely (see also Gustus & Cloney 1972 and Orrhage 1973). Some of the phylogenies based on DNA sequencing place the brachiopods and annelids as sister groups (within the molluscs) (Field *et al.* 1988), and one wonders whether this could be the result of this convergence and the similarity in the shell structure.

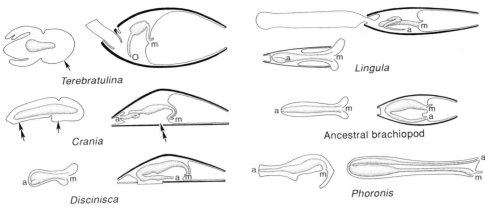

Fig. 45.3. Larvae and adults of the four main types of brachiopods, a hypothetical brachiopod ancestor, and a phoronid. The guts are shaded with an 'm' at the mouth and an 'a' at the anus (the circle in *Terebratulina* indicates the end of the intestine); in *Terebratulina* and *Crania* the single arrows indicate the position of the stomodaeum, and the double arrows indicate the position of the closed blastopore. (From Nielsen 1991.)

The deuterostome nature of the brachiopods can hardly be questioned. The apparent lack of a protocoel in the adults can be interpreted as a specialization in a direction different from that of the pterobranchs, echinoderms and enteropneusts which have the protocoel associated with blood vessels forming a complicated excretory organ, the axial organ or glomerulus; the axial organ is here interpreted as a synapomorphy of these three phyla (see Chapter 46). The question is then whether the phoronids and brachiopods, which show no sign of an axial organ, are sister groups. The ontogeny of the phoronids shows that the U-shaped gut becomes curved by an enormous elongation of the ventral side, while the ontogeny of *Crania* shows the curvature to be the result of an elongation of the dorsal side (Fig. 45.3). It appears natural to interpret these two life cycles as derived independently from the life cycle of a common ancestor having a straight gut like that of notoneuron (Nielsen 1991; Fig. 45.3) and having a lophophore with ciliated tentacles. This results in a phylogeny with a trichotomy (see Fig. 43.8), but this can only be resolved when more information becomes available.

Interesting subjects for future research

1. Embryology of discinids.
2. Development and structure of nervous systems.
3. Origin of the large arm sinus.

References

Atkins, D. 1958. A new species and genus of Kraussinidae (Brachiopoda) with a note on feeding. – Proc. zool. Soc. Lond. 131: 559-581, 1 pl.

Atkins, D. 1961. A note on the growth stages and structure of the adult lophophore of the brachio-
    pod *Terebratella (Waltonia) inconspicua* (G.B. Sowerby). – Proc. zool. Soc. Lond. **136**: 255-271.
Bemmelen, J.F. van 1883. Untersuchungen über den anatomischen und histologischen Bau der Bra-
    chiopoda Testicardinia. – Jena. Z. Naturw., N.F. **16**: 88-161, pls 5-9.
Blochmann, F. 1892-1900. Untersuchungen über den Bau der Brachiopoden I-II. Text 124 pp, atlas
    18 pls. – Gustav Fischer, Jena.
Chuang, S.-H. 1977. Larval development in *Discinisca* (Inarticulate brachiopod). – Am. Zool. **17**:
    39-53.
Conklin, E.G. 1902. The embryology of a brachiopod, *Terebratulina septentrionalis* Couthouy. –
    Proc. Am. phil. Soc. **41**: 41-76, 10 pls.
Field, C.G., G.J. Olsen, D.J. Lane, S.J. Giovanni, M.T. Ghiselin, E.C. Raff, N.R. Pace & R.A. Raff
    1988. Molecular phylogeny of the animal kingdom. – Science **239**: 748-753.
Gilmour, T.H.J. 1981. Food-collecting and waste-rejecting mechanisms in *Glottidia pyramidata*
    and the persistence of lingulacean inarticulate brachiopods in the fossil record. – Can. J. Zool.
    59: 1539-1547.
Gustus, R.M. & R.A. Cloney 1972. Ultrastructural similarities between setae of brachiopods and
    polychaetes. – Acta zool. (Stockh.) **53**: 229-233.
Hoverd, W.A. 1985. Histological and ultrastructural observations of the lophophore and larvae of
    the brachiopod, *Notosaria nigricans* (Sowerby 1846). – J. nat. Hist. **19**: 831-850.
Jeuniaux, C. 1982. Composition chimique comparée des formations squelettiques chez les Lopho-
    phoriens et les Endoproctes. – Bull. Soc. zool. Fr. **107**: 233-249.
Long, J.A. 1964. The embryology of three species representing three superfamilies of articulate
    brachiopods. – Ph.D. thesis, Univ. Washington.
Long, J.A. & S.A. Stricker 1991. Brachiopoda. – *In* A.C. Giese, J.S. Pearse & V.B. Pearse (eds): Re-
    production of Marine Invertebrates, vol. 6, pp 47-84. Blackwell Scientific Publs., Boston/ Box-
    wood Press, Pacific Grove.
Lowenstam, H.A. & S. Weiner 1989. On Biomineralization. – Oxford Univ. Press, New York.
Nielsen, C. 1987. Structure and function of metazoan ciliary bands and their phylogenetic signifi-
    cance. – Acta zool. (Stockh.) **68**: 205-262.
Nielsen, C. 1991. The development of the brachiopod *Crania (Neocrania) anomala* (O.F. Müller)
    and its phylogenetic significance. – Acta zool. (Stockh.) **72**: 7-28.
Orrhage, L. 1973. Light and electron microscope studies of some brachiopod and pogonophoran
    setae. – Z. Morph. Tiere **74**: 253-270.
Percival, E. 1944. A contribution to the life-history of the brachiopod, *Terebratella inconspicua*
    Sowerby. – Trans. R. Soc. N.Z. **74**: 1-23, pls 1-7.
Pross, A. 1980. Untersuchungen zur Gliederung von *Lingula anatina* (Brachiopoda). – Archimerie
    bei Brachiopoden. – Zool. Jb., Anat. **103**: 250-263.
Reed, C.G. & R.A. Cloney 1977. Brachiopod tentacles: ultrastructure and functional significance of
    the connective tissue and myoepithelial cells in *Terebratalia*. – Cell Tissue Res. **185**: 17-42.
Rudwick, M.J.S. 1970. Living and Fossil Brachiopods. – Hutchinson, London.
Schumann, D. 1973. Mesodermale endoskelette terebratulider Brachiopoden. I. – Paläont. Z. **47**:
    77-103.
Storch, V. & U. Welsch 1972. Über Bau und Entstehung der Mantelstacheln von *Lingula unguis* L.
    (Brachiopoda). – Z. wiss. Zool. **183**: 181-189.
Storch, V. & U. Welsch 1975. Elektronenmikroskopische und enzymhistochemische Untersuch-
    ungen über die Mitteldarmdrüse von *Lingula unguis* L. (Brachiopoda). – Zool. Jb., Anat. **94**:
    441-452.
Storch, V. & U. Welsch 1976. Elektronenmikroskopische und enzymhistochemische Untersuch-
    ungen über Lophophor und Tentakeln von *Lingula unguis* L. (Brachiopoda). – Zool. Jb., Anat.
    96: 225-237.
Strathmann, R. 1973. Function of lateral cilia in suspension feeding of lophophorates (Brachio-
    poda, Phoronida, Ectoprocta). – Mar. Biol. (Berl.) **23**: 129-136.
Stricker, S.A. & C.G. Reed 1985. The ontogeny of shell secretion in *Terebratalia transversa*
    (Brachiopoda, Articulata) I. Development of the mantle. – J. Morph. **183**: 233-250.

Williams, A. 1973. The secretion and structural evolution of the shell of thecideidine brachiopods. – Phil. Trans. R. Soc. B **264**: 439-478, pls 40-53.

Williams, A. & A.J. Rowell 1965. Evolution and phylogeny. – *In* R.C. Moore (ed.): Treatise on Invertebrate Paleontology, part H, vol. 1, pp 164-214. Geol. Soc. Am., Lawrence, Kansas.

Yatsu, N. 1902. On the development of *Lingula anatina*. – J. Coll. Sci. imp. Univ. Tokyo **17**(4): 1-112, 8 pls.

NEORENALIA

Pterobranchs, echinoderms and enteropneusts share many features, the most characteristic organ probably being their excretory organ, the axial complex (Fig. 46.1). The ontogenetic origin and the structure/function of this organ in the three groups are so similar that their homology can hardly be questioned, and I have chosen to erect a new supraphyletic group for these three phyla (and the sister group of the enteropneusts, see below) and to name it after this new kidney.

The axial complex is situated in the prosome and comprises a heart consisting of a blood vessel (without an endothelium), partially surrounded by a pericardium, and a specialized area with podocytes along the efferent branch. Primary urine is believed to be formed through ultrafiltration from the blood vessel through the basement membrane covered with podocytes to the protocoel, which functions as a nephridial capsule. The urine flows from the protocoel through the short coelomoduct to a median coelomopore/nephridiopore. The prosome cannot be recognized in the adult echinoderms, but the ontogenetic origin of the respective components of their axial organ clearly demonstrates its homology with those of the other two phyla. The pericardium can be recognized as a pulsatile vesicle already in the larvae of echinoderms and enteropneusts (Ruppert & Balser 1986); the pterobranch larvae have not been studied. The origin of the vesicle is not clear in all cases, but development from the protocoel has been indicated in some cases (Chapters 47, 48, 50).

Fig. 46.1. Morphology of the anterior coelomic cavities in larvae and adults of non-chordate neorenalians with emphasis on the transformations of parts of the larval coelomic sacs into the adult's axial complex. – Pterobranchs: adult *Cephalodiscus gracilis* based on Lester (1985) and Dilly, Welsch & Rehkämper (1986). – Echinoderms: larva of *Asterias forbesi* and adult of *Asterias* sp. based on Nichols (1962) and Ruppert & Balser (1986). – Enteropneusts: larva of *Schizocardium brasiliense* and adult of *Saccoglossus kowalevskii* based on Balser & Ruppert (1990). – The larvae are drawn on the basis of photographs in Ruppert & Balser (1986), while the adults are diagrammatic sections with some of the structures shifted slightly to get the necessary details into the plane of the drawing. The lines with dots indicate a coelomic layer consisting of podocytes. The small arrows indicate the direction of the presumed or proven ultrafiltration of primary urine.

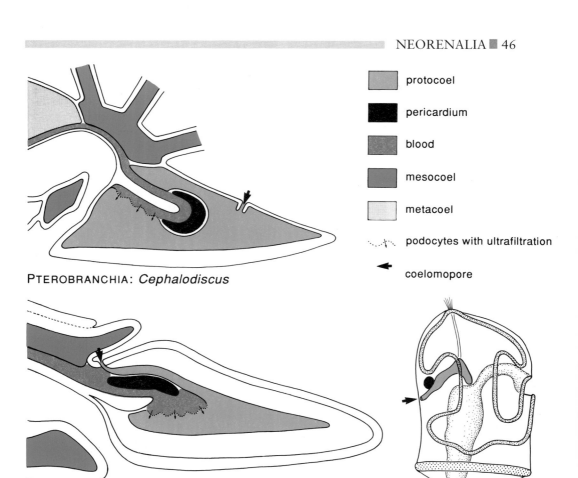

protocoel

pericardium

blood

mesocoel

metacoel

podocytes with ultrafiltration

coelomopore

PTEROBRANCHIA: *Cephalodiscus*

ENTEROPNEUSTA: *Saccoglossus*

Schizocardium

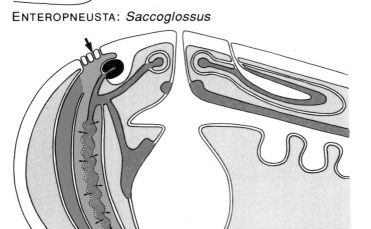

ECHINODERMATA: *Asterias*

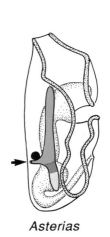

Asterias

357

This new excretory organ has apparently made the excretory function of the metanephridia in the metasome of the ancestors redundant, and such organs are not known from any of the neorenalian phyla (the vertebrate capsule nephridia are a vertebrate apomorphy, see Chapter 54). The metanephridia of phoronids and brachiopods have the additional function of gonoducts, the gametes being liberated from the gonads into the metacoels, but the neorenalians all have gonads with separate gonoducts and lack coelomoducts from the metacoels.

The chordates lack the axial organ but are included in the new group because they are interpreted as the sister group of the enteropneusts (see Chapters 49 and 51).

References

Balser, E.J. & E.E. Ruppert 1990. Structure, ultrastructure, and function of the preoral heart-kidney in *Saccoglossus kowalevskii* (Hemichordata, Enteropneusta) including new data on the stomochord. – Acta zool. (Stockh.) 71: 235-249.

Dilly, P.N., U. Welsch & G. Rehkämper 1986. Fine structure of heart, pericardium and glomerular vessel in *Cephalodiscus gracilis* M'Intosh, 1882 (Pterobranchia, Hemichordata). – Acta zool. (Stockh.) 67: 173-179.

Lester, S.M. 1985. *Cephalodiscus* sp. (Hemichordata: Pterobranchia): observations of functional morphology, behavior and occurrence in shallow water around Bermuda. – Mar. Biol. (Berl.) 85: 263-268.

Nichols, D. 1962. Echinoderms. – Hutchinson University Library, London.

Ruppert, E.E. & E.J. Balser 1986. Nephridia in the larvae of hemichordates and echinoderms. – Biol. Bull. Woods Hole 171: 188-196.

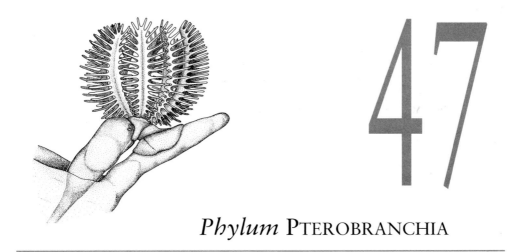

Phylum PTEROBRANCHIA

Pterobranchs form one of the smallest animal phyla, comprising only two or three genera of marine, benthic organisms: *Rhabdopleura* with four species forming small adnate colonies; *Cephalodiscus* with 15-20 species forming sometimes quite extensive aggregations of tubes, called coenecia, housing solitary individuals with lively budding; and *Atubaria*, which has only been recorded once, resembles *Cephalodiscus*, but is supposedly not tube-building. The fossil record of *Rhabdopleura*-like forms goes back to the Middle Cambrian (Bengtsson & Urbanek 1986), and also the highly diverse fossil group Graptolithina (Cambrian-Carboniferous) is now believed to be closely related to the rhabdopleurids (Armstrong, Dilly & Urbanek 1984).

The individual zooids are rather similar with an archimeric body (Chapter 43) consisting of a preoral shield (prosome), used in creeping and in secreting the tubes, a short perioral collar (mesosome), carrying ciliated tentacles, and an elongated globular body (metasome), posteriorly extended into a narrow tail or stolon from which the budding takes place. The gut is U-shaped with a short oesophagus, a globular stomach and a narrow rectum passing dorsally from the posterior side of the stomach to the anus, which is situated a short distance behind the lophophore. Many ectodermal, endodermal and peritoneal cells are ciliated, and there is always only one cilium per cell (Dilly, Welsch & Rehkämper 1986a, 1986b, Nielsen 1987); this is also the case with the cells of the intraepithelial ganglion (Rehkämper, Welsch & Dilly 1987).

The prosome is a flat shield with a rather narrow neck. It is used as a creeping sole and its thick, ventral epithelium is ciliated with many mucus cells. There is a pigmented transverse stripe without cilia on the ventral side, and the prosome can be folded along this zone when material secreted for tube-building is being added to the edge of the tube (Dilly 1988). The zooids can move around in the tubes and *Cephalodiscus* may even leave the coenecium and start to build a new one if conditions become too adverse (Lester 1985). The unpaired protocoel is lined by a monolayered peritoneum and opens to the exterior through a pair of dorsal, ciliated ducts. Middorsally, the protocoel is filled by the heart, which is an anterior extension of a

Chapter vignette: *Cephalodiscus gracilis* in feeding position. (Redrawn from Lester 1985.)

median, U-shaped vessel, in the usual position between basement membranes, surrounded by a pericardial sac; the dorsal vessel probably carries blood to the heart and the ventral vessel leads the blood posteriorly. The pericardial sac consists of a monolayered myoepithelium, which develops from a cluster of mesenchymal cells just behind the protocoel in *Rhabdopleura* (Lester 1988b), but which may be interpreted as an isolated pocket of protocoelomic peritoneum. The ventral part of the ventral vessel, called glomerulus, has convoluted ventrolateral walls with podocytes and is believed to be a site for ultrafiltration of primary urine (Dilly, Welsch & Rehkämper 1986b; Fig. 46.1). The protocoel thus functions like a Bowman's capsule in the vertebrate kidney. It could be expected that the primary urine in the protocoel should be modified during the passage through the coelomoducts, but this has not been investigated.

The mesosome is quite short, forming a collar surrounding the mouth and the foregut and carrying one to nine pairs of dorsal tentacles. The tentacles are feather-shaped with a row of pinnules on each side. In feeding specimens of *Cephalodiscus*, the tentacles are held in a curved position so that an almost spherical shape is formed (Lester 1985; see the chapter vignette). The pinnules and tentacles are ciliated, with a double row of ciliated cells on each side of the pinnules (Dilly, Welsch & Rehkämper 1986c, Nielsen 1987); Gilmour (1979) reported a row of laterofrontal, probably sensory, cells along the frontal side of the double row of cilia. The lateral cilia form an upstream-collecting system which pumps water into the sphere and out through a distal opening between the tentacle tips (Lester 1985). Particles strained from the water are passed to the mouth along the frontal side of the pinnules/tentacles, i.e. the side of the tentacles at the outside of the sphere (Lester 1985). The paired mesocoelic cavities extend into the tentacles and are surrounded by a monolayered peritoneum. Dilly (1972) observed a blood sinus in the basement membrane along the frontal side of the tentacles in *Cephalodiscus*, while Gilmour (1979) described a small vessel in each tentacle in the shape of a longitudinal fold of the frontal part of the peritoneum in *Rhabdopleura*; it is possible that the peritoneum and basement membrane are only folded into the shape of a separate vessel in the proximal part of the tentacles or that differences exist between species.

There is a pair of dorsal, ciliated coelomoducts which open posterolaterally. The walls of the coelomoducts contain groups of cells with cross-striated myofilaments, and the function of these muscles appear to be an opening of the duct, perhaps in connection with rapid retractions of the zooids (Dilly, Welsch & Rehkämper 1986c). The main nervous concentration is situated at the dorsal side of the mesosome (see below). *Cephalodiscus* and *Atubaria* have a pair of lateral gill pores from the pharynx to the anterior part of the metasome, just behind the mesosomal coelomopores; the ciliated coelomic canals continue posteriorly in shallow furrows (Schepotieff 1907c, Dilly, Welsch & Rehkämper 1986c). Gilmour (1979) was of the opinion that the gill pores developed to allow the escape of excess water from the filter-feeding process entering the oesophagus, but direct observations are lacking.

A dorsal extension from the pharynx, the stomochord, runs anteriorly between the pharynx and the peritoneum of the protocoel. It is compact or has a central cavity, consists of vacuolated cells and is surrounded by a thickened extracellular sheath

(Schepotieff 1907a,c, Balser & Ruppert 1990). Both during metamorphosis and budding, the stomochord develops as a specialization of the anterior part of the endodermal pharynx (see below). The function of this structure is uncertain, but it appears to support the neck region between prosome and metasome, and perhaps also to support the muscular heart (Balser & Ruppert 1990).

The large metasome contains the major part of the gut, which is suspended in mesenteria formed by the median walls of the paired metacoelomic sacs. One or two gonads are situated at the dorsal side of the metacoel, covered by the peritoneum. The two metacoelomic sacs have no connection to the exterior, i.e. no metanephridia, and the gametes are shed directly through short canals separate from the coelom. The sexes of the individual zooids are usually separate, but some species of *Cephalodiscus* have one gonad of each sex. The metacoelomic sacs extend into the stalk region, where the septum between the two cavities is lacking in *Cephalodiscus*; the blood vessels and the nerves have the same arrangement in the stalks of the two genera (Schepotieff 1907a,c).

The posterior end of the stalk has somewhat different structure in the two main genera. In *Cephalodiscus*, the tip has a small attachment organ, and the budding takes place from this area (see below). In *Rhabdopleura*, the stalk ends in a genuine branched stolon from which new buds arise. The stolon is a narrow string of vacuolated tissue surrounded by a black tube. The zooids bud off from the growing tips of the stolon and form stalk and main body, which become contained in larger chambers which are partially erect. The stalk region appears to be homologous with the stalk of *Cephalodiscus*. The stalks are highly contractile and may retract the zooids to the bottom of their chambers (Stebbing & Dilly 1972, Lester 1985).

The central nervous system is located at the dorsal side of the mesosome between the tentacle crown and the anus. It is situated basally in an oval area of thickened epithelium (Rehkämper, Welsch & Dilly 1987). The peripheral nervous system comprises nerves to the tentacles, a median nerve to the prosome, and a pair of connectives around the oesophagus to a midventral nerve which continues along the ventral side to the stalk in *Rhabdopleura* (Schepotieff 1907a) and along the ventral side and back along the dorsal side of the whole stalk in *Cephalodiscus* (Schepotieff 1907c).

The haemal system (Schepotieff 1907a,c) comprises the above-mentioned vessels associated with the heart and the tentacle vessels, but important parts of the system are more lacunar than vessel-like, especially around the gut. A sinus extends from the median glomerulus along the dorsal wall of the preoral shield to the pharynx, which is surrounded by a pair of vessels uniting again behind the pharynx into a mid-ventral vessel; this vessel extends all the way to the tip of the stalk where it curves around and follows the dorsal side of the body almost to the tentacle area.

The coenecium of *Cephalodiscus* consists of individual tubes held together in a mass of interwoven fibres so that characteristic, species-specific structures are formed. *Rhabdopleura* has naked branching tubes extending from a hemispherical ancestral chamber. The tubes are made up of a double row of alternating U-shaped pieces which are secreted by special glandular areas of the cephalic shield. The tube-material contains keratin and collagen (Armstrong, Dilly & Urbanek 1984). Chitin has not been found (Jeuniaux 1982).

361

The budding of *Cephalodiscus* has been studied by Masterman (1898), Harmer (1905) and Schepotieff (1908). The buds develop from the small attachment plate at the tip of the stalk where the first stage is an outgrowth from the stalk consisting of ectoderm and peritoneum of the paired colomic cavities of the stalk; its tip becomes the very large oral shield, and the gut develops from an ectodermal invagination starting orally. At a later stage, the anus breaks through at the dorsal side and the bud attains the proportions of the adult. The metacoels of the stalk pinch off the pro- tocoel, and possibly the pericardium, in the oral shield and the mesocoels in the col- lar with the tentacles. The stomochord is formed as a small, anteriorly directed diver- ticulum from the epithelium of the gut. The buds remain attached to the parent until an advanced stage, and large clusters of buds of varying ages are often found.

The buds of *Rhabdopleura* develop from the tips or along the sides of the branching stolons (Schepotieff 1907b, see above). The early buds are simple evagina- tions of the stolon consisting of ectoderm and extensions of the paired coelomic com- partments of the stolon. The origin of the gut is in need of further investigation, but the coelomic compartments develop much like those of *Cephalodiscus*. A special type of buds is dormant with a thick cuticle (Stebbing 1970, Dilly 1975).

The gonads are mesodermally derived and have a pair of separate ducts formed from ectodermal invaginations (Masterman 1898). Sperm and eggs are shed through the narrow gonoducts, not through the metacoels via the metanephridia as in phoro- nids and brachiopods. The sperm of *Rhabdopleura* is of the specialized type with a spindle-shaped head and a most unusual mitochondrial filament (Lester 1988a). Fer- tilization has not been observed; Andersson (1907) observed sperm within the ovaries of a *Cephalodiscus*, which is indicative of internal fertilization, but it is not known if this is the case for other species.

All species appear to deposit the fertilized, yolk-rich eggs in the tubes, usually behind the zooids, where the early part of the development takes place (Harmer 1905, John 1932, Lester 1988a).

Early studies of the development of *Cephalodiscus* (Andersson 1907, Schepo- tieff 1909, John 1932) were rather incomplete but showed gastrulation by invagina- tion and formation of mesoderm and coelomic pouches, which can be interpreted in accordance with the following description of *Rhabdopleura*. Schepotieff (1909) also observed a number of quite different larvae which, as pointed out by Hyman (1959), are so similar to ectoproct larvae that it must be taken for granted that a confusion had taken place.

Rhabdopleura (Dilly 1973, Lester 1988a) shows total, equal cleavage which leads to the formation of a spherical, completely ciliated larva; later stages are elon- gate and show a shallow, anteroventral concavity. The spherical stage consists of a layer of ectodermal cells around a mass of endodermal cells with much yolk; a nar- row cavity at one side has a thin layer of cells interpreted as mesoderm covering the outer, ectodermal side. The elongate larvae have a thin layer of mesoderm covering the inside of the whole ectodermal sheet, and the cavity is now enlarged and situated at the anterodorsal side of the endoderm. The mesoderm is monolayered except in two pairs of lateral areas where flat coelomic sacs are found; these sacs are interpreted as mesocoel and metacoel (Fig. 47.1).

362

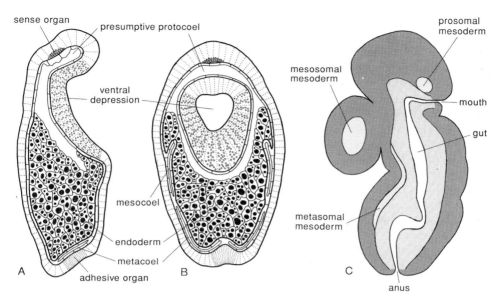

Fig. 47.1. Development of pterobranchs. – A-B, median and horizontal sections of swimming larvae of *Rhabdopleura normani*; the mesocoelomic and metacoelomic cavities are well defined, but the extension of the protocoel is uncertain, because parts of the anterior cavity between mesoderm and endoderm may represent fixation artifacts (modified from Lester 1988). – C, almost median section of a juvenile of *Cephalodiscus nigrescens* in the stage with a complete, straight gut (based on John 1932).

Settlement and metamorphosis of *Rhabdopleura* have been studied by Lester (1988b). She observed that the larvae started to test the substratum creeping on the ventral side after a short period of swimming; the larva then settled with the ventral depression in close contact with the substratum and secreted a thin surrounding cap-sule. The metamorphosis involves the development of the adult body regions with the tentacles and the early stages of the pinnules; the endoderm is at first a compact mass of cells, but a lumen develops first in the region which becomes the intestine and later in the stomach, while the pharynx originates as an invagination from the ectoderm. The pericardial vesicle forms from a small mass of mesodermal cells be-tween the dorsal epidermis and the pharynx. The cocoon breaks open after a few days and the first zooid of the colony, the ancestrula, starts feeding shortly after.

Pterobranchs and enteropneusts are usually treated together as the group Hemi-chordata, but a scrutiny of the characters purported to characterize the group reveals mostly characters shared with other deuterostome phyla, and the characters must be interpreted as plesiomorphies; the only possible synapomorphy which could support the idea of a sister-group relationship is the stomochord. This enigmatic structure has been discussed in great detail in the literature because it has been interpreted as a homologue of the chorda of the chordates. Balser & Ruppert (1990) summarized the discussions and concluded that stomochord and urochord/notochord have very similar structure and possibly also function; the differences between the two structures, especially with respect to degree of isolation from the endoderm and position relative to the gut, could perhaps be interpreted as two steps in the evolution of a supporting rod. On the other hand, the stomochord is situated at the anterior end of the gut and is not associated with any nervous concentration or tube, and it could instead be interesting to look for similar structures associated with the axial complex of echinoderms.

The archimeric body-plan is shared with most other deuterostomes, and the structure, function and development of the mesocoel and lophophore show that these structures must be homologous in phoronids, brachiopods, pterobranchs, and echinoderms. I have earlier (Nielsen 1985, 1987) united these four phyla in the group Brachiata, but in light of new evidence about the structure and function of the axial complex in several phyla, I have now come to the conclusion that the Brachiata must be paraphyletic (see Fig. 43.8). Phoronids and brachiopods have no axial complex, and there is no sign of that they have ever had one. Pterobranchs, echinoderms, and enteropneusts have the axial complex, and constitute a monophyletic taxon, together with the chordates, which I have here given the name Neorenalia (Chapter 46). The presence of the characteristic U-shaped gill-slits unite the enteropneusts and the chordates in the monophyletic group Cyrtotreta (Fig. 43.8), and this leaves pterobranchs and echinoderms as possible sister groups. However, I have not been able to identify a character which could be interpreted as a synapomorphy of the two phyla, and the phylogenetic tree is therefore left with a trichotomy (see further discussion in Chapter 46).

Interesting subjects for future research

1. Development of *Rhabdopleura* and *Cephalodiscus*.
2. Structure and biology of *Atubaria*.

References

Andersson, K.A. 1907. Die Pterobranchier der schwedischen Südpolarexpedition. – Wiss. Ergebn. schwed. Südpolarexped., vol. 5, no. 10, pp. 1-122, pls 1-8.

Armstrong, W.G, P.N. Dilly & A. Urbanek 1984. Collagen in the pterobranch coenecium and the problem of graptolite affinities. – Lethaia 17: 145-152.

Balser, E.J. & E.E. Ruppert 1990. Structure, ultrastructure, and function of the preoral heart-kidney in *Saccoglossus kowalevskii* (Hemichordata, Enteropneusta) including new data on the stomochord. – Acta zool. (Stockh.) 71: 235-249.

Bengtsson, S. & A. Urbanek 1986. *Rhabdotubus*, a Middle Cambrian rhabdopleurid hemichordate. – Lethaia 19: 293-308.

Dilly, P.N. 1972. The structures of the tentacles of *Rhabdopleura compacta* (Hemichordata) with special reference to neurociliary control. – Z. Zellforsch. 129: 20-39.

Dilly, P.N. 1973. The larva of *Rhabdopleura compacta* (Hemichordata). – Mar. Biol. (Berl.) 18: 69-86.

Dilly, P.N. 1975. The dormant buds of *Rhabdopleura compacta* (Hemichordata). – Cell Tissue Res. 159: 387-397.

Dilly, P.N. 1988. Tube building by *Cephalodiscus gracilis*. – J. Zool. (Lond.) 216: 465-468.

Dilly, P.N., U. Welsch & G. Rehkämper 1986a. On the fine structure of the alimentary tract of *Cephalodiscus gracilis* (Pterobranchia, Hemichordata). – Acta zool. (Stockh.) 67: 87-95.

Dilly, P.N., U. Welsch & G. Rehkämper 1986b. Fine structure of heart, pericardium and glomerular vessel in *Cephalodiscus gracilis* M'Intosh, 1882 (Pterobranchia, Hemichordata). – Acta zool. (Stockh.) **67**: 173-179.

Dilly, P.N., U. Welsch & G. Rehkämper 1986c. Fine structure of tentacles, arms and associated coelomic structures of *Cephalodiscus gracilis* (Pterobranchia, Hemichordata). – Acta zool. (Stockh.) **67**: 181-191.

Gilmour, T.H.J. 1979. Feeding in pterobranch hemichordates and the evolution of gill slits. – Can. J. Zool. **57**: 1136-1142.

Harmer, S.F. 1905. The Pterobranchia of the Siboga-Expedition. – Siboga Exped. **26**bis: 1-132, pls 1-14.

Hyman, L.H. 1959. Class Pterobranchia. – The Invertebrates, vol. 5, pp. 155-191. McGraw-Hill, New York.

Jeuniaux, C. 1982. La chitine dans le règne animal. – Bull. Soc. zool. Fr. **107**: 363-386.

John, C.C. 1932. On the development of *Cephalodiscus*. – 'Discovery' Rep. **6**: 191-204, pls 43-44.

Lester, S.M. 1985. *Cephalodiscus* sp. (Hemichordata: Pterobranchia): observations of functional morphology, behavior and occurrence in shallow water around Bermuda. – Mar. Biol. (Berl.) **85**: 263-268.

Lester, S.M. 1988a. Ultrastructure of adult gonads and development and structure of the larva of *Rhabdopleura normani* (Hemichordata: Pterobranchia). – Acta zool. (Stockh.) **69**: 95-109.

Lester, S.M. 1988b. Settlement and metamorphosis of *Rhabdopleura normani* (Hemichordata: Pterobranchia). – Acta zool. (Stockh.) **69**: 111-120.

Masterman, A.T. 1898. On the further anatomy and the budding process of *Cephalodiscus dodecalophus* (M'Intosh). – Trans. R. Soc. Edinb. **39**: 507-527, 5 pls.

Nielsen, C. 1985. Animal phylogeny in the light of the trochaea theory. – Biol. J. Linn. Soc. **25**: 243-299.

Nielsen, C. 1987. Structure and function of metazoan ciliary bands and their phylogenetic significance. – Acta zool. (Stockh.) **68**: 205-262.

Rehkämper, G., U. Welsch & P.N. Dilly 1987. Fine structure of the ganglion of *Cephalodiscus gracilis* (Pterobranchia, Hemichordata). – J. comp. Neurol. **259**: 308-315.

Schepotieff, A. 1907a. Die Pterobranchier. Die Anatomie von *Rhabdopleura normanii* Allmann. – Zool. Jb., Anat. **23**: 463-534, pls 25-33.

Schepotieff, A. 1907b. Die Pterobranchier. Knospungsprozesse und Gehäuse von *Rhabdopleura*. – Zool. Jb., Anat **24**: 193-238, pls 17-23.

Schepotieff, A. 1907c. Die Pterobranchier. Die Anatomie von *Cephalodiscus*. – Zool. Jb., Anat. **24**: 553-600, pls 38-47.

Schepotieff, A. 1908. Die Pterobranchier. Knospungsprozess von *Cephalodiscus*. – Zool. Jb., Anat. **25**: 405-486, pls 12-14b.

Schepotieff, A. 1909. Die Pterobranchier des Indischen Ozeans. – Zool. Jb., Syst. **28**: 429-448, pls 7-8.

Stebbing, A.R.D. 1970. Aspects of the reproduction and life cycle of *Rhabdopleura compacta* (Hemichordata). – Mar. Biol. (Berl.) **5**: 205-212.

Stebbing, A.R.D. & P.N. Dilly 1972. Some observations on living *Rhabdopleura compacta* (Hemichordata). – J. mar. biol. Ass. U.K. **52**: 443-448.

365

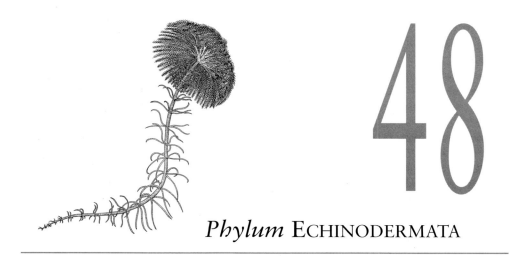

Phylum ECHINODERMATA

Living echinoderms form one of the most well-defined animal phyla, characterized by the unique specialization of one coelomic compartment into a water-vascular system comprising a perioesophageal ring and five radial canals, usually with tube feet (sometimes called podia). These five radial canals indicate a pentameric symmetry, which can be recognized in all adult, living echinoderms, even though some holothurians and echinoids externally appear bilateral. That the pentamery is in itself secondary is indicated by the bilaterality of early larval stages, and also by some of the early fossils which are not pentameric. The stereomic, calcareous, mesodermal skeleton is another characteristic which is present in all echinoderms, although the skeletal elements may be inconspicuous, for example in some holothurians; however, similar mesodermal skeletons are found in some articulate brachiopods (Chapter 45). A peculiarity is that the echinoderms lack a brain; there are nerve cords around the mouth and along the ambulacra, but there is no central, coordinating nervous centre. All echinoderms are marine, and almost all are benthic in the adult stage, but planktotrophic larvae are known in almost all groups, and the pelago-benthic life cycle appears to be the primitive developmental type. About 7000 living species are recognized, and there is a very extensive fossil record comprising about 13 000 species. There is no consensus about the phylogeny of fossil and living echinoderms (see below), and especially some of the earliest fossils are open to re-interpretation. There was obviously an enormous radiation in the Cambrian, and a number of extinct classes are known from the Ordovician; already the Cambrian forms had ambulacral plates indicating the presence of a water-vascular system with tube feet (Paul & Smith 1984). The living classes (except the holothuroids) are known from the early Ordovician, while all the other classes of the Cambrian radiation died out during the Palaeozoic. The phylogenetic interrelationships of the five living classes are better known, also through the fossil record, but there is still no agreement about the shape of the phylogenetic tree. The newly detected small deep-sea echinoderm *Xyloplax* (Rowe, Baker & Clark 1988) is by its discoverers placed in a class of its own,

Chapter vignette: The stalked crinoid *Cenocrinus asterias*. (Based on Rasmussen 1977.)

the Concentricycloidea; however, the authors regard the class as sister group of certain valvatid asteroids (op. cit., p. 452) and this is incompatible with the cladistic method adopted here, and I will therefore regard them as asteroids (see also Smith 1988a). If the so-called circumoral water-vascular canal is a misinterpreted circumoral haemocoel, *Xyloplax* would be much less enigmatic.

The alternative interpretation of many of the fossils and the completely different phylogeny of the deuterostomes advocated by Jefferies (1986) are discussed below.

The body is of quite different shape in the five living phyla, flower-shaped, star-shaped, globular, or worm-like, and there is considerable variation in the shape of the gut, which may be a straight tube between mouth and anus, coiled with several loops, a large sack without an anus, or a small sack with radial extensions. All epithelia are monolayered, and ciliated cells, always with only one cilium, occur both in ectoderm, endoderm, peritoneum (Fig. 3.9), and some myocytes (Walker 1979, Rieger & Lombardi 1987).

The calcareous endoskeleton is formed in the mesoderm and has a lattice-like structure, called stereomic. Each ossicle consists of numerous microcrystals with parallel orientation (Emlet 1982); they are secreted by primary mesenchyme cells in many of the larvae (see below) and some of the skeletal plates, for example four of the genital plates of echinoids, develop directly from larval skeletal elements (Ubisch 1913a, Emlet 1985). The plates make firm contact for example in the test of the sea-urchins, but elsewhere they mostly form looser connections or joints held together by collagenous material or muscles.

The presence of prosome, mesosome and metasome cannot be recognized externally, but the development of the coelomic compartments from coelomic pouches of the bilateral larvae clearly reveals the archimery.

The left protocoel, usually called the axocoel, is connected to the exterior through the madreporite in most groups. The coelomic cavity just below the madreporite, the madreporic chamber, is in open connection with the stone canal, which is a part of the left mesocoel (see below). The canal from the madreporic chamber to a ring surrounding the mouth is specialized as the axial gland or axial complex, which is a nephridium in many species (Fig. 46.1). It consists of a haemal space lined by a basement membrane covered by peritoneal podocytes of the axocoel (Bargmann & Hehn 1968, Welsch & Rehkämper 1987). Several elements in the axial complex contain muscle cells and various parts of the organ have been observed to pulsate. It is believed that primary urine is formed through pressure ultrafiltration from the blood through the basement membrane and between the podocytes to the axocoel; the urine may become modified in the axial complex on the way to the madreporite which functions as nephridiopore (Ruppert & Balser 1986). The axial complex is clearly recognized in asteroids, ophiuroids and echinoids. In crinoids and holothurians, the axial canal has lost contact with the madreporic chamber, and the organ is not a filtrating nephridium (Holland 1970, Erber 1983); these groups use other organs/tissues for excretion, and excretion for example from gills, papulae, respiratory trees, and especially from the gut is known from many species of all classes (Jangoux 1982).

The right protocoel and metacoel degenerate except for the small contractile sac, the dorsal sac, which surrounds a blind haemal space (MacBride 1896).

The left mesocoel, usually called the hydrocoel, forms the water-vascular system with a stone canal along the axial complex from the circumoral ring canal to the madreporic chamber; radial canals extend from the ring canal along the body wall, giving rise to a double row of tube feet. These zones are called radial areas or ambulacra, and the areas between them are called interradial or interambulacral. In several holothurians, the stone canal has lost connection with the madreporic chamber and thus the connection with the exterior (Erber 1983). In most crinoids, the stone canal loses the connection with the madreporic chamber, and many accessory stone canals develop from the ring canal; they all open into the metacoel (Seeliger 1893, Chadwick 1907).

The metacoel consists of the larger, perivisceral coeloms, which surround gut, gonads and most of the other coelomic canals, and various narrow coelomic canals along the ambulacra. The asteroids have the most complicated system of metacoelomic canals comprising both an oral ring with radial canals and an aboral ring with extensions surrounding the gonads (MacBride 1896; the most instructive illustrations of the structure and origin of the coelomic cavities are still to be found in Delage & Hérouard 1903). The perivisceral coeloms are quite narrow, for example in ophiuroids, but spacious, for example in asteroids (Fig. 46.1).

The above-mentioned haemal system is partially a spongiose mass and it is clear that other organs are involved in transport of nutrients and oxygen. Peritoneal cilia in several of the coelomic compartments (Ferguson 1982, Walker 1982; Fig. 3.9) create circulation, for example around the gonads, in the coelomic fluid of the tube feet, and in gill-like structures with extensions of various coelomic compartments.

The nervous system is complicated and quite unusual in that a coordinating centre which could be called a brain is lacking (Cobb 1988). There are ring nerves around the oesophagus and radial nerves along the ambulacra. The ring and radial nerves are internalized by a neurulation-like process in ophiuroids, echinoids and holothuroids; the infolding, which forms the so-called epineural canals, can be directly observed during the development (Ubisch 1913b). The nerves consist of an ectodermal, intraepithelial part, called the ectoneural nerve, and a mesodermal part, called the hyponeural nerve, separated by a basement membrane; the hyponeural nerve is differentiated from the peritoneum of the narrow radial oral canal of the metacoel. In types with epineural canals, the ectoneural nerve is situated in the aboral ectodermal lining of the canal. The ectoneural part of the nervous system is apparently mainly sensory while the hyponeural is motory; the communication between the two systems is via synapses across the basement membrane (Cobb 1985, 1988). Some muscle cells of podia and pedicellariae have a long thin extension reaching to the the basement membrane just opposite to ectoneural nerve endings (Cobb 1967, 1986).

The gonads are formed from mesodermal elements of the metacoel (MacBride 1896). They have separate gonoducts which open through pores in the genital plates (the madreporite is a genital plate too). Most species spawn the gametes freely in the water, where fertilization takes place, but a few have brood protection.

The polar bodies indicate the apical end of the apical-blastoporal (animal-vegetative) axis, which is established before fertilization both in echinoids and asteroids (Hörstadius 1927, Schroeder 1980, Kominami 1983); the first two cleavages follow this axis so that four cells of equal size are formed; the plane of the first cleavage is predetermined and not induced by the site of sperm entrance. The question about the planes of these cleavages relative to the symmetry of the larvae is more difficult to assess. Kominami (1983) found that the first cleavage is median in some of the embryos of the starfish *Asterina*, but that considerable variation exists. Observations on the development of the sea urchin *Strongylocentrotus* (Cameron *et al.* 1987, 1989, 1990, Davidson 1989) have shown that the first two cleavages divide the embryo into oral, right, aboral, and left quadrants, but again some variation was observed. These studies established a generalized cell-lineage for the embryos so that the origin of most of the larval organs are known. In the direct developing sea urchin *Heliocidaris erythrogramma*, Henry, Wray & Raff (1990) have shown that the dorsoventral axis is specified before the first cleavage, which divides the egg perpendicular to the median plane of the larva; a comparative study of several species (Henry, Kluege & Raff 1992) has shown that the first cleavage is median in species of *Strongylocentrotus* and *Lytechinus*.

The development of echinoderms has been studied in detail by many authors over more than a century (see also Delage & Hérouard 1903), and especially echinoids, such as *Echinus*, *Strongylocentrotus* and *Paracentrotus* (MacBride 1903, Ubisch 1913b), and asteroids, such as *Asterina*, *Asterias* and *Leptasterias* (MacBride 1896, Gemmill 1914, Chia 1968), have been studied intensively. These forms, which have small eggs and planktotrophic larvae, are easy to keep and to spawn, and the embryos and larvae are easy to handle both in studies of normal development and in experimental setups. For the sake of clarity, the following description of development in the main first deals with the types having planktotrophic larvae and the variations within this type, and then deals with the species with lecithotrophic larvae and direct development.

A ciliated coeloblastula is formed after several cleavages, usually with a tuft of longer cilia at the apical pole indicating the first nervous centre, the apical organ (Burke 1983). The often rather small cells at the blastoporal pole move into the blastocoel where they form the primary mesenchyme. The larger of these micromeres give rise to the cells which form the larval skeleton, while the smaller micromeres later become incorporated in the developing coelomic sacs (Pehrson & Cohen 1986). The blastomere sizes vary a great deal and small cells at the apical pole and large cells at the blastoporal pole have been observed in several species.

The blastoporal end of the embryo now invaginates so that a typical gastrula is formed.

A pair of small lateral pockets develop from the apical end of the archenteron and these pockets are the origin of the secondary mesoderm, which thus is formed by enterocoely. The right and left coelomic compartments have quite different fates with the water-vascular system developing from a coelomic compartment of the left side, but abnormalities with the water-vascular system developing from the right side or with a system on each side with subsequent development of a set of tube feet on each side have been observed several times, especially in asteroids (Gemmill 1915, Newman 1925). This clearly shows that the pentamery has originated from a normal bilaterality by the asymmetrical development of the two sides.

369

A deviating mode of forming coelom and hydropore is seen in holothurians such as *Labidoplax* (Selenka 1883) and *Stichopus* (Rustad 1938), in which the apical end of the archenteron first curves towards the aboral side and forms a connection with the exterior, thus forming the hydropore; the distal part of the archenteron then pinches off as the coelomic pouch and the archenteron curves towards the oral side.

The larva now assumes a more prismatic shape with the first stages of the arms developing; the apical end of the archenteron makes contact with the future oral epithelium, and the larval mouth breaks through. The cells of the apical organ spread laterally and form a small transverse ciliary band which becomes the preoral part of the circumoral ciliary band; individual nerve cells in the transversely elongated preoral ciliary band can be made visible with immunofluorescence (Burke 1983, Bisgrove & Burke 1987).

The left coelomic pouch elongates and divides into an anterior part, which becomes axocoel and hydrocoel, and a posterior metacoel, which becomes part of the large main body cavity. The anterior coelomic sac sends out a small canal which meets an invagination from the dorsal ectoderm, and the hydropore is formed. The narrow canal from the coelomic pouch to the exterior is ciliated and these cilia slowly transport coelomic fluid towards the hydropore; this is interpreted as a primitive excretory organ, the primordium of the adult axial gland (Ruppert & Balser 1986). The anterior part of this complex becomes the axocoel with the axial gland and an oral ring in close contact with the metacoelomic oral ring described below. The posterior part becomes the hydrocoel when the posterior end elongates and its posterior part curves into a circle with five small buds; this is the first stage of the stone canal and the ring canal with radial canals of the water-vascular system.

The right coelomic pouch likewise divides into an anterior axo-hydrocoel and a posterior metacoel which becomes the main body cavity. The anterior part disappears except for a small pulsating pouch, which can be seen already in the larva (Ruppert & Balser 1986), and it later becomes the small, pulsating dorsal sac (Fig. 46.1). In asteroid larvae, the axocoels fuse anteriorly and remain large during the whole larval period (Gemmill 1914). It is often claimed that the axocoel in crinoids develops an extension into the stalk, but this has been refuted by Grimmer, Holland & Hayami (1985).

The left and right metacoelomic pouches develop differently. The left pouch curves around the oral invagination of the adult rudiment and becomes the larger, oral, body cavity. Both the metacoelomic oral ring with radial canals and (in asteroids) the aboral ring with extensions surrounding the gonads (Fig. 46.1) develop as five small extensions from the main coelom; these extensions become Y-shaped and the branches fuse into a ring.

There are many variations in the origin and development of the coelomic pouches (Hyman 1955; Fig. 43.2). An anteromedian coelomic pouch which gives rise to all the coelomic cavities is found in several holothurians and asteroids. The anterior two pairs of coeloms develop from the anterior part of the archenteron while the posterior part of the archenteron gives rise to the metacoel in some crinoids and sea stars.

The general ciliation with an apical tuft characteristic of the gastrula differentiates simultaneously with the development of the coelomic compartments and the gut. A band of narrower cells is formed around a perioral area, and the cilia of this band become an upstream-collecting system, the neotroch, which is both particle-collecting and locomotory in the planktotrophic larvae. This early larval stage, which is usually called dipleurula (Chapter 43), can be recognized in representatives of all the living classes except the crinoids, which apparently all have lecithotrophic larvae (see below). This supposedly ancestral shape of the neotroch becomes differentiated in the planktotrophic larval types characteristic of the classes (Fig. 48.1). Asteroid larvae develop soft, flexible larval arms with loops of the neotroch; the earlier stages are called auricularia larvae, while older stages which have developed three short, preoral arms without ciliary bands but with small attachment organs are called brachiolaria larvae. Echinoid and ophiuroid larvae likewise develop long arms, but these larvae, called echinopluteus and ophiopluteus larvae, respectively, have stiff arms containing characteristic calcareous skeletons. Many of these larvae specialize parts of the neotroch into locomotory structures called epaulettes, which are much wider bands with strong ciliation. Holothuroid larvae resemble asteroid larvae but the arms are short and lobe-like, and the planktotrophic stage soon develops into a lecithotrophic stage with five circular ciliary bands (Fig. 48.2).

The gut with the blastopore transformed into the anus and with the new mouth is retained in the adult holothurians, but important modifications are found in the other classes. The larval mouth is retained after metamorphosis in ophiopluteus larvae, but the intestine and anus are lost. Brachiolaria and echinopluteus larvae develop a new mouth on the left side of the larva (or almost medially, see Fig. 48.3) outside the neotroch, and the larval mouth and oesophagus become cast off or resorbed together with the larval body at metamorphosis. The ring canal of the water-vascular system thus surrounds different openings in these two types. The oral side of the adult rudiment is formed from an invagination of the larval body wall, the vestibule or amnion, in echinoids and ophiuroids.

The first adult organs which can be recognized externally are often five tube feet around the future mouth, each containing the tip of one of the radial canals of the water-vascular system.

The planktotrophic larvae of many echinoids, asteroids and ophiuroids go through a more or less catastrophic metamorphosis where important parts of the larval body are cast off or resorbed by the juvenile (Chia & Burke 1978). In the pluteus larvae, the larval arms with the skeleton become resorbed, with exception of the basal parts of the skeletal arms, which become transformed into four of the genital plates (Ubisch 1913a). This indicates that also the adult skeletal system is formed by the descendants of the large micromeres. The soft arms of the brachiolaria are resorbed, except in *Luidia sarsi* (Fig. 48.3), where the juvenile detaches from the larval body.

The blastocoel of the planktonic larvae is not simply a fluid-filled space; a gelatinous material occupies the cavity and is clearly of great importance both for the development and the stability of the body form (Strathmann 1989).

The many species with lecithotrophic larvae and especially those with direct development show interesting deviations from the general developmental pattern described above, but only those which appear to contribute to the understanding of the phylogeny will be discussed here.

Lecithotrophic larvae may be superficially similar to planktotrophic larvae of related species, but there are many examples of characteristic lecithotrophic larval types (Fig. 48.1). The doliolaria, which is barrel-shaped with four or five ciliated rings, is known in holothurians and is the only known larval type in the crinoids. It is clearly derived from the planktotrophic type, and a development from a planktotrophic auricularia through a doliolaria to the juvenile bottom stage has been described, for example in the holothurian *Synapta* (Fig. 48.2). This indicates that the crinoid larva is a derived type too, and the development of *Florometra* shows indications of ciliary rings developing from one circumoral ring (Lacalli & West 1986). Some of the lecithotrophic ophiuroid larvae have ciliary bands looking like the fragment-stage in the development of *Synapta*, so this type also appears to be derived from the planktotrophic larvae. The origin of the gut, mouth and coelomic compartments of the lecithotrophic larvae resembles that of the planktotrophic larvae to various degrees.

Direct development, often associated with brooding and in certain cases with ovovivipary (Komatsu *et al.* 1990), occurs in all classes. In the still unidentified ophiuroid called Kirk's ophiuroid, the rather large eggs are found deposited on stones and the development is completely direct (Fell 1941). The early development shows an epiboly-like formation of an inner mass of cells with a 'vestigial archenteron'; the embryo is slightly bilateral at this stage and the opening to the invagination could be interpreted as the blastopore. However, the opening later becomes the adult mouth, which in the ophiuroids with planktotrophic development forms as a new opening from the stomach to the left side of the larva. The stomach forms later as a new invagination from the same area, and the coelomic cavities all form through schizocoely in the compact mass of mesendodermal cells.

Direct-developing asteroids go through a stage where the three brachiolaria arms can be recognized as small stubs and the development of the coelomic sacs takes place in the above described manner (MacBride 1896, Komatsu *et al.* 1990).

Fig. 48.1. Planktotrophic and lecithotrophic larval types of the five echinoderm classes. – Crinoidea: lecithotrophic larva of *Florometra serratissima* (based on Lacalli & West 1986; planktotrophic larvae have not been reported). – Asteroidea: full-grown bipinnaria larva and brachiolaria larva with developing sea star of *Luidia* sp. (plankton, off Nassau, The Bahamas, October 1990); lecithotrophic larva of *Astropecten latespinosus* (redrawn from Komatsu, Murase & Oguro 1988). – Ophiuroidea: ophiopluteus of *Amphiura filiformis* (redrawn from Mortensen 1931), metamorphosing larva of *Amphiura* sp. (plankton, off Kristineberg Marine Biological Station, Sweden, October 1984); lecithotrophic ophiuroid larva (possibly *Ophioderma squamulosa*, see Mortensen 1921) (plankton, off San Salvador Island, The Bahamas, October 1990). – Echinoidea: full-grown echinopluteus larva and larva with developing sea urchin of *Psammechinus miliaris* (redrawn from Czihak 1960); lecithotrophic larva of *Heliocidaris erythrogramma* (redrawn from Williams & Anderson 1975). – Holothuroidea: planktotrophic auricularia larva of *Parastichopus californicus* (Stimpson) (Friday Harbor Laboratories, WA, USA, July 1992), metamorphosing doliolaria larva of a holothuroid (plankton, off Phuket Marine Biological Center, Thailand, March 1982); lecithotrophic larva of *Cucumaria elongata* (redrawn from Chia & Buchanan 1969).

Planktotrophic larvae   Lecithotrophic larvae

full-grown   metamorphosing

Crinoidea

Asteroidea

Florometra

Luidia

Astropecten

Ophiuroidea

Amphiura

Ophioderma?

Echinoidea

Psammechinus

Heliocidaris

Holothuroidea

Parastichopus   doliolaria   *Cucumaria*

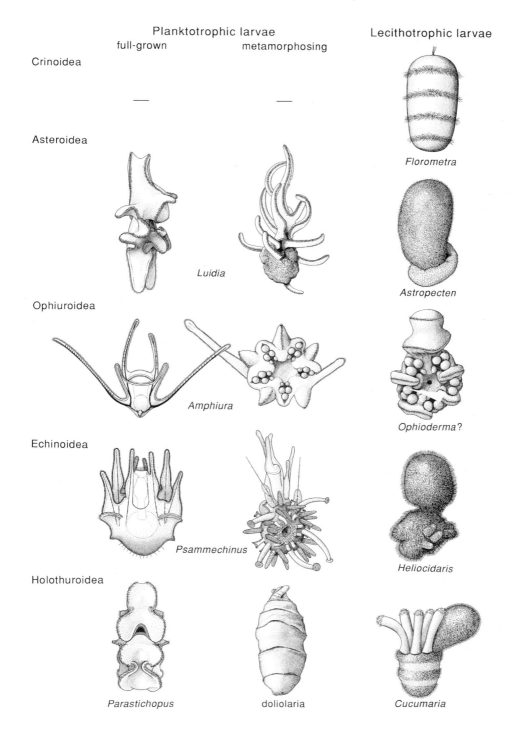

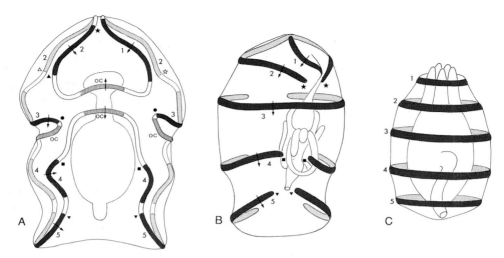

Fig. 48.2. Transformation of the ciliary bands of the larva of *Synapta digitata* from an auricularia stage through an intermediate stage to an old doliolaria stage which has developed the first tube feet. (From Nielsen 1987.)

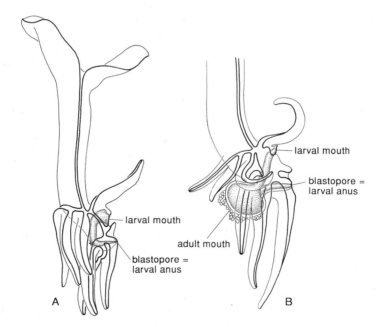

Fig. 48.3. Right views of larvae of the starfish *Luidia sarsi* Düben & Koren (plankton, off Kristineberg Marine Biological Station, Sweden, October 1989). – A, young stage just before the adult rudiment begins to develop. – B, older stage where the adult mouth is being formed but the whole larval gut with larval mouth and blastopore (anus) is still retained (the anterior part of the larval body is omitted).

374

The embryology of the direct-developing echinoid *Heliocidaris erythrogramma* has been compared with the planktotrophic type described above (Wray & Raff 1990). The cleavage pattern is different, with *H. erythrogramma* having equal cleavage up to the 32-cell stage, and the fates of the corresponding blastomeres of the two types of embryos show conspicuous differences. However, a comparison of the fate maps of the two types shows complete identity of the spatial relationships between areas developing into identical organs in the larva (Wray & Raff 1990, fig. 13).

The planktotrophic larval types described above from four of the classes are clearly variations over a common theme: the dipleurula, which has had the ciliary bands extended on various types of arms with or without skeleton or convoluted and broken up to form for example the circular bands of the doliolaria larvae (Fig. 48.2). Most of these larvae go through a complex metamorphosis during which the larval structures are abandoned. There are, however, numerous intermediate types between the planktotrophic larvae and various types of lecithotrophic larvae or direct development in all the five classes (Fig. 48.1). It is clear that the change away from planktotrophy has taken place numerous times within all the classes (Strathmann 1974), and there are even examples of genera in which one species has planktotrophic larvae and another has a brief, lecithotrophic larval stage (for example *Heliocidaris*; Williams & Anderson 1975). It seems exceedingly improbable that the evolution has gone in the opposite direction and given rise to so many larval forms with identical ciliary feeding systems (Strathmann 1978, 1988). The variations in cell-lineage observed between different echinoids which have identical fate maps has a parallel in the annelids (Chapter 17) and it is clear that the cleavage patterns are much more labile than the spatial relationships between the different areas of the embryos.

Larval types can sometimes give a clue to the understanding of phylogenetic relationships within a larger systematic category, but it appears that the larval types of the five classes of living echinoderms each show so wide a variation that it is impossible to see a pattern (Strathmann 1988). The skeletons of echinoplutei and ophioplutei look very similar, but may not be homologous (Raff *et al.* 1987). In echinoid and ophiuroid larvae, the small adult is formed from the bottom of an epidermal invagination of the left side of the larva, the vestibule or amniotic invagination. This has been interpreted as a synapomorphy of the two groups, but the elaborate, planktotrophic larva of *Eucidaris* has a metamorphosis without trace of a vestibule (Emlet 1988), so the importance of this character is uncertain indeed. At present it seems impossible to identify one of the planktotrophic developmental types as the most 'primitive', and the apparent lack of planktotrophic larvae in the crinoids, which are often regarded as the outgroup of the other four living classes, only makes the problem deeper. On the other hand, the early larval stage with apical organ, complete gut, slightly curved perioral ciliary band, and probably three pairs of coelomic pouches, i.e. a dipleurula, can be identified as common to all echinoderms. This is indeed important for the understanding of the phylogenetic position of the phylum.

Regeneration and autotomy occur in all classes and asexual reproduction by fission is known in asteroids, ophiuroids and holothurians (Emson & Wilkie 1980). It has sometimes been presumed that the large larval body left after the liberation of the juvenile in *Luidia sarsi* should be able to survive and form a new sea star, but this has never been observed directly. However, a completely different type of asexual reproduction has been observed in larvae of another species of *Luidia*, in which the posterior larval arms may form a new gut through a gastrulation process and then become pinched off so that a complete pluteus larva is formed (Bosch, Rivkin & Alexander 1989).

Phylogenetic analyses of the living echinoderm classes, with emphasis on molecular data, larval and adult characters and fossil evidence, result in different trees (Raff *et al.* 1988, Smiley 1988, Smith 1988b) and this does to some extent reflect the choice of characters or the different interpretations of the same characters. I find it impossible to make a choice between the various trees at this point, but my guess is that one way out of this impasse could be to study the phylogeny of each of the five classes in more detail to establish the characters of their ancestral forms; this should make it easier to make educated comparisons between the classes.

Fortunately, it appears that an established phylogeny of the echinoderm classes is not crucial to the analysis of the relationships of the echinoderms to other phyla because both larval type and morphology of the adult ancestor can be deduced from the combined knowledge of the five classes.

The bilaterally symmetrical larvae and also some of the non-pentameric fossil types indicate that the ancestor of the echinoderms was bilateral, and the evolution of pentamery from bilaterality has fascinated many authors. Grobben (1923) summarized earlier ideas and gave a detailed account of the theory which derives the echinoderm ancestor from a *Cephalodiscus*-like form which settled with the prosome, held the right side towards the substratum and curved the left row of tentacles around the mouth while the right row degenerated. The tentacles with the left hydrocoel should then have become the primary tube feet with the early water-vascular system. This ancestral form was probably a ciliary filter feeder as *Cephalodiscus*, but this aspect was not discussed. Grobben's theory finds support in the attachment of the crinoid larva by the anterior pole and in the presence of three small tentacles with attachment organs in the anterior end of certain bipinnaria larvae, but both structures may just as well be interpreted as specializations.

Jefferies (1986) proposed that a *Cephalodiscus*-like ancestor turned over on the right side, became sessile and lost the stalk; the tentacles facing the bottom degenerated while those of the left side continued to function as ciliary feeding organs. His considerations on the further evolution of echinoderms and other deuterostomes from this ancestor are discussed below.

The alternative explanation proposed by Holland (1988, with a useful table summarizing earlier theories) involves almost the same spatial transformations, but is much more attractive because it links the morphological changes with changes in the feeding biology. The adult ancestor was believed to be *Cephalodiscus*-like but with a straight gut like that of the enteropneusts, in all much like the deuterostome ancestor called notoneuron (Fig. 43.6). The first step leading towards the echinoderms was that the adults of the ancestor turned over on the left side and began to collect particles from the substratum with the tentacles of the left side; the tentacles of the right side could then degenerate when the organism switched completely to deposit feeding. This early ancestor (called the stem laeothete by Holland) had no pentamery and could thus be compatible with some of the early echinoderm fossils. Holland linked the chordates also to the stem laeothetes, but this interpretation is not accepted here (Chapter 51).

A totally different phylogeny of the deuterostomes, the calcichordate theory, has been advocated in a series of recent publications by Jefferies (summarized in 1986). It is based on a very detailed interpretation of the Palaeozoic group Carpoidea or Homalozoa, which by most palaeontologists, such as Ubaghs (1975) and Philip (1979), is regarded as non-pentameric echinoderms. The more traditional view is that the carpoids had an asymmetrical body with a series of openings, which could have housed retractile gills, and one articulated arm possibly with tube feet and possibly a feeding organ. Jefferies' interpretation of the series of openings is that they were gill slits like those of amphioxus and that the long, jointed appendage was a tail with chorda and neural tube; the body should have housed a spacious branchial chamber like that of tunicates and cephalochordates. This group of organisms, called calcichordates, should then have given rise to echinoderms and chordates. The theory completely disregards the similarities between the gill slits of enteropneusts and chordates. The interpretation of the series of openings as gill slits is proposed almost axiomatically: 'Since the openings suggest outlet valves, they can plausibly be seen as gill slits.' (Jefferies 1986, p. 197). However, it is very difficult to see how the pharynx of the reconstructed cornutes (for example *Cothurnocystis*; Jefferies 1986, fig. 7.6) can be compared to the gill chamber of a living tunicate or cephalochordate; both of these types have a gill chamber with large areas of gills, which both carry the ciliary bands creating the water currents and support the mucus net which is the filtering device. If the mucus filter extended only across the row of gill slits in the cornute, the filtering area would have been disproportionately small both in relation to the size of the pharynx and to the size of the whole animal when the living organisms are considered. A possibility not considered by Jefferies is of course that the 'gill slits' were merely exit openings for the filtered water and that the filter was a more extensive structure somewhere else in the 'pharynx', but this brings the speculations into the realm of fantasy. A functionally even more improbable explanation is the interpretation of the closely related *Scotiaecystis*, which had a long series of closely fitting, chevron-shaped ossicles in the same position as the 'gill slits' of *Cothurnocystis*. These ossicles would appear to close the gill slits, but the following explanation was offered: 'When water pressure was high inside the head the dorsal integument would inflate upwards. The chevron complex, bisecting the integument and therefore situated along the line of maximal stretching, would itself be stretched and gaps between the chevrons would open, allowing water to escape' (Jefferies 1986, p. 207). It is unclear how the pressure inside the head/pharynx would be created. In living tunicates, the pharynx/gill chamber is kept expanded by the elastic tunic and the ciliated gill bars pump the water through the mucus filter out of the pharynx; there is accordingly a slightly higher pressure at the exhalant siphon than inside the filter (Riisgård 1988). If the pressure were to have been higher inside the pharynx of *Scotiaecystis*, a mechanism unknown in living tunicates or cephalochordates would have been present, and the discussion again becomes mere fantasy. The reconstructions of notochord and spinal chord with ganglia in the articulated extremity and of nerves, ganglia with eyes, even, in the head/body of the Middle-Ordovician *Mitrocystella* and other mitrates, interpreted as early vertebrate ancestors, appear as extreme examples of wishful thinking. Consequently, I reject the calcichordate theory on functional grounds, in accordance with a number of other authors who have rejected it on other grounds (Philip 1979, Ubaghs 1975, Jollie 1982).

The theory of Holland (1988) mentioned above for the evolution of the echinoderms from a bilateral, pterobranch/notoneuron-like organism is just as valid without the calcichordate story. To me it appears that the derivation of the echinoderms from a notoneuron with a dipleurula larva gives a very coherent story. The scenario outlined by Holland (1988), as modified slightly above, with a notoneuron which takes up a deposit-feeding habit by turning the left side towards the substratum and using the tentacles of that side for deposit feeding, and later losing the tentacles of the right side, links the structural changes to changes in feeding-biology. I believe that such linked changes in structure and function have taken place in connection with the establishment of all the major divergences in the animal kingdom.

The relationship between echinoderms, pterobranchs and enteropneusts is discussed in Chapter 47.

Interesting subjects for future research

1. Origin of the adult nervous system.
2. Crinoid embryology.

References

Bargmann, W. & G.v. Hehn 1968. Über das Axialorgan ('mysterious gland') von *Asterias rubens* L. – Z. Zellforsch. **88**: 262-277.

Bisgrove, B.W. & R.D. Burke 1987. Development of the nervous system of the pluteus larva of *Strongylocentrotus droebachiensis*. – Cell Tissue Res. **248**: 335-343.

Bosch, I., R.B. Rivkin & S.P. Alexander 1989. Asexual reproduction by oceanic echinoderm larvae. – Nature **337**: 169-170.

Burke, R.D. 1983. Development of the larval nervous system of the sand dollar, *Dendraster excentricus*. – Cell Tissue Res. **229**: 145-154.

Cameron, R.A., B.R. Hough-Ewans, R.J. Britten & E.H. Davidson 1987. Lineage and fate of each blastomere of the eight-cell sea urchin embryo. – Genes Dev. **1**: 75-84.

Cameron, R.A., S.E. Fraser, R.J. Britten & E.H. Davidson 1989. The oral-aboral axis of a sea urchin embryo is specified by first cleavage. – Development **106**: 641-647.

Cameron, R.A., S.E. Fraser, R.J. Britten & E.H. Davidson 1990. Segregation of oral from aboral ectoderm precursors is completed at fifth cleavage in the embryogenesis of *Strongylocentrotus purpuratus*. – Dev. Biol. **137**: 77-85.

Chadwick, H.C. 1907. L.M.B.C. Memoir 15: *Antedon*. – Proc. Trans. Lpool biol. Soc. **21**: 371-417, 7 pls.

Chia, F.-S. 1968. The embryology of a brooding starfish, *Leptasterias hexactis* (Stimpson). – Acta zool. (Stockh.) **49**: 321-364.

Chia, F.-S. & J.B. Buchanan 1969. Larval development of *Cucumaria elongata* (Echinodermata: Holothuroidea). – J. mar. Biol. Ass U.K. **49**: 151-159, 2 pls.

Chia, F.-S. & R.D. Burke 1978. Echinoderm metamorphosis: fate of larval structures. – In F.-S. Chia & M.E. Rice (eds): Settlement and Metamorphosis of Marine Invertebrate Larvae, pp 219-246. Elsevier, New York.

Cobb, J.L.S. 1967. The innervation of the ampulla of the tube foot in the starfish Astropecten irregularis. – Proc. R. Soc. Lond. B 168: 91-99, pls 10-14.

Cobb, J.L.S. 1985. The neurobiology of the ectoneural/hyponeural synaptic connection in an echinoderm. – Biol. Bull. Woods Hole 168: 432-446.

Cobb, J.L.S. 1986. Neurobiology of the Echinodermata. – In M.A. Ali (ed.): Nervous Systems in Invertebrates (NATO ASI, Ser. A 141), pp 483-525. Plenum Press, New York.

Cobb, J.L.S. 1988. A preliminary hypothesis to account for the neural basis of behaviour in echinoderms. – In R.D. Burke, P.V. Mladenov, P. Lambert & R.L. Parsley (eds): Echinoderm Biology, pp 565-573. A. A. Balkema, Rotterdam.

Czihak, G. 1960. Untersuchungen über die Coelomanlagen und die Metamorphose des Pluteus von Psammechinus miliaris (Gmelin). – Zool. Jb., Anat. 78: 235-256, pls 1-2.

Davidson, E.H. 1989. Lineage-specific gene expression and the regulative capacities of the sea urchin embryo: a proposed mechanism. – Development 105: 421-445.

Delage, Y. & E. Hérouard 1903. Les Échinodermes. – Traité de Zoologie Concrète, vol. 3. Paris.

Emlet, R.B. 1982. Echinoderm calcite: a mechanical analysis of larval spicules. – Biol. Bull. Woods Hole 163: 264-275.

Emlet, R.B. 1985. Crystal axes in recent and fossil adult echinoids indicate trophic mode in larval development. – Science 230: 937-940.

Emlet, R.B. 1988. Larval form and metamorphosis of a 'primitive' sea urchin, Eucidaris thouarsi (Echinodermata: Echinoidea: Cidaroida), with implications for developmental and phylogenetic studies. – Biol. Bull. Woods Hole 174: 4-49.

Emson, R.H. & I.C. Wilkie 1980. Fission and autotomy in echinoderms. – Oceanogr. mar. Biol. ann. Rev. 18: 155-250.

Erber, W. 1983. Zum Nachweis des Axialkomplexes bei Holothurien (Echinodermata). – Zool. Scr. 12: 305-313.

Fell, B.B. 1941. The direct development of a New Zealand ophiuroid. – Q. Jl microsc. Sci., N.S. 82: 377-441, pls 20-22.

Ferguson, J.C. 1982. Nutrient translocation. – In M. Jangoux & J.M. Lawrence (eds): Echinoderm Nutrition, pp 373-393. A.A. Balkema, Rotterdam.

Gemmill, J.F. 1914. The development and certain points in the adult structure of the starfish Asterias rubens, L. – Phil. Trans. R. Soc. B 205: 213-294.

Gemmill, J.F. 1915. Double hydrocoele in the development and metamorphosis of the larva of Asterias rubens, L. – Q. Jl microsc. Sci., N.S. 61: 51-60.

Grimmer, J.C., N.D. Holland & I. Hayami 1985. Fine structure of the stalk of an isocrinoid sea lily (Metacrinus rotundus) (Echinodermata, Crinoidea). – Zoomorphology 105: 39-50.

Grobben, K. 1923. Theoretische Erörterungen betreffend die phylogenetische Ableitung der Echinodermen. – Sber. Akad. Wiss. Wien, 1. Abt. 132: 263-290.

Henry, J.J., K.M. Kluege & R.A. Raff 1992. Evolutionary dissociation between cleavage, cell lineage and embryonic axes in sea urchin embryos. – Development 114: 931-938.

Henry, J.J., G.A. Wray & R.A. Raff 1990. The dorsoventral axis is specified prior to first cleavage in the direct developing sea urchin Heliocidaris erythrogramma. – Development 110: 875-884.

Holland, N.D. 1970. The fine structure of the axial organ of the feather star Nemaster rubiginosa (Echinodermata: Crinoidea). – Tissue Cell 2: 625-636.

Holland, N.D. 1988. The meaning of developmental asymmetry for echinoderm evolution: a new interpretation. – In C.R.C. Paul & A.B. Smith (eds): Echinoderm Phylogeny and Evolutionary Biology, pp 13-25. Oxford Univ. Press, Oxford.

Hörstadius, S. 1927. Studien über die Determination von Paracentrotus lividus Lk. – Wilhelm Roux Arch. Entwicklungsmech. Org. 112: 239-246.

Hyman, L.H. 1955. Echinodermata. The Invertebrates, vol. 4. – McGraw-Hill, New York.

Jangoux, M. 1982. Excretion. – In M. Jangoux & J.M. Lawrence (eds): Echinoderm Nutrition, pp 437-445. A.A. Balkema, Rotterdam.

Jefferies, R.P.S. 1986. The Ancestry of the Vertebrates. – British Museum (Nat. Hist.), London.

Jollie, M. 1982. What are the 'Calcichordata'? and the larger question of the origin of the chordates. – Zool. J. Linn. Soc. **75**: 167-188.

Komatsu, M., Y.T. Kano & C. Oguro 1990. Development of a true ovoviviparous sea star, *Asterina pseudoexigua pacifica* Hayashi. – Biol. Bull. Woods Hole **179**: 254-263.

Komatsu, M., M. Murase & C. Oguro 1988. Morphology of the barrel-shaped larva of the sea-star, *Astropecten latespinosus*. – *In* R.D. Burke, P.V. Mladenov, P. Lambert & R.L. Parsley (eds): Echinoderm Biology, pp 267-272. Balkema, Rotterdam.

Kominami, T. 1983. Establishment of embryonic axes in larvae of the starfish, *Asterina pectinifera*. – J. Embryol. exp. Morph. **75**: 87-100.

Lacalli, T.C. & J.E. West 1986. Ciliary band formation in the doliolaria larva of *Florometra*. – J. Embryol. exp. Morph. **96**: 303-323.

MacBride, E.W. 1896. The development of *Asterina gibbosa*. – Q. Jl microsc. Sci., N.S. **38**: 339-411, pls 18-29.

MacBride, E.W. 1903. The development of *Echinus esculentus*, together with some points in the development of *E. miliaris* and *E. acutus*. – Phil. Trans. R. Soc. B **195**: 285-327, pls 7-16.

Mortensen, T. 1921. Studies on the development and larval forms of echinoderms. – G.E.C. Gad, Copenhagen.

Mortensen, T. 1931. Contributions to the study of the development and larval forms of echinoderms. I-II. – K. danske Vidensk. Selsk. Skr., Mat. nat. Afd., 9. Rk. **4**(1): 1-39, 7 pls.

Newman, H.H. 1925. On the occurrence of paired madreporic pores and pore-canals in the advanced bipinnaria larvae of *Asterina (Patiria) miniata* together with a discussion of the significance of similar structures in other echinoderm larvae. – Biol. Bull. Woods Hole **40**: 118-125.

Nielsen, C. 1987. Structure and function of metazoan ciliary bands and their phylogenetic significance. – Acta zool. (Stockh.) **68**: 205-262.

Paul, C.R.C. & A.B. Smith 1984. The early radiation and phylogeny of echinoderms. – Biol. Rev. **59**: 443-481.

Pehrson, J.R. & L.H. Cohen 1986. The fate of the small micromeres in sea urchin development. – Dev. Biol. **113**: 522-526.

Philip, G.M. 1979. Carpoids – echinoderms or chordates? – Biol. Rev. **54**: 439-471.

Raff, R.A., J.A. Anstrom, J.E.Chin, K.G. Field, M.T. Ghiselin, D.J. Lane, G.J. Olsen, N.R. Pace, A.L. Parks & E.C. Raff 1987. Molecular and developmental correlates of macroevolution. – *In* R.A. Raff & E.C. Raff (eds): Development as an Evolutionary Process, pp 109-138. Alan R. Liss, New York.

Raff, R.A., K.G. Field, M.T. Ghiselin, D.J. Lane, G.J. Olsen, N.R. Pace, A.L. Parks, B.A. Parr & E.C. Raff 1988. Molecular analysis of distant phylogenetic relationships in echinoderms. – *In* C.R.C. Paul & A.B. Smith (eds): Echinoderm Phylogeny and Evolutionary Biology, pp 29-41. Oxford Univ. Press, Oxford.

Rasmussen, H.W. 1977. Function and attachment of the stem in Isocrinidae and Pentacrinidae: review and interpretation. – Lethaia **10**: 51-57.

Rieger, R.M. & J. Lombardi 1987. Ultrastructure of coelomic lining in echinoderm podia: significance for concepts in the evolution of muscle and peritoneal cells. – Zoomorphology **107**: 191-208.

Riisgård, H.U. 1988. The ascidian pump: properties and energy cost. – Mar. Ecol. Prog. Ser. **47**: 129-134.

Rowe, F.W.E., A.N. Baker & H.E.S. Clark 1988. The morphology, development and taxonomic status of *Xyloplax* Baker, Rowe and Clark (1986) (Echinodermata: Concentricycloidea), with description of a new species. – Proc. R. Soc. Lond. B **233**: 431-459.

Ruppert, E.E. & E.J. Balser 1986. Nephridia in the larvae of hemichordates and echinoderms. – Biol. Bull. Woods Hole **171**: 188-196.

Rustad, D. 1938. The early development of *Stichopus tremulus* (Gunn.) (Holothuroidea). – Bergens Mus. Årb. 1938, Naturv. R. 8: 1-23, pls 1-2.

Schroeder, T.E. 1980. Expressions of the prefertilization polar axis in sea urchin eggs. – Dev. Biol. **79**: 428-443.

Seeliger, O. 1893. Studien zur Entwicklungsgeschichte der Crinoiden (Antedon rosacea). – Zool. Jb., Anat. 6: 161-444, pls 12-22.

Selenka, E. 1883. Studien über Entwicklungsgeschichte der Thiere. II. Die Keimblätter der Echinodermen. – C.W. Kreidel, Wiesbaden.

Semon, R. 1888. Die Entwicklung der Synapta digitata und die Stammesgeschichte der Echinodermen. – Jena. Z. Naturw. 22: 1-135, pls 1-6.

Smiley, S. 1988. The phylogenetic relationships of holothurians: a cladistic analysis of the extant echinoderm classes. – In C.R.C. Paul & A.B. Smith (eds): Echinoderm Phylogeny and Evolutionary Biology, pp 69-84. Oxford Univ. Press, Oxford.

Smith, A.B. 1988a. To group or not to group: the taxonomic position of Xyloplax. – In R.D. Burke, P.V. Mladenow, P. Lambert & R.L. Parsley (eds): Echinoderm Biology, pp 17-23. A.A. Balkema, Rotterdam.

Smith, A.B. 1988b. Fossil evidence for the relationships of extant echinoderm classes and their times of divergence. – In C.R.C. Paul & A.B. Smith (eds): Echinoderm Phylogeny and Evolutionary Biology, pp 85-97. Oxford Univ. Press, Oxford.

Strathmann, R. 1974. Introduction to function and adaptation in echinoderm larvae. – Thalassia Jugosl. 10: 321-339.

Strathmann, R.R. 1978. The evolution and loss of feeding larval stages of marine invertebrates. – Evolution 32: 894-906.

Strathmann, R.R. 1988. Larvae, phylogeny, and von Baer's law. – In C.R.C. Paul & A.B. Smith (eds): Echinoderm Phylogeny and Evolutionary Biology, pp 53-68. Oxford Univ. Press, Oxford.

Strathmann, R.R. 1989. Existence and function of a gel filled primary body cavity in development of echinoderms and hemichordates. – Biol. Bull. Woods Hole 176: 25-31.

Ubaghs, G. 1975. Early Paleozoic echinoderms. – Ann. Rev. Earth Planet. Sci. 3: 79-98.

Ubisch, L.v. 1913a. Die Anlage und Ausbildung des Skeletsystems einiger Echiniden und die Symmetrieverhältnisse von Larve und Imago. – Z. wiss. Zool. 104: 119-156, pls 6-7.

Ubisch, L.v. 1913b. Die Entwicklung von Strongylocentrotus lividus (Echinus microtuberculatus, Arbacia pustulosa). – Z. wiss. Zool. 106: 409-448, pls 5-7.

Walker, C.W. 1979. Ultrastructure of the somatic portion of the gonads in asteroids, with emphasis on flagellated-collar cells and nutrient transport. – J. Morph. 162: 127-162.

Walker, C.W. 1982. Nutrition of gametes. – In M. Jangoux & J.M. Lawrence (eds): Echinoderm Nutrition, pp 449-468. A.A. Balkema, Rotterdam.

Welsch, U. & G. Rehkämper 1987. Podocytes in the axial organ of echinoderms. – J. Zool. (Lond.) 213: 45-50.

Williams, D.H.C. & D.T. Anderson 1975. The reproductive system, embryonic development, larval development and metamorphosis of the sea urchin Heliocidaris erythrogramma (Val.) (Echinoidea: Echinometridae). – Aust. J. Zool. 23: 371-403.

Wray, G.A. & R.A. Raff 1990. Novel origins of lineage founder cells in the direct-developing sea urchin Heliocidaris erythrogramma. – Dev. Biol. 141: 41-54.

CYRTOTRETA

The group Cyrtotreta comprises the phyla Enteropneusta, Urochordata, Cephalochordata, and Vertebrata, and is characterized by a pharynx with a series of gill slits on each side (Fig. 49.1). Each gill slit develops as an opening from the endodermal pharynx directly to the ectoderm, but these primary openings almost always become protected inside ectodermal chambers formed through invaginations of the ectoderm around one or more slits or by various ectodermal folds. The gill slits are present in all enteropneusts and cephalochordates, and in all but a few, highly specialized, urochordates. In the vertebrates, gill slits are present in the 'fishes' but absent in the tetrapods, where they can only be recognized during ontogeny.

The gill slits are U-shaped in enteropneusts and cephalochordates (Chapters 50, 53), and a similar shape is possibly indicated in early developmental stages of the gill slits of some ascidians (Chapter 52); the vertebrate gill slits are simple. The first stages of the gill slits are almost circular, and the U-shape is attained through the development of an extension from the dorsal side. The original bars between the gill slits are called gill bars, while the secondary bars are called tongue bars. The gill openings remain U-shaped in several of the enteropneusts, but in the ptychoderids and in amphioxus, horizontal structures, synapticles, develop across the slits, so that the openings become divided into rows of gill pores. The bars surrounding the gill slits have a basal structure which can be recognized in members of all the four phyla. The partially ciliated epithelium is supported by a skeleton, which is a strengthened collagenous basement membrane; the structure is often described as cartilaginous, but only the vertebrates have true cartilage in the skeleton (Chapter 54). There are one or more blood vessels in the core of the larger skeletal bars. Amphioxus has a coelomic canal along the outer side of each gill bar and no coelom in the tongue bars, whereas the enteropneusts have a blind coelomic extension in the tongue bar but no coelomic canal in the gill bars. The gill skeleton develops as separate thickenings of the basement membrane along the sides of the gill slits, and these structures may in later stages fuse into a more continuous skeleton.

The gill slits are ciliated and the cilia create a flow of water from the pharynx through the gill slits to the exterior, with new water entering the pharynx through the mouth. Only the vertebrates (and the salps) have the water current set up by mus-

Transverse sections of gill bars

Development of gill slits

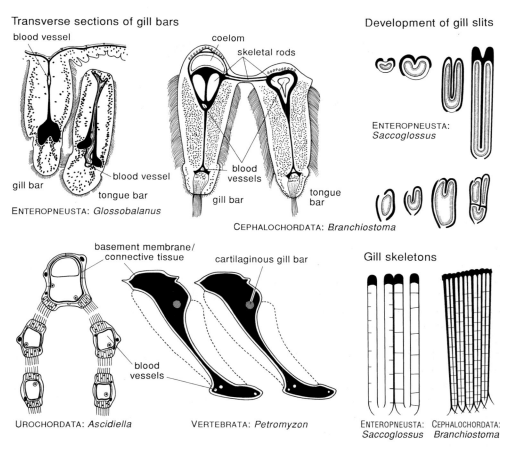

ENTEROPNEUSTA: *Saccoglossus*

CEPHALOCHORDATA: *Branchiostoma*

Gill skeletons

UROCHORDATA: *Ascidiella*

VERTEBRATA: *Petromyzon*

ENTEROPNEUSTA: *Saccoglossus* CEPHALOCHORDATA: *Branchiostoma*

Fig. 49.1. Structure and development of the gill slits in the the cyrtotrete phyla. – Transverse sections of gill bars in Enteropneusta: *Glossobalanus cambrensis* (based on Knight-Jones 1953, with blood vessels added after the diagrams of Pardos & Benito 1984), Urochordata: *Ascidiella aspersa* (drawn from a section of a Danish specimen, with cytological details after Mackie *et al.* 1974: *Corella willmeriana*), Cephalochordata: *Branchiostoma lanceolatum* (based on Franz 1927 and Rähr 1982), and Vertebrata: *Petromyzon marinus* (based on Mallatt 1984). – Development of gill slits and skeletons in Enteropneusta: *Saccoglossus kowalevskii* (based on Bateson 1886a and Dawydoff 1948) and Cephalochordata: *Branchiostoma lanceolatum* (based on Lönnberg 1901-1905 and Drach 1948). – Gill skeletons: Enteropneusta: skeletons of two gill slits of *Glossobalanus minutus* (redrawn from Spengel 1893) and Cephalochordata: skeletons of gill slits of *Branchiostoma lanceolatum* (redrawn from Krause 1923).

cular contractions instead. The whole structure appears primarily to be associated with feeding through particle collection, but respiratory functions are associated in most cases (Chapters 50, 51). The gill slits of the enteropneusts appear to be a ciliary filter-feeding structure (Chapter 50) whereas the three chordate phyla are mucociliary filter feeders with the filter produced by the endostyle (Chapter 51). The presumed evolution of these structures is discussed in Chapter 51.

The similarities between the complicated gill structures of amphioxus and the ptychoderid enteropneusts are striking, but since there are much simpler shapes of gill slits and gill skeletons in other enteropneusts a convergence appears likely. This is also indicated by the differences in the extent of the coelomic cavities in the two groups. On the other hand, the gill slits have similar position, function and origin in the four phyla, and these similarities indicate homology.

The ciliation shows much variation among the four phyla. Adult enteropneusts have multiciliary cells in both ectoderm, endoderm and in some of the special organs; the larvae have monociliate cells everywhere in the ectoderm except for the perianal ciliary ring where each cell has a large compound cilium (Chapter 50; Fig. 43.5). Urochordates and vertebrates (Chapters 52, 54) have mainly multiciliate epithelia but monociliate cells are found in certain organs. Cephalochordates are exclusively monociliate (Chapter 53).

The multiciliate condition is regarded as an apomorphy of the cyrtotretes, and it may be significant that multiciliate epithelia with an accessory centriole at the base of each cilium is found only in the enteropneusts, where it may be regarded as a transitory stage from the monociliate stage, which always has the accessory centriole, to the multiciliate stage which generally lacks them (see also discussion of multiciliarity in Chapter 9). The complete lack of multiciliate cells in amphioxus must be interpreted as a return to the primitive condition.

Maisey (1986) mentioned that the presence of giant nerve cells in all the cyrtotrete phyla could be interpreted as a synapomorphy, but giant nerve cells have been described in *Phoronis* (Chapter 44) and in several protostomes, such as annelids, so the presence alone cannot indicate a synapomorphy.

Another character which has been drawn into the discussion is the dorsal, mesosomal nerve tube of the enteropneusts. It does resemble the neural tube of the chordates in position, shape and ontogeny, but functionally it is apparently not brain-like and there is no sign of an associated chorda (Chapter 50).

Bateson (1886b) was probably the first to recognize the the close relationships between the enteropneusts and the chordate phyla, and he drew a tree showing the tunicates as 'sister group' to the enteropneusts + notochordates (= cephalochordates + vertebrates, see Chapter 51). The emphasis was clearly on the serial gill slits (and on the dorsal nerve tube and the 'notochord'). The anatomy of the pterobranchs was hardly known and their affinities with the enteropneusts were not proposed until the following year (Harmer 1887). Cladograms of the chordate relationships have been presented and discussed in a number of recent papers, for example Maisey (1986), Schaeffer (1987) and Gans (1989), but they have all discussed the relationships between the chordates and the 'hemichordates', a group which must be regarded as polyphyletic (Chapter 47). The cladograms have been identical to that in Fig. 49.2 if the word Hemichordata is replaced by Enteropneusta, and the discussions have all been centred around enteropneust characters, generally disregarding the pterobranch characters in the matrices. The disagreement between the two types of cladograms is therefore more a matter of nomenclature than of opinion, but it should be stressed that only the cladogram in Fig. 49.2 follows the precise cladistic argumentation.

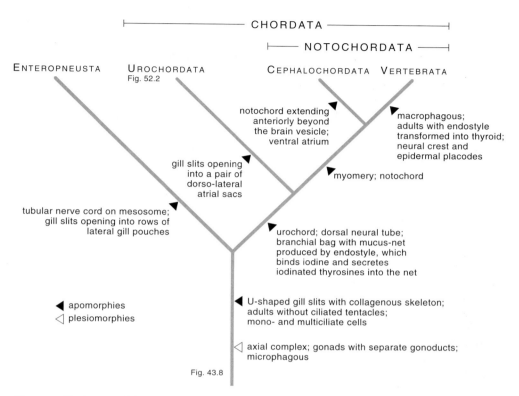

Fig. 49.2. Phylogeny of the Cyrtotreta.

The above discussion has demonstrated that the cyrtotretes must be regarded as a monophyletic group, and the discussion in Chapter 51 will show that the enteropneusts and the chordates must be sister groups with relationships as those shown in Fig. 49.2. The position of the cyrtotretes among the neorenalians is clearly shown by the presence of the characteristic axial complex in the enteropneusts, but the chordates have evolved alternative types of kidneys.

References

Bateson, W. 1886a. Continued account of the later stages in the development of *Balanoglossus kowalevskii*, and of the morphology of the Enteropneusta. – Q. Jl microsc. Sci., N.S. **26**: 511-533, pls 28-33.

Bateson, W. 1886b. The ancestry of the Chordata. – Q. Jl microsc. Sci., N.S. **26**: 535-571.

Dawydoff, C. 1948. Classe des Entéropneustes. – Traité de Zoologie, vol. **11**, pp 369-453. Masson, Paris.

Drach, P. 1948. Embranchement des Céphalochordés. – Traité de Zoologie, vol. **11**, pp 931-1037. Masson, Paris.

Franz, V. 1927. Morphologie der Akranier. – Ergebn. Anat. Entwicklungsgesch. **27**: 464-692.

Gans, C. 1989. Stages in the origin of vertebrates: analysis by means of scenarios. – Biol. Rev. **64**: 221–268.

Harmer, S.F. 1887. Report on *Cephalodiscus dodecalophus* M'Intosh, a new type of Polyzoa, procured on the Voyage of H.M.S. Challenger during the years 1873–76. Appendix. – Challenger Rep., Zool. **20**(62): 39–47.

Knight-Jones, E.W. 1953. Feeding in *Saccoglossus*. – Proc. zool. Soc. Lond. **123**: 637-654, pl. 1.

Krause, R. 1923. Mikroskopische Anatomie der Wirbeltiere in Einzeldarstellungen, vol. 4 (pp 609-906). – Walter de Gruyter, Berlin.

Lönnberg, E. 1901-1905. Leptocardii. – Bronn's Klassen und Ordnungen des Tierreichs, 6. Band, 1. Abt., 1. Buch, pp 99-249. Akademische Verlagsgesellschaft, Leipzig.

Mackie, G.O., D.H. Paul, C.M. Singla, M.A. Sleigh & D.E. Williams 1974. Branchial innervation and ciliary control in the ascidian *Corella*. – Proc. R. Soc. Lond. B **187**: 1-35.

Maisey, J.G. 1986. Heads and tails: a chordate phylogeny. – Cladistics **2**: 201-256.

Pardos, F. & J. Benito 1984. Estudio histológico de la faringe de *Glossobalanus minutus* (Enteropneusta, Ptychoderidae). – Boln R. Soc. esp. Hist. nat., Biol. **80**: 101-118.

Rähr, H. 1982. Ultrastructure of gill bars of *Branchiostoma lanceolatum* with special reference to gill skeleton and blood vessels (Cephalochordata). – Zoomorphology **99**: 167-180.

Schaeffer, B. 1987. Deuterostome monophyly and phylogeny. – Evol. Biol. **21**: 179-235.

Spengel, J.W. 1893. Die Enteropneusten des Golfes von Neapel. – Fauna Flora Golf. Neapel **18**: 1-758, 37 pls.

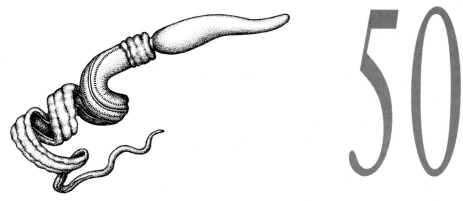

Phylum ENTEROPNEUSTA

Enteropneusts or acorn worms form a well-defined phylum of about 70 burrowing or creeping, marine, worm-like species, usually referred to four families. There is no reliable fossil record. Many species have a characteristic planktotrophic larval stage called a tornaria. The spherical, centimetre-large, planktotrophic organism with very complicated ciliary bands known as *Planctosphaera* (Spengel 1932, van der Horst 1927-39) is possibly an enteropneust larva. The monograph of van der Horst (1927-39) is still an indispensable source of information about anatomy and development, and if no special reference is given in the following, the information can be found there.

The body is archimeric (Chapter 43) with three clearly defined regions: an almost spherical to elongate conical proboscis, a short collar and a long main body which has two lateral rows of U-shaped gill slits in the anterior part. The gut is a straight tube from the ventral mouth at the base of the proboscis to the posterior tip of the body.

The epithelia of the adult enteropneusts consist mainly of multiciliate cells, but monociliate secretory cells have been observed in the preoral ciliary organ of *Saccoglossus* (Welsch & Welsch 1978). In the multiciliate ectodermal cells of *Saccoglossus*, each cilium has an accessory centriole (Hrauda 1987), but such centrioles are apparently lacking in *Glossobalanus* (Saita, Castellani & Tripepi 1978). In the pharynx of *Glossobalanus*, only the lateral cells of the gill-bars have cilia with accessory centrioles (Pardos 1988). Also the epithelial cells at the lumen of the stomochord are multiciliate (Welsch & Storch 1970). The lateral and frontal ciliary bands of the gill bars of *Saccoglossus* consist of very narrow cells (Knight-Jones 1953) and it should be checked whether these epithelia are actually monociliate. In the larvae, the general epithelium and the perioral ciliary band, the neotroch, consist of monociliate cells, whereas the perianal ciliary ring consists of multiciliate cells (Strathmann & Bonar 1976).

The proboscis or prosome has a coelomic cavity derived from the protocoel (see below), but the cavity is rather narrow because the mesoderm forms a thick layer of muscles. There is a small ciliated duct connecting the left side of the protocoel with the

---

Chapter vignette: *Saccoglossus kowalevskii*. (Redrawn from Sherman & Sherman 1976.)

surface at the posterodorsal side of the proboscis. The dorsal side of the pharynx and the underside of the proximal part of the proboscis is supported by a thickened basement membrane, the so-called proboscis skeleton. An anterior extension of the pharynx, the stomochord, is another structure which is apparently supporting the proximal part of the proboscis. It is hollow with ciliated epithelium and surrounded by a strengthened basement membrane (Welsch & Storch 1970, Balser & Ruppert 1990).

The protocoel, stomochord and associated blood vessels form an axial complex which is an ultrafiltration kidney (Balser & Ruppert 1990; Fig. 46.1). A median blood vessel carries the blood anteriorly on the dorsal side of the stomochord and posteriorly on the ventral side. The anterior and ventral parts of the vessel have strongly folded walls consisting of peritoneal cells in the shape of podocytes resting on a basement membrane, and the structure, called the glomerulus, is the site of formation of primary urine. A pericardial sac of myoepithelial cells and paired muscular extensions from the mesocoels surround the dorsal side of the blood vessel. A modification of the urine before it leaves the proboscis pore has been suggested but not proven.

The collar region has a pair of mesocoelomic cavities, each with an anterodorsal ciliated coelomoduct, but the median septum is incomplete in many species. The coelomopores are situated in an ectodermal invagination in connection with the sac of the first gill slit. The function of the coelomoducts is unknown.

The main body or metasome is long and comprises an anterior region with numerous gill slits and gonads and a long tail with a terminal anus. There is an undivided metacoelomic sac on each side, without any metanephridium or coelomopore.

The two rows of gill slits develop as series of circular openings between endoderm and ectoderm; they soon become U-shaped by the development of a so-called tongue bar, an outgrowth from the dorsal side of the opening (Bateson 1886); the bars between adjacent gill slits are called gill bars or septa. The system becomes supported by a skeleton, which is a strengthened basement membrane with collagen-like fibrils in an amorphous matrix (Pardos & Benito 1988a). The skeleton develops as curved bars or sheets along the anterior and posterior sides of the original gill slits and remain in this undifferentiated shape in *Protoglossus* (Caullery & Mesnil 1904). In most other species the skeletal parts differentiate further and send extensions into the tongue bars; the elements from the two sides finally fuse so that structures of very characteristic shapes are formed (Fig. 49.1). The tongue bars are free and can be moved in some species, but for example the ptychoderids have transverse bars, synapticles, containing extensions of the skeletal system across the gill slits, so that a more rigid gill basket is formed. The main skeletal element of the tongue bars contains a coelomic extension whereas the gill bars are solid.

The gill slits open into ectodermal invaginations called gill pouches; each U-shaped gill slit has its own gill pouch in some species whereas several gill slits open into a common pouch in other species.

The function of the gill slits is not well understood and their role in respiration has been questioned (Pardos & Benito 1988a, Northcutt & Gans 1983). Knight-Jones (1953) studied the feeding mechanism and the function of the cilia of two species of *Saccoglossus*. He found that particles are captured in mucus secreted from various gland cells on the proboscis and transported to the mouth and through the gut by a dense ciliation. The ciliation of the gill slits was studied on sections and on live specimens cut open and observed from the inside of the pharynx. The gill bars have characteristic ciliary bands resembling the bands on the gill filaments of bivalves. A wide lateral ciliary band transports water through the gill slits to the gill pouches from where the water leaves through the narrower gill pores, but particles are retained on the frontal side of the bars and transported dorsally along the grooves between the bars; at some point, the particles are taken over by the general ciliation which transports particles posteriorly towards the intestine. It seems probable that the enteropneusts can be both deposit-feeders and suspension-feeders; the presence of sand in gut and castings indicates deposit feeding, but some of this may be associated with burrowing; on the other hand, the structure and beating of the ciliary bands of the gill slits resemble those of the bivalve gills, which secrete and transport mucus when disturbed, but which are now generally thought to be ciliary feeders when undisturbed (Jørgensen 1990).

Burdon-Jones (1962) studied *Balanoglossus* and found similar ciliary patterns and observed that a suspension of 1-2 μm graphite particles was cleared by the pharynx. He concluded that deposit feeding both by engulfing of substratum and by ciliary mechanisms of the proboscis is the most important type of feeding, but that filter feeding may be of some importance too.

The nervous system has been studied in detail by Bullock (1946), Silén (1950) and Knight-Jones (1952). There is a net of fine nerve fibres everywhere in the basis of the epithelia, with distinctive concentrations in all body regions. The proboscis has many longitudinal nerves connected to a thicker ring at the posterior side; this ring is extended dorsally and continues posteriorly through the narrow proboscis stalk to the underside of the neural tube of the collar region. The most conspicuous component of the nervous system is the dorsal neural tube or collar cord, which is an invagination of the dorsal epithelium (see below). It is hollow in a few species, but the lumen becomes partly or completely obliterated in several species. The collar cord seems mainly to be a nervous pathway and much of the autonomous nerve activity appears to be centred in the proboscis. A middorsal nerve cord extends from the collar cord to the anus and a number of lateral cords curve around the anterior side of the branchial region to a conspicuous midventral cord which extends to the anus. Giant unipolar nerve cells form a small group of cell bodies in the anterior end and a larger group in the posterior end of the collar cord; their axons cross over to the opposite side and run anteriorly or posteriorly, respectively, with the posterior axons following the the prebranchial ring nerves to the midventral nerve cord. There is an extensive system of nerve cells in the epithelium of all regions of the gut with a midventral concentration which is most prominent in the anterior part; a weaker nerve is found middorsally. A nerve plexus is present in the ventral mesentery, with nerve fibres traversing the basement membranes both to the endoderm and the ectoderm. There are apparently no special sense organs in the adults, but sensory cells are found everywhere in the epithelia.

The haemal system (Pardos & Benito 1984) is situated between the basement membranes and comprises the above-mentioned anterior heart complex, which pumps the blood from the mediodorsal vessel to the ventral vessels surrounding the pharynx and uniting into a ventral longitudinal vessel. The blood leaks through a subintestinal plexus in the ventral part of the metasome to a pair of lateral longitudinal vessels from which smaller vessels run dorsally along the gill bars to the dorsal end of the gill slits, where they divide and send one branch ventrally along the anterior and one along the posterior side of the neighbouring tongue bars. Numerous anastomosing narrow blood spaces connect these blind-ending vessels with a median vessel leading the blood to the median dorsal vessel. Many of these blood spaces are located inside the skeletal rods of the gills, and a respiratory function of these areas seems unlikely (Pardos & Benito 1988a). On the other hand, there are apparently quite extensive blood lacunae under the thin epithelium of the parts of both tongue bars and gill bars facing the branchial sac (Pardos & Benito 1988a, figs 1, 12), so gas exchange may after all take place in this region. A more enigmatic feature of the haemal system is the presence of podocytes in the coelomic lining of some of the blood spaces in the branchial sacs of *Glossobalanus* (Pardos & Benito 1988b). The function of such structures is usually believed to be ultrafiltration from the blood and the formation of primary urine, but the metacoels have no ducts in the enteropneusts, so the function of this structure is uncertain.

The gonads occur in rows along the sides of the metasome. Each gonad is sac-shaped with a short gonoduct; they are apparently formed from the peritoneum, but the exact origin is unknown. Fertilization is external in most species.

The first cleavage is median and the second cleavage divides the embryo into a dorsal and a ventral half; the third cleavage separates the future prosomal and mesosomal ectoderm from the future metasomal ectoderm and mesoderm + endoderm (Colwin & Colwin 1951; Fig. 51.1). A coeloblastula is soon formed and gastrulation is by invagination. In the planktotrophic forms, the ciliated embryo hatches at this stage and the whole epithelium consists of monociliate cells. The mouth breaks through from the apical end of the archenteron, and a convoluted ring of ciliated cells around the new mouth becomes specialized as the neotroch, which is an upstream-collecting ciliary band (Strathmann & Bonar 1976, Nielsen 1987). This band becomes highly complex in many larvae, forming loops on small tentacles (Fig. 43.4). The neotroch is both the feeding and the locomotory organ of the youngest larvae, but a large, perianal ring of compound cilia soon develops and takes over as the swimming organ (Fig. 50.1). The compound cilia are formed on a band of multiciliate cells, which is several cells wide with the cells arranged in a regular rhomboid pattern in some species (Spengel 1893; Fig. 43.5). The compound cilia are at first not much longer than the single cilia of the neotroch (about 20-25 μm), but with increasing size of the larva the compound cilia increase in length, and Strathmann & Bonar (1976) estimated the length of compound cilia of large tornaria larvae to be more than 200 μm. The lecithotrophic larvae remain uniformly ciliated except for the apical tuft and the perianal ring of compound cilia (Burdon-Jones 1952).

An apical organ with a ciliary tuft is found in all larvae, and some species have a pair of eyes with a pigment cup and a lens at the sides of the apical organ. Each photosensitive cell has both a modified cilium and an array of microvilli, which makes this type of cell unique among the photoreceptor cells (Brandenburger, Woollacott & Eakin 1973).

The development of the coelomic sacs (Fig. 43.2) has attracted much attention for over a century, but it appears that some of the developmental types which are repeatedly illustrated in the textbooks are not well documented. The protocoel is apparently always pinched off as a spacious pouch from the anterior part of the archenteron. The meso- and metacoels have different origins in different species, but some of the reports seem to be based on observations of only a few stages and should be checked.

The following species with direct development have been reared from eggs and studied in more detail, and the reports appear reliable: *Saccoglossus pusillus*, *S. kowalevskii* and *Balanoglossus clavigerus*. Davis (1908) studied *D. pusillus* and found that the lateral parts of the protocoel extended posteriorly along the sides of the gut and that meso- and metacoel became established when these long extensions became isolated from the protocoel and broke up into an anterior and a posterior compartment. Bateson (1884) studied *S. kowalevskii* and observed that both the meso- and the metacoelomic sacs developed as separate pockets from the lateral walls of the gut. Stiasny (1914a,b) made new studies and discussed earlier reports on *B. clavigerus* and concluded that meso- and metacoel developed together from a pair of lateral pockets from the posterior part of the gut, and that the two elongate pockets afterwards divided into meso- and metacoel. Tornaria larvae from plankton of several oceans have revealed that the feeding tornaria larvae generally form meso- and metacoel in a slightly different way. Morgan (1891, 1894) noted that these coelomic compartments form when the larvae have already begun to feed, and that the coeloms are therefore compact instead of hollow. Dawydoff (1944, 1948) described similar types and it can be concluded that two additional types of formation of meso- and metacoel occur: 1) by isolated cells given off from the lateral parts of the protocoel (corresponding to the hollow pouches in *S. pusillus*) and 2) by compact cell groups given off from two zones of the gut (corresponding to the pouches in *S. kowalevskii*). A number of smaller deviations from these types have been mentioned but need not be considered here, except for the type described by Dawydoff (1944) in a tornaria from Indochina, in which the coelomic pouches were observed to form from a flat sheet of cells which then folded together surrounding a small cavity; this is remarkably similar to the process of coelom formation observed in the brachiopod *Crania* (Chapter 45).

The heart vesicle is variously reported to originate from ectoderm (Stiasny 1914b) or mesoderm (Morgan 1894, Ruppert & Balser 1986). It soon develops muscle cells, and can then be seen making small pulsating movements.

The primary body cavity of older larvae is filled by a gelatinous material which is of importance both for the stability of the shape and for several of the developmental processes (Strathmann 1989).

At metamorphosis, the planktotrophic larvae lose the neotroch very rapidly, while the perianal ring can be recognized for a while (Fig. 50.1). The metamorphosis of the lecithotrophic larvae is more gradual, and the perianal ring becomes pulled out onto a small, temporary 'tail' ventral to the anus (Burdon-Jones 1952).

Large tornaria larvae show two to three pairs of gill pockets on the pharynx near the mouth. At metamorphosis the gut becomes pulled backwards so that the

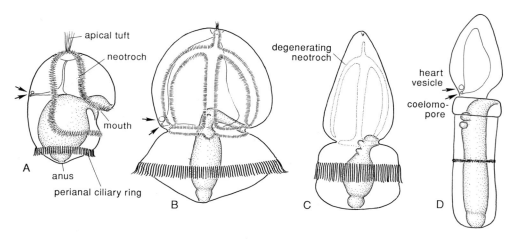

Fig. 50.1. Developmental stages of enteropneusts with tornaria larvae seen from the left side. – A, a young larva (*Saccoglossus clavigerus*; after Stiasny 1914a). – B, a full-grown larva. – C, a larva which has just commenced metamorphosis; the neotroch is degenerating and the gut is being pulled posteriorly. – D, a newly settled specimen; the collar with the collar cord has been formed and the gill pouches are now located in the anterior end of the metasome but have not yet broken through to the exterior (B-D, unidentified larvae from plankton; mainly based on Morgan 1894).

pockets become situated in the anterior part of the metasome where the gill slits break through (Fig. 50.1). Additional gill slits develop in a series behind the first few openings. The newly formed gill openings are oval, but a tongue bar soon develops from the dorsal side of the pore. The metacoel finally grows around the canals and forms the musculature, and the skeletal basement membrane develops between the cell layers.

The basiepithelial nervous system forms concentrations along the median zones and the collar cord is formed by regular infolding in some species and through a de-lamination-like process in others.

The collar cord resembles the neural tube of the chordates, but since it has none of the functions of a brain it could perhaps better be compared to the infolded nerves of some of the echinoderms. There is no specific indication of homology with the neural tube complex of the chordates.

The possession of both an axial complex and the series of gill slits place the en-teropneusts in a key position within the deuterostomes. The homology of the axial complexes of pterobranchs, echinoderms and enteropneusts can hardly be doubted, so these phyla must belong to a monophyletic group. The gill slits are regarded as a synapomorphy of the cyrtotretes, i.e. enteropneusts and chordates, and these two groups together must form a monophyletic group. The cephalochordates and verte-brates have paired rows of gill slits which are considered homologous to those of the enteropneusts, but the urochordates have gill slits where the U-shape is only weakly indicated in the development of certain species (Chapter 52). However, the three chordate phyla are characterized by the possession of the chorda/neural plate complex and a number of other apomorphies (Chapter 51) which demonstrate the mono-phyletic character of the chordates. Taken together these character complexes indicate that the enteropneusts and chordates are sister groups, and have a phylogeny as that shown in Fig. 51.2.

It can be questioned whether the gill slits of the cyrtotretes are homologous to the gill pores of *Cephalodiscus*, as suggested by the trochaea theory, but the position in the anteriormost part of the metasome and the development of the openings appear identical. Gilmour (1982) suggested that the gill slits of the enteropneusts were an adaptation providing an outlet of excess water from the ciliary feeding process of the larva, but the gill slits do not open until after the metamorphosis of the larva when the larval ciliary bands have ceased to function. The gill openings can therefore not be interpreted as adaptations to larval ciliary feeding; their functions in feeding and respiration of the adults have been discussed above. The single gill pores of the pterobranch *Cephalodiscus* (Chapter 47) may have a function such as that suggested by Gilmour (1982, see also 1979).

There is no indication of an endostyle in the enteropneusts, and the mucus cells in the oesophagus of the tornaria larvae (Gilmour 1982) cannot be accepted as an endostyle homologue until more specific similarities have been demonstrated. The enteropneusts are capable of binding iodine and of secreting mono-iodotyrosine, one of the compounds secreted by the endostyles, but the function is apparently not restricted to special glands (Barrington & Thorpe 1963).

Interesting subjects for future research

1. Function of the gill slits.
2. Development of the nervous system.

References

Balser, E.J. & E.E. Ruppert 1990. Structure, ultrastructure, and function of the preoral heart-kidney in *Saccoglossus kowalevskii* (Hemichordata, Enteropneusta) including new data on the stomochord. – Acta zool. (Stockh.) 71: 235-249.

Barrington, E.J.W. & A. Thorpe 1963. Comparative observations on iodine binding by *Saccoglossus horsti* Brambell and Goodhart, and by the tunic of *Ciona intestinalis*. – Gen. comp. Endocrinol. 3: 166-175.

Bateson, W. 1884. The early stages in the development of *Balanoglossus* (sp. incert.). – Q. Jl microsc. Sci., N.S. 24: 208-236, pls 18-21.

Bateson, W. 1886. Continued account of the later stages in the development of *Balanoglossus kowalevskii*, and the morphology of the Enteropneusta. – Q. Jl microsc. Sci., N.S. 26: 511-533, pls 28-33.

Brandenburger, J.L., R.M. Woollacott & R.E. Eakin 1973. Fine structure of eyespots in tornarian larvae. – Z. Zellforsch. 142: 89-102.

Bullock, T.H. 1946. The anatomical organization of the nervous system of enteropneusts. – Q. Jl microsc. Sci., N.S., 86: 55-111, pls 2-8.

Burdon-Jones, C. 1952. Development and biology of the larva of *Saccoglossus horsti* (Enteropneusta). – Phil. Trans. R. Soc. B 236: 553-590.

Burdon-Jones, C. 1962. The feeding mechanism of *Balanoglossus gigas*. – Bolm Fac. Filos. Ciênc. Univ. S Paulo, Zool. 24: 255-280.

Caullery, M. & F. Mesnil 1904. Contribution à l'étude des Entéropneustes. *Protobalanus* (n.g.) *koehleri* Caull. et Mesn. – Zool. Jb., Anat. **20**: 227-256, pls 12-13.

Colwin, A.L. & L.H. Colwin 1951. Relationships between the egg and larva of *Saccoglossus kowalevskii* (Enteropneusta): axes and planes; general prospective significance of the early blastomeres. – J. exp. Zool. **117**: 111-137.

Davis, B.M. 1908. The early life-history of *Dolichoglossus pusillus* Ritter. – Univ. Calif. Publs Zool. **4**: 187-226, pls 4-8.

Dawydoff, C. 1944. Formation des cavités coelomiques chez les tornaria du plancton indochinois. – C. r. hebd. Séanc. Acad. Sci., Paris **218**: 427-429.

Dawydoff, C. 1948. Classe des Entéropneustes. – Traité de Zoologie, vol. 11, pp 369-453. Masson, Paris.

Gilmour, T.H.J. 1979. Feeding in pterobranch hemichordates and the evolution of gill slits. – Can. J. Zool. **57**: 1136-1142.

Gilmour, T.H.J. 1982. Feeding in tornaria larvae and the development of gill slits in enteropneust hemichordates. – Can. J. Zool. **60**: 3010-3020.

Hrauda, G. 1987. Licht- und elektronenmikroskopische Untersuchungen an der Körperdecke von *Saccoglossus kowalewskii* und *Saccoglossus mereschkowskii* (Enteropneusta, Hemichordata). – Zool. Jb., Anat. **116**: 399-408.

Jørgensen, C.B. 1990. Bivalve Filter Feeding: Hydrodynamics, Bioenergetics, Physiology and Ecology. – Olsen & Olsen, Fredensborg.

Knight-Jones, E.W. 1952. On the nervous system of *Saccoglossus cambrensis* (Enteropneusta). – Phil. Trans. R. Soc. B **236**: 315-354, pls 32-35.

Knight-Jones, E.W. 1953. Feeding in *Saccoglossus* (Enteropneusta). – Proc. zool. Soc. Lond. **123**: 637-654, 1 pl.

Morgan, T.H. 1891. The growth and metamorphosis of tornaria. – J. Morph. **5**: 407-458, pls 24-28.

Morgan, T.H. 1894. The development of *Balanoglossus*. – J. Morph. **9**: 1-86, pls 1-6.

Nielsen, C. 1987. Structure and function of metazoan ciliary bands and their phylogenetic significance. – Acta zool. (Stockh.) **68**: 205-262.

Northcutt, R.G. & C. Gans 1983. The genesis of neural crest and epidermal placodes: a reinterpretation of vertebrate origins. – Q. Rev. Biol. **58**: 1-28.

Pardos, F. 1988. Fine structure and function of pharynx cilia in *Glossobalanus minutus* Kowalewsky (Enteropneusta). – Acta zool. (Stockh.) **69**: 1-12.

Pardos, F. & J. Benito 1984. Blood circulatory system in the pharynx of an enteropneust *Glossobalanus minutus* (Ptychoderidae). – Riv. Biol. **77**: 69-85.

Pardos, F. & J. Benito 1988a. Blood vessels and related structures in the gill bars of *Glossobalanus minutus* (Enteropneusta). – Acta zool. (Stockh.) **69**: 87-94.

Pardos, F. & J. Benito 1988b. Ultrastructure of the branchial sacs of *Glossobalanus minutus* (Enteropneusta) with special reference to podocytes. – Arch. Biol. (Brux.) **99**: 351-363.

Ruppert, E.E. & E.J. Balser 1986. Nephridia in the larvae of hemichordates and echinoderms. – Biol. Bull. Woods Hole **171**: 188-196.

Saita, A., L.C. Castellani & S. Tripepi 1978. The integument of *Glossobalanus minutus* Kowalevsky (Enteropneusta Ptychoderidae). Ultrastructural analysis. – Monitore zool. ital., N.S. **12**: 155-179.

Sherman, I.W. & V.G. Sherman 1976. The Invertebrates: Function and Form. A Laboratory Guide. – Macmillan, New York.

Silén, L. 1950. On the nervous system of *Glossobalanus marginatus* Meek (Enteropneusta). – Acta zool. (Stockh.) **31**: 149-175.

Spengel, J.W. 1893. Die Enteropneusten des Golfes von Neapel. – Fauna Flora Golf. Neapel **18**: 1-758, pls 1-37.

Spengel, J. 1932. *Planctosphaera pelagica*. – Rep. scient. Results Michael Sars N. Atlant. deep Sea Exped. **5**(5): 1-28, 1 pl.

Stiasny, G. 1914a. Studien über die Entwicklung des *Balanoglossus clavigerus* Delle Chiaje.– I. Die Entwicklung der Tornaria. – Z. wiss. Zool. **110**: 36-75, pls 4-6.

Stiasny, G. 1914b. Studien über die Entwicklung des *Balanoglossus clavigerus* Delle Chiaje. II. Darstellung der weiteren Entwicklung bis zur Metamorphose. – Mitt. zool. Stn Neapel **22**: 255-290, pls 6-9.

Strathmann, R.R. 1989. Existence and functions of a gel filled primary body cavity in development of echinoderms and hemichordates. – Biol. Bull. Woods Hole **176**: 25-31.

Strathmann, R. & D. Bonar 1976. Ciliary feeding of tornaria larvae of *Ptychodera flava* (Hemichordata: Enteropneusta). – Mar. Biol. (Berl.) **34**: 317-324.

van der Horst, C.J. 1927-39. Hemichordata. – Bronn's Klassen und Ordnungen des Tierreichs, 4. Band, 4. Abt., 2. Buch, 2. Teil, pp 1-737. Akademische Verlagsgesellschaft, Leipzig.

Welsch, U. & V. Storch 1970. The fine structure of the stomochord of the enteropneusts *Harrimania kupfferi* and *Ptychodera flava*. – Z. Zellforsch. **107**: 234-239.

Welsch, L.T. & U. Welsch 1978. Histologische und elektronenmikroskopische Untersuchungen an der präoralen Wimpergrube von *Saccoglossus horsti* (Hemichordata) und der Hatschekschen Grube von *Branchiostoma lanceolatum* (Acrania). Ein Beitrag zur phylogenetischen Entwicklung der Adenohypophyse. – Zool. Jb., Anat. **100**: 564-578.

51

CHORDATA

The three living phyla Urochordata, Cephalochordata and Vertebrata, which are classified as chordates, have been subjects of comparative anatomical and embryological studies for over a century, and it is now almost unanimously agreed that the group is monophyletic and that cephalochordates and vertebrates are sister groups. Several alternative, mostly rather fantastic ideas have been proposed over the years. The calcichordate theory, which derives the chordate phyla from various fossils which most zoologists and palaeontologists interpret as echinoderms, has been discussed in Chapter 48. The theory which derives the chordates from nemertines was proposed in the last century and taken up again by Jensen (1963) and Willmer (1974); it is based on very little direct evidence and will not be discussed here.

The group consisting of cephalochordates and vertebrates is referred to in various ways, for example the somitic chordates (Schaeffer 1987), the cephalochordate-vertebrate assemblage (Gans 1989), but the group is apparently without a formal name. I find it practical for the discussion to have a name for the group, and I therefore propose the name Notochordata for the supposedly monophyletic group consisting of the phyla Cephalochordata and Vertebrata. In this connection it appears clearer to use the term chorda as a collective term for the dorsal, stiffening rod formed from the archenteron in all the chordates, the term urochord for the chorda in the tail of urochordate larvae and adult larvaceans, and to restrict the term notochord to cover the structure along the whole dorsal side of cephalochordates and vertebrates.

Several characters unite the three phyla: 1) The chorda, a stiffening rod of tissue formed from the roof of the archenteron. 2) The dorsal neural tube, formed from a longitudinal mid-dorsal fold of ectoderm mostly overlying the chorda. 3) The longitudinal muscles along the chorda, used in locomotion by creating wagging or undulatory movements of a finned tail. 4) The ciliated pharyngeal gill slits, which support a mucus filter secreted by the endostyle; this last character is found in the vertebrates only in the somewhat modified filtration system of the ammocoetes larva of the lampreys, but the various cell types of the endostyle can be recognized in the thyroid of all the adult vertebrates. 5) The mouth develops on the 'dorsal' side of the apical pole (Fig. 51.1) which makes it difficult to homologize it with the mouth of the 'lower' deuterostomes. Characters 1-3 are not independent, since the chorda-mesoderm in-

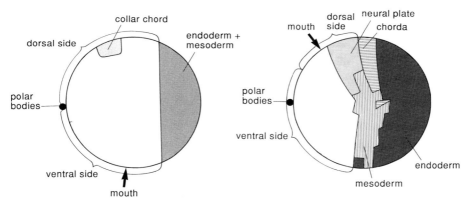

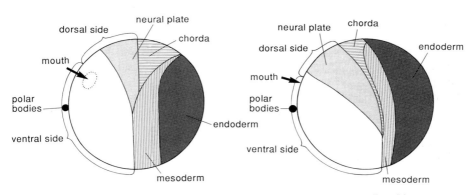

Fig. 51.1. Fate maps of enteropneusts and chordates; fertilized eggs or blastula stages seen from the left side. – Enteropneusta: *Saccoglossus kowalevskii* (based on Colwin & Colwin 1951). – Urochordata: *Halocynthia roretzi* (based on Nishida 1987). – Cephalochordata: *Branchiostoma lanceolatum* (based on Tung, Wu & Tung 1962). – Vertebrata: *Bombinator pachypus* (based on Vogt 1929).

duces the formation of the neural tube during ontogeny and the lateral musculature of the tail (or longer parts of the body) is innervated from the neural tube, and this probably means that the three structures evolved together.

The chorda is a cylinder of cells surrounded by a strengthened basement membrane. However, the cellular elements of the chorda are not uniform in the chordate phyla. In the urochordates, the early larval chorda is a row of cylindrical cells, but in most of the older larvae it becomes transformed to a hollow tube of cells surrounding a core of elastic material, and this is also seen in the adult larvaceans (Chapter 52). In amphioxus, the chorda consists of a stack of flat muscle cells (Chapter 53). In the vertebrates, the chorda is a rod of irregularly arranged cells with large vacuoles in the agnathes, but this structure is modified by the development of the vertebrae in the gnathostomes.

The neural tube is formed as a longitudinal infolding of the middorsal ectoderm, the neural plate, overlying the chorda. The lateral edges of the neural plate fuse medially so that the neural tube becomes situated below the ectoderm. Posteriorly, the folds continue around the blastopore, which becomes enclosed by the fusion of the folds so that the archenteron and the neural canal become connected by a narrow canal, the neurenteric canal. Anteriorly, the infolding follows the notochord in cephalochordates and vertebrates, but in the urochordates, the short urochord of the tail apparently induces also the infolding of the anterior part of the nerve tube along the body (Chapter 52). Detailed structural similarities of the neural tube in the three phyla underline the homology of the structure, for example the presence of the infundibular organ/flexural organ which secretes Reissner's fibre (Olsson 1956).

There is an unsegmented longitudinal musculature in the urochordate tail (Chapter 52) and a segmented longitudinal musculature along most of the body in notochordates. The homology of these muscles is mainly based on ontogeny; the muscles develop from longitudinal zones of the archenteron lateral to the chorda in all three groups, but there is no sign of coelomic sacs or segmentation in the urochordates, while the muscle segments of the two other groups originate from mesodermal pockets from the archenteron (Chapters 53, 54). However, these muscles are all primarily involved in swimming, and their innervation from the neural tube and functional connection with the notochord strongly indicate that the whole structure has evolved as a functional unit.

The enteropneusts have a pharynx with gill slits of a structure very similar to that of the cephalochordates and tunicates, but without a mucous filter. The homology of these structures is partially based on the assumption that the ciliary filtering system of the enteropneusts can be modified to a mucociliary filter system.

A parallel example of such a transformation can be seen in the molluscs where the gills of bivalves and gastropods, which are considered homologous, have similar structures but different functions. The unspecialized gill is a respiratory organ with ciliary tracts which clean the gill. This ciliary system is modified to a filter feeding system of the upstream-collecting type without mucus in most bivalves; mucus is secreted in quantities only when the bivalves are stressed, and observations on stressed mussels have given rise to the older theories which regarded the bivalves as mucociliary feeders (Jørgensen 1990). Most gastropods have the unmodified gill type, but some of the prosobranchs, as for example *Crepidula*, have modified the gill to a filter-feeding structure which resembles that of the bivalves, but which uses a mucous net secreted by an endostyle-like gland at the base of the gill as the filtering structure (Werner 1953). The structure of the gill filaments of the filter-feeding gastropods and bivalves is very similar, although laterofrontal cilia have not been described in the gastropods.

It seems unquestionable that the ciliary filter-feeding gill slits of the enteropneusts, the mucociliary filter-feeding gill slits of cephalochordates and tunicates, and the respiratory gill slits of the vertebrates are homologous organs. The homology of the gill pores of *Cephalodiscus* seems more uncertain.

The endostyles of urochordates, cephalochordates and vertebrates comprise characteristic bands of cells with identical position and function. The ancestral chordate probably had some rows of glandular and ciliated cells around the entrance to the pharynx, or perhaps only along its ventral part, and these cells secreted and shaped the mucous net used in feeding. This complex may then have become folded posteriorly into a U-shape with the anteriormost cells situated along the bottom of a ventral groove with the more posterior bands forming parallel lines more dorsally. This is indicated in the ontogeny of amphioxus (Chapter 53). The cells of the endostyle secrete the net of mucus with proteins and additionally bind iodine which becomes secreted into the net or released into the blood as hormonal iodothyrosines (Goodbody 1974). Thyroxine, which is one of the well-known hormones of the vertebrate thyroid, is also formed in small quantities in amphioxus. The different bands of glandular and ciliated cells in the endostyles of urochordates, cephalochordates and the ammocoetes larva of the lampreys cannot be homologized directly, since the functions are not distributed uniformly in the various forms, but all forms apparently have a midventral zone of monociliate cells with very long cilia. The ascidians appear to have the most complicated endostyle, with 8 bands of cells on each side, while the appendicularian *Oikopleura* has only 3 bands with the major band performing both major secretory functions (Olsson 1963).

The complex of the gill slits and the endostyle forms a beautiful evolutionary transformation series: the first step is the ciliary filter-feeding slits of the enteropneusts, which lack an endostyle, but which have iodine-binding cells in several epithelia (Chapter 50). The second step is the mucociliary filter-feeding systems of urochordates and cephalochordates (Chapters 52, 53), which have an endostyle with bands of cells secreting the mucous filter and of cells which bind iodine and secrete iodinated proteins into the net and into the blood stream. The third step is the mucous filter-feeding system with the water pumped through the filter by contractions of the pharynx as seen in the ammocoetes larva (Chapter 54), which has an endostyle with only little importance for the secretion of the filter but with iodine-binding and -secreting cells. The fourth step is the exclusively respiratory gill system of adult lampreys (Chapter 54) and many other aquatic vertebrates, which have an endostyle modified into an endocrine gland, which accumulates iodine and secretes tyroxine (Chapter 54).

The last chordate apomorphy to be discussed here is the position of the mouth in relation to the apical pole. Fate maps of a number of chordates show that the mouth develops from an area between the apical pole and the neural plate (Fig. 51.1). In some ascidians, it even appears that the mouth originates from the anterior part of the neural plate (Chapter 52). This is in contrast to the conditions in the other deuterostomes, and since there is no indication of a shift of the 'old' mouth to the dorsal side, it must be supposed that the chordate mouth is not homologous with that of the other deuterostomes. This does not appear unacceptable because none of the chordates have feeding primary larvae, and the development of a new anus in a position ventral to the blastopore is an unavoidable consequence of the enclosure of the blastopore at neurulation.

The axial complex so characteristic of pterobranchs, echinoderms and enteropneusts is not found in any of the chordates, but various small coelomic pouches and ciliated ectodermal pits or pockets have been homologized with different parts of this complex. Ruppert (1990) contributed to our knowledge about structure and function of some of these structures and made a very informative summary and interpretation of the literature about the various structures (see Table 51.1). He pointed out that two completely separate organ complexes should be distinguished: the anterior coelomopore complex and the anterior neuropore. The anterior coelomopore complex consists of the left protocoel and the ectodermal invagination forming part of its coelomoduct, while the anterior neuropore is derived from the anterior opening of the neural tube. The anterior coelomopore complex is obviously part of the axial complex, while the anterior neuropore is restricted to the cyrtotrete phyla, which have a dorsal infolding of the nervous system. The collar chord of the enteropneusts is situated in the same position as part of the neural tube of the chordates, but its homology with the neural tube is doubtful (Chapter 50). The neural gland complex of the tunicates has a cavity derived from the neural canal, but the neuropore closes and the ciliated funnel develops as a new opening from the anterior part of the pharynx (Chapter 52); this makes a homology of the ciliated funnel and the neuropore very questionable. The presence and variation of these organs support the monophyly of the supraphyletic groups, but their role in our understanding of the interrelationships of the chordate phyla is far from clear, and they will not be discussed further.

The three pairs of coelomic sacs so characteristic of all the deuterostome phyla discussed above cannot all be recognized in the chordates. The protocoel and its derivatives have been mentioned above and are listed in Table 51.1. The mesocoel has to my knowledge not been identified with certainty in any chordate. The metacoel is perhaps represented by the musculature in the urochordate tail. The homologies of the coelomic sacs (except the first pair) of the cephalochordates and vertebrates are very uncertain, and it appears that attempts to use them in a phylogenetic analysis will only lead to circular reasoning.

The apomorphies of the various groups are summarized in Figs 49.2 and 51.2.

One theory for the origin of the chordates emphasizes the similarities of the gill systems of enteropneusts and chordates and regards the sessile ascidians as a evolutionary dead-end group (see for example Bateson 1886 and Garstang 1928). Garstang (1928, p. 51) directly stated that his interpretation 'implies an evolutional

Table 51.1. The structure/function of the anterior coelomopore complex (= left protocoel + ectodermal invagination) (and the right protocoel) in the neorenalian phyla; based on Ruppert (1990).

Pterobranchia	protocoel + proboscis pore (right protocoel = ?pericard)
Echinodermata	axocoel + hydropore (right protocoel = dorsal sac)
Enteropneusta	protocoel + proboscis pore (right protocoel = ?pericard)
Urochordata	not identified
Cephalochordata	larva: left anterior head coelom + preoral pit adult: absent + Hatschek's pit (right protocoel = right head coelom)
Vertebrata	embryo: premandibular somite + Rathke's pouch adult: eye muscles + adenohypophysis (right protocoel = eye muscles)

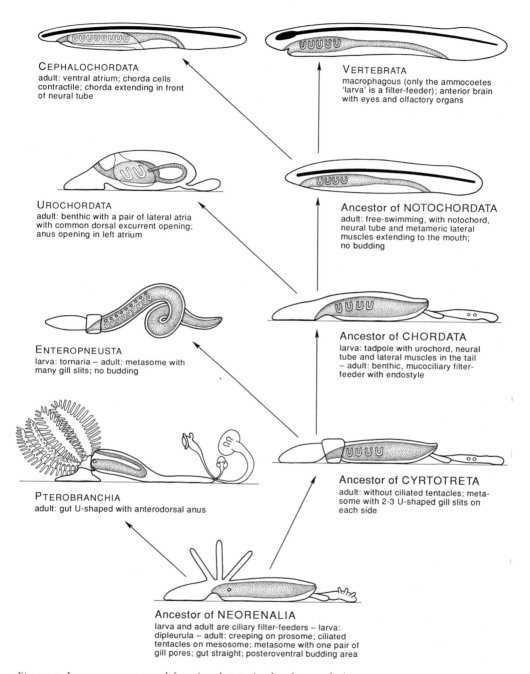

CEPHALOCHORDATA
adult: ventral atrium; chorda cells contractile; chorda extending in front of neural tube

VERTEBRATA
macrophagous (only the ammocoetes 'larva' is a filter-feeder); anterior brain with eyes and olfactory organs

UROCHORDATA
adult: benthic with a pair of lateral atria with common dorsal excurrent opening; anus opening in left atrium

Ancestor of NOTOCHORDATA
adult: free-swimming, with notochord, neural tube and metameric lateral muscles extending to the mouth; no budding

ENTEROPNEUSTA
larva: tornaria – adult: metasome with many gill slits; no budding

Ancestor of CHORDATA
larva: tadpole with urochord, neural tube and lateral muscles in the tail – adult: benthic, mucociliary filter-feeder with endostyle

PTEROBRANCHIA
adult: gut U-shaped with anterodorsal anus

Ancestor of CYRTOTRETA
adult: without ciliated tentacles; metasome with 2-3 U-shaped gill slits on each side

Ancestor of NEORENALIA
larva and adult are ciliary filter-feeders – larva: dipleurula – adult: creeping on prosome; ciliated tentacles on mesosome; metasome with one pair of gill pores; gut straight; posteroventral budding area

Fig. 51.2. Important structural-functional steps in chordate evolution.

progress of plankton-feeding organisms from a fixed condition with external ciliated tentacles and food-grooves to an eventually free and motile state with endopharyngeal apparatus of gill-slits and endostyle'. This is in complete agreement with the conclusions drawn in the preceding chapters about the evolution of the deuterostomes from a notoneuralian ancestor with ciliated tentacles to the enteropneusts with pharyngeal gill slits, and with the above-mentioned derivation of the mucociliary feeding structures of the chordates from the ciliary feeding structures of the enteropneusts.

Garstang (1928) further proposed that the ancestral urochordate had a free-living larval stage resembling a dipleurula. The circumoral ciliary band with associated nerve cells should then have become dislocated towards the dorsal side and developed into the edges of the infolding dorsal nervous system of the chordates (op. cit., fig. 3). This looks very appealing geometrically, with the posterior loop passing around the anus so that a neurenteric canal would be formed by the fusion of the bands. However, if we compare the body regions of this hypothetical ancestor with those of the metamorphosing tornaria a number of important differences become apparent: 1) In the tornaria, the neotroch is situated around the mouth with its main parts situated on the prosome; this is clearly shown at metamorphosis when the degenerating ciliary band can be recognized on the proboscis of the young bottom stage (Fig. 50.1). 2) The small endodermal primordia of gill pores formed already in the latest tornaria stages break through after metamorphosis on the metasome, i.e. behind the area of the neotroch and not in the perioral field as in Garstang's diagram.

The interpretation of the urochordates as an evolutionary dead-end group, i.e. as a monophyletic group, makes considerable room for interpretations of their phylogeny. It is possible to regard the ascidians and salps as a monoplyletic group characterized by the presence of the tunic, and the doliolarians and appendicularians as another monophyletic group characterized by their habit of moulting the cuticle or changing their house (which is a specialized cuticle) (see Chapter 52).

An alternative interpretation of the urochordates, and further of the chordates, regards the sessile ascidians as the ancestral group of urochordates and the pelagic urochordates, and further the notochordates as derived from neotenic tadpole larvae. Views of this nature, which are found both in a number of review papers and several textbooks, are especially clearly expressed by Berrill (1955). The ascidian tadpole larva with the urochord is interpreted as a specialized dispersal stage evolved within the ascidians. Berrill (1955, p. 11) explicitly rejected the homology of enteropneust and chordate gill system: 'the hemichordates represented by the pterobranchs and enteropneusts together are not in any way relevant to the story and may exhibit no more than a convergent resemblance to chordate organization in so far as gill slits and a dorsally placed nerve centre may have been independently acquired'. The origin of the ancestral ascidian is usually not discussed. A consequence of this interpretation is that the holopelagic tunicate groups are regarded as specialized, and that the doliolids and larvaceans should have lost the tunic.

Berrill's theory is not explicit about the closest relatives of the chordates, but more recent papers, such as Romer (1972), Maisey (1986), Schaeffer (1987), and

Gans (1989), appear almost unanimously to regard the enteropneusts as the closest and the pterobranchs as the second-closest sister group, although several of the papers use the less explicit collective term hemichordates. This 'softens' the rejection of the homology of the gill systems of enteropneusts and chordates, but the phylogenetic consequences are usually not discussed.

It is probably clear that I find Garstang's ideas much more probable than Berrill's, and I will try to outline a number of the most important evolutionary steps leading from the common ancestor of enteropneusts and chordates to the living chordate phyla (Fig. 51.2).

The ancestral cyrtotrete (Chapter 49) was a pelago-benthic organism with a dipleurula larva and an adult which crept or burrowed by means of the proboscis. Each side of the pharynx had a number of U-shaped gill slits which were ciliary feeding structures; there was no endostyle. This ancestor was derived from the common neorenalian ancestor, which had ciliated tentacles as the feeding apparatus and resembled a pterobranch with straight gut. Asexual reproduction took place as budding from a posteroventral area as in the pterobranchs.

The ancestral chordate specialized the pharyngeal gill apparatus to a mucociliary feeding structure with a mucous net produced by a ventral endostyle; the mucous net retained particles from the water current set up by the gill slits and was transported dorsally by the action of special cilia; the mucus with the captured particles was then transported to the intestine and digested. The gill slits were probably opening directly to the exterior. The other important innovation of this ancestor was seen in the larva, which elongated the posterior part of the body including the posterior ends of the metacoelomic sacs into a laterally compressed tail (Fig. 51.3). There was probably a median fin around the tail and the larva could move by wiggling the tail from side to side in addition to the slow movements caused by the cilia of the neotroch. This new swimming movement was created by longitudinal muscles in the metacoelomic sacs and with the coelomic sacs themselves functioning as antagonistic hydrostatic skeletons. Since none of the living chordates have planktotrophic larvae with ciliary bands, it seems probable that the larvae of the ancestor soon became lecithotrophic and lost the neotroch, and this is also in agreement with the closure of the blastopore by the formation of the neurenteric canal in the embryos.

An analogous larval specialization can be observed in larvae of the asteroid *Luidia* (Fig. 51.3; see also Fig. 48.3). In these larvae, the two protocoels are fused anteriorly and the preoral lobe is greatly elongated with two large anterior fins; the larvae swim by wiggling this anterior arm in the median plane. The muscles are found on the oral side of the arm and the protocoel is probably functioning as a hydrostatic skeleton (Tattersall & Sheppard 1934). The larval body including the muscular larval arm becomes resorbed at metamorphosis in *L. ciliaris* while the whole larval body is cast off in *L. sarsi*. This larval arm is of course not homologous with the tail of the hypothetical chordate larva, but it shows that a locomotory organ can easily be formed by the musculature of the coelomic sacs.

The ancestral chordate larva then developed a stiffening rod (the urochord) in the tail, i.e. the posterior part of the metasome, and a special nervous system to co-

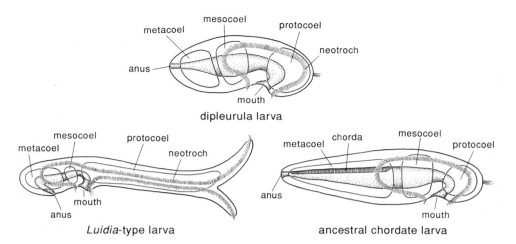

Fig. 51.3. Larvae of notoneuron (dipleurula), a *Luidia*-type starfish (simplified after Fig. 48.3) and the hypothetical chordate ancestor.

ordinate the swimming movements of the 'new' tail. The urochord developed from a median row or zone of cells along the roof of the tail region of the gut. This stiffening structure could then take over the function of keeping a constant length of the tail when the lateral muscles contracted alternatively, producing the swimming movements. The stomochord of pterobranchs and enteropneusts (Chapters 47, 50) is a stabilizing extension of the anterior part of the gut and fulfils some of the same functions as the proposed urochord. Again here, the two structures are obviously not homologous, but the possibility of developing a stiffening structure from the dorsal wall of the gut is demonstrated. The neural tube developed as an infolding of the middorsal ectoderm with the motor neurons associated with the tail muscles. An anterior extension of the neural tube with centres coordinating the tail movements and containing new sensory organs developed simultaneously. The first stage of the development of the tail was probably a rather stiff structure which could be wagged from side to side, possibly more as an escape reaction than a directed swimming, while the undulatory swimming movements necessitated a more complicated nervous coordination (Gans 1989).

In the living chordates there is an intimate connection between chorda and neural tube: the chorda induces the formation of the neural tube chemically. I believe that the complex chorda/neural-tube/tail-musculature evolved simultaneously and that the neural tube is in reality a new nervous system developed to control the movements of the new tail. The dorsal ectodermal area which became infolded above the urochord extended anteriorly along the dorsal midline enclosing the 'old' nervous centre on the mesosome. The new locomotory nervous system, the neural tube, with sensory organs like eye and statocyst, thus incorporated the already existing more vegetative system in the bottom of the anterior expansion. This interpretation is suggested by the metamorphosis of the ascidian larvae, which have a

404

visceral ganglionic area developed from one side of the anterior swelling of the neural tube, and this is the only part of the central nervous system which is retained after metamorphosis, when the whole locomotory system of the tail is resorbed (Chapter 52). The close connection of the three elements of the tail, chorda, neural tube and muscles, has been demonstrated experimentally in all the chordate phyla (Chapters 52-54).

None of the living chordate phyla has a terminal anus, and it must be assumed that a new anus at the base of the tail became established at an early evolutionary stage; this is also indicated by the closure of the blastopore by the formation of the neurenteric canal, which must have taken place after the larvae became lecithotrophic. I believe that the chordate ancestor also acquired a new mouth, as indicated by the fate maps of all the chordate types (Fig. 51.1), but the mechanism behind this is obscure.

The ancestral deuterostomes discussed so far were all pelago-benthic, and the adults were supposedly creeping on the prosome. It has been suggested that the ancestral chordate was more permanently attached, and that this is indicated by the asymmetries of amphioxus (Berrill 1987). Emphasis has been on the presence of three small, possibly adhesive papillae in the anterior end of amphioxus (Chapter 53) which have been homologized with the adhesive papillae of the ascidian larvae. However, the function of the papillae of amphioxus has not been demonstrated with certainty and an innervation like that found in the ascidian tadpoles (Chapter 52) is apparently absent (Chapter 53), so the proposed homology is completely unsupported.

From this creeping chordate ancestor, two evolutionary lines can be envisaged: one towards the urochordates and the other towards the notochordates (cephalochordates and vertebrates).

The urochordate line retained the life cycle with the lecithotrophic tadpole larva and a tail-less adult, which developed a pair of ectodermal invaginations around the external gill openings, the atria, which fused dorsally, forming an excurrent siphon. The anus became incorporated into the left atrium, so that the faeces became expelled with the excurrent jet. This interpretation accepts the generally held belief that the appendicularians are neotenous.

In the notochordate line, the urochord and the larval tail musculature extended anteriorly and the musculature became segmented as the adults adopted a swimming habit, using the undulatory movements of the whole body including the tail.

The cephalochordates have retained the mucociliary filter feeding and their branchial bag has become protected by the development of a ventral atrium. This new structure is formed by paired folds of the lateral body wall which extend ventrally and fuse below the branchial bag, leaving only the posterior atrial pore (Chapter 53).

The vertebrates changed their feeding habit from filter feeding to macrophagy; the branchial bag became respiratory with the muscular ventilation and the endostyle became an endocrine gland, the thyroid. The ammocoetes larva of the lampreys still has a filtratory branchial bag, but the mucous filter is secreted mainly by the gill bars and the water is moved by ventilatory movements. At metamorphosis the endostyle directly becomes the thyroid (Chapter 54). Numerous new characters evolved in connection with the new life style.

References

Bateson, W. 1886. The ancestry of the Chordata. – Q. Jl microsc. Sci., N.S. **26**: 535-571.

Berrill, N.J. 1955. The Origin of the Vertebrates. – Oxford Univ. Press, Oxford.

Berrill, N.J. 1987. Early chordate evolution Part 2. Amphioxus and ascidians. To settle or not settle. – Int. J. Invert. Reprod. Dev. **11**: 15-28.

Colwin, A.L. & L.H. Colwin 1951. Relationships between the egg and larva of *Saccoglossus kowalevskii* (Enteropneusta): axes and planes; general prospective significance of the early blastomeres. – J. exp. Zool. **117**: 111-137.

Gans, C. 1989. Stages in the origin of vertebrates: analysis by means of scenarios. – Biol. Rev. **64**: 221-268.

Garstang, W. 1928. The morphology of the Tunicata, and its bearings on the phylogeny of the Chordata. – Q. Jl microsc. Sci., N.S. **72**: 51-187.

Goodbody, I. 1974. The physiology of ascidians. – Adv. mar. Biol. **12**: 1-149.

Jensen, D.D. 1963. Hoplonemertines, myxinoids, and vertebrate origins. – *In* E.C. Dougherty (ed.): The Lower Metazoa, pp 113-126. Univ. California Press, Berkeley.

Jørgensen, C.B. 1990. Bivalve Filter Feeding: Hydrodynamics, Bioenergetics, Physiology and Ecology. – Olsen & Olsen, Fredensborg.

Maisey, J.G. 1986. Heads and tails: a chordate phylogeny. – Cladistics **2**: 201-256.

Nishida, H. 1987. Cell lineage analysis in ascidian embryos by intracellular injection of a tracer enzyme III. Up to the tissue restricted stage. – Dev. Biol. **121**: 526-541.

Olsson, R. 1956. The development of Reissner's fibre in the brain of the salmon. – Acta zool. (Stockh.) **37**: 235-250.

Olsson, R. 1963. Endostyles and endostylar secretions: a comparative histochemical study. – Acta zool. (Stockh.) **44**: 299-328.

Romer, A.S. 1972. The vertebrate as a dual animal – somatic and visceral. – Evol. Biol. **6**: 121-156.

Ruppert, E.E. 1990. Structure, ultrastructure and function of the neural gland complex of *Ascidia interrupta* (Chordata, Ascidiacea): clarification of hypotheses regarding the evolution of the vertebrate anterior pituitary. – Acta zool. (Stockh.) **71**: 135-149.

Schaeffer, B. 1987. Deuterostome monophyly and phylogeny. – Evol. Biol. **21**: 179-235.

Tattersall, W.M. & E.M. Sheppard 1934. Observations on the bipinnaria of the asteroid genus *Luidia*. – *In* James Johnstone Memorial Volume, pp 35-61. Univ. Press of Liverpool.

Tung, T.C., S.C. Wu & Y.Y.F. Tung 1962. The presumptive areas of the egg of amphioxus. – Scientia sin. **11**: 629-644

Vogt, W. 1929. Gestaltungsanalyse am Amphibienkeim mit örtlicher Vitalfärbung. II. Teil. Gastrulation und Mesodermbildung bei Urodelen und Anuren. – Arch. Entwicklungsmech. Org. **120**: 384-706.

Werner, B. 1953. Über den Nahrungserwerb der Calyptraeidae (Gastropoda Prosobranchia). – Helgoländer wiss. Meeresunters. **4**: 260-315.

Willmer, E.N. 1974. Nemertines as possible ancestors for the vertebrates. – Biol. Rev. **49**: 321-363.

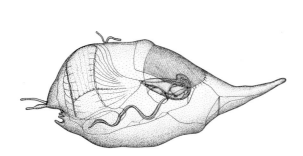

Phylum UROCHORDATA

Urochordates or tunicates are a very distinct, marine phylum of about 1250 species, often arranged in three classes, Ascidiacea, Thaliacea and Larvacea (sometimes called appendicularians); the sorberaceans and *Octacnemus*, which lack the branchial basket, are probably best regarded as specialized ascidians. The Thaliacea are divided into three quite different orders, Pyrosomida (only one genus, *Pyrosoma*), Salpida (with 9 genera) and Doliolida (with 3 genera), but the pyrosomes are so similar to colonial ascidians, as for example *Cyathocormus*, that they are here included in the Ascidiacea. The characteristic tadpole larva with a chorda only in the tail (here called a urochord) is found in all major groups except the salps and is usually regarded as the most conspicuous apomorphy of the phylum, but most of the features usually regarded as characteristic may actually be plesiomorphic (see below). The presence of a thick cuticle or tunic, hence the alternative name Tunicata, with a cellulose-like compound called tunicin is often listed as another characteristic of the tunicates, but it is not found in larvaceans and doliolids. The ancestry of the urochordates is discussed below, and one of the conclusions is that the most reliable apomorphy of the phylum must be the presence of a pair of branchial atria, which develop as invaginations of the body wall and fuse at the dorsal side to form one excurrent siphon; also the anus opening into the left atrium may be an important apomorphy. A revised classification based on the cladistic method is discussed below.

There is a meagre fossil record, but larvaceans have been reported from Early Cambrian shales of China (Zhang 1987).

The shape of the adult urochordates is highly variable and many species form colonies or other aggregates of highly characteristic shapes. An archimeric regionation cannot be recognized, and the mesoderm does not form the coelomic pouches characteristic of the other deuterostomes.

The four groups ascidians, salps, doliolids and larvaceans are so different in several respects that they will be given separate descriptions in many of the following sections to avoid confusing generalizations.

Chapter vignette: The appendicularian *Oikopleura dioica* in its house. (Redrawn from Lohmann 1903.)

All the urochordates have monolayered epithelia. The ectoderm is unciliated, but many areas of the endoderm consist of multiciliate cells. Monociliate epithelial cells are found in certain zones of the endostyle (Goodbody 1974) and of the stomach (Ermak 1982), and sensory cells may be monociliate or multiciliate (Bone 1986). Myoepithelial cells have been found in the adhesive papillae of ascidian tadpole larvae (Cloney 1977).

The ascidians have a thick tunic, primarily secreted by the epithelial cells. Ectodermal extensions from a zone near the heart form a more or less extensive system of blood vessels in the tunic, and mesodermal cells from the blood wander through the ectoderm or through the walls of the vessels into the tunic (Seeliger 1893b), where they take part in the formation of additional tunic material. The tunic is only adherent to the epithelium in the areas around the siphons and at the blood vessels. The tunic contains cellulose, mucopolysaccharides and protein, and may accumulate unusual substances such as vanadium. Colonial forms have a common tunic, in some cases with interzooidal blood vessels. The tunic of didemnids contains a network of myocytes which can cause contractions of the tunic for example around the common excurrent openings (Mackie & Singla 1987).

The salps have a hyaline, sometimes coloured tunic with cellulose but with only few cells, which have been observed to move from the blood vessels through the epithelium (Seeliger 1893b).

The doliolids have a thin ectoderm which secretes a thin cuticle without cellulose and without cells; large areas of the cuticle are moulted regularly, keeping the surface of the animal free from detritus (Uljanin 1884).

The larvacean ectoderm comprises a number of very specialized cell groups, the oikoplast epithelium, which secretes the various parts of a more or less complicated filter or house; this structure is used as a filtering device which concentrates the plankton particles in the water pumped through the filter by the undulating movements of the tail. The filter/house is discarded periodically and a new one secreted.

The gut consists of a spacious pharynx, often called the branchial basket, with gill slits and a ventral endostyle, a narrow oesophagus, a stomach with various digestive diverticula and glands, and an intestine which opens in the left atrial chamber (except in the larvaceans, which lack atrial chambers and have a ventral anus). The adult larvaceans have a urochord consisting of a central hyaline mass surrounded by flattened cells and a strengthened basement membrane (Welsch & Storch 1969).

Almost all ascidians have a branchial basket with a high number of gill slits or stigmata. The finest gill bars consist of one layer of ectodermal cells with a thin basement membrane surrounding a blood vessel, but the longitudinal gill bars are thicker and have unciliated zones with muscle cells surrounded by basement membrane (Pennachetti 1984; Fig. 49.1). Each stigma is surrounded by 7 (6-8) rings of very narrow cells each with a row of cilia (Mackie et al. 1974, Monniot 1979). The cilia of the gill slits transport water out of the branchial basket and new water with particles is sucked in through the mouth, which forms the incurrent siphon. The particles are caught by a fine mucous net produced continuously by the endostyle and transported along the wall of the basket by the cilia of the peripharyngeal bands and of various structures on the gill bars to a middorsal structure, which rolls the net together and passes it posteriorly to the oesophagus. The filtered water passes through the stigmata into the lateral atrial chambers and out through the middorsal excurrent siphon. The filter consists of proteins and polysaccharides and is organized as a rectangular meshwork; the filaments of the two directions are sometimes not of the same thickness (Flood & Fiala-Medioni 1981). The filter is secreted by bands of special cells of the endostyle, possibly by a mechanism resembling that of the gastropod *Crepidula* (Werner 1953, Werner & Werner 1954), but the details have not been worked out.

The salps have an apparently highly specialized gill filter with only one enormous gill opening on each side. Nevertheless the normal mucous filter is formed by the endostyle and pulled dorsally by the peripharyngeal bands and then posteriorly to the oesophagus, but the current through the filter is created by the muscular swimming movements of the salp; the transverse muscles constrict the body and the flow of water is regulated by one-way valves consisting of a system of flaps at each siphon (Bone, Braconnot & Ryan 1991).

Oozooids of most doliolids have only four gill slits on each side while all other types of zooids have a vertical row of 8-200 gill slits on each side. The endostyle secretes a mucous filter, which is transported dorsally by the peripharyngeal bands; however, the net is not transported along the gill bars as in the ascidians but coiled together into a funnel-shape by the cilia at the entrance to the oesophagus (Deibel & Paffenhöfer 1988).

The larvaceans have only one, circular gill opening on each side, and a first concentration of the food particles takes place in the filter structure formed by the oikoplast epithelium (Deibel 1986). The final capture of the particles takes place in a fine, normal mucous filter formed by the endostyle (Deibel & Powell 1987). Only *Kowalevskaia* lacks the endostyle completely, and the food particles are apparently captured by cilia on some elaborate folds surrounding the unusually large gill pores (Fol 1872).

The endostyle (Chapter 51) has a similar structure in ascidians and salps with three longitudinal bands of glandular cells on each side, while the doliolids have only two rows of gland cells (Neumann 1935). The gland cells secrete the mucous net which consists primarily of proteins, and they also bind iodine which is again secreted to the filter as mono- and di-iodothyrosines (Goodbody 1974). The larvaceans have a short endostyle which is usually curved, in *Fritillaria* so strongly that there is only a narrow opening from the endostyle to the pharynx. The endostyle of *Oikopleura* has only one row of gland cells, which produce both the mucous net and the iodotyrosines (Olsson 1963).

The central nervous system of ascidians, salps and doliolids consists of a cerebral ganglion or brain at the dorsal side of the pharynx just behind the peripharyngeal ciliary bands. In front of the brain, a convoluted ciliated funnel leads from the pharynx via a ciliated duct to the neural gland, and water is pumped through this neural gland complex to the haemal system; a glandular function is not indicated (Ruppert 1990). The gland is situated dorsal or ventral to the brain, or in a few cases at its right side (Elwyn 1937). The connection of this complex with the nervous system is described below.

The larvaceans have a more complicated central nervous system which resembles that of the tadpole larvae. It consists of a neural tube with an extended brain vesicle, a ganglion dorsal to the gut and a neural tube with a Reissner's fibre in the tail (Olsson 1969). The brain vesicle contains a large statocyst (Holmberg 1984). The two rows of lateral muscle cells in the tail are innervated from cells in the neural cord (Bone 1989); these cells may be arranged in groups resembling ganglia, but considerable variation between specimens has been observed so a true segmentation is not indicated (Seeliger 1900).

The mesoderm does not form the coelomic cavities usually recognized in deuterostomes. Some of the ascidians have a pair of more or less extensive cavities, called epicardial sacs, which are extensions from the caecum of the endostyle (Damas 1900). These two pockets could be interpreted as mesodermal sacs formed by enterocoely, and they surround parts of the gut as the metacoels. However, they are not formed until a late ontogenetic stage, and their homology with the coelomic sacs of such forms as enteropneusts appears more than questionable. A heart situated ventral to the posterior part of the pharynx pumps the blood alternately in two directions. It consists of a mesodermal sac folded around a blood lacuna; its inner wall forms the circular musculature of the heart and the lumen forms the pericardial sac. The haemal system is a complex of channels or lacunae surrounded by basement membrane and without endothelium; the ectodermal extensions in the tunic form well-defined vessels. The blood contains various cell types, such as lymphocytes, macrophages, vanadocytes, which accumulate vanadium, and nephrocytes (Berrill 1950).

Appendicularian musculature is restricted to the heart and the two longitudinal tail muscles, which each consist of one row of flat cells. There is a species-specific number of muscle cells in the tail, which is apparently the number laid down during the embryonic development.

There is no excretory system like metanephridia or an axial complex, but waste products accumulate in special blood cells which may become deposited in the tunic or in special regions of the body as renal organs.

One or more mesodermal gonads are situated near the stomach and their common or separate ducts open into the atrial chamber. In some appendicularians the ripe eggs disrupt the maternal body wall and the animal dies (Berrill 1950).

There are numerous studies of ascidian embryology (review in Cloney 1990), from classical studies using direct observations of the embryos (as those of van Beneden & Julin (1887) on *Clavelina*, and Conklin (1905) on *Cynthia* and *Ciona*) to recent studies using various marking techniques, such as the elegant studies on cell lineage by Nishida (1987, on *Halocynthia*); there seems to be only little variation between the species.

The unfertilized egg has an apical-blastoporal axis containing the position of the polar bodies, and fertilization triggers a characteristic reorganization of the contents of the egg, called ooplasmic segregation, followed by the formation of the polar bodies (Bates & Jeffery 1988). It is not known if the point of sperm entry determines the plane of the first cleavage. A bilateral organization, often indicated by variously coloured areas and with the polar bodies in the median plane, can be recognized after a few minutes. The fertilized egg is highly determined, and isolated cells cannot regenerate whole embryos (Reverberi 1971).

The first cleavage is median, the second transverse, and the third equatorial; the polar bodies indicating the apical pole can usually be followed through the early development (Fig. 52.1). The fate map (Fig. 51.1) shows that the whole ectodermal epithelium originates from the apical quartet and the whole endoderm from the blastoporal quartet. The nervous system develops from a crescentic area at the anterior part of the equatorial zone of both quartets, while the mesodermal organs develop from an circumequatorial zone consisting almost exclusively of areas of the blastoporal quartet (Nishida 1987).

The following development goes through a coeloblastula and an embolic gastrula in which the blastocoel becomes obliterated. The posterior, dorsal strip of endo-

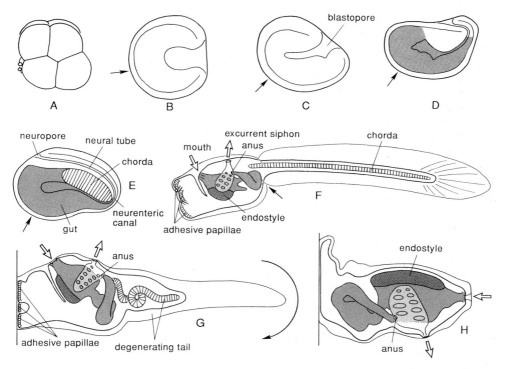

Fig. 52.1. Ontogeny of ascidians with the primitive larval type (A, *Styela partita*; B-H: *Clavelina lepadiformis*), all stages seen from the left; the position of the polar bodies is seen in A and their approximate position is indicated by black arrows in the later stages (based on the observations on *Halocynthia roretzi* by Nishida 1987). – A, 8-cell stage. – B, gastrula. – C, late gastrula, the blastopore has turned slightly to the dorsal side. – D, beginning of neural tube formation. – E, early tailbud stage with chorda, mesoderm and neural tube. – F, newly hatched larva with attachment papillae, endostyle, atrium with two rows of gill openings, and the rectum opening into the left atrium. – G, newly settled larva, the tail is almost completely retracted and the gut rotation (indicated by the long arrow) has just begun. – H, juvenile ascidian with completed rotation of the gut. (A redrawn from Conklin 1905; B-E redrawn from van Beneden & Julin 1887; F redrawn from Julin 1904; G-H redrawn from Seeliger 1893-1907.)

derm differentiates into the cylindrical urochord and the zones lateral to the urochord become isolated as a pair of lateral mesodermal strips; the remaining ventral endoderm becomes tubular. The rather large mesodermal cells along the urochord become arranged in one layer while a mass of smaller cells, the lateral trunk mesoderm, can be recognized on each side at the anterior end of the urochord (van Beneden & Julin 1887). The urochord induces the ectoderm to form a neural plate over the urochord and extending to the anterior end of the embryo (Reverberi, Ortolani & Farinella-Ferruzza 1960). The lateral sides of the neural plate fold together dorsally, forming a narrow neural tube; posteriorly, the folds encompass the blastopore so that a neurenteric canal is formed when the folds fuse; anteriorly, the tube is extended to form a small brain vesicle which for a period retains a narrow anterior canal to the exterior, the neuropore.

The embryo now begins to curve ventrally while the tail elongates, and this becomes very pronounced during the late embryonic stages where the tail encircles the body. The apical pole becomes ventrally displaced and is finally situated ventrally at the base of the tail, so that the whole epithelium of the body is derived from the 'dorsal' pair of cells of the apical quartet, while the tail has ectoderm derived from the 'ventral' (Fig. 52.1). The urochord cells slowly rearrange and form a single row, and the posterior endodermal tube becomes a narrow strip, the endodermal chord. The neural tube elongates and the cells become arranged in four longitudinal rows (Nicol & Meinertzhagen 1988).

The embryo now consists of a globular body and a slim tail with a cuticular dorsal-ventral fin. There is a considerable interspecific variation in the internal organization, and the following description of the large embryos is mainly based on a type like *Clavelina* or *Ciona*, with a rather complete gut but without the precocious development of adult structures and buds seen in many species.

The body contains the rather voluminous gut, which may have a cavity, but which has neither mouth nor anus. The neuropore is closed and the anterior part of the neural tube has become divided longitudinally with the left part separated from the main neural tube. The left vesicle becomes the adult cerebral ganglion or brain and the neural gland complex, whereas the right part enlarges as the larval brain with an anterior brain vesicle and a more posterior visceral ganglion.

The right brain vesicle may contain three types of sense organs, an ocellus, a statocyte and a group of modified cilia which may be pressure receptors, but only few species have all three types (Eakin & Kuda 1971). The ocellus and the statocyte develop from a pair of cells (a8.25 and *a*8.25 in the cell lineage), but there is no fixed right/left origin of the two organs (Nishida 1987). The ocellus has ciliary photoreceptor cells which show some resemblance to those of vertebrates, but the membrane folds are arranged parallel to the cilium instead of perpendicular to it.

The left brain vesicle elongates anteriorly as a ciliated duct and an opening through the anterodorsal wall of the pharynx becomes established; the ciliated funnel originates from the anterior cells of the duct. A group of cells of the dorsal or the ventral side of the vesicle differentiates as the nerve cells of the cerebral ganglion, which becomes separated from the vesicle; the remaining part of the vesicle becomes the neural gland (Elwyn 1937).

Three frontal adhesive papillae are the attachment organs used in settling. In addition to the adhesive cells they contain two types of primary sensory cells, both monociliate with the cilium entering the tunic. The axons from these cells unite, forming one nerve from each papilla, and the three nerves merge into a nerve which winds through the anterior haemocoel to the visceral ganglion (Torrence & Cloney 1983). The papillae or palps are reported to develop from the anterior part of the neural plate (Nicol & Meinertzhagen 1988, in *Ciona*) or from an area just in front of this (Nishida 1987, in *Halocynthia*); this discrepancy may indicate differences between species, but the possibility of migration of cells from the neural plate to the papillae cannot be ruled out.

The oral siphon develops as a shallow dorsal, ectodermal invagination just behind the adhesive papillae (Nishida 1987).

The gut already has the general regions of the adult gut. In the pharynx, the endostyle has differentiated from a ventral longitudinal groove, and the epicardial sacs develop as evaginations from the ventral side of the pharynx behind the endostyle (Berrill 1950). The origin of the pericardium has been much discussed, but it now appears that it develops from paired ventral parts of the lateral trunk mesoderm which migrate anteriorly, develop a cavity at each side, and fuse around a haemal sinus (Berrill 1950). Other parts of this mesoderm have migrated laterally to the anterior part of the body in the region of the adhesive papillae, the pregastral mesoderm; this region swells up at metamorphosis and forms a rather large cavity with scattered mesodermal cells, which do not form a mesodermal sac (Willey 1893). These cells also give rise to blood cells and to the cells in the tunic (Nishide, Nishikata & Satoh 1989).

The atrium develops from a pair of ectodermal invaginations which later on fuse dorsally forming the median anal siphon (Seeliger 1893a). The gill openings or stigmata break through between the pharynx endoderm and the ectoderm of the atrial invaginations. In many species there are two or three primary gill openings, protostigmata, on each side, and these openings then divide and form the sometimes very complicated patterns of round, elongate or curved stigmata (Julin 1904, Brien 1948), and the whole branchial basket may form complicated folds. The first stage of the protostigmata is often circular and they then become elongate and in some species curve into a J-shape and become divided into series of stigmata. The J-shape has sometimes been interpreted as homologous to the U-shape of the gill slits with tongue bars of enteropneusts and cephalochordates (Garstang 1928), but this resembles wishful thinking.

The tail has the urochord as the most conspicuous structure. In a few genera, such as *Dendrodoa* (Welsch & Storch 1969), it consists of a stack of coin-shaped cells with many yolk globules and surrounded by a strengthened basement membrane as in the older embryos; in most other genera, extracellular matrix is secreted between the cells, and these matrix lenses finally fuse into a central rod surrounded by a continuous layer of flat urochordal cells (Cloney 1964). The two lateral bands of muscle cells are arranged in 2-4 rows according to the species, but the cells in the rows are not aligned (Nishida & Satoh 1985), so there is no indication of segmentation. The neural tube (spinal cord) consists of four longitudinal rows of cells which retain their epithelial structure with cilia, but axons from nerve cells in the visceral

413

ganglion extend along the lateral sides of the tube and innervate the muscle cells across the basement membrane (Torrence & Cloney 1982). In *Dendrodoa*, which has three rows of muscle cells, the dorsal row of cells and the anteriormost cells of the ventral row are innervated directly, and all the cells are coupled through gap junctions (Bone 1989). A dorsal and sometimes also a ventral sensory nerve is found in the basement part of the ectoderm below the cuticular tail fins; these nerves are bundles of axons from primary sense organs, each with a cilium extending into the fin; these sensory cells can be recognized in the tail at an early stage and their axons pass to the visceral ganglion without contact with the motor axons of the cord (Torrence & Cloney 1982). The neural canal contains a Reissner's fibre, but its point of origin has not been ascertained (Olsson 1972). The only trace of the original gut is a row of endodermal cells along the ventral side of the chorda. A layer of ectodermal cells surrounds the tail and secretes a thin tunic which extends into a thin dorsal, posterior and ventral fin. The tail is rotated 90° to the right in many species.

The type described above is considered ancestral and has been found in a number of families of mainly solitary forms, but many modifications involving delayed development of the gut or precocious development of buds have been described (see Millar 1971). Some of the molgulids have embryos without tails, but the chorda and the neural tube develop as in the normal tadpoles (Berrill 1931). The closure of the neural tube is especially clearly shown in *Molgula*, where the neural folds fuse along the midline leaving both the neural pore and the blastopore open for a short time; the blastopore finally becomes constricted and the neurenteric canal is formed.

The embryos hatch from the egg membrane and have a short pelagic stage.

At settling, the larva attaches by the adhesive papillae, the larval cuticle is shed, and the tail becomes retracted by methods which vary between species (Cloney 1978, 1990). The cellular material of the tail and of the whole neural tube including the sensory vesicle with its sense organs becomes resorbed. Just after metamorphosis, the zone between the attachment and the mouth expands strongly so that the gut rotates 90-180° and the oral siphon finally points away from the substratum (Fig. 52.1).

Sexual reproduction and development of the oozooid of salps are complicated and not well known. The eggs become fertilized in the gonoduct where the development takes place, each embryo being nourished by a placenta, which is very different from the mammalian placenta in that its two layers are both of maternal origin (Bone, Pulsford & Amoroso 1985). The embryology is complicated through the invasion of follicular cells, called calymmocytes, which conceal some of the developing organs (Brien 1948, Sutton 1960). It is unclear whether a urochord is actually present during ontogeny, and the morphology of the newborn oozooid is undescribed.

The development of *Doliolum* has been studied by Uljanin (1884), Neumann (1906), Godeaux (1958), and Braconnot (1970). The eggs are spawned freely into the water where fertilization takes place. The embryo goes through a blastula and a gastrula stage, and the next stage which has been observed already has a body and a tail, which is bent dorsally in the middle like a hairpin; the urochord consisting of one row of cells can be recognized but the other structures are not well described. A later stage has the tail stretched out and the urochord separated from the developing gut by a mass of mesodermal cells; a short anterior neural tube is formed by the infolding of a neural plate. The tail has three rows of muscle cells on each side, and the embryo is able to make weak swimming movements, but the movements seem mainly to be sharp bendings of the base of the tail while the tail itself is kept stiff (Braconnot 1970). There is apparently no neural tube in the tail, and nerves from body to tail have not been described. The large atrium develops from a pair of dorsal ectodermal invaginations which fuse medially. The endoderm has become a hollow tube which breaks through anteriorly and posterodorsally, forming mouth and anus. The anterior part of the gut becomes the spacious pharynx with the ventral endostyle; four pairs of gill openings break through at each side. The tail is resorbed and the young oozooid hatches from the egg membrane.

Knowledge of the development of larvaceans is limited, but Delsman (1910, 1912) described the cell lineage of the cleavages until the sixth cleavage and the development of the juvenile of *Oikopleura* (see also Fenaux 1976). Polar bodies were not observed and the orientation of the first stages of the embryos could not be determined; gastrulation is epibolic. The older embryos are curved and the first stages of the formation of the tail with urochord and nerve tube could be observed. The unciliated juvenile hatches at a stage resembling that of an ascidian embryo of similar age; it has a nerve tube with an anterior sensory vesicle with a statocyst. The tail has a urochord consisting of a row of 20 cells, a row of endodermal cells, a nerve tube situated along the left side of the urochord, and a row of 10 muscle cells on the dorsal and a similar row on the ventral side. The urochord cells are at first rather flat and in full contact but the cells elongate and small vacuoles form between the cells and finally fuse centrally so that the urochord consists of a central rod of elastic material surrounded by few, thin cells inside the basement membrane (see also Welsch & Storch 1969). The cells of the endodermal cord were observed to wander anteriorly in the tail and all except two finally become situated in the region of the endostyle. Mouth, anus and the two gill openings break through from the pharynx without any signs of ectodermal invaginations. The juvenile thus resembles an ascidian tadpole, but the tail soon bends ventrally and the epithelium becomes differentiated, enabling the secretion of the first house.

Asexual reproduction is found in a number of ascidians and is obligatory in salps and doliolids, which alternate between sexual and asexual generations, whose individuals are morphologically different.

The colonial ascidians show a whole series of different types of budding, which may take place both in the adult stage and precociously in the larval stage (Nakauchi 1982). The buds may develop through a strobilation-like process involving parts of the gut, but buds may also originate from epicardial or peribranchial areas; stolonial budding involving only vascular elements in the tunic is found in several types. The budding gives rise to characteristic colonies with many identical zooids. A special type of budding is associated with dormant buds, which may similarly develop from various types of tissues.

The salps have more complicated life cycles: the zygotes develop into solitary oozooids, which lack gonads but have a ventral budding zone which gives rise to a stolon; this stolon gives rise to a chain of blastozooids, which have gonads but lack the stolon (Ihle 1935). The doliolids have still more intricate cycles with several types of individuals formed through budding (Neumann 1906, Braconnot 1971): The zygote develops into an oozooid which forms hundreds of small buds from a ventral stolon; the buds separate from the stolon and wander around the right side of the oozooid to a long posterodorsal appendage where they differentiate according to their position on the appendage. The laterally positioned buds become small feeding gastrozooids, which are generally stated to provide nourishment to the growing oozooid (old nurse), which loses its gut and develops wide muscle bands. The buds situated along the dorsal midline of the appendage develop into feeding phorozooids, which detach from the oozooid after having received another small wandering bud, which has attached to a midventral outgrowth; the small bud gives off several new buds which in turn develop into feeding gonozooids (with gonads), which then detach.

The buds of salps and doliolids are formed through processes which resemble various types known in the ascidians, and it seems difficult to find any phylogenetic pattern in the distribution of the types.

The most generally held opinion about the phylogeny of the urochordates is that the ancestor was sessile with a tadpole larva, that the holopelagic thaliaceans are derived from the ascidians, and that the larvaceans are neotenic (see for example Garstang 1928 and Berrill 1955). The discussion of the phylogeny of the chordates (Chapter 51) came to the conclusion that the ancestral urochordate was pelago-benthic with a tadpole larva and with an adult having a pair of lateral atria which fused dorsally and an anus opening into the left atrium, and this does not give any hint about the intraphyletic evolution.

The usual interpretation implies that the doliolids and larvaceans have lost the tunic with mesodermal cells and cellulose; this is not supported by direct observations. An alternative phylogeny (Fig. 52.2, also suggested by Garstang 1928), which emphasizes the two different types of cuticle, interprets the ascidians and salps as representatives of one evolutionary line characterized by a thick, permanent tunic with mesodermal cells and cellulose, while the doliolids and larvaceans represent the sister group characterized by a cuticle which lacks mesodermal cells and cellulose; this cuticle is moulted regularly over a large region of the body, being discarded in the doliolids and used as the 'house' in the larvaceans. The budding process and the enormously complicated life cycle of the doliolids show no specific similarities with those of the ascidians or salps, so the budding from a midventral area could be a plesiomorphy, shared with the pterobranchs, and the holopelagic life cycle could have evolved independently in the two lines. Ultrastructural studies of the sperm (Holland 1989) support the idea that the salps are closely related to the ascidians, especially the family Didemnidae; on the other hand the larvaceans appear to have the most primitive type of sperm, and a sister-group relationship with the doliolids is not indicated. It is possible that the salps are a specialized ascidian type, so that the group Ascidiacea is in fact paraphyletic, but this is without importance for the present discussion.

The phylogeny proposed in Fig. 52.2 implies a urochordate ancestor with a simple cuticle and a paired atrium with an unpaired, dorsal anal siphon. The larvaceans are regarded as secondarily simplified (neotenic), a view held by most modern authors. This ancestor could have been creeping and would fit well into the phylogeny of the chordates proposed in Fig. 51.2.

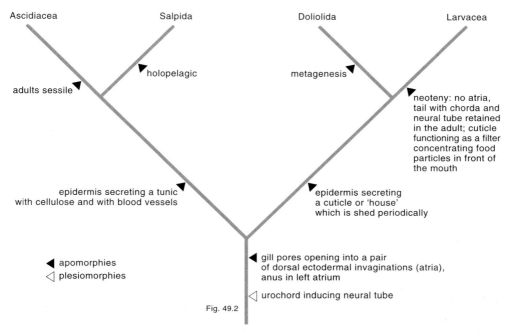

Fig. 52.2. The proposed phylogeny of the urochordates.

The neural tube in the tail region of the tadpole larvae has basiepithelial bundles of neurons but lacks nerve cells. This could be an adaptation to the absorption of the tail at metamorphosis since the adult larvaceans have ganglionic cells in the neural tube of the tail. The nerve cells in the larvacean tail innervate the individual muscle cells along the tail and thus enable the tail to make undulatory movements. The first stage in the evolution of the tail may have been a short, rather stiff structure, only capable of jerking movements which could have been an escape mechanism (Gans 1989); the more organized swimming by means of undulatory movements required a more complicated innervation, for example like that seen in the larvaceans. This could be interpreted as a synapomorphy between the larvaceans and the 'higher' chordates, but the differences between the muscle systems of the two types makes this rather unlikely.

Willey (1893) interpreted the 'pregastral' mesoderm as homologous with the prosomal mesoderm of the 'lower' deuterostomes and amphioxus; this would fit the phylogeny presented here perfectly, but except for the position of the cells in the anteriormost part of the body, there is unfortunately nothing which supports the idea.

The mouth (oral siphon) is definitely formed at the dorsal side of the ascidian embryo, as indicated by its position between the adhesive organs and the anterior part of the brain. Cell lineage studies further document this by showing that the shallow stomodaeum develops from the anteriormost area of the neural plate or just in front of it, and that the apical pole of the young larva is situated ventrally near the base of the tail (see also the position of the mouth on the fate map in Fig. 51.1). The

417

mouth of larvaceans is situated at the morphological anterior pole, but there are no signs of adhesive papillae and cell lineage studies are lacking, so it is impossible to make comparisons with the ascidian larvae. The ascidian mouth is generally accepted as homologous with the mouth of for example enteropneusts and amphioxus, but I have found only very few discussions of the origin and significance of its dorsal position (Bone 1958). It appears that only two possibilities exist: either the ventral mouth has been shifted laterally to end up in the dorsal position, or the dorsal mouth is a new structure, non-homologous with the mouth for example of the enteropneusts. The mouth of amphioxus develops on the left side of the body and later on moves to a midventral position (Chapter 53), but the stomodaeum of all the ascidians studied so far appear to be formed middorsally without any sign of asymmetry. There are small asymmetries in the anatomy of the adult ascidians, for example the position of the anus in the left atrium, and also during ontogeny, for example in the development of the brain, but none of these indicate a lateral displacement of the mouth. It appears to me that the ascidian mouth, and probably the mouth of all urochordates, must be interpreted as a new opening to the archenteron. The anterior lobe with adhesive organs of the ascidians can still be interpreted as homologous with the prosome for example of the enteropneusts, although the apical pole is situated at the posteroventral part of the body in the tadpole larva.

Eakin (1973) speculated that the ocellus of the ascidian larva could be homologous to the vertebrate median eye, but the receptor cells of the tadpoles have their lamellar folds stacked parallel to the main axis of the cilium, whereas all the vertebrate eyes have the lamellae arranged perpendicular to the cilium; this makes a homology very unlikely.

The urochordates share so many important apomorphies, such as the complex of chorda, neural tube and tail mesoderm, with the notochordates that the monophyletic nature of the chordate group can hardly be doubted. On the other hand, the urochordates have a series of characters which set them apart from the other two phyla, such as the lack of segmentation of the mesoderm, which is here interpreted as a plesiomorphy, the presence of a pair of lateral atria which fuse dorsally forming an unpaired anal siphon, and the position of the anus in the left atrium. The larvaceans are regarded as neotenic, so that their retention of the tail, which has a more complicated innervation than that of the tadpole larvae, is interpreted as an apomorphy. The sister-group relationship between urochordates and notochordates appears well founded, and the series of events which I believe have taken place during the evolution of these phyla from their ancestors are described in Chapter 51.

Interesting subjects for future research

1. Formation of the mucous net by the endostyle
2. Embryology of doliolids and larvaceans.

References

Bates, W.R. & W.R. Jeffery 1988. Polarization of ooplasmic segregation and dorsal-ventral axis determination in ascidian embryos. – Dev. Biol. 130: 98-107.

Berrill, N.J. 1931. Studies in tunicate development. Part II. Abbreviation of development in the Molgulidae. – Phil. Trans. R. Soc. B **219**: 225-346.

Berrill, N.J. 1950. The Tunicata with an Account of the British Species. – Ray Society, London.

Berrill, N.J. 1955. The Origin of the Vertebrates. – Oxford Univ. Press, Oxford.

Bone, Q. 1958. The asymmetry of the larval amphioxus. – Proc. zool. Soc. Lond. **130**: 289-293.

Bone, Q. 1986. Tunicates. – *In* M.A. Ali (ed.): Nervous Systems in Invertebrates (NATO ASI, Ser. A 141), pp 527-557. Plenum Press, New York.

Bone, Q. 1989. Evolutionary patterns of axial muscle systems in some invertebrates and fish. – Am. Zool. **29**: 5-18.

Bone, Q., J.-C. Braconnot & K.P. Ryan 1991. On the pharyngeal feeding filter of the salp *Pegea confoederata* (Tunicata: Thaliacea). – Acta zool. (Stockh.) **72**: 55-60.

Bone, Q., A.L. Pulsford & E.C. Amoroso 1985. The placenta of the salp (Tunicata: Thaliacea). – Placenta **6**: 53-64.

Braconnot, J.-C. 1970. Contribution a l'étude des stades successifs dans le cycle des Tuniciers pélagiques Doliolides I. Les stades larvaire, oozooide, nourrice et gastrozoide. – Archs Zool. exp. gén. **111**: 629-668.

Braconnot, J.-C. 1971. Contribution a l'étude des stades successifs dans le cycle des Tuniciers pélagiques Doliolides II. Les stades phorozoide et gonozoide des doliolides. – Archs Zool. exp. gén. **112**: 5-31.

Brien, P. 1948. Embranchement des Tuniciers. Morphologie et reproduction. – Traité de Zoologie, vol. **11**, pp 553-930. Masson, Paris.

Cloney, R.A. 1964. Development of the ascidian notochord. – Acta Embryol. Morph. exp. **7**: 111-130.

Cloney, R.A. 1977. Larval adhesive organs and metamorphosis in ascidians. I. Fine structure of the everting papillae of *Distaplia occidentalis*. – Cell Tissue Res. **183**: 423-444.

Cloney, R.A. 1978. Ascidian metamorphosis: review and analysis. – *In* F.-S. Chia & M.E. Rice (eds): Settlement and Metamorphosis of Marine Invertebrate Larvae, pp 255-282. Elsevier, New York.

Cloney, R.A. 1990. Urochordata – Ascidiacea. – *In* K.G. Adiyodi & R.G. Adiyodi (eds): Reproductive Biology of Invertebrates, vol. 4B, pp 391-451. Oxford & IBH Publishing, New Delhi.

Conklin, E.G. 1905. The organization and cell lineage of the ascidian egg. – J. Acad. nat. Sci. Philad., 2. ser. **13**: 1-119, pls 1-12.

Damas, D. 1900. Les formations épicardiques chez *Ciona intestinalis* (L.). – Archs Biol. **16**: 1-25, pls 1-3.

Deibel, D. 1986. Feeding mechanism and house of the appendicularian *Oikopleura vanhoeffeni*. – Mar. Biol. (Berl.) **93**: 429-436.

Deibel, D. & G.-A. Paffenhöfer 1988. Cinematographic analysis of the feeding mechanism of the pelagic tunicate *Doliolum nationalis*. – Bull. mar. Sci. **43**: 404-412.

Deibel, D. & C.V.L. Powell 1987. Ultrastructure of the pharyngeal filter of the appendicularian *Oikopleura vanhoeffeni*: implications for particle size selection and fluid mechanics. – Mar. Ecol. Prog. Ser. **35**: 243-250.

Delsman, H.C. 1910. Beiträge zur Entwicklungsgeschichte von *Oikopleura dioica*. – Verh. Rijks-inst. Onderz. Zee **3**(2): 3-24, 3 pls.

Delsman, H.C. 1912. Weitere Beobachtungen über die Entwicklung von *Oikopleura dioica*. – Tijd-schr. ned. dierk. Vereen., 2. ser. **12**: 199-205, pl. 8.

Eakin, R.M. 1973. The Third Eye. – Univ. California Press, Berkeley.

Eakin, R.M. & A. Kuda 1971. Ultrastructure of sensory receptors in ascidian tadpoles. – Z. Zell-forsch. **112**: 287-312.

Elwyn, A. 1937. Some stages in the development of the neural complex in *Ecteinascidia turbinata*. – Bull. neurol. Inst. N.Y. **6**: 163-177.

Ermak, T.H. 1982. The renewing cell populations of ascidians. – Am. Zool. **22**: 795-805.

Fenaux, R. 1976. Cycle vital, croissance et production chez *Fritillaria pellucida* (Appendicularia), dans le baie de Villefranche-sur-Mer, France. – Mar. Biol. (Berl.) **34**: 229-238.

Flood , P.R. & A. Fiala-Medioni 1981. Ultrastructure and histochemistry of the food-trapping mucous film in benthic filter-feeders (Ascidians). – Acta zool. (Stockh.) **62**: 53-65.

419

Fol, H. 1872. Études sur les Appendiculaires du détroit de Messine. – Mém. Soc. Phys. Hist. nat. Genève **21**: 445-499, 11 pls.

Gans, C. 1989. Stages in the origin of vertebrates: analysis by means of scenarios. – Biol. Rev. **64**: 221-268.

Garstang, W. 1928. The morphology of the Tunicata, and its bearings on the phylogeny of the Chordata. – Q. Jl microsc. Sci., N.S. **72**: 58-187.

Godeaux, J. 1958. Contribution à la connaissance des Thaliacés (Pyrosome et Doliolum). – Annls Soc. r. zool. Belg. **88**: 5-285.

Goodbody, I. 1974. The physiology of ascidians. – Adv. mar. Biol. **12**: 1-149.

Holland, L.Z. 1989. Fine structure of spermatids and sperm of *Dolioletta gegenbauri* and *Doliolum nationalis* (Tunicata: Thaliacea): implications for tunicate phylogeny. – Mar. Biol. (Berl.) **101**: 83-95.

Holmberg, K. 1984. A transmission electron microscopical investigation of the sensory vesicle in the brain of *Oikopleura dioica* (Appendicularia). – Zoomorphology **104**: 298-303.

Ihle, J.E.W. 1935. Desmomyaria. – Handbuch der Zoologie, 5. Band, 2. Hälfte, pp 401-532. Walter de Gruyter, Berlin.

Julin, C. 1904. Recherches sur la phylogenèse des Tuniciers. – Z. wiss. Zool. **76**: 544-611.

Lohmann, H. 1903. Neue Untersuchungen über den Reichtum des Meeres an Plankton und über die Brauchbarkeit der verschiedenen Fangmethoden. – Wiss. Meeresunters., Kiel N.F. **7**: 1-86, pls 1-4.

Mackie, G.O., D.H. Paul, C.M. Singla, M.A. Sleigh & D.E. Williams 1974. Branchial innervation and ciliary control in the ascidian *Corella*. – Proc. R. Soc. Lond. B **187**: 1-35.

Mackie, G.O. & C.L. Singla 1987. Impulse propagation and contraction in the tunic of a compound ascidian. – Biol. Bull. Woods Hole **173**: 188-204.

Millar, R.H. 1971. The biology of ascidians. – Adv. mar. Biol. **9**: 1-100.

Monniot, F. 1979. Microfiltres et ciliatures branchiales des ascidies littorales en microscope électronique. – Bull. Mus. natn. Hist. nat., Paris 4. sér., 1A: 843-859.

Nakauchi, M. 1982. Asexual development of ascidians: its biological significance, diversity, and morphogenesis. – Am. Zool. **22**: 753-763.

Neumann, G. 1906. Doliolum. – Wiss. Ergebn. dt. Tiefsee-Exped. 'Valdivia' **12**: 93-243. pls 11-25.

Neumann, G. 1935. Cyclomyaria. – Handbuch der Zoologie, 5. Band, 2. Hälfte, pp 324-400. Walter de Gruyter, Berlin.

Nicol, D. & I.A. Meinertzhagen 1988. Development of the central nervous system of the larva of the ascidian, *Ciona intestinalis* L. II. Neural plate morphogenesis and cell lineages during neurulation. – Dev. Biol. **130**: 737-766.

Nishida, H. 1987. Cell lineage analysis in ascidian embryos by intracellular injection of a tracer enzyme III. Up to the tissue restricted stage. – Dev. Biol. **121**: 526-541.

Nishida, H. & N. Satoh 1985. Cell lineage analysis in ascidian embryos by interacellular injection of a tracer enzyme II. The 16- and 32-cell stage. – Dev. Biol. **110**: 440-454.

Nishide, K., T. Nishikata & N. Satoh 1989. A monoclonal antibody specific to embryonic trunk-lateral cells of the ascidian *Halocynthia roretzi* stains coelomic cells of juvenile and basophilic blood cells. – Dev. Growth Differ. **31**: 595-600.

Olsson, R. 1963. Endostyles and endostylar secretions: a comparative histochemical study. – Acta zool. (Stockh.) **44**:299-328.

Olsson, R. 1969. Phylogeny of the ventricle system. – *In* G. Sterba (ed.): Zirkumventrikuläre Organe und Liquor, pp 291-305. Gustav Fischer, Jena.

Olsson, R. 1972. Reissner's fiber in ascidian tadpole larvae. – Acta zool. (Stockh.) **53**: 17-21.

Pennachetti, C.A. 1984. Functional morphology of the branchial basket of *Ascidia paratropa* (Tunicata, Ascidiacea). – Zoomorphology **104**: 216-222.

Reverberi, G. 1971. Ascidians. – *In* G. Reverberi (ed.): Experimental Embryology of Marine and Freshwater Invertebrates, pp 507-550. North-Holland, Amsterdam.

Reverberi, G., G. Ortolani & N. Farinella-Ferruzza 1960. The causal formation of the brain in the ascidian larva. – Acta Embryol. Morph. exp. **3**: 296-336.

Ruppert, E.E. 1990. Structure, ultrastructure and function of the neural gland complex of *Ascidia interrupta* (Chordata, Ascidiacea): Clarification of hypotheses regarding the evolution of the vertebrate anterior pituitary. – Acta zool. (Stockh.) **71**: 135-149.

Seeliger, O. 1893a. Über die Entstehung des Peribranchialraumes in den Embryonen der Ascidien. – Z. wiss. Zool. **56**: 365-401, pls 19-20.

Seeliger, O. 1893b. Einige Beobachtungen über die Bildung des äusseren Mantels der Tunicaten. – Z. wiss. Zool. **56**: 488-505, pl. 24.

Seeliger, O. 1893-1907. Die Appendicularien und Ascidien. – Bronn's Klassen und Ordnungen des Tierreichs, 3. Band (Suppl.), 1. Abt., pp 1-1280. Akademische Verlagsgesellschaft, Leipzig.

Seeliger, O. 1900. Einige Bemerkungen über den Bau des Ruderschwanzes der Appendicularien. – Z. wiss. Zool. **67**: 361-400, pls 21-23.

Sutton, M.F. 1960. The sexual development of *Salpa fusiformis* (Cuvier). – J. Embryol. exp. Morph. **8**: 268-290.

Torrence, S.A. & R.A. Cloney 1982. Nervous system of ascidian larvae: caudal primary sensory neurons. – Zoomorphology **99**: 103-115.

Torrence, S.A. & R.A. Cloney 1983. Ascidian larval nervous system: primary sensory neurons in adhesive papillae. – Zoomorphology **102**: 111-123.

Uljanin, B. 1884. Die Arten der Gattung *Doliolum* im Golfe von Neapel. – Fauna Flora Golf. Neapel **10**: 1-140, 12 pls.

van Beneden, E. & C. Julin 1887. Recherches sur la morphologie des Tuniciers. – Archs Biol. **6**: 237-476, pls 7-16.

Welsch, U. & V. Storch 1969. Zur Feinstruktur der Chorda dorsalis niederer Chordaten (*Dendrodoa grossularia* (v. Beneden) und *Oikopleura dioica* Fol). – Z. Zellforsch. **93**: 547-559.

Werner, B. 1953. Über den Nahrungserwerb der Calyptraeidae (Gastropoda Prosobranchia). – Helgoländer wiss. Meeresunters. **4**: 260-315.

Werner, E. & B. Werner 1954. Über den Mechanismus des Nahrungserwerbs der Tunicaten, speziell der Ascidien. – Helgoländer wiss. Meeresunters. **5**: 57-92.

Willey, A. 1893. Studies in the Protochordata. – Q. Jl microsc. Sci., N.S. **34**: 317-369, pls 30-31.

Zhang, A. 1987. Fossil appendicularians in the Early Cambrian. – Scientia sin. B **30**: 888-896.

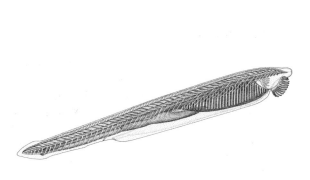

Phylum CEPHALOCHORDATA

Cephalochordates or lancelets are a very small phylum, comprising only about 25 species representing two families. All species are marine, mostly living more or less buried in coarse sand. The larvae, perhaps more appropriately called juveniles, are pelagic, and some sexually mature planktonic forms (*Amphioxides*) may be neotenic. The amphioxus-like fossil *Pikaia* from the Burgess Shale has not been described in detail (Conway Morris & Whittington 1979), but the preservation suggests the presence of a well-defined cuticle, which makes the relationship with cephalochordates improbable (Butterfield 1990).

The most common genus, *Branchiostoma*, more generally known as amphioxus, has been in the focus of evolutionary studies for more than a century, and only selected topics from the literature will be touched upon here.

The body is lanceolate and an archimeric regionation can be distinguished neither from the outer shape nor from the embryology or morphology of the mesoderm.

All epithelia are monolayered, and all the ectoderm of the juvenile stages as well as many areas of the gut of the adult stages are ciliated; occasional cilia have been observed also on the peritoneum; only monociliate cells have been observed (Hatschek 1881, Welsch & Storch 1969, Tjoa & Welsch 1974, Welsch 1975, Holland & Holland 1990).

The mouth is situated at the bottom of an anteroventral invagination, the vestibule. The opening of the vestibule is surrounded by a horseshoe of cirri, and there is a system of ciliated ridges, the wheel organ, at the bottom of the vestibule in front of the mouth opening. A ciliated structure at the dorsal side of the vestibule, called Hatschek's pit, secretes mucus which flows over the wheel organ; the mucus with trapped particles becomes ingested. The mouth is surrounded by a ring of ciliated velar tentacles. The gut consists of a spacious pharynx with numerous parallel gill slits, the branchial basket, an oesophagus, an intestine with an anterior, digestive diverticulum, and a short rectum. The gill slits open into a lateroventral atrial chamber which has a posterior, midventral opening some distance in front of the

Chapter vignette: *Branchiostoma lanceolatum* or amphioxus. (Based on Drach 1948 and Pearse *et al.* 1987.)

anus. The anus is situated ventrally a short distance from the posterior end (Drach 1948).

The U-shaped gill slits are subdivided into vertical rows of gill pores. The system is supported by a skeleton formed by a thickened basement membrane with stiffening rods (Rähr 1982; Fig. 49.1). The basement membrane contains collagen, but this is absent in the rods, which consist of structural proteins and acid mucopolysaccharides, probably including chondroitin sulphate and thus possibly representing a cartilage-like composition; chitin is apparently absent. There are several blood vessels both in the basement membrane and in the rods, but no special respiratory areas have been reported (Rähr 1982). The gill slits represent one type of segmentation, branchiomery, which is reflected in the excretory organs and in parts of the haemal system; it is not synchronous with the myomeric segmentation defined by the coelomic compartments giving rise to the longitudinal muscles.

The branchial basket is the feeding organ. The cilia of the gill pores create the water current which enters the mouth, passes through the pores to the atrial chamber and leaves through the atriopore. A ventral endostyle secretes a fine mucous filter which is transported dorsally along the gill bars to the dorsal side where the filters of the two sides with the captured particles are rolled together and transported posteriorly to the oesophagus. The endostyle is a longitudinal groove with parallel bands of different cell types (Barrington 1958, Olsson 1963). One pair of bands secretes the muco-proteinaceous filter and another pair binds iodine and secretes iodinated thyrosines to the filter.

The almost cylindrical chorda extends anteriorly beyond the mouth and the dorsal nerve tube and posteriorly to the tip of the tail. It consists of a stack of coin-shaped cells with transverse, striated myofibrils and is surrounded by a thick basement membrane. This chordal sheath has two rows of laterodorsal pits where the membrane is very thin, and extensions from the notochordal cells extend to these thin areas where synapses with cells in the apposing spinal chord are formed (Flood 1970). The function of the muscle cells is debated, but it is believed that the contraction stiffens the chorda, so that it becomes more efficient as an antagonist to the lateral muscles when the animal swims and burrows (Guthrie & Banks 1970, Webb 1973).

The central nervous system consists of a neural tube or spinal chord with a small anterior brain vesicle (Bone 1960). Several cells have retained the epithelial character and are monociliate. A midventral group of such cells at the posterior side of the brain vesicle form the infundibular organ, which secretes Reissner's fibre, which extends through the neural canal to its posterior end (Olsson & Wingstrand 1954). The neural tube contains three types of putative photoreceptors, two types belonging to the rhabdomeric type and one probably to the ciliary type (Coomans 1981). None of them bear special resemblance to other chordate photoreceptors. There are no other multicellular sense organs, such as ears or nose. Several giant (Rohde) cells are found in the anterior and the posterior dorsal parts of the chord; the axons of the anterior cells cross over ventrally to the opposite side and extend posteriorly, while the posterior cells have anteriorly extending axons. There are alternating dorsal nerve roots and alternating ventral nerve roots; the dorsal roots pass between the myomeres and contain both sensory and viscero-motory elements, while the ventral roots go to the myomeres and consist of somatic-motory elements (Bone 1960).

The coelomic compartments are quite complicated, and their morphology is best explained in connection with the ontogeny (see below).

The principal muscles are the segmented lateral muscles used in swimming and burrowing. Each segment has developed from one coelomic sac, whose median side has become transformed into a large longitudinal muscle; the muscle segments of the two sides alternate. The septa between the segments are conical so that the characteristic interlocking muscular lamellae also known for example from teleosts are formed, although the shape is simpler. These muscles consist of three types of muscle cells, one type perhaps being a developmental stage of one of the others. Each muscle cell sends a dorsal tail to the spinal cord where a synapse is formed (Flood 1968).

There is a haemal system (Rähr 1979, 1981, 1982) which resembles the vertebrate haemal system in general layout, and a number of vessels have been homologized with vertebrate blood vessels. The vessels are surrounded by basement membranes and lack endothelia. Some of the vessels are contractile, but the muscle cells are part of the surrounding peritoneum and there is no heart. The contractile vessels create a slow circulation of the blood in constant directions.

A row of nephridia is found on each side in the branchial region; each organ has a nephridiopore in the atrial chamber at the base of a tongue bar. The nephridia are made up of cells of a unique type, the cyrtopodocyte, consisting of one part forming a usual podocytic lining of a blood vessel and another part which resembles a protonephridial solenocyte, i.e. a ring of long microvilli surrounding a cilium (Brandenburg & Kümmel 1961). The solenocyte part of the cells traverses the subchordal coelom and the tips of the microvilli and the cilium penetrates between the cells of the nephridial canals. The cyrtopodocytes appear to be of coelomic origin (see below), and the ring of microvilli surrounding a cilium could perhaps be interpreted as a specialization of the corresponding structures seen for example in peritoneal cells of echinoderms (Fig. 3.5). The structure superficially resembles a protonephridial cell, but the basement membrane surrounding almost all the weirs in the ectodermal protonephridia is not present and a homology is not indicated.

The gonads are arranged segmentally along the branchial basket. They develop from the ventral parts of the myocoels, and the ripe gametes break through the thin body wall to the atrial cavity; fertilization is external (Drach 1948).

The development of amphioxus has been the subject of several studies. The fundamental descriptions of the embryology are by Hatschek (1881), Cerfontaine (1906) and Conklin (1932), and these papers are the sources of the following description when nothing else is stated.

The first cleavage is median and the two first blastomeres are able to develop into small but apparently completely normal larvae if isolated. The second cleavage separates the anterodorsal and posteroventral halves of the embryo, and isolated blastomeres of this stage are not capable of forming complete embryos (Conklin 1932, Tung, Wu & Tung 1962a). Cell-lineage studies have made it possible to construct fate maps (Fig. 51.1), which show that the ectoderm is formed from the apical quartet, the endoderm from the blastoporal quartet, the mesoderm and chorda from a ring-shaped area at the apical half of the blastoporal quartet, and the neural tube from a dorsal crescent at the apical quartet. The following development goes through a coeloblastula and an embolic gastrula in which the blastocoel soon becomes obliterated. The ectodermal cells develop one cilium each at the stage when the blastopore narrows, and the endodermal cells become monociliate at a later stage. The second polar body is situated inside the fertilization membrane and can be used as a marker of the apical pole until a late gastrula stage, when the embryo hatches. The gastrula becomes bilaterally symmetrical with a flattened dorsal side (Fig. 53.1). The dorsomedian zone of the endoderm becomes thick and induces the overlying ectoderm to form the neural plate, which becomes overgrown by a pair of lateral ectodermal folds (Tung, Wu & Tung 1962b). These folds leave an anterior opening, the neuropore, but at the posterior end they continue around the blastopore which thus becomes enclosed and the neurenteric canal formed. The lateral parts of the neural plate soon fold up and fuse so that the neural tube is formed. The mediodorsal zone of endoderm then folds up forming the notochord, and the laterodorsal zones of endoderm form longitudinal folds, which soon break up into rows of coelomic sacs. The coelomic sacs differentiate first at the anterior end, and new sacs become added at the posterior end, though not from teloblasts as supposed by Hatschek (1881). The developing anterior coelomic sacs have small cavities in open connection with the archenteron while the following sacs are more compact (Conklin 1932).

The anterior part of the archenteron forms a pair of small lateral pockets which become pinched off as the anteriormost pair of coelomic sacs. This is very similar to the formation of the protocoelomic sacs in the other deuterostomes (Fig. 43.2) and

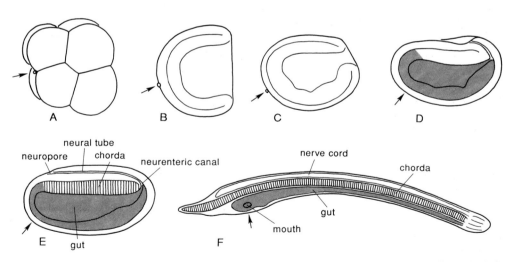

Fig. 53.1. Development of amphioxus (*Branchiostoma lanceolatum*); all stages seen from the left side with the blastoporal side oriented towards the right; the position of the polar bodies/apical pole is indicated by black arrows. – A, 8-cell stage. – B, blastula. – C, early gastrula. – D, late gastrula. – E, neurula with mesodermal sacs. – E, juvenile with mouth and the first gill slit. (A redrawn from Conklin 1933; B-D redrawn from Conklin 1932; E redrawn from Tung, Wu & Tung 1962a).

the two sacs are generally believed to represent the protocoel (Table 51.1). The more posterior coelomic sacs apparently represent meso- and metacoel, but direct homology has not been proved. The coelomic sacs of the two sides are symmetrically aligned until a stage of about 7-8 pairs of sacs, but at this stage the left anterior diverticulum remains small while the right becomes a larger, thin-walled sac and the somites of the right side thus become situated a little more posteriorly than those of the left side so that the somites of the two sides alternate.

The wall of the coelomic sacs in contact with the chorda differentiates into muscle cells while the remaining parts remain as a thin peritoneum. Each sac then divides into a dorsal and a ventral sac. The dorsal sacs, somites, comprise the muscles and a thin peritoneal part which covers the body wall and the dorsal part of the gut (Hatschek 1888). The ventral sacs are situated lateral to the ventral parts of the gut and they soon fuse to a pair of longitudinal sacs and finally to one sac surrounding the gut except mid-dorsally (it is not quite clear if the ventral parts of the sacs have been separated completely from the archenteron and fuse afterwards or if the ventral cavity develops as one long sac at each side; there may be differences between the anterior and posterior sacs). The muscular part of the dorsal sacs extends dorsally around the neural tube and ventrally around the gut, but the ventral part of the thin peritoneum develops an extension, the sclerocoel, separating the median side of the muscles from the chorda; the coelomic compartments lateral to the longitudinal muscles are called myocoels. This arrangement is seen in the region between the atrial pore and the anus; the anterior part of the body becomes more complicated due to the development of the gill pores and the atrium. The gill slits (see below) divide the ventral coelom into a number of narrower spaces: paired longitudinal epibranchial (or subchordal) coeloms dorsal to the gill slits, narrow channels through the gill bars and a midventral hypobranchial (or subendostylar) coelom.

A pair of lateral metapleural or atrial folds grow ventrally covering the branchial basket and finally fuse in the midline leaving only a posterior atrial pore; these folds contain coelomic compartments which are originally parts of the ventral coelom (Lankester & Willey 1890).

The mouth breaks through on the left side of the anterior region, and the first gill pore breaks through on the right side near the ventral midline well behind the mouth. Additional gill pores develop in a series behind the first pore; this series slowly moves to the left side while a new series of pores develops on the right side. The pores soon become heart-shaped and then U-shaped by the development of a dorsal extension which becomes the tongue bar; each vertical slit becomes divided into a row of gill pores by the development of transverse bars, synapticles, between the gill bars and tongue bars. The branchial skeleton develops first along the anterior and posterior sides of the gill pores; in later stages the two curved rods fuse dorsally and extend into the tongue bar, and finally transverse bars are formed in the synapticles (Lönnberg 1901-1905; Fig. 49.1).

The endostyle develops as a circular endodermal thickening inside the mouth; it soon develops an anterior indentation and then becomes V-shaped (Willey 1891). Finally, the endostyle moves to a mid-ventral position behind the mouth and its two branches move close together so that the original anterior indentation becomes situated along the bottom of the endostylar groove, with the more posterior rows of cells forming parallel bands along the sides of the groove.

The cyrtopodocytes apparently develop from the peritoneum of the subchordal coelom (Goodrich 1934), but the differentiation of the cells and the origin of the nephridial ducts are in need of further investigation.

At the stage of one gill pore, three small papillae are found in the mouth region (van Wijhe 1926, Berrill 1987). One papilla is situated on the left side in front of the mouth just below the preoral pit (Table 51.1); another papilla lies on the right side opposite from the mouth; and the third papilla is situated midventrally just ventral to the gill opening. The right and median papillae become incorporated into the anterior parts of the atrial folds, while the anteriormost, left papilla soon disappears. Some of the earlier authors supposed that the papillae are adhesive and proposed a homology to the adhesive papillae of ascidian larvae. However, the ultrastructure of the oral papilla shows no signs of secretory activity, cilia or innervation (Andersson & Olsson 1989).

Cell-lineage studies (Tung, Wu & Tung 1962a) have shown that the apical pole is now situated midventrally, almost directly under the lateral mouth (Fig. 53.1).

A series of asymmetries in the juvenile stages of amphioxus have been the subject of many phylogenetic discussions. These features must be interpreted in one of two ways: They are either adaptational specializations of the juvenile, probably associated with feeding (i.e. apomorphies), or they reflect the ancestry of the phylum. The latter view was held by for example Willey (1891), who derived amphioxus from sessile, ascidian-like ancestors which had tadpole larvae with a dorsal mouth. When the notochord of the cephalochordate ancestor extended further forward, the dorsal mouth 'came in the way' and became displaced to the left side. The asymmetries of several other anterior organs should have evolved as a consequence of this change. The origin of the cephalochordates (and the vertebrates) from sessile, ascidian-like ancestors has been rejected above (Chapters 50, 51). Bone (1958) made new observations on the feeding of juvenile amphioxus and found that the epithelial cilia create a posteriorly directed current along the body, and that a part of this current enters the mouth and leaves through the single gill slit, propelled by the cilia of the gill slit. The lateral position of the juvenile mouth was interpreted as an adaptation which made an enlargement of the mouth possible, which should enhance the flow of water through the filtering structure of the juvenile. A similar expansion of a ventral (or dorsal) mouth on a laterally compressed, fish-like organism was considered mechanically unsound. In conclusion it was suggested that the lateral position of the juvenile mouth of amphioxus 'is an adaptation related to the mode of feeding, and has no phylogenetic significance; it is not inherited from an asymmetrical ancestor, nor is it a result of an adult adaptation' (Bone 1958, p. 293). This is in complete agreement with the phylogeny proposed here (Fig. 51.2) and is further supported by the fact that no asymmetries are found in the pharyngeal region of primitive vertebrates.

The longitudinal muscles are used in swimming just like those of the lower vertebrates, and their structure with two main types of muscles cells is very similar to that observed especially in the lampreys, although the type of innervation is different (Bone 1989).

427

The position of the adult mouth on the ventral side 'in front of' the apical pole (Fig. 53.1) is identical to that of the vertebrates, and the fate maps of urochordates, cephalochordates and vertebrates show significant agreement (Fig. 51.1). The asymmetries observed during development have been interpreted as larval specializations.

I think that it is documented beyond all doubt that the cephalochordates and vertebrates form a monophyletic group. They share such fundamental apomorphies as the segmented lateral musculature developing from lateral coelomic sacs and extending almost to the anterior end of the body, the retention of the chorda, sometimes in a much modified form, in the adult stage, haemal systems of similar general shape, and innervation of segmented organs from dorsal and ventral nerve roots from the notochord.

On the other hand, the cephalochordates show a series of apomorphies, such as the notochord consisting of muscle cells innervated directly from the neural tube, the ventral atrium and the asymmetries of the juveniles; these apomorphies make it very unlikely that the vertebrates are derived from cephalochordates or vice versa. Amphioxus does indeed show many features which appear to be characteristic also of the ancestral vertebrate and which can be used to illustrate the early evolution of this phylum, but the just-mentioned apomorphies of the cephalochordates make it necessary to regard the two phyla as sister groups.

Some of the classical treatises of the animal kingdom included the cephalochordates in the vertebrates (where it was usually placed as the group Leptocardii within the fishes, see for example Lönnberg 1901-05), but it is more in accordance with recent concepts of the phyla to regard Cephalochordata and Vertebrata as separate phyla.

Interesting subjects for future research

1. Development of nephridia and cyrtopodocytes.
2. Secretion of the mucous filter by the endostyle.

References

Andersson, A. & R. Olsson 1989. The oral papilla of the lancelet larva (*Branchiostoma lanceolatum*) (Cephalochordata). – Acta zool. (Stockh.) **70**: 53-56.

Barrington, E.J.W. 1958. The localization of organically bound iodine in the endostyle of Amphioxus. – J. mar. biol. Ass. U.K. **37**: 117-126.

Barrington, E.J.W. 1965. The biology of Hemichordata and Protochordata. – Freeman, San Francisco.

Berrill, N.J. 1987. Early chordate evolution Part 2. Amphioxus and ascidians. To settle or not to settle. – Int. J. Invertebr. Reprod. Dev. **11**: 15-28.

Bone, Q. 1958. The asymmetry of the larval amphioxus. – Proc. zool. Soc. Lond. **130**: 289-293.

Bone, Q. 1960. The central nervous system in amphioxus. – J. comp. Neurol. **115**: 27-64.

Bone, Q. 1989. Evolutionary patterns of axial muscle systems in some invertebrates and fish. – Am. Zool. **29**: 5-18.

Brandenburg, J. & G. Kümmel 1961. Die Feinstruktur der Solenocyten. – J. ultrastruct. Res. **5**: 437-452.

Butterfield, N.J. 1990. Organic preservation of non-mineralizing organisms and the taxonomy of the Burgess Shale. – Paleobiology **16**: 272-286.

Cerfontaine, P. 1906. Recherches sur le développement de l'*Amphioxus*. – Archs Biol. **22**: 229-418, pls 12-22.

Conklin, E.G. 1932. The embryology of *Amphioxus*. – J. Morph. **54**: 69-118.

Conklin, E.G. 1933. The development of isolated and partially separated blastomeres of amphioxus. – J. exp. Zool. **64**: 303-375.

Conway Morris, S. & H.B. Whittington 1979. The animals of the Burgess Shale. – Scient. Am. **240**: 122-133.

Coomans, A. 1981. Phylogenetic implications of the photoreceptor structure. – Atti Conv. Lincei **49**: 23-68.

Drach, P. 1948. Embranchement des Céphalochordés. – Traité de Zoologie, vol. **11**, pp 931-1040. Masson, Paris.

Flood, P.R. 1968. Structure of the segmental trunk muscles in amphioxus. – Z. Zellforsch. **84**: 389-416.

Flood, P.R. 1970. The connections between spinal chord and notochord in *Amphioxus (Branchiostoma lanceolatum)*. – Z. Zellforsch. **103**: 115-128.

Goodrich, E.S. 1934. The early development of the nephridia in *Amphioxus*: Part II, The paired nephridia. – Q. Jl microsc. Sci., N.S. **76**: 655-674, pls 37-40.

Guthrie, D.M. & J.R. Banks 1970. Observations on the function and physiological properties of a fast paramyosin muscle – the notochord of amphioxus *(Branchiostoma lanceolatum)*. – J. exp. Zool. **52**: 125-138.

Hatschek, B. 1881. Studien über Entwicklung des *Amphioxus*. – Arb. zool. Inst. Univ. Wien **4**: 1-88, pls 1-9.

Hatschek, B. 1888. Über den Schichtenbau von *Amphioxus*. – Anat. Anz. **3**: 662-667.

Holland, N.D. & L.Z. Holland 1990. Fine structure of the mesothelia and extracellular materials in the coelomic fluid of the fin boxes, myocoels and sclerocoels of a lancelet, *Branchiostoma floridae* (Cephalochordata = Acrania). – Acta zool. (Stockh.) **71**: 225-234.

Lankester, E.R. & A. Willey 1890. The development of the atrial chamber of *Amphioxus*. – Q. Jl microsc. Sci., N.S. **31**: 445-466, pls 29-32.

Lönnberg, E. 1901-1905. Leptocardii. – Bronn's Klassen und Ordnungen des Tierreichs, 6. Band, 1. Abt., 1. Buch, pp 99-249. Akademische Verlagsgesellschaft, Leipzig.

Olsson, R. 1963. Endostyles and endostylar secretions: a comparative histochemical study. – Acta zool. (Stockh.) **44**: 299-328.

Olsson, R. & K.G. Wingstrand 1954. Reissner's fibre and the infundibular organ in amphioxus. – Univ. Bergen Årb. 1954, Nat. R. **14**: 1-15, 3 pls.

Pearse, V., J. Pearse, M. Buchsbaum & R. Buchsbaum 1987. Living Invertebrates. – Blackwell, Palo Alto, CA.

Rähr, H. 1979. The circulatory system of amphioxus (*Branchiostoma lanceolatum* (Pallas)). – Acta zool. (Stockh.) **60**: 1-18.

Rähr, H. 1981. The ultrastructure of the blood vessels of *Branchiostoma lanceolatum* (Pallas)(Cephalochordata). – Zoomorphology **97**: 53-74.

Rähr, H. 1982. Ultrastructure of gill bars of *Branchiostoma lanceolatum* with special reference to gill skeleton and blood vessels (Cephalochordata). – Zoomorphology **99**: 167-180.

Tjoa, L.T. & U. Welsch 1974. Electron microscopical observations on Kölliker's and Hatschek's pit and on the wheel organ in the head region of amphioxus (*Branchiostoma lanceolatum*). – Cell Tissue Res. **153**: 175-187.

Tung, T.C., S.C. Wu & Y.Y.F. Tung 1962a. The presumptive areas of the egg of amphioxus. – Scientia sin. **11**: 629-644.

Tung, T.C., S.C. Wu & Y.Y.F. Tung 1962b. Experimental studies on the neural induction in amphioxus. – Scientia sin. **11**: 805-820.

van Wijhe, J.W. 1926. On the temporary presence of the primary mouth-opening in the larva of amphioxus, and the occurrence of three postoral papillae, which are probably homologous with those of the larva of ascidians. – Proc. Sect. Sci. K. ned. Akad. Wet. **29**: 286-295.

Webb, J.E. 1973. The role of the notochord in forward and reverse swimming and burrowing in the amphioxus *Branchiostoma lanceolatum*. – J. Zool. (Lond.) **170**: 325-338.

Welsch, U. 1975. The fine structure of the pharynx, cyrtopodocytes and digestive caecum of *Amphioxus (Branchiostoma) lanceolatum*. – Symp. zool. Soc. Lond. **36**: 17-41.

Welsch, U. & V. Storch 1969. Zur Feinstruktur und Histochemie des Kiemendarmes und der 'Leber' von *Branchiostoma lanceolatum* (Pallas). – Z. Zellforsch. 102: 432-446.

Welsch, L.T. & U. Welsch 1978. Histologische und elektronenmikroskopische Untersuchungen an der präoralen Wimpergrube von *Saccoglossus horsti* (Hemichordata) und der Hatschekschen Grube von *Branchiostoma lanceolatum* (Acrania). Ein Beitrag zur phylogenetischen Entwicklung der Adenohypophyse. – Zool. Jb., Anat. 100: 564-578.

Willey, A. 1891. The later larval development of *Amphioxus*. – Q. Jl microsc. Sci., N.S. 32: 183-234, pls 13-15.

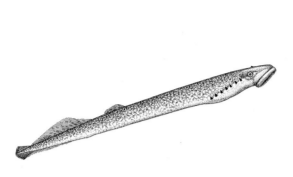

Phylum VERTEBRATA (= CRANIATA)

The literature about our own phylum is absolutely overwhelming, and the list of subjects which could be taken up in a phylogenetic discussion seems endless. I have therefore chosen not to follow the outline used in the preceding chapters, but to arrange selected characters in three main groups: 1) general chordate characters, 2) synapomorphies of the notochordates, and 3) autapomorphies of the vertebrates. Most of the information can be checked in the common textbooks, so references have been kept to a minimum.

It is almost universally agreed that hagfishes and lampreys show a whole suite of characters which must be regarded as ancestral to the vertebrates, although the few living representatives are at the same time specialized for unusual feeding modes. These groups are therefore of special interest for the phylogenetic discussion, but it should be remembered that many of the characters which show similarities between vertebrates and the other chordates are found in the ammocoetes larva of the lampreys, and in embryonic stages of the various vertebrate groups. The conodont animals have now been shown to have been vertebrates, and a probable phylogeny shows the hagfishes as the sister group of conodont animals plus the remaining vertebrates (Briggs 1992); this extends the vertebrate record back into the Cambrian.

In Chapter 51, the following character complexes were listed as important apomorphies of the chordates: chorda, dorsal neural tube, longitudinal muscles along the chorda, ciliated pharyngeal gill slits functioning as a mucociliary filtering structure with the mucous net secreted by the ventral endostyle, and the embryology with the mouth developing from an area dorsal to the apical pole. The expressions of these characters in the vertebrates will be commented on briefly.

The notochord (chorda) develops from the dorsal side of the archenteron or from the primitive streak, also when the development is complicated through large amounts of yolk or a placenta. In the lampreys, the notochord is formed as median fold from the roof of the archenteron (see below); the first stages show an irregular arrangement of cells, but the later larval stages have a chorda consisting of one row

Chapter vignette: *Petromyzon marinus*. (Redrawn from Muus 1964.)

of flat cells. The adults again show an irregular arrangement of cells with large vacuoles. The chorda is surrounded by a thickened basement membrane.

The neural tube is formed from the ectoderm in contact with the chorda, and the chemical induction of the ectoderm from the chorda cells is documented through numerous studies already from the beginning of the century. The adult central nervous system consists of a dorsal nerve tube, the spinal cord, and a highly complex, anterior brain. Comparisons with amphioxus show that the spinal cord is in principle very similar in the two groups, and that the brain is an enormously enlarged and specialized brain vesicle. It is difficult to find homologous areas or organs in vertebrates, cephalochordates and urochordates (see below).

The mesoderm develops from the invaginated archenteron (see below) or from corresponding masses of cells from the dorsal blastopore lip.

The pharynx of the ammocoetes larva has a row of gill slits on each side and a ventral endostyle, but although the general morphology is rather similar to that of amphioxus (except that the slits are not U-shaped), a number of differences have been pointed out (Mallat 1981). An important difference is that the water flow through the pharynx is set up by ventilatory movements of the pharynx musculature rather than by the beat of the cilia of the gill slits. Another important difference is that although the endostyle has bands of cells secreting proteinaceous mucus and iodinated compounds to the pharynx, it does not organize the mucous filter which is the filtering structure. The mucus appears to be secreted mainly from goblet cells on the gill bars. The adult lampreys and all other vertebrates lack an endostyle, but the transformation of the endostyle of the ammocoetes larva into the thyroid of the adult lamprey has been well documented; the other vertebrates have a thyroid comprising the same cell types as those of the lampreys, and it seems unquestionable that endostyle and thyroid are homologous organs.

The first cleavage is median in anurans (Klein 1987). In *Lampetra* (Balfour 1881, Damas 1944), cleavage leads to a coeloblastula with a narrow blastocoel at the apical side; gastrulation is through invagination, and the narrow, tubular archenteron lies close to the dorsal ectoderm. The dorsal cells of the archenteron become the chorda and the two plates of somewhat smaller cells lateral to the chorda become the mesoderm. This general type of development can with more or less modification be recognized in all vertebrates.

The main axis of the egg is determined already in the ovary and detailed fate maps of the various regions of the egg have been constructed for many species (Fig. 51.1). The fate maps of the three chordate phyla show identical spatial relations between the various areas of the egg, and the position of the mouth on the dorsal side behind the apical pole is documented also in a number of vertebrates.

All the general chordate characters can thus be recognized in the vertebrates.

The mesoderm usually develops through ingression of cells from the dorsal blastopore lip or corresponding areas, but the supposedly primitive formation of mesodermal sacs as pockets from the roof of the archenteron lateral to the chorda has been observed in a few groups, for example lampreys (Koltzoff 1902, Damas 1944). Their mesoderm is at first a pair of compact longitudinal cell masses lateral to notochord and neural tube, but the anterior parts of the mesodermal plates become divided into about 20 segments, which subsequently divide into dorsal somites and ventral sacs (lateral plates); the posterior part of the mesodermal bands forms segmental somites, but the ventral part remains undivided. The primitive vertebrate nephridia develop from the narrow stalk (nephrotome) between the somites and the ventral sacs. In other vertebrates, the somites can always be recognized, but are formed as compact cell masses without connection with the archenteron, and the ventral mesoderm is undivided. So although the origin of the mesoderm is in most cases quite different from that found in amphioxus, the early morphology of the coeloms is quite similar, and the segmentation of the mesoderm in the two phyla must be regarded as homologous.

The fate maps (Fig. 51.1) show that the mouth develops from a 'dorsal' area behind the apical pole in all the chordate phyla, but the development shows that the mouth remains dorsal in the urochordates (Fig. 52.1) while both the apical pole and the mouth area move to the ventral side so that the apical pole becomes situated behind the ventral mouth in both cephalochordates and vertebrates (Figs 53.1 and 54.1). The origin of this character seems completely enigmatic.

There are important differences between the anterior parts of the neural tubes of amphioxus and the vertebrates, but the posterior parts are more similar, and Bone (1960) found 'rather striking' resemblances between the arrangements of neurons in amphioxus and the young ammocoetes larvae, with amphioxus representing a more primitive type.

The synapomorphies of cephalochordates and vertebrates clearly demonstrate the monophyly of the notochordates.

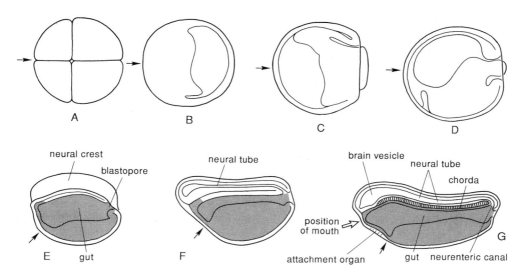

Fig. 54.1. Development of *Bombinator pachypus*; all stages are seen from the left side with the blastoporal pole to the right; the position of the polar bodies/apical pole is indicated by black arrows. – A, 8-cell stage. – B, blastula. – C, early gastrula. – D, late gastrula. – E, neural crest formation. – F, early neural tube stage. – G, fully developed neural tube stage. (Redrawn from Vogt 1929.)

433

The apomorphies of the vertebrates are numerous, and have been discussed in several recent papers (for example Maisey 1986 with a list of 25 apomorphies, Schaeffer 1987 with a list of 10 apomorphies, Gans 1989); only some of the more conspicuous characters will be mentioned.

One of the most important complexes appears to be related to the evolution of the neural crest and the epidermal placodes. These structures develop at the edges of the neural plate (Moury & Jacobson 1990) and give rise to a number of structures which are unknown in the non-vertebrates (Gans 1989). Many of the structures are related to the macrophagous habits of the vertebrates and the related evolution of a more 'active' life style with complicated sense organs and a complex brain. Cells of the neural crest give rise to sensory nerves with ganglia, peripheral motor ganglia and higher-order motor neurons, and cells of the placodes form sense organs like eyes (not the sensory tissues which are parts of the brain), ears, lateral line organs, and gustatory organs. Also important parts of the skeleton, for example the cartilage in the gill bars, are derived from the neural crest.

The vertebrate haemal system has an inner layer of cells, the endothelium, a character which is only found in a few scattered 'invertebrate' groups, where it is regarded as apomorphic within the phyla (Ruppert & Carle 1983).

The transformation of the endostyle into the thyroid in all vertebrates but the larval lampreys is discussed above.

Another vertebrate apomorphy is the multilayered ectoderm, where cell division in the basement layer gives rise to new cells which can replace old cells which are worn off at the surface. Similar epithelia are only known from certain body regions of chaetognaths (Chapter 31), where a homology is not indicated.

The anterior end of the neural tube is greatly enlarged and specialized as a brain in all vertebrates. It is difficult to make comparisons with the very simple 'brain' of amphioxus, but the position of the cells secreting Reissner's fibre, the infundibular organ at the posteroventral side of the brain vesicle in amphioxus and the flexural organ in larval salmon (Olsson 1956), indicates that it is the areas in front of this region which have become the vertebrate brain with the several new multicellular sense organs and centres for processing the information from the new sense organs and the coordination of more complex movements.

The fully developed vertebrate nephridia are of the glomerular type with podocyte-lined blood vessels surrounded by a small coelomic compartment, the Bowmann's capsule; there is a common nephridial duct on each side. The nephridia develop from approximately the same position as those of amphioxus, but the vertebrate nephridia are situated behind the branchial region and aligned with the somites, i.e. myomeric, while those of amphioxus are aligned with the gill slits, i.e branchiomeric. This makes it unlikely that the two types of nephridia are homologous.

The transformation of the endostyle to an endocrine gland, the thyroid, in all vertebrates except the larval lampreys has been mentioned above.

The few character complexes just mentioned should suffice to demonstrate the (unquestioned) monophyly of the vertebrates.

References

Balfour, F.M. 1881. A Treatise of Comparative Embryology, vol. 2. – Macmillan, London.

Bone, Q. 1960. The central nervous system in amphioxus. – J. comp. Neurol. 115: 27-64.

Briggs, D.E.G. 1992. Conodonts: a major extinct group added to the vertebrates. – Science 256: 1285-1286.

Damas, H. 1944. Recherches sur le développement de *Lampetra fluviatilis* L. – Archs Biol. 55: 1-284, pls 1-3.

Gans, C. 1989. Stages in the origin of vertebrates: analysis by means of scenarios. – Biol. Rev. 64: 221-268.

Klein, S.L. 1987. The first cleavage furrow demarcates the dorsal-ventral axis in *Xenopus* embryos. – Dev. Biol. 120: 299-304.

Koltzoff, N.K. 1902. Entwickelungsgeschichte des Kopfes von *Petromyzon planeri*. – Bull. Soc. imp. Nat. Moscou 15: 259-589, pls 1-7.

Maisey, J.G. 1986. Heads and tails: a chordate phylogeny. – Cladistics 2: 201-256.

Mallat, J. 1981. The suspension feeding mechanism of the larval lamprey *Petromyzon marinus*. – J. Zool. (Lond.) 194: 103-142.

Moury, J.D. & A.G. Jacobson 1990. The origins of neural crest cells in the axolotl. – Dev. Biol. 141: 243-253.

Muus, B. 1964. Havfisk og fiskeri i Nordvesteuropa. – Gad, Copenhagen.

Olsson, R. 1956. The development of the Reissner's fibre in the brain of the salmon. – Acta zool. (Stockh.) 37: 235-250.

Ruppert, E.E. & K.J. Carle 1983. Morphology of metazoan circulatory systems. – Zoomorphology 103: 193-208.

Schaeffer, B. 1987. Deuterostome monophyly and phylogeny. – Evol. Biol. 21: 179-235.

Vogt, W. 1929. Gestaltungsanalyse am Amphibierkeim mit örtlicher Vitalfärbung. II. Teil. Gastrulation und Mesodermbildung bei Urodelen und Anuren. – Arch Entwicklungsmech. Org. 120: 384-706.

Five enigmatic taxa

In the preceding chapters, I have tried to assign phylogenetic positions to all groups of metazoans, but five taxa have not been discussed because their position is completely unresolved. Two of the taxa are the 'mesozoans', now usually regarded as two separate phyla, Dicyemida and Orthonectida, and two are the monospecific genera *Xenoturbella* and *Buddenbrockia*. The fifth taxon is the completely enigmatic *Salinella salve*, which was obtained by Frenzel (1892) from a saline culture of material from Córdoba, Argentina. The description shows a tube of cells with cilia both on the inner and the outer side and with special cilia around both openings. Various developmental stages and an encystation after 'conjugation' were also described. The organism has not been refound, and since ultrastructural details are obviously unknown, it seems futile to discuss its phylogenetic position.

Dicyemids and orthonectids are well-established 'Problematica', which are discussed in all major textbooks and encyclopedias. Their position is sometimes regarded as close to the metazoan stem (as indicated by the name Mesozoa), whereas other authors regard them as specialized parasitic flatworms. I cannot contribute anything new to the discussion.

The status of the two remaining taxa is different. They are omitted from several multiauthor texts, probably because no-one has wanted to deal with them. They should be discussed shortly here to point out that they are worth studying and that they are both readily obtainable.

Xenoturbella bocki is a centimetre-long, whitish, totally ciliated 'worm' which was described from the Gullmarfjord, Sweden, by Westblad (1949), who had also found one specimen near Millport, Scotland. It has subsequently been collected from a number of localities in the Gullmarfjord, where it can routinely be obtained from muddy and mixed bottoms at depths of 20-100 m. Franzén (1956), Reisinger (1960), Pedersen & Pedersen (1986, 1988), Franzén & Afzelius (1987), and Ehlers (1991) have contributed to the knowledge of the spermatozoa, the general anatomy, the ultrastructure of the epidermis and of the statocyst. The development is unknown. The systematic-phylogenetic position has been discussed by several of these authors, and two main opinions have prevailed: one is that *Xenoturbella* is a very primitive platyhelminth, but it appears to be so primitive that none of the (somewhat uncertain) platyhelminth apomorphies can be recognized; the other is that the epidermis and the statocyst show enteropneust-echinoderm affinities, and that *Xenoturbella* therefore could be a very primitive deuterostome, but without any trace of coeloms. Studies of the development could probably throw light on the systematic position.

Buddenbrockia plumatellae is a cylindrical organism with rounded ends and attains a length of about 3 mm; it occurs in the coelomic cavity of freshwater bryozoans. It was described by Schröder (1910a, b, 1912a, b) from *Plumatella* from Germany and has been found also in *Stolella*, *Hyalinella* and *Lophopodella* in Belgium, Bulgaria, Turkestan, Japan, and Brazil (Dumortier & van Beneden 1843, Braem 1911, Marcus 1941, Grancarova 1968, Oda 1972, 1978). Structure, reproduction and development have been studied on the light microscopic level, and it has been suggested that the organism is related to mesozoans or nematodes, but without convincing evidence. Electron microscopical studies and studies on the development are needed before this taxon can be placed in the system.

References

Braem, F. 1911. Beiträge zur Kenntniss der Fauna Turkestans. VII. Bryozoen und deren Parasiten. – Trav. Soc. imp. Natur. S.-Peterb. **42** (Lief. 2, Teil 1): 1-56.

Dumortier, B.C. & P.J. van Beneden 1850. Histoire naturelle des polypes composés d'eau douce. – Nouv. Mém. Acad. R. Sci. Bruxelles **16**: 33+96 pp, 6 pls.

Ehlers, U. 1991. Comparative morphology of statocysts in the Plathelminthes and the Xenoturbellida. – Hydrobiologia **227**: 263-271.

Franzén, Å. 1956. On spermiogenesis, morphology of the spermatozoon, and biology of fertilization among invertebrates. – Zool. Bidr. Uppsala **31**: 355-480.

Franzén, Å. & B.A. Afzelius 1987. The ciliated epidermis of *Xenoturbella bocki* (Platyhelminthes, Xenoturbellida) with some phylogenetic considerations. – Zool. Scr. **16**: 9-17.

Frenzel, J. 1892. Untersuchungen über die mikroskopische fauna Argentiniens. *Salvinella salve* nov. gen. nov. spec. – Arch. Naturgesch. **58**(1): 66-96, pl. 7.

Grancarova, T. 1968. Neue Bryozoen in der Bulgarischen Fauna. 1. *Urnatella gracilis* Leidy (Bryozoa, Entoprocta); *Hyalinella punctata* (Hancock) (Bryozoa Ectoprocta). – Izv. zool. Inst. (Sofia) **28**: 197-204. (In Russian, German summary.)

Marcus, E. 1941. Sôbre Bryozoa do Brasil.– Bolm Fac. Filos. Ciênc. S Paulo, Zool. **5**: 3–208.

Oda, S. 1972. Some problems on *Buddenbrockia*. – Zool. Mag. (Tokyo) **81**: 173-183 (In Japanese).

Oda, S. 1978. A note on Bryozoa from Lake Shoji, Japan. – Proc. Jap. Soc. syst. Zool. **15**: 19-23.

Pedersen, K.J. & L.R. Pedersen 1986. Fine structural observations on the extracellular matrix (ECM) of *Xenoturbella bocki* Westblad, 1949. – Acta zool. (Stockh.) **67**: 103-113.

Pedersen, K.J. & L.R. Pedersen 1988. Ultrastructural observations on the epidermis of *Xenoturbella bocki* Westblad, 1949; with a discussion of epidermal cytoplasmic filament systems of invertebrates. – Acta zool. (Stockh.) **69**: 231-246.

Reisinger, E. 1960. Was ist *Xenoturbella*? – Z. wiss. Zool. **164**: 188-198.

Schröder, O. 1910a. Eine neue Mesozoenart (*Buddenbrockia plumatellae* n.g. n.sp.) aus *Plumatella repens* L. und *Pl. fungosa* Pall. – Sber. heidelb. Akad. Wiss., Math.-nat. Kl. **1910** (6): 1-8.

Schröder, O. 1910b. *Buddenbrockia plumatellae*, eine neue Mesozoenart aus *Plumatella repens* L. und *Pl. fungosa* Pall. – Z. wiss. Zool. **96**: 525-537, pls 23-23a.

Schröder, O. 1912a. Zur Kenntniss der *Buddenbrockia plumatellae* Ol. Schröder. – Z. wiss. Zool. **102**: 79-91.

Schröder, O. 1912b. Weitere Mitteilungen zur Kenntnis der *Buddenbrockia plumatellae* Ol. Schröder. – Verh. naturh.-med. Ver. Heidelb., N.F. **11**: 230-237.

Westblad, E. 1949. *Xenoturbella bocki* n.g., n.sp. a peculiar, primitive turbellarian type. – Ark. Zool., 2. ser. **1**: 11-29, 4 pls.

Concluding remarks

The starting point for this whole book has been my earlier work with the trochaea theory (Chapter 2), and although the main points in the theory have turned out to stand the more detailed scrutiny, a number of modifications have been necessary. The following short review emphasizes these modifications.

The first part of the theory, which is identical with Hatschek's gastraea theory, invokes the evolution of gastrula-like organisms such as cnidarians from colonies of choanoflagellates, and this idea is now becoming more and more accepted; the new data, for example on ultrastructure, have added further support to the theory and made it possible to assign precise phylogenetic positions to sponges and placozoans.

The evolution of a trochaea with a circumblastoporal ring of compound cilia on multiciliate cells from the gastraea, and the further evolution and radiation of the protostomes from the trochaea through the ancestral protostome, gastroneuron with a trochophora larva, have turned out to be in excellent agreement with almost all the available facts. Only the ectoprocts are still difficult to place in the phylogeny, but their 'deuterostome' characters have almost all been shown to be based on misinterpretations, and their position as the sister group of the entoprocts seems to be the most reasonable one at present.

The derivation of the deuterostomes from the trochaea has turned out to be more questionable. A perianal ring of compound cilia is only found in actinotrocha and tornaria larvae, and the compound cilia are formed from monociliate cells in the actinotrochs. It appears just as probable that these perianal bands have developed independently in the two phyla and that they are therefore not homologous with the archaeotroch of the trochaea. This would imply that the ancestral deuterostome, notoneuron, had a dipleurula larva instead of a tornaria larva. It would also imply that the protornaea had no archaeotroch and that the protornaeozoans were derived directly from the gastraea. The more detailed discussion of the deuterostome phyla has led to the conclusion that notoneuron was not a deposit-feeder but a suspension-feeder with the larval neotroch pulled out on lateral tentacles.

It must still be admitted that the origin of the larval mouth surrounded by the neotroch in the deuterostomes has no functional explanation. The trochaea theory proposes that four secondary openings developed from the apex of the archenteron in a gastraea (trochaea), and that these openings became the mouth, the hydropore and two primary gill pores; parts of the general ciliation of the ectoderm should then become specialized as the neotroch. This theory lacks the desirable additional explanation, but I have not seen any other attempts at explaining how a new mouth can evolve.

The changed ideas about the origin of the deuterostomes have not resulted in any change in the shape of the phylogenetic tree. The revised concept of both larval and adult morphology of notoneuron has made the group Brachiata paraphyletic, and a new phylogeny has been proposed.

A number of the character complexes which have been used in constructing the phylogeny have been discussed earlier (Nielsen 1985, 1987). Complexes such as the pelago-benthic life cycles, the fate of the blastopore, the types of larval ciliary bands, the position and gross morphology of the central nervous systems, and the cleavage patterns have been integral parts of the argumentation, and the importance and interpretation of these complexes have not been brought into question by the more comprehensive coverage of taxonomic groups and details. Only a few new or modified conclusions need to be mentioned.

The evolution of the third germ layer, mesoderm, has been the subject of much discussion over the last century, but it appears that the various types of mesoderm formation observed in the living organisms, such as ingression from the ectoderm (ectomesoderm), 4-d mesoderm from a cell at the posterior edge of the blastopore, and endomesoderm from the archenteron, are not homologous; I have not found anything indicating that 4-d mesoderm is ancestral to endomesoderm or vice versa. The phylogeny proposed here is more in accordance with the theory that at least the two last-mentioned types of mesoderm have evolved independently.

The idea of independent origins of the coeloms in spiralians and deuterostomes is an important consequence of the phylogeny proposed here. Many earlier authors, prominently among them Hyman (1951), have divided the Bilateria into three groups, Acoelomata, Pseudocoelomata and Coelomata, with the more or less tacit understanding that the coelom has evolved only once. This made it necessary to regard the enterocoelic coelom formation of the deuterostomes as derived from the schizocoelic type found in many protostomes, or vice versa, which has proved very difficult (Rieger 1985). Discussions of this problem have almost always been coupled with considerations about the nature of the ancestral bilaterian: was it an organism with gastral pouches which turned into coelomic sacs (a type of protocoeloma or bilaterogastraea), or was it a compact, turbellarian-like organism? Such considerations may perhaps have obscured some of the investigations, because the search has always been for intermediate forms (Rieger 1986). I have not found any morphological or embryological support for the idea that the bilaterian ancestor was coelomate, and that the acoelomate and pseudocoelomate groups therefore should be derived or simplified in this respect. This does not exclude the possibility that the coelomic cavities may become secondarily reduced, as for example in some interstitial annelids. The origins of the various types of secondary body cavities found for example in annelids, ectoprocts, chaetognaths, and deuterostomes are obscure, and it cannot be stated that the coelomate condition evolved either from the acoelomate or from the pseudocoelomate condition. Consequently, I believe that introduction of the character states: coelom +/–, automatically makes a phylogenetic analysis suspect, be it manual or computer generated.

It seems now to be generally accepted that the early metazoans had monociliate cells, but this does not necessarily mean that a monociliate epithelium is an ancestral character in all cases. A few observations of ciliogenesis in multiciliate cells indicate that at least some of these cells pass through a monociliate stage, and the centrioles are an integral part of almost all dividing cells; this means that almost all cells possess the basal apparatus characteristic of monociliate cells at some stage of the differentiation; a reversal from the multiciliate to the monociliate condition could simply be the result of a loss of the ability to organize additional basal bodies.

Also the myoepithelial cell has been interpreted as a plesiomorphic character in a number of groups, but it is evident that actin is present in all dividing cells, and the specialization as contractile cells of various types can have evolved independently many times.

A more general statement could be that all animal cells have the genetic information needed for organizing organelles such as a cilium with an accessory centriole, contractile elements formed by actin fibres, and perhaps also microvilli, and that these possibilities are only suppressed in many differentiated cells. It should be clear that such characters cannot be used in a phylogenetic analysis.

My final conclusion is that it is indeed possible to construct a robust phylogenetic tree on the base of morphological/embryological characters alone, but only after a careful analysis of all the characters.

References

Hyman, L.H. 1951. The Invertebrates, vol. 2. – MacGraw-Hill, New York.
Nielsen, C. 1985. Animal phylogeny in the light of the trochaea theory. – Biol. J. Linn. Soc. 25: 143-299.
Nielsen, C. 1987. Structure and function of metazoan ciliary bands and their phylogenetic significance. – Acta. zool. (Stockh.) 68: 205-262.
Rieger, R.M. 1985. The phylogenetic status of the acoelomate organization within the Bilateria: a histological perspective. – In S. Conway Morris, J.D. George, F. Gibson & H.M. Platt (eds): The Origins and Relationships of Lower Invertebrates, pp 101-122. Oxford Univ. Press, Oxford.
Rieger, R.M. 1986. Über den Ursprung der Bilateria: die Bedeutung der Ultrastrukturforschung für ein neues Verstehen der Metazoenevolution. – Verh. dt. zool. Ges. 79: 31-50.

Postscript

It goes without saying that I do not regard my speculations about animal phylogeny as finished with the publication of this book. I have felt very tempted to try to answer some of the questions which have turned up during the process of writing, but I have felt a need to stop and make a review of what we know (or believe to know) and a synthesis, which should in turn focus the attention on questions of special phylogenetic interest.

I would therefore be most grateful if my readers will let me know about mistakes or oversights in this book, and I will of course be delighted to be informed about new results which could cast light on phylogeny – be it results which corroborate my ideas or results which will make it necessary for me to change them.

Archirrhinos haeckelii. (Redrawn from Stümpke, H. 1962. Bau und Leben der Rhinogradentia. – Gustav Fischer, Stuttgart.)

Glossary

actinotrocha larva – planktonic phoronid larva.

adoral ciliary zone – band of single cilia around the mouth and between proto- and metatroch in protostome larvae (trochophores) and adult rotifers.

animal – see apical-blastoporal.

apical organ – group of ciliated cells at the anterior (apical) pole of eumetazoan larvae; the cells are probably sensory and some of these cells and their neighbouring ectodermal cells sink in and form part of the brain in protostomes.

apical-basal – orientation of cells in an epithelium; the apical side faces the outside of the epithelium, either the exterior, the gut lumen or a coelomic space; the basal side usually rests on a basement membrane.

apical-blastoporal – orientation of an embryo; the apical pole is the pole of the apical organ, the blastoporal pole the pole of the blastopore; the apical pole is the same as the animal pole and the blastoporal pole the same as the vegetative pole in protostomes and deuterostomes, but the orientation is reversed in cnidarians and ctenophores.

apomorphy – derived or specialized character.

archaeotroch – ring of compound cilia surrounding the blastopore of the bilaterian ancestor tornaea.

archenteron – the gut of the gastrula.

archimery – body plan with three main regions, pro-, meso- and metasome, containing proto-, meso- and metacoel, respectively; characteristic of deuterostomes

basement membrane – a layer of extracellular matrix secreted by the basal side of an epithelium.

blastopore – the mouth opening of the gastrula.

coelom – inner cavity lined by mesoderm, which often forms a peritoneum.

dipleurula larva – hypothetical deuterostome larval type having a perioral band of single cilia on monociliate cells functioning as an upstream-collecting system but lacking a 'telotroch/archaeotroch'.

ectoderm – the outer cell layer of the gastrula and cell layers retaining the position at the outside of the organism.

endoderm – the inner cell layer of the gastrula and cell layers of the gut originating from the endoderm of the gastrula.

enterocoely – coelomic pouches formed as pockets from the endoderm.

episphere – the upper part of a trochophore, i.e. apical to the prototroch.

epithelium – layer of cells with uniform apical-basal orientation, joined with cell junctions of the septate or tight type or zonulae adhaerentes, and usually with basement membrane at the basal side.

gastrotroch – band of single cilia from the posterior side of the mouth along the ventral side to the anus in protostome larvae; functions as a rejection band in filter feeding or as a locomotory structure when the larva creeps.

homology – organs/structures are homologous when they are derived from the same organ/structure in the common ancestor.

hyposphere – the lower part of a trochophore, i.e. below the prototroch.

mesoderm – the secondary germ layer, situated between ectoderm and endoderm; it originates from ectoderm or endoderm.

mesothelium – see peritoneum.

metatroch – band of compound cilia behind the mouth in protostome larvae (and adult rotifers); functions as a downstream-collecting band in filter feeding.

monophyletic group – taxonomic unit consisting of an ancestral species and all its extinct and living descendants.

neotroch – ring of separate cilia on monociliate cells around the mouth of deuterostome larvae or adults; functions as a locomotory structure or as an upstream-collecting band in filter feeding.

neurotroch – see gastrotroch.

paraphyletic group – taxonomic unit comprising an ancestral species and only some of its extinct and living descendants.

pericalymma larva – lecithotrophic larva which has the whole hyposphere or parts of it covered by thin extensions from various parts of the episphere or the hyposphere; the thin extension is sometimes called serosa.

peritoneum – mesodermal epithelium surrounding a coelomic cavity; also called mesothelium.

planula larva – gastrula larva of cnidarians; can have an archenteron or be compact.

plesiomorphy – ancestral or primitive character.

polar bodies – the abortive cells of the meiotic division of an oocyte; the polar bodies are situated at the apical pole of the embryo in spiralians and deuterostomes, but at the blastoporal pole in cnidarians and ctenophores, and equatorially in some sponges.

polyphyletic group – taxonomic assemblage consisting of a number of groups, but not including their latest common ancestor and all its extinct and living descendants.

primitive character (state) – ancestral or plesiomorph character (state).

primitive group – a small, usually rather uniform sister group of a large, highly diverse group. It should be stressed that the 'primitive' group has the ancestral state of certain characters which are considered phylogenetically important, but that it is often highly derived in other characters. It is definitely misleading for

443

example to claim that 'primitive' groups should have the ancestral larval type. The term should mostly be avoided.

prototroch – horseshoe or ring of compound (with few exceptions) cilia in front of the mouth of protostome larvae (and adult rotifers), functioning as a locomotory organ and often as a downstream-collecting band in filter feeding.

proctodaeum – ectodermal invagination to the posterior opening between ectoderm and gut.

schizocoely – coelomic spaces formed as cavities in compact masses of mesodermal cells.

serosa – see pericalymma larva.

sister groups – two groups (taxa) resulting from one speciation event in the ancestral species.

stomodaeum – ectodermal invagination to the anterior opening between ectoderm and gut.

symplesiomorphy – character shared by a number of groups but regarded as inherited from ancestors older than the latest common ancestor.

synapomorphy – apomorphy shared by two or more groups and therefore believed to have been present in their latest common ancestor.

telotroch – horseshoe or ring of compound cilia surrounding the anus in protostome larvae.

tornaria larva – planktotrophic enteropneust larva.

trochophora larva – planktotrophic protostome larva with prototroch and metatroch of compound cilia, surrounding the adoral ciliary zone of separate cilia; gastrotroch and telotroch may be present.

vegetative – see apical-blastoporal.

veliger larva – trochophore of gastropods and bivalves with the prototroch at the edge of a pair of wing-like expansions.

Systematic index

Acanthobdella 79
Acanthocephala 76, 234, 235, 248–52
  body cavity 235–6, 249
  and rotifers 244, 245, 251–2
Acanthocephalus 248
Acanthochitona 117
Acoela 216, 217, 218, 219, 221–2
  cleavage pattern 92, 219
  spermatozoa 140, 218
Acoelomata 65, 189, 231, 439
Acoelomorpha 215, 216, 217, 221
acorn worms, *see* Enteropneusta
Acotylea 218, 219
Acyclus 239
'Adenopoda' 112
Aglantha 54, 70, 313
Aglaspis 163
Agnostus 163
Aiptasia 47
Alalcomenaeus 163
Alcyonidium 196, 197
Ambulacralia 61
Amiskwia 254
Amphioxides 422
amphioxus, *see Branchiostoma*
Amphiporus 213
Amphitrite 99
Amphiura 373
Ancalagon 250–1
Animalia (Metazoa) 2, 18–29, 33–4
Annelida 76, 99, 100, 109, 124, 126–44, 352
  ancestral 126, 144
  blastopore fate 77, 132
  ciliated cells 70, 85, 128, 141
  cleavage patterns 90–1, 92, 93, 94, 130, 136–8,
    141–2
  fate maps 95, 130–1
  gastrulation 46, 131–2
  haemal system 67, 129–30

  larvae 81, 84, 96, 130, 132–4
  nervous system 78, 79, 129, 133, 142
Anopla 225
Anostraca 172
Antedon 322
Anthozoa 54, 55, 56–7, 58, 70
Aplacophora 111, 113, 114, 118–21
appendicularians, *see* Larvacea
Arachnida 160, 167
Arachnidium 204
Archiannelida 126
Archirrhinos 441
'Archoophora' 218
Arenicola 91, 93, 131–2
Arenicolidae 127
Argulus 164
Arhynchonemertes 225, 226, 231
arrowworms, *see* Chaetognatha
Artemia 172
Arthropoda 76, 149, 150, 151, 159–73
  blastopore fate 77, 170
  cleavage patterns 92, 93, 94, 125, 167–70, 173
  fate maps 95, 159–60
  haemal system 67, 167
  larvae 95, 163–4, 171, 172
  nervous system 78, 79, 164, 166
  and tardigrades 179–80
Arthrotardigrada 176
Articulata (brachiopod group) 63, 346, 347, 348,
  350
Articulata (segmented protostomes) 99, 100–1,
  109, 333
Ascaris 275
Aschelminthes 62, 76, 234–7, 252, 258–9
  ancestral 12
  cleavage pattern 71, 92, 234
  larvae 83, 96
  mesoderm formation 64, 96–7, 234–5
  and tardigrades 179

Ascidiacea 407, 408–9, 415, 416, 417, 418
  chordate evolution and 400, 402
  embryology 410–14
  endostyle 399, 409
Ascidiella 383
Aspidobothrii 216
Asplanchna 239, 240, 241, 242, 243
Asterias 357, 369
Asterina 21, 325, 369
Asteroidea 367, 368, 375–6
  development 319, 368, 370, 371, 372, 373
Astropecten 327, 373
Atubaria 359, 360
Aurelia 46
Aysheaia 149, 150, 152

Babinka 118, 120
Balanoglossus 47, 325, 389, 391
Balanophyllia 23
Balanus 92, 160, 168, 169, 170, 172
Baltoeurypterus 163
Barentsia 185, 186, 187
Barentsiidae 185, 189
Barnea 82, 84, 119
Bathybelos 255
Batillipes 178
Bdelloidea 238, 239, 241, 243, 244
Bdellonemertini 225, 228
Beroe 315
Beroida 311, 315
Bilateria 20, 29, 61–72
  cell junctions 50
  coelomic cavities 65, 439
  embryology 47, 63, 64–5, 71–2
  neurotransmitters 22, 49, 69
  phylogenies 61–2, 222, 236, 308, 332
Bivalvia 111, 112, 116, 117–18, 119, 398
Bombinator 397, 433
Bothrioplana 221
Bowerbankia 196, 200, 201, 206
Brachiata 332, 364, 439
Brachiopoda 318, 346–53
  apical organs 320, 352
  archimery 321
  blastopore fate 318, 350, 352
  chaetae 142, 143, 348, 350, 352
  ciliary bands 324, 327, 328, 347
  ciliated cells 70, 326, 347
  cleavage pattern 319, 349, 352
  coelom formation 325, 351
  coelomic compartments 321, 322, 323, 348,
    352, 353
  gut 63, 347
  metanephridia 323, 349, 358
  nervous system 320, 348, 352
  phylogenetic position 205, 206, 333, 344,
    352–3
Branchiocaris 163

Branchiocerianthus 54
Branchiopoda 165, 172
Branchiostoma (amphioxus) 422
  adhesive papillae 405, 427
  apical tuft 320
  chorda 397, 423
  development 46, 325, 397, 424–7
  endostyle 399, 423, 426–7
  gill structures 382, 383, 384, 422–3, 426
Bredocaris 163
Bryozoa 97, 182–4, 190, 204, 205, 214
  gizzard teeth 142, 143, 194–5, 205
  larval apical organ 78, 183
Buddenbrockia 436, 437
Bugula 199
  embryology 93, 94, 197–8
  metamorphosis 200, 201
Burgessia 163
Busycon 91

Caenorhabditis 275–6, 278, 284
  embryology 279, 281, 282
  nervous system 78, 79, 261, 277
Calcarea 35, 36, 37
Calliactis 23
Callianira 308
Callyspongia 26
Cambropachycope 163
Cambropodus 162
Canadaspis 163
Capitella 134
Capitellidae 128, 129–30, 134–6
Capitomastus 134
Cardoarachnidium 204
Carpoidea 377
Carybdea 55
Catenulida 140, 215, 216, 218, 219, 221
Caudofoveata 111, 112, 113, 117, 119
Cellularia 35, 37, 38
Cenocrinus 366
Cephalocarida 164, 165, 172
Cephalochordata 384, 385, 396, 422–8
  ciliated cells 70, 384, 422
  cleavage 319, 424
  coelomic compartments 325, 400, 424, 425–6
  endostyle 398–9, 423, 426–7
  evolution 401, 405
  fate map 397, 424–5
  gastrulation 46, 425
  gill slits 382–4, 422–3
  nervous system 398, 423–4, 425
Cephalodiscus 357, 359, 360, 376
  budding 362
  ciliation 327, 360
  development 362, 363
  gill pores 360, 393, 398
Cephalopoda 111, 112, 115, 142, 143
  cleavage pattern 91, 116

Cephalorhyncha 262, 272, 291–2
Cercomeromorpha 216
Cerebratulus 95, 227
Cestoda 215, 216, 217
Chaetoderma 117, 119, 120
Chaetogaster 94, 170
Chaetognatha 76, 234, 235, 245, 254–9, 329
  ciliated cells 70
  coelomic cavities 236, 255, 258
  mesoderm formation 65, 256–8
  nervous system 78, 79, 255, 258–9
  origin of germ cells 66, 256, 257
Chaetonotoida 264, 265, 269
Chaetopterus 106
Chamaesipho 160
Cheilostomata 192, 197, 204
Chelicerata 151, 156, 159, 160, 163
  appendages 161, 162
Childia 219
chitons 142, 143
Chlorogonium 24
Choanoflagellata 18, 19, 20, 29, 32–4
  cilia 22, 23, 32–3
  and metazoans 25–7, 33–4
  mitochondria 23, 33
Chordata 334, 358, 384, 385, 392–3, 396–405
  ancestral 401, 403–4, 405
  blastopore fate 319
  coelomic compartments 321, 400
  nervous system 321, 396, 398, 404–5
Chordonia 61
Chthamalus 160
Ciona 410, 412, 413
Cirripedia 160, 172
  cleavage patterns 92, 94, 167–8, 170, 173
Cladocera 94, 170
Clavelina 410, 411, 412
Clitellata 126, 134–8, 143, 157
Cnidaria 20, 29, 53–8, 308
  cell junctions 50–1, 54, 58
  cilia 23, 54, 70
  and ctenophores 61, 307, 315–16
  embryology 46, 47, 48, 56
  gastraea theory and 10, 11, 57, 58
  larvae 56–7, 95
  nervous system 49, 55
Codosiga 23, 32
Coelenterata 61, 62, 315–16
Coelomata 61, 65, 236, 439
Coeloplana 311, 316
comb jellies, *see* Ctenophora
Concentricycloidea 366–7
Conchifera 111, 112, 113
Conochilus 82, 239, 240
conodont animals 431
Conopeum 197
Conulata 57, 58
Copepoda 94
Cordylophora 50

Corella 383
Corticium 23
Corynotrypa 204
Cothurnocystis 377
Cotylea 218, 219
Crania (*Neocrania*) 320, 346, 347, 348, 353
  coelomic compartments 323, 325, 348, 351, 352
  development 318, 349–51, 352
Craniata, *see* Vertebrata
Craniidae 346, 348, 350
Crepidula 81, 89, 93, 119, 398, 409
Crinoidea 367, 368, 370, 372, 373
Crisia 50, 194, 195, 201
Cristatella 196, 202
Crustacea 156, 159–60, 163, 164–70, 334
  appendages of cephalic region 161, 164, 165
  cleavage patterns 94, 167–70, 173
  fossil 162–4
  larvae 163, 171, 172
  limbs 162, 165
  moulting 165
Ctenophora 62, 307–9, 310–16, 332
  biradial symmetry 64, 310
  ciliated cells 70, 311
  and cnidarians 61, 307, 315–16
  gonads 66, 313
  gut 63, 312–13
  mesoderm formation 307, 308, 315
  neurotransmitters 69, 312
Ctenostomata 192, 195
Cubozoa 54, 55–6, 57, 58
Cucumaria 373
Cupelopagis 239
Cyathocormus 407
Cycloneuralia 234, 235, 236, 261–3
Cycloporus 93
Cyclorhagida 299
Cyclostomata 192, 193, 195, 201, 202, 204
Cydippida 310, 313, 315
Cynthia 410
Cyrtotreta 333, 364, 382–5, 401, 403
  ciliated cells 70, 331, 384

Dactylocalyx 36
'Dalyellioida' 216, 217
Decapoda 172
Demospongiae 22, 35, 37–8
Dendraster 328
Dendrodoa 413–14
Dentalium 114, 119
Deuterostomia (Notoneuralia) 14, 61–2, 307–9,
    318–35, 438–9
  ancestor 11, 13, 329–30, 334–5, 438–9
    adult 15, 330, 331
    larva 14, 323, 324, 329–30, 331
  blastopore fate 63, 318–19
  chaetognaths and 258
  ciliated cells 70, 324, 330–2

cleavage pattern 92, 319–20
and ectoprocts 205
larvae 95, 324–6, 328
mesoderm formation 64, 65, 307, 323
nervous system 49, 80, 320–1, 326
vs. protostomes 14, 332
Diaphanoeca 32, 33
Dicranophorus 239
Dicyemida 436
Didemnidae 408, 416
Digenea 67, 215, 216
Digononta 238
Dinomiscus 185
Dinophilus 140, 141
Discinidae 346, 348, 350
Discinisca 327, 346, 347, 348–9, 353
larvae 350, 351, 352
Doliolida 407, 408, 409, 415–16, 417
Doliolum 414
Doliopharyngiophora 216
Dorylaimida 276
Draconema 275
Drepanophorus 228

Echiniscoidea 176
Echinoderes 273, 292, 299, 300–1
Echinodermata 332, 333, 334, 366–78
ancestral 330, 376–8
anterior coelomopore complex 400
apical organs 320, 370
axial complex 356–8, 367, 370
blastopore fate 318–19
ciliary bands 324, 327, 328, 370–1, 372
ciliated cells 70, 326, 367
cleavage 319, 368–9, 375
coelomic compartments 321, 322, 325, 367–8, 369–70
larvae 330, 370–5
mesoderm formation 308, 369
nervous system 320, 366, 368
water-vascular system 66, 67, 366, 368, 369
Echinoidea 319, 367, 371, 373, 375
Echinus 369
Echiura 126, 141–2, 143, 333
Echiurus 141, 142
Ecterocoelia 61
Ectoprocta 76, 97, 182–4, 190, 192–207, 329
ciliated cells 70, 193, 194
cleavage pattern 92, 93, 94, 197–9
coelomic cavities 100, 195–6, 205
haemal system 67, 195
larvae 83, 96, 197, 199–200, 201, 202–3
nervous system 78–9, 195, 205
pharynx 193–4, 262, 274
Eisenia 94
Electra 194, 195, 196, 197
larvae 96, 197, 199
metamorphosis 201, 202
Emeraldella 163

Emplectonema 227
Enopla 225, 227
Enterocoelia 61
Enteropneusta 363, 384, 385, 387–93, 401
anterior coelomopore complex 400
archimery 321, 387
axial complex 356–8, 388, 392
blastopore fate 319
ciliary bands 324, 327, 328, 389, 390
ciliated cells 70, 384, 387, 390
cleavage 319, 390
coelomic compartments 321, 322, 323, 325, 387–8, 391
endostyle 393, 399
fate map 397
gill slits 382–4, 388–9, 391–2, 393, 398
larval apical organ 47, 320, 390
nervous system 321, 384, 389, 392, 400
Entoprocta 76, 97, 182, 185–90, 206, 234
ciliation 70, 185–6
cleavage pattern 92, 93, 94, 186–7, 189
larvae 47, 81, 82, 96, 187, 189
larval eyes 183–4, 187, 206
nervous system 78–9, 186, 189–90
Eoperipatus 155
Epimenia 117, 119
Epineuralia 61
Epiperipatus 155
Epistomia 197
Erpobdella 138, 170
Euarthropoda 149
Euarticulata 100, 109, 124–5
Eubilateria 62, 222
Eucidaris 375
Euglena 24
Eukaryotae 18
Eumetazoa (Gastraeozoa) 22, 24–5, 29, 45–51
Eunice 131, 132
Eunicida 128
Euplatyhelminthes 216
Euplectella 26, 35
Euplokamis 313
Eupomatus 46, 131, 132
Eurotatoria 238
Euspiralia 222
Eutardigrada 176
Exogone 126

Farrea 35
Farrella 192, 197
Fecampiidae 217
flatworms, *see* Platyhelminthes
Florometra 372, 373
Floscularia 239
Flustrellidra 200
Fredericella 196, 202
Frenulata 138, 139–40, 143
Fritillaria 409
Fungi 18

Gastraeozoa (Eumetazoa) 22, 24–5, 29, 45–51
Gastroneuralia, *see* Protostomia
Gastropoda 111, 112, 116, 117–18, 119, 398
Gastrotricha 76, 234, 261, 262, 264–70
  body cavities 236, 265
  ciliated cells 70, 85–6, 269–70
  cleavage pattern 92, 265–7, 269
  nervous system 78, 79, 265
  pharynx 262, 264, 269, 273–4
Genitoconia 114
Geonemertes 228
Glossobalanus 383, 387, 390
Glottidia 328, 350
Glyceridae 127, 129–30
Gnathostomula 63
Gnathostomulida 85, 126, 140–1, 143, 215
  ciliation 85, 141
  gut 63, 141
Golfingia 93–4, 105, 106
Gordioidea 287–9
Gordius 273, 287, 288
Goticaris 163
Gryphus 348
Gymnolaemata 192, 193, 194, 195, 196–7
  metamorphosis 201, 203

Habelia 163
Haeckelia 311, 315, 316
hagfishes 431
hair worms, *see* Nematomorpha
Haliclona 23, 39
Haliclystus 49
Halicryptus 294, 296
Hallucigenia 149
Halocynthia 397, 410, 411, 413
Haplognathia 63, 141
Heliocidaris 72, 369, 373, 375
Helobdella 136, 137
Hemichordata 334, 335, 363, 384, 402
Hemithyris 351–2
Henningsmoenia 163
Hesionides 129
Heteronemertini 225, 226, 227–8
Heterotardigrada 176
Hexactinellida 35, 36–7, 38
Hexarthra 238, 239, 240
Hirudinea (leeches) 66, 126, 134–8, 143, 157
Holopedium 93, 94, 170
Holothuroidea 368, 375–6
  coelomic compartments 367, 368, 370
  larvae 371, 372, 373
Homalorhagida 299
Homalozoa 377
Hoplonemertini 225, 226, 228
Hoploplana 218
Hubrechtella 227–8
Hutchinsoniella 163, 164, 166, 172
Hyalinella 437
Hydra 21, 28, 54, 57

Hydroctena 315–16
Hydrozoa 50–1, 54, 55–7, 58
Hyponeuralia 61
Hypophorella 197

Ilyanassa 90
Inarticulata 346, 347
Insecta 124, 156, 159, 160, 165, 167
Introverta 262, 272–4

Kinonchulus 272, 273, 276
Kinorhyncha 76, 234, 262, 272, 291, 299–301
  cuticle 272, 300
  introvert 273, 292, 299
  nervous system 78, 79, 261, 300
  pharynx 262, 299
  and priapulans 296–7
Kowalevskaia 409

Labidoplax 324, 370
Lampetra 432
lampreys 399, 405, 431–2
lamp-shells, *see* Brachiopoda
lancelets, *see* Cephalochordata
Larvacea (appendicularians) 407, 408, 409–10,
    416, 417
  development 415, 418
  neoteny 405, 416
Leanchoilia 163
Lecane 241
Lecithoepitheliata 215
leeches, *see* Hirudinea
Lepas 160
Lepidochitona 114, 117, 118
Lepidodasys 262, 264, 265
Lepidodermella 269
Lepidopleurus 120
Leptasterias 369
Leptocardii 428
Limifossor 79
Limulus 163, 167
Lineus 71, 95, 227, 228, 230
Lingula 346, 347, 353
  development 349, 351, 352
  nervous system 348
Lingulidae 346, 348, 350
Littorina 114
Lobata 315
Lobatocerebridae 126, 140, 141, 143
Lobatocerebrum 140
Lophophorata 182, 204, 205, 206, 332, 342
Lophopodella 437
Loricifera 76, 234, 262, 272, 291, 303–6
  cuticle 272, 304
  introvert retractor muscles 292, 305
  nervous system 78, 261, 305

pharynx 262, 273–4, 303
and priapulans 296–7
Lottia 119
Loxokalypus 189
Loxosoma 182, 189, 204
larvae 47, 81, 82, 96, 189
Loxosomatidae 185, 189
Loxosomella 92, 185, 186, 187–9
Lucilia 77
Luidia 371, 373, 374, 376, 403, 404
Luolishania 149
Lyrodus 83, 84, 117, 119
Lytechinus 369

Maccabeus 294, 295
Macracanthorhynchus 250, 251, 252
Macrobiotus 79
Macrodasyoida 264, 265, 269–70
Macroperipatus 152, 153
Macrostomida 140, 216, 218, 219
Magelona 69, 128, 139
Malacobdella 93, 95, 227, 228, 230
Malacostraca 172
Mandibulata 151, 160, 161
Marella 163
Martinssonia 163
Maxillopoda 165, 172
Medusozoa 57, 58
Meiopriapulus 294, 295, 296
Melicerta 81
Membranipora 194, 195, 196, 197, 199, 203
Mermithoidae 289
Merostomata 160
Mesotardigrada 176
Mesozoa 436
Metazoa (Animalia) 2, 18–29, 33–4
Micrura 228, 229–30
Minona 219
Mitrocystella 378
Mnemiopsis 308, 312
Molaria 163
Molgula 414
Mollusca 76, 99, 100, 110–21, 142, 332, 334
ancestral form 111–13
ciliation 70
cleavage pattern 90–1, 92, 93, 94, 116
gills 115, 398
haemal system 67, 115
larvae 81, 82, 83, 84, 95, 96, 117–18, 119
nervous system 78, 79, 111, 114, 116–17
segmentation 109, 118–21
Monocelis 219
Monogenea 215, 216
Monogononta 238, 239, 241–3
Monoplacophora 111, 112, 115
Mopalia 119
moss-animals, *see* Ectoprocta
mud dragons, *see* Kinorhyncha
Myriapoda 156, 159, 160

Myriochele 71, 85, 128
Mytilus 118
Myzostomida 126, 134, 143, 333

Nanaloricus 261, 292, 303, 304, 305
Naraoia 163
Nautilus 114, 115, 118
Neanthes 132
Nebalia 238
Nectonema 287, 288, 289
Nectonematoidea 287
Nemathelminthes 234, 236–7, 261
Nematoda 76, 234, 262, 272, 275–84
blastopore fate 77, 281, 282
cuticle 272, 275, 276
mermithoid 289
nervous system 78, 79, 261, 277–8, 281–2, 283
origins of germ cells 66, 279–81
pharynx 262, 273–4, 276
Nematomenia 120
Nematomorpha 76, 234, 262, 272, 287–9, 291
cuticle 272, 287, 289
introvert 273, 288, 289
nervous system 78, 287–8
Nemertini 76, 97, 211–13, 225–31, 396
ciliated cells 70, 71
cleavage pattern 92, 93, 95, 213, 227
coelomic cavities 100, 212–13, 226, 231
larvae 83, 95, 211, 212, 214, 227–31
nervous system 79–80, 213, 214, 226
Nemertodermatida 140, 216, 218
Neocrania, see Crania
Neodasys 264, 265, 269
Neodermata 215, 216, 217
Neogossea 269
Neomenia 83, 84, 117, 119
Neoophora 216, 218, 219
Neopilina 110, 114, 115, 118, 120
Neorenalia 333, 356–8, 364
axial complex 67, 321, 356–8
evolution 401, 403
Nereidae 127, 128
Nereis 128, 132, 133
Notochordata 384, 385, 396, 401, 402
Notommata 239, 240
Notoneuralia, *see* Deuterostomia
Notoplana 218

Octacnemus 407
Odaraia 163
Oerstedia 228
Oikopleura 399, 407, 409, 415
Olenellidae 163
Olenoides 163
Oligacanthorhynchidae 249
Oligobrachia 139
Oligochaeta 126, 134–8, 143

Oligomera 333
Onychophora 76, 149, 151, 152–7, 159, 173
  blastopore fate 77
  cleavage pattern 93, 124–5, 154
  metanephridia 150, 153–4, 156
  nervous system 78, 79, 153
  segmentation 124, 152
Ophioderma 373
Ophiuroidea 319, 367, 371, 373, 375–6
Orthonectida 436
Ostrea 96
Owenia 85, 132, 133
  ciliated cells 70, 71, 85, 128, 270
  larvae 84, 130
Oweniidae 85, 128, 133–4
Oxychomatus 325

Pachygrapsus 49
Palaeonemertini 225, 226, 227–8
Paludina 94
Panagrellus 279
Panarthropoda 99, 100, 109, 149–51, 156–7,
    160
  ectoderm 101, 149
  embryology 124, 150
  segmentation 124, 149
Pan-Chelicerata 163
Pan-Crustacea 163
Pantopoda 160
Paracentrotus 308, 369
Parascaris 279, 280
Paraspadella 254
Parastichopus 373
Paratomella 217
Parenchymia 95, 97, 190, 211–14, 231
Paucijaculum 254
Pectinaria 129, 133
Pectinariidae 127
Pectinatella 202
Pedicellina 182, 183, 186, 188, 204
  cleavage 93, 94, 186
Pedicellinidae 189
Pelagia 57
Penaeus 172
Pentastomida 149, 159, 164
Peripatoides 79
Peripatopsis 77, 154, 155
Perspicaris 163
Petromyzon 383, 431
Phascolopsis 104
Phascolosoma 104, 106
Pholas 115
Phoronida 318, 337–44
  apical organs 320, 340
  archimery 321, 337–8, 342–3
  blastopore fate 318, 342, 343
  ciliary bands 324, 327, 328, 337–8, 340, 343
  ciliated cells 70, 326, 343
  cleavage pattern 319, 339, 342

coelom formation 325, 339–40
  coelomic compartments 321, 322, 323, 337–8,
      342–3
  metanephridia 323, 339, 358
  nervous system 320, 338, 340, 342, 343
  phylogenetic position 204, 205, 206, 333,
      342–4, 353
Phoronis 337, 338, 339, 353
  actinotrocha larva 83, 329, 339, 340
  coelomic compartments 322, 325, 339–40
  nervous system 338, 384
Phoronopsis 337, 341
Phylactolaemata 192, 195, 196
  development 201, 202–3
  tentacles 193, 194, 203–4
Phyllocarida 165
Phyllochaetopterus 139
Phyllodocida 128
Physa 99, 116
Pikaia 422
Pilidium 229, 230
Placozoa 20, 22, 23, 24–5, 29, 42–4
Planctosphaera 387
Plantae (plants) 2, 18, 23
Plathelminthes 216
Plathelminthomorpha 62, 222
Platyhelminthes 76, 97, 211–14, 215–22, 231
  ciliated cells 70, 71
  circulatory system 67
  cleavage patterns 92, 93, 94–5, 213, 218–19
  and gnathostomulids 141, 215
  gut 63, 217, 221
  larvae 83, 95, 211, 212, 214, 219–21
  nervous system 79–80, 213, 214, 217, 220
  reproduction 140, 218
Plenocaris 163
Pleurasiga 26
Pleurobrachia 310
Pliciloricus 273, 303, 305
Ploesoma 241
Plumatella 194, 196, 201, 202, 437
Podarke 130–1, 132
Podocoryne 55
Poeobius 127
Pogonophora 126, 138–40, 143
Polychaeta 124, 126–34, 138, 143
Polycladida 215, 216, 218, 219
Polygordius 129
  embryology 77, 131, 132
  larvae 80, 81, 83, 84, 132, 133–4, 135, 329
Polyphemus 170
Polyplacophora 111, 113, 116, 117, 118, 119
Pomatoceros 128
Pontonema 77
Porichthys 49
Porifera (sponges) 20, 24–5, 29, 35–40
  cell junctions 21, 22, 36, 39
  choanocytes/choanosyncytia 25–7, 35, 37, 38
  cilia 23, 36, 37–8
  neurotransmitters 22, 37

Priapula 76, 234, 262, 272–3, 291, 294–7
  and acanthocephalans 250–1
  cuticle 262, 294
  introvert retractor muscles 292, 295
  nervous system 78, 79, 261, 295
Priapulus 79, 272, 273, 294, 295, 296
Proarthropoda 149
Prochaetoderma 114
Prokaryotae 18
Prolecithophora 215
Proseriata 219
Prosorhochmus 228
Prostheceraeus 215
Prostoma 228
Protista 2, 18
Protoctista (Protista) 2, 18
Protodrilus 127, 129
Protoglossus 388
Protornaeozoa 62, 307–9
Protostomia (Gastroneuralia) 14, 61–2, 75–86,
    308, 438
  ancestral 11, 12, 13, 15
  bilaterality 63–4
  ciliated cells 69, 70, 85–6, 270
  and ectoprocts 205
  mesoderm formation 65, 75, 86, 307
  nervous system 49, 75, 76–80
  vs. deuterostomes 14, 332
Psammechinus 373
Psammodriloides 127
Pseudoceros 212
Pseudocoelomata 65, 182, 234, 235, 439
Pterobranchia 206, 333, 359–64, 401
  anterior coelomopore complex 400
  apical organs 320
  archimery 321, 359–60
  axial complex 356–8, 359–60
  ciliary bands 324, 327, 360
  ciliated cells 70, 326, 359
  coelomic compartments 321, 322, 325, 359–60,
    363, 364
  nervous system 320, 360, 361
Ptychoderidae 388
Pumilus 346
Pyrosomida 407

Radiata 315
Raillietiella 164
Remipedia 164, 165, 167
Rhabditophora 215, 216, 221
Rhabdocalyptus 35
Rhabdocoela 216
Rhabdopleura 359, 360, 361
  coelomic compartments 322, 325
  development 320, 362–3
Rhynchaschelminthes 291
Rhynchohelminthes 291
Rhynchoscolex 219
ribbon worms, *see* Nemertini

Ridgeia 138–9
Rostroconcha 112
Rotaria 240
Rotifera 76, 234, 235, 238–45, 259
  and acanthocephalans 244, 245, 251–2
  body cavity 235
  ciliary bands 81, 82, 83, 234, 239, 240
  cleavage pattern 92, 241–2
  nervous system 79, 241, 245
Rugiloricus 303–4, 305

Sabellidae 127
Saccoglossus 357, 387
  coelomic compartments 322, 325, 391
  development 391, 392, 397
  gill slits 383, 388–9
Sagitta 79, 256, 257
Salinella 436
Salpida 407, 408, 409, 414, 415–16, 417
Sanctacaris 163
Sarotrocercus 163
Scalibregmatidae 127
Scaphopoda 111, 112, 116, 117, 119
Schizocardium 357
Scolecida 204
Scoloplos 77, 129, 131, 132
Scotiaecystis 377
Scutopus 113
Scypha 26, 38
Scyphozoa 54, 55–6, 57, 58
Seison 238, 239, 241, 243, 244
Seisonidea 238
Seriata 215, 216
Serpula 82, 132, 133
Siboglinum 139
Sidneyia 163
Simrothiella 114
Siniopelta 197
Siphonophora 54
Sipuncula 76, 99, 100, 102–7, 333–4
  and annelids 126, 142
  ciliation 70, 106
  circulatory system 66, 67, 103, 106, 109
  cleavage pattern 92, 93–4, 103–4
  epithelium 101, 102
  larvae 81, 96, 104, 105
  nervous system 78, 79, 102, 106
Sipunculus 79, 99, 103, 104
  larvae 83, 96, 104, 105
Skara 163
Solaster 325
Solemya 117
Solenogastres 111, 112, 117, 119
Sorberacea 407
Spadella 254, 256
Speleonectes 159, 163, 164, 165
Sphaerium 116
Sphaeroeca 32
Spiochaetopterus 139

Spionidae 127
Spiralia 12, 62, 76, 89–97, 206, 214
   ciliated cells 70
   cleavage patterns 71, 89–95
   entoprocts and 189–90
   mesoderm formation 64, 96–7
Spirobranchus 96
Spisula 115
sponges, see Porifera
Stenolaemata 192, 193, 194, 195–6, 203, 204
Stichopus 324
Stolella 437
Strongylocentrotus 28, 325, 369
Styela 411
Stylaria 94, 138
Sycon 21
Symplasma 35, 38
Synapta 372, 374

Tardigrada 76, 100, 125, 150, 176–80
   cleavage pattern 93, 179
   nervous system 78, 79, 178, 179–80
   pharynx 177, 262
   phylogenetic relations 149, 151, 179–80
Teloblastica 97, 99–101, 190, 214
Tentaculata 182, 204, 333, 342
Terebratalia 325, 350
Terebratella 322
Terebratulina 353
Tesnusocaris 162
Testaria 111, 112, 114, 118–20
Tetraclita 160
Tetrastemma 228
Thalassema 142
Thaliacea 407, 416
Themiste 103
Thermozodium 176
Tjalfiella 310
Tracheata 156, 159, 160, 161, 162
Trachylina 57
Travisia 101, 128, 258
Trematoda 67, 216
Trepaxonemata 140, 216, 218
Triarthrus 163
Trichoplax 20, 21, 23, 36, 42–4
Trilobita 156, 159, 163, 164
Trilobitomorpha 161
Triops 166
Tripedalia 53, 55
Triploblastica 61, 64
Triticella 197
Trochaeozoa 61
Trochus 91
Tubificida 138
Tubiluchus 261, 292, 294, 295–6
Tubulanus 95, 225, 227
Tubuliporata 192
tunicates, see Urochordata

Turbanella 79, 261, 264, 265–8, 269
Turbellaria 215, 216, 217
'Typhloplanoida' 216

Uniramia 150, 156, 157, 159
Urechis 141–2
Urnatella 186
Urochordata (tunicates) 384, 385, 396, 405, 407–18
   body regions 321, 407
   ciliated cells 70, 384, 408
   cleavage 319, 411
   coelomic compartments 321, 400, 410
   endostyle 398–9, 409
   evolution 400–2, 416–17
   fate map 397, 411
   gill slits 382–4, 408–9, 413
   musculature 398, 410
   nervous system 398, 409–10, 412, 413–14
   neural gland complex 400, 409, 412
   tight junctions 21
   urochord 396, 397, 407, 408, 412, 413

Vema 118
Vertebrata 332, 384, 385, 396, 428, 431–4
   cell junctions 21
   chorda 397, 431–2
   ciliated cells 70, 384
   cleavage 319, 432
   coelomic sacs 400
   endostyle 398–9, 432, 434
   evolution 401, 405
   fate maps 397, 432, 433
   gill slits 382–4, 432
   nervous system 398, 432, 433, 434
Vestimentifera 126, 138–9, 143
Victorella 183
Viviparus 116
Volvox 9, 23, 24

Waptia 163
water bears, see Tardigrada
Wingstrandarctus 176

Xenodasys 264, 265
Xenoprorhynchus 219
Xenoturbella 215, 436–7
Xyloplax 366–7

Yohoia 163
Yoldia 83, 118, 119, 120

Zygoneura 61

Subject index

acanthella 248, 249, 250
acanthor 248, 250
acetylcholine/cholinesterase 20, 22, 55
  bilaterians 49, 69, 72
  ctenophores 312
  sponges 37
actinotrocha larva 328, 329, 339, 340–1, 442
  vs. cyphonautes larva 204, 343
  vs. trochophora larva 83, 324–5, 329
adhesive papillae, *see* papillae (adhesive)
adhesive sacs 199, 200, 201, 202, 203, 206
adoral ciliary zone 12, 80–1, 82, 442
  annelids 132
  entoprocts 186
  molluscs 117
  rotifers 238, 239, 240
ammocoetes larva 399, 405, 432
amphiblastula 38
amphids 278, 284
anal siphon, urochordates 413
ancestrula 193, 200, 363
animal pole, *see* apical pole
antennae 161, 164, 165, 171
antennal glands 167, 171
antennules 161, 164, 165, 171
antero-posterior axis 11, 45
  annelid embryos 130
  eumetazoans 45
  gastrotrich embryos 265, 269
anus 63
  annelids 140, 141
  chordates 405
  deuterostomes 318–19
  lack 63, 222
    platyhelminths 63, 217, 221
  molluscs 116
  trochaea theory 10, 13, 14
  urochordates 407, 408, 416
apical–basal (orientation) 442
apical–blastoporal axis 10, 11, 442
  annelids 130

arthropod embryos 168
cnidarians 54
ctenophores 310
nematode embryos 279
spiralian embryos 89
apical organs 10, 11, 442
  acanthocephalans 249
  annelids 130, 131
  brachiopods 320, 352
  ctenophores 311
  deuterostomes 320, 326
  echinoderms 320, 370
  ectoprocts 183, 199, 202–3
  enteropneusts 47, 320, 390
  entoprocts 183, 187, 189
  eumetazoans 45, 47
  nemertines 227, 228, 229
  parenchymians 213–14
  phoronids 320, 340
  protostomes 78
  pterobranchs 320
  spiralians 95–6
apical (animal) pole 89, 442
  amphioxus 417
  ascidian tadpoles 412
  molluscan embryos 115
apodemes 166
apomorphy 442
appendages, arthropod cephalon 161, 163, 164,
    165, 171, 172
archaeocytes 36, 37, 38
archaeotroch 10–11, 330, 442
archenteron 10–11, 442
  bilaterians 47, 65
  brachiopods 350, 351, 352
  chaetognaths 256
  cnidarians 47, 54
  nemertines 228, 229
  rotifers 242
archicoelomate concept 222, 330
archimery 321, 326, 442

echinoderms 367
enteropneusts 321, 387
phoronids 321, 337–8, 342–3
pterobranchs 359–60
asexual reproduction
choanoflagellates 33
cnidarians 54, 56
echinoderms 375–6
placozoans 43
sponges 38
urochordates 415
atrium/atria
cephalochordates 422–3
urochordates 407, 413, 416
auricularia larva 327, 371, 373, 374
axial complex 67, 321, 356–8, 364
chordates 358, 399–400
echinoderms 356–8, 367, 370
enteropneusts 356–8, 388, 392
phoronids and brachiopods 344, 353, 358
pterobranchs 356–8, 359–60
axocoel 367, 370

bacteria, symbiotic sulphur 139, 140
basal lamina, sponges 36, 39, 40
basement membrane 20, 45, 51, 442
chaetognaths 254–5
cnidarians 54
gill skeleton 423
platyhelminths 216–17
belt desmosomes 19, 20, 21, 22
cnidarians 54
placozoans 21, 43
biciliate cells 165
bilaterality 54, 63–4
bilaterogastrea 222
binary fission, choanoflagellates 33
bipinnaria larva 373
biradial cleavage 94, 197–9
biradial symmetry 64, 310
blastaea 9, 10, 40
blastocoel 9, 66, 67–8, 124
echinoderm larvae 371
nematodes 278, 281
blastomeres
determination of fate 95, 197
spiralians 89–90, 91
see also cell lineages; fate maps
blastoporal pole 89, 442
see also apical–blastoporal axis
blastopore 10, 14, 47, 442
cnidarians 54
fate 63
annelids 77, 132
arthropods 77, 170
brachiopods 318, 350, 352
chordates 319
deuterostomes 63, 318–19
echinoderms 318–19

enteropneusts 319
molluscs 116
nematodes 77, 281, 282
onychophorans 154–5
phoronids 318, 342, 343
protostomes 63, 75–7, 78
protostomian ancestor 12
blastozooids 415
blastula 24, 320
cnidarians 56
eumetazoans 45, 47, 48
sponges 37, 38, 40
blood vessels 66–7
molluscs 115
nemertines 212–13, 226, 231
urochordates 408
see also haemal systems
body cavities
acanthocephalans 235–6, 249
aschelminthes 235–6
entoprocts 185
gastrotrichs 236, 265
kinorhynchs 300
nematodes 278
nematomorphs 288
priapulans 295
see also coelom(ic cavities); haemocoel
body coelom 102, 103
Bowmann's capsule 434
brachiolaria larva 371, 373
brain 69
acanthocephalans 249, 252
annelids 133, 142
arthropods 166
cycloneuralians 261, 263
loriciferans 261, 305
molluscs 116
nematodes 261, 277
nemertines 226
onychophorans 153
panarthropods 150
protostomes 76, 78
rotifers 241
sipunculans 106
tardigrades 178
urochordates 412
vertebrates 432, 434
branchial basket 408–9, 413, 423
branchiomery 423
bristles, natatory 287, 289
brooding, ectoproct embryos 196–7, 204
budding 38, 189, 202, 203
cnidarians 54, 56
pterobranchs 361, 362
urochordates 415, 416

calcichordate theory 377–8, 396
catecholamines 22, 338
caudal ganglion 241, 243

cell junctions 19–22
  see also desmosomes; gap junctions; septate
    junctions
cell lineages
  arthropods 168
  clitellates 136–8
  ctenophores 314, 3..
  ectoprocts 198
  gastrotrichs 268
  nematodes 279, 281–3
  nemertines 227
  platyhelminths 218
  rotifers 243
cells
  constant numbers 275
  de-differentiation/re-differentiation 35, 55
  determination of fate 95, 320
centrioles, accessory 20, 23, 71, 387
cephalon (arthropod head region) 159, 163–4
  appendages 161, 163, 164, 165, 171, 172
chaetae (setae) 142–3
  annelids 128, 133, 134, 138
  brachiopods 142, 143, 348, 350, 352
  echiurans 141, 142
chelicerae 161
chitin 142–3, 291
  annelids 128, 139–40
  arthropods 164, 165
  brachiopods 348
  chaetognaths 254, 259
  ectoprocts 193
  entoprocts 185
  introverts 272
  molluscs 113
  nemertines 211–12, 225
  platyhelminths 211–12, 216
  rotifers 239
choanocyte-like (collar) cells 27, 28, 51
choanocytes/choanosyncytia 20, 22, 39, 51
  collared units 25–7, 36
  sponges 25–7, 35, 37, 38
cholinesterase, see acetylcholine/cholinesterase
chorda 396, 397, 404
  cephalochordates 423
  vertebrates 397, 431–2
  see also notochord; urochord
cilia 22–3
  basal structures 20, 23, 71
  bilaterians 64
  choanoflagellates 22, 23, 32–3
  cnidarians 23, 54, 70
  ectoprocts 193, 194
  entoprocts 70, 185–6
  introverts 273
  placozoans 42, 43
  spermatozoa 22, 23, 70
  sponges 23, 36, 37–8
  see also monociliate cells; multiciliate cells
ciliary bands 69, 70
  brachiopods 324, 327, 328, 347

cnidarians 54
  deuterostomes 324–6, 327, 328, 329
  echinoderms 324, 327, 328, 370–1, 372
  ectoproct larvae 199, 205
  enteropneusts 324, 327, 328, 389, 390
  Götte's larva 211, 212
  Müller's larva 219–20
  phoronids 324, 327, 328, 337–8, 340, 343
  pilidium larva 211, 212, 229
  protostome larvae 80–3, 324, 329
  pterobranchs 324, 327, 360
  rotifers 81, 82, 83, 234, 239, 240
  sipunculan larvae 106
  trochaea theory 10–11, 12, 13, 15–16
  see also adoral ciliary zone; archaeotroch;
    metatroch; neotroch; prototroch; telotroch
cingulum 238
circulatory systems 66–8
  see also coelomic circulatory systems; haemal
    systems
cladistics 3–4
claws, tardigrades 177
cleavage 24, 71–2
  acanthocephalans 250, 252
  annelids 90–1, 92, 93, 94, 130, 136–8, 141
  arthropods 92, 93, 94, 125, 167–70, 173
  aschelminthes 71, 92, 234
  biradial 94, 197–9
  brachiopods 319, 349, 352
  cephalochordates 319, 424
  chaetognaths 256
  cnidarians 56
  ctenophores 314, 319
  deuterostomes 92, 319–20
  discoidal 91, 116
  duet 92, 219
  echinoderms 319, 368–9, 375
  enteropneusts 319, 390
  gastrotrichs 92, 265–7, 269
  monet 92, 94
  nematodes 279, 280, 283, 284
  onychophorans 93, 124–5, 154
  panarthropods 150
  phoronids 319, 339, 342
  priapulans 296
  protostomes 75, 76
  pterobranchs 362
  radial 319, 339, 349
  rotifers 92, 241–2
  spiral, see spiral cleavage
  sponges 37
  tardigrades 93, 179
  urochordates 319, 411
  vertebrates 319, 432
cleptocnidia 54, 311
clitellum 134
cnidae, see nematocysts
cnidocil 57
cnidocytes (nematoblasts) 53–4, 57
coeloblastula 37, 45, 48, 56

coelom(ic cavities) 4, 65, 235, 439–40, 442
  annelids 127, 133, 138, 140, 141, 142
  arthropods 166, 170, 171
  brachiopods 321, 322, 323, 348, 352, 353
  cephalochordates 400, 424
  chaetognaths 236, 255, 258
  chordates 321, 400
  deuterostomes 321–3
  echinoderms 321, 322, 367–8
  ectoprocts 100, 195–6, 205
  enteropneusts 321, 322, 323, 387–8
  entoprocts 186
  formation 86, 439
    brachiopods 325, 351
    cephalochordates 325, 425–6
    deuterostomes 323–4, 325
    echinoderms 325, 369–70
    enteropneusts 325, 391
    euarticulates 124–5
    phoronids 325, 339–40
    pterobranchs 325, 362, 363
    see also enterocoely; schizocoely
  gills 382
  lining cells 27, 28
  molluscs 115, 116, 121
  nemertines 100, 212–13, 226, 231
  neorenalians 357
  onychophorans 153, 155–6
  panarthropods 150
  phoronids 321, 322, 323, 337–8, 342–3
  platyhelminths 212, 222
  protostomes 86
  pterobranchs 321, 322, 359–60, 364
  sipunculans 102–3
  teloblasticans 99, 100
  urochordates 321, 400, 410
  see also blastocoel; body cavities; mesocoel;
    metacoel; protocoel
coelomic circulatory systems 66, 67
  sipunculans 66, 67, 103, 106, 109
coelomocytes 102
coelomoducts 65, 66, 68–9, 323, 360
coelomopore complex, anterior 400
coenecia 359, 361
collagen 20, 23–4, 36, 291
  annelids 128–9
  choanoflagellates 33
  ctenophores 313
  nematodes 276
collar, enteropneusts 388
collar cells 27, 28, 51
collar cord (neural tube of enteropneusts) 321,
    384, 389, 392, 400
colloblasts 54, 310, 311, 312
colonies 19
  choanoflagellate 32, 33
  ectoproct 192, 193, 195
  urochordates 408
comb plates 310–11
compensation sacs 102–3

contractile cells, sponges 36, 37
corona
  chaetognaths 255
  rotifers (wheel organ) 238, 239, 240, 244–5
coronate larva 199–200, 206
crural glands 152
Cuénot, L. 61–2
cuticle
  annelids 128
  arthropods 164, 165
  cephalorhynchs 291
  chaetognaths 254, 259
  ectoprocts 193, 202
  entoprocts 185
  gastrotrichs 264, 269
  intracellular 239, 244, 248–9, 252
  introverts 272–3
  kinorhynchs 272, 300
  loriciferans 272, 304
  molluscs 113
  nematodes 272, 275, 276
  nematomorphs 272, 287, 289
  nemertines 212, 225
  onychophorans 152
  panarthropods 149
  platyhelminths 212, 216
  priapulans 262, 294
  rotifers 239
  sipunculans 102
  tardigrades 177
  urochordates (tunic) 407, 408, 416
cyphonautes larva 194, 197, 199, 205
  settling/metamorphosis 200, 201
  vs. actinotrocha larva 204, 343
cyrtocytes 27
cyrtopodocytes 68, 424, 427
cystacanth 251

Darwin, C. 1
desmosomes 19, 22
  see also belt desmosomes; hemidesmosomes
Desor's larva 228, 230
determination, cell fate 95, 320
deutocerebrum 161, 166, 178, 179
development, see embryology
dipleurula larva 330, 442
  ancestral chordate 402, 403, 404
  ancestral deuterostome 323, 330, 331
  echinoderms 371, 375
diploidy 20, 24
discoidal cleavage 91, 116
DNA sequencing 4–5
doliolaria larva 372, 373, 374, 375
dorso-ventral axis 89, 115
duet cleavage 92, 219

ecdysis, see moulting
ecdysones/ecdysteroids 165, 278

echinopluteus larva 371, 373, 375
ectoderm 10, 45, 442
　acanthocephalans 248–9, 251
　annelids 127, 128
　arthropods 165
　chaetognaths 254
　cnidarians 54
　ctenophores 311
　gastrotrichs 264
　kinorhynchs 300
　loriciferans 304
　nematodes 276
　nematomorphs 287
　nemertines 225
　onychophorans 152
　priapulans 294
　rotifers 238–9
　tardigrades 178
　urochorda   408
　ve
　　　　　　　　...s; epithelia
e　　　　　m 64, 65
　annelids 131, 132
　molluscs 116
ectoteloblasts131, 136–8
eggs
　onychophorans 154
　sipunculans 103
　spiralians 91–2
embryology 24, 71–2
　acanthocephalans 250, 251, 252
　annelids 130–2, 136–8, 139, 141–2
　arthropods 167–71
　bilaterians 47
　brachiopods 349–52
　cephalochordates 424–7
　chaetognaths 256–8
　cnidarians 46, 47, 48, 56
　ctenophores 314–15
　echinoderms 368–75
　ectoprocts 196–9, 202, 203, 204
　eumetazoans 45, 46
　gastrotrichs 265–9
　kinorhynchs 300–1
　loriciferans 305
　molluscs 115–16
　nematodes 279–83
　nematomorphs 288–9
　nemertines 227, 228
　onychophorans 154–6
　panarthropods 124, 150
　phoronids 339–41
　platyhelminths 218–19
　priapulans 296
　protostomes 76
　pterobranchs 362–3
　rotifers 241–3
　sipunculans 103–4
　spiralians 89–97
　sponges 37

tardigrades 178–9
teloblasticans 99–100
urochordates 410–15
vertebrates 432–3
　see also blastula; cleavage; gastrulation
embryonic discs 229, 230
endoderm 10, 45, 443
　cnidarians 54
endomesoderm 116
endopodites 165
endostyle 393, 396, 398–9
　cephalochordates 398–9, 423, 426–7
　enteropneusts 393, 399
　urochordates 398–9, 409
　vertebrates 398–9, 432, 434
endothelium 66, 115, 434
enterocoely 86, 222, 439, 443
　chaetognaths 258
　deuterostomes 307, 323
　echinoderms 369
epaulettes 371
epicardial sacs 410
epidermal placodes 434
epidermis
　acanthocephalans 248–9
　cnidarian 54
　molluscs 114
　platyhelminths 216–17
　see also ectoderm; epithelia
episphere 443
epistome 337
epithelia 443
　annelids 128, 141
　brachiopods 347
　cephalochordates 422
　chaetognaths 254–5, 258
　enteropneusts 387
　oikoplast 408, 409
　onychophorans 152
　pallial 197, 200, 201
　rotifers 238–9
　sipunculans 101, 102
　teloblasticans 101
　urochordates 408
　see also ectoderm; endoderm
epithelio-muscular cells 55
eutely 179
excretory organs
　nematodes 278, 283
　nemertines 226
　neorenalians 321, 356–8
　urochordates 410
　see also nephridia
exopodite 162
exoskeleton 149, 159, 303
　see also shells
eyes
　arthropods 161, 166
　chaetognaths 255
　compound (faceted) 161, 166

ectoproct larvae 183–4, 206
enteropneust larvae 390
entoproct larvae 183–4, 187, 206
onychophorans 153
tardigrades 178, 179
see also photoreceptors/photoreceptor cells

fate maps
annelids 95, 130–1
arthropods 95, 159–60
cephalochordates 397, 424–5
chordates 397, 399
enteropneusts 397
urochordates 397, 411
vertebrates 397, 432, 433
feeding biology
arthropods 165, 171
ectoprocts 193
enteropneusts 388–9
kinorhynchs 299
loriciferans 306
placozoans 42
protostome larvae 80–1
rotifers 240–1
tardigrades 176
fertilization
ectoprocts 196
internal 339
membrane 130
sipunculans 103
spiralians 89
fibre cells 42
filter/house, larvaceans 408
fins, chaetognaths 255
fission 38, 43
binary, choanoflagellates 33
echinoderms 375–6
flagella 22
FMRFamide 55, 338, 340
foot
larval entoprocts 187, 188, 189
molluscan 110–11, 112, 113, 118
rotifers 239–40
foot glands 112, 177, 183, 206
foot retractor muscles 118, 120
fossils
annelids 126
arthropods 159, 162–4
brachiopods 346
echinoderms 366
ectoprocts 204
molluscan 110
panarthropods 149
frontal organs 183, 187, 189, 206
funiculus 67, 195

ganglia
annelids 129, 133, 142

arthropods 161, 162, 164, 166, 170–1
chaetognaths 255
molluscs 114, 117
tardigrades 178, 179–80
gap junctions 19, 20
cnidarians 50–1, 54, 58
eumetazoans 49–51
gastraea theory 9–10, 51, 63, 438
cnidarian structure and 10, 11, 57, 58
gastral pores 10
gastrodermis 54
gastroneuron 12, 13
gastrotroch 12, 80, 81, 82, 443
annelids 131, 132
entoprocts 187
gastrovascular system 312–13
gastrozooids 416
gastrula 320
cnidarians 54, 56
ctenophores and 310
eumetazoans 45–7, 51
onychophorans 154
gastrulation 46
annelids 46, 131–2, 136
arthropods 169
cephalochordates 46. 425
cnidarians 48
molluscs 116
nematodes 279, 280, 282
nemertines 227, 228
protostomes 76
rotifers 242–3
sipunculans 104
gemmulae 38
germ cell determinant 256, 257
germ cells, primordial 235, 256, 279–81
germovitellarium 241, 242
giant nerve cells 383, 389, 423
gill bars 382, 388, 408
gill pores 360, 423, 426
gill pouches 388
gills
arthropods 165
molluscs 115, 398
gill slits 364, 377, 382–4
cephalochordates 382–4, 422–3
chordates 396, 398, 399
enteropneusts 382–4, 388–9, 391–2, 393, 398
urochordates (stigmata) 382–4, 408–9, 413
vertebrates 382–4, 432
girdle hairs 142, 143
gizzard teeth 142, 143, 194–5, 205
glandular pit 112
glomerulus 360, 388, 434
glycocalyx 212, 216, 225
glycoproteins, structural 23–4
gnathobases 160, 161, 165, 171
gonads 65, 66
acanthocephalans 249, 250
annelids 130

brachiopods 349
cephalochordates 424
chaetognaths 256
ctenophores 66, 313
echinoderms 368
ectoprocts 196
enteropneusts 390
entoprocts 186
gastrotrichs 265
kinorhynchs 300
loriciferans 305
molluscs 115, 116
nematodes 279
nematomorphs 288
nemertines 213, 226–7
onychophorans 154
phoronids 339
platyhelminths 218
priapulans 296
pterobranchs 361, 362
rotifers 241, 244
sipunculans 103
tardigrades 178
urochordates 410
gonoducts 66, 68–9, 323
gonozooids 202, 204
Götte's larva 211, 212, 214, 219
Grobben, K. 61
growth, ectoproct colonies 195
gut 47
    annelids 132, 138–9, 140
    arthropods 164–5
    brachiopods 63, 347
    cephalochordates 422–3
    chaetognaths 255
    ctenophores 63, 312–13
    echinoderms 367, 371
    ectoprocts 193–5, 202
    enteropneusts 387
    entoprocts 186
    gastrotrichs 265
    kinorhynchs 299–300
    loriciferans 303–4
    molluscs 114
    nematodes 276–7, 284
    nemertines 225, 228
    onychophorans 152–3
    phoronids 337
    platyhelminths 63, 217, 221
    priapulans 295
    pterobranchs 359, 361
    rotifers 241
    tardigrades 176, 179
    trochaea theory 12, 13
    tube-shaped 63
    urochordates 408, 411, 413, 414

Haeckel, E. 1, 2, 9, 18, 45
haemal systems 66–8

annelids 67, 129–30
arthropods 67, 167
articulates 109
brachiopods 348–9
cephalochordates 424
closed and open 67
ectoprocts 67, 195
enteropneusts 389–90
entoprocts 186
molluscs 67, 115
onychophorans 153
phoronids 338–9
pterobranchs 360, 361
teloblasticans 100
urochordates 410
vertebrates 434
see also blood vessels; water-vascular system
haemocoel 150
    arthropods 166, 171
    onychophorans 153
    tardigrades 178
haemocytes 102, 103
Hatschek, B. 61, 62, 244
Hatschek's pit 422
head
    annelids 127
    arthropods, see cephalon
    chaetognaths 254
    tardigrades 176
heart 67
    arthropods 167
    articulates 109
    enteropneusts 388, 391
    molluscs 115
    neorenalians 321
    onychophorans 153
    panarthropods 150
    pterobranchs 359–60
    urochordates 410
hemidesmosomes 19, 20, 22
    cnidarians 50, 51, 55
    eumetazoans 50, 51
hermaphroditism 218
Higgins larva 305
homology 1, 443
hood 254
hooks
    acanthocephalans 249
    gastrotrichs 264–5
Hüllglocken larva, see pericalymma larva
hydrocoel 368, 370
hydropore 10, 370
hydrostatic skeleton 65
hyposphere 443

infundibular organ 398, 423, 434
intracellular skeletal lamina 239, 244, 248–9, 252
introvert 272, 276, 291
    kinorhynchs 273, 292, 299

loriciferans 303
nematomorphs 273, 288, 289
priapulans 294
retractor muscles 292, 295, 300, 305
iodine-binding cells 393, 399, 409
Iwata's larva 228, 229–30

jaws
annelids 128–9, 141
onychophorans 152

Kirk's ophiuroid 372
Kölliker's tufts 142, 143

lacunar system 249
lamina, intracellular skeletal 239, 244, 248–9, 252
larvae 24
annelids 81, 84, 96, 130, 132–4
arthropods 95, 163–4, 171, 172
aschelminthes 83, 96
brachiopods 350–1, 352, 353
cnidarians 56–7, 95
crustaceans 163, 171, 172
deuterostomes 95, 324–6, 328
echinoderms 330, 370–5
ectoprocts 83, 96, 197, 199–200, 201, 202–3
enteropneusts 390–2
entoprocts 47, 81, 82, 96, 187, 189
eumetazoan 45–7
loriciferans 304, 305
molluscs 81, 82, 83, 84, 95, 96, 117–18, 119
nematomorphs 288, 289
nemertines 83, 95, 211, 212, 214, 227–31
phoronids, see actinotrocha larva
planktotrophic 15–16
platyhelminths 83, 95, 211, 212, 214, 219–21
priapulans 296
protostomes 75, 76, 80–4
pterobranchs 362–3
sipunculans 81, 96, 104, 105
sponges 37–8, 39
urochordates, see tadpole larva, urochordates
see also specific types
lemnisci 244, 250, 252
ligament sacs 235–6, 249
limbs
arthropods 160–1, 162, 164, 165, 171
onychophorans 152, 156
panarthropods 150–1
tardigrades 176–7, 179
Linnaeus, C. 18
lobopodia 150, 152
lophophore
brachiopods 347, 348
ectoprocts 192, 193, 203–4, 205

phoronids 337, 338
lorica 303, 305
'Luther's larva' 219

macrocilia 311
madreporic chamber 367
Malpighian tubules 167, 178
mandibles, arthropods 160, 161, 164, 165, 171
mantle 110, 112, 113, 118
mantle fold 346
mastax 239, 240–1
mastax ganglion 240–1, 243, 245
maxillae, arthropods 161, 164, 165, 171
maxillary glands 167, 171
medusae 54, 56, 57
meiosis 20, 24
mesenchyme, primary 307
mesoblasts 141–2
mesocoel 321, 322
brachiopods 348, 351, 352
chordates 400
echinoderms 367, 368
enteropneusts 388, 391
phoronids 338, 340, 342
pterobranchs 360, 362, 363
mesoderm 47, 64–5, 443
ctenophores (mesogloea) 307, 310, 313, 316
formation 64–5, 439
annelids 129, 132, 139, 142
arthropods 159–60, 169–70, 171, 173
aschelminthes 64, 96–7, 234–5
ctenophores 307, 308, 315
deuterostomes 64, 65, 307, 323
echinoderms 308, 369
euarticulates 124–5
molluscs 116
platyhelminths 217–18
phoronids 339, 342
protostomes 65, 75, 86, 307
spiralians 64, 96–7
vertebrates 432–3
pregastral, urochordates 413, 417
mesodermal bands 99, 116, 124
mesogloea
cnidarians 54, 55, 316
ctenophores 307, 310, 313, 316
mesohyl 36
mesosome 321, 359, 360
mesoteloblasts 99, 100, 104, 125, 136, 171
mesothelium 443
metacoel 321, 323
brachiopods 348, 351, 352
chordates 400
echinoderms 367, 368, 370
enteropneusts 388, 391, 392
phoronids 338, 340, 342
pterobranchs 361, 362, 363
metamorphosis
annelids 133, 134

brachiopods 351
echinoderms 371, 373, 375
ectoprocts 183, 200, 201, 206
enteropneusts 391–2
entoprocts 183, 187–9
phoronids 341, 342
pterobranchs 363
urochordates 414
metanauplius larva 163–4, 171, 172
metanephridia 4, 68, 69, 100
annelids 130
arthropods 166–7
brachiopods 323, 349, 358
molluscs 115
onychophorans 150, 153–4, 156
panarthropods 150, 156–7
phoronids 323, 339, 358
sipunculans 103, 106
metasomal sac 340, 341
metasome 321
enteropneusts 388
pterobranchs 361
metatroch 14, 80–1, 82, 83, 443
ancestral forms 12
annelids 132
molluscs 117
rotifers 238, 239, 240
sipunculans 106
mitochondria 20, 23, 33
mitraria larva 134
mixocoel 67, 150
arthropods 166, 171
onychophorans 153, 156
tardigrades 178
molecular methods 4–5
monet cleavage 92, 94
monociliate cells 23, 69–71, 263, 440
annelids 70, 85, 128, 141
brachiopods 70, 326, 347
cephalochordates 70, 384, 422
cyrtotretes 70, 331, 384
deuterostomes 70, 324, 330–2
echinoderms 70, 326, 367
ectoprocts 70, 193, 194
enteropneusts 70, 384, 387
gastrotrichs 70, 85–6, 269–70
phoronids 70, 326, 343
protostomes 69, 70, 85–6, 270
pterobranchs 70, 326, 359
urochordates 70, 384, 408
mono-iodotyrosine 393, 409
monophyletic group 1, 5, 7, 443
moulting (ecdysis) 149, 152
arthropods 165
introverts 272–3
nematodes 278, 283
priapulans 296
tardigrades 177
mouth 10, 13, 63
cephalochordates 422, 426, 428

chordates 396, 399, 405
cnidarians 47, 54
cycloneuralians 263
deuterostomes 318–19, 438–9
introverts 272
molluscs 116
nematodes 276
nematomorphs 287
onychophorans 152
priapulans 294–5
tardigrades 176
urochordates (oral siphon) 413, 417–18
vertebrates 434
mouth-anus, onychophorans 154–5
mouth parts, arthropods 160–1
Müller's larva 219–21
multicellularity 19
multiciliate cells 69–71, 309, 330–2
annelids 85, 128
aschelminthes 234
ctenophores 311
cyrtotretes 384
ectoprocts 193
enteropneusts 384, 387, 390
gastrotrichs 85–6, 269–70
protostomes 69, 70, 85–6
rotifers 239
urochordates 384, 408
muscle cells
gnathostomulids 141
nematodes 276
muscles
arthropods 166
cephalochordates 427
chaetognaths 255–6
chordates 396, 398
ctenophores 312, 313
gastrotrichs 265
kinorhynchs 300
loriciferans 305
nematomorphs 288, 289
nemertines 226
onychophorans 153
platyhelminths 217–18
priapulans 295
rotifers 241
sipunculans 104
tardigrades 177, 178
urochordates 398, 410
myocoel 426
myocytes
cnidarians 55
sponges 37
myoepithelial cells 236, 262–3, 440
annelids 128
brachiopods 347
gastrotrichs 269
priapulans 296
urochordates 408
see also pharynx, myoepithelial

nauplius larva 163–4, 171, 172
nematoblasts (cnidocytes) 53–4, 57
nematocysts (cnidae) 27, 53–4, 57, 316
    acquired (cleptocnidia) 54, 311
neoteny
    larvaceans (appendicularians) 405, 416
    rotifers 244, 245
neotroch 10, 13, 14, 443
    deuterostomes 327, 328
    echinoderms 371
    enteropneusts 390, 392
nephridia 68–9
    arthropods 166–7, 171
    cephalochordates 424
    echinoderms 367
    ectoprocts 196
    enteropneusts 388
    vertebrates 433, 434
    see also excretory organs; metanephridia;
        protonephridia
nephridiopores 115
nephromixia 68
nerve cells 20
    giant 383, 389, 423
    sponges 36–7
nerve cords, ventral, see ventral nerve cords
nervous system
    acanthocephalans 249, 252
    annelids 78, 79, 129, 133, 142
    arthropods 78, 79, 164, 166
    aschelminthes 234
    bilaterians 69
    brachiopods 320, 348, 352
    cephalochordates 398, 423–4, 425
    chaetognaths 78, 79, 255, 258–9
    cnidarians 49, 55
    ctenophores 311–12
    deuterostome ancestor 14, 15
    deuterostomes 49, 80, 320–1, 326
    echinoderms 320, 366, 368
    ectoprocts 78–9, 195, 205
    enteropneusts 321, 384, 389, 392, 400
    entoprocts 78–9, 186, 189–90
    eumetazoans 49
    gastrotrichs 78, 79, 265
    kinorhynchs 78, 79, 261, 300
    loriciferans 78, 261, 305
    molluscs 78, 79, 111, 114, 116–17
    nematodes 78, 79, 261, 277–8, 281–2, 283
    nematomorphs 78, 287–8
    nemertines 79–80, 213, 214, 226
    onychophorans 78, 79, 153
    panarthropods 150
    phoronids 320, 338, 340, 342, 343
    platyhelminths 79–80, 213, 214, 217, 220
    priapulans 78, 79, 261, 295
    protostomes 49, 75, 76–80
    protostomian ancestor 12, 13, 14, 15
    pterobranchs 320, 360, 361
    rotifers 79, 241, 245

sipunculans 78, 79, 102, 106
    tardigrades 78, 79, 178, 179–80
    urochordates 398, 409–10, 412, 413–14
    vertebrates 398, 432, 433, 434
    see also brain; ganglia; neural tube; sensory
        organs; ventral nerve cords
neural crest 434
neural gland complex 400, 409, 412
neural plate 412, 425
neural tube
    cephalochordates (amphioxus) 423, 425, 434
    chordates 321, 396, 398, 404–5
    enteropneusts (collar cord) 321, 384, 389, 392,
        400
    urochordates 409–10, 412, 413–14, 417
    vertebrates 432, 434
neurenteric canal 398, 405
neuropore 400, 425
neurotransmitters 22
    bilaterians 22, 49, 69
    cnidarians 55
    ctenophores 69, 312
    eumetazoans 49
    phoronids 338, 340
    sponges 22, 37
neurotroch, see gastrotroch
notochord 396, 423, 428, 431–2
notoneuron 13, 330, 376–7, 378, 438
nuclei, constant numbers 275

occluding junctions 19
    see also septate junctions; tight junctions
ocelli 166, 412, 418
oikoplast epithelium 408, 409
ommatidia 161, 166
ontogeny 1
ooplasmic segregation 410
oozooids 409, 415, 416
ophiopluteus larva 327, 371, 373, 375
opisthosome 138, 139
oral shield 112
oral siphon 413, 417–18
orthogon theory 80
ovoviparous development 154

pallial epithelium 197, 200, 201
papillae (adhesive)
    amphioxus 405, 427
    urochordates 411, 413, 414
paraphyletic group 443
parapodia 128, 129, 135, 136, 150–1
parthenogenesis 238, 241
pedal glands 239, 240
pedal shield 112
pediveliger larva 117, 119
pelago-benthic ancestor 11–12, 13
pelagosphaera larva 104, 105, 106
pentameric symmetry 366, 376

pericalymma larva 83–4, 189, 443
   annelids 133–4, 135
   molluscs 117, 119
   sipunculans 104, 105
pericardium
   molluscs 116, 121
   neorenalians 356
   pterobranchs 360
   urochordates 413
perinotum 113
periostracum 113, 348
peristomium 127
peritoneum 65, 102, 129, 195–6, 255, 443
pharynx
   annelids 128, 141
   ctenophores 312
   cycloneuralians 262, 263
   cyrtotretes 382–4
   ectoprocts 193–4, 262, 274
   gastrotrichs 262, 264, 269, 273–4
   kinorhynchs 262, 299
   loriciferans 262, 273–4, 303
   myoepithelial 262, 269, 273–4, 292
   nematodes 262, 273–4, 276
   platyhelminths 217
   priapulans 294–5
   tardigrades 177, 262
   triradiate 262, 273–4
phorozooids 416
photoreceptors/photoreceptor cells 49
   cephalochordates 423
   entoproct and ectoproct larvae 183–4, 206
   rotifers 241
   see also eyes; ocelli
phylogenetic trees
   current 3–5
   history 1, 2
   proposed 5–7
phylogeny 1
phylum 5
pigment spots, ectoprocts 199–200
pilidium larva 211, 212, 214, 227–31
pinacocytes 36
pinacoderm 35–6
placental structures 76, 154, 202
planktotrophic ancestral forms 9–10, 13
planktotrophic larvae 15–16
planula larva 47, 56, 58, 221–2, 443
plesiomorphy 443
podocytes 103, 338–9, 390
polar bodies 24, 443
   acanthocephalans 250, 252
   arthropods 168, 169
   cnidarians 56
   ctenophores 313, 314
   nematodes 279
   phoronids 339
   spiralians 89
polyembryony 202
polyphyletic group 7, 443

polypides 192
polyps 54, 56, 57
pp60*v-src* kinase 20, 24
preancestrula (primary disc) 200, 201, 202
primary disc (preancestrula) 200, 201, 202
primitive character (state) 443
primitive group 443–4
primordial germ cells 235, 256, 279–81
proboscis
   acanthocephalans 248, 249, 251, 252
   enteropneusts 387–8
   nematomorphs 288
   nemertines 225–6, 228, 230, 231
proctodaeum 170, 281, 444
prosome 321, 359–60
   enteropneusts 387–8
   phoronids 337
prostomium 127
prothoracic glands 165
protocerebrum 161, 166, 178, 179
protocoel 321, 324, 353, 356
   brachiopods 321, 348, 351, 352
   chordates 321, 400
   echinoderms 367
   enteropneusts 387–8, 391
   neorenalians 321
   phoronids 321, 337, 340, 342
   pterobranchs 359–60
protonephridia 68, 69, 72
   acanthocephalans 249
   annelids 130, 132, 134
   entoprocts 186, 187
   gastrotrichs 265
   kinorhynchs 300
   larval 81, 83
   loriciferans 305
   molluscs 118
   nemertines 226
   phoronid larvae 340–1
   platyhelminths 220, 221
   priapulans 295–6
   rotifers 241
protornaea 10, 13
prototroch 12, 14, 80–1, 82, 444
   annelids 130, 131, 132
   ectoprocts 183, 197–8, 199, 203, 206
   entoprocts 183, 187, 188
   molluscs 116, 117
   rotifers 238, 239, 240
   sipunculans 103
   spiralians 91
pseudocoel 182, 235, 288
pseudocolloblasts 311
pyriform organ 183, 199, 200, 206

radial cleavage 319, 339, 349
radula 111, 113, 114
Reissner's fibre 398, 414, 423, 434
renopericardial canal 115

reproduction
  acanthocephalans 250
  annelids 130, 140
  arthropods 167
  asexual, *see* asexual reproduction
  brachiopods 349
  chaetognaths 256
  choanoflagellates 33
  cnidarians 55–6
  ctenophores 313–14
  ectoprocts 196–7, 202
  entoprocts 186
  nemertines 226–7
  phoronids 339
  placozoans 43
  platyhelminths 140, 218
  priapulans 296
  pterobranchs 362
  sponges 37, 38
resilin 165
respiratory organs
  arthropods 162, 165–6
  enteropneusts 390
  onychophorans 152
retrocerebral organ/complex 239, 252, 255
RFamide 20, 22, 49
rhabdites 216, 221
rhynchocoel 212–13, 231
rhynchodaeum 228
RNA sequencing 4–5
rostrum 244

sacculus 166–7
salivary glands 154, 156, 177–8
scalids 291, 292, 294, 299, 303–4, 305
schizocoely 86, 99, 100, 439, 444
  deuterostomes 323
  euarticulates 124
  nemertines 212–13
  phoronids 340
  sipunculans 104
Schmidt's larva 228, 230
scleroblasts 55
sclerocoel 426
sclerocytes 36
segmentation
  annelids 126–7, 134, 136, 140, 141, 142
  arthropods 159, 171, 173
  chordates 398
  euarticulates 124
  kinorhynchs 301
  molluscs 109, 118–21
  nemertines 213
  onychophorans 124, 152
  panarthropods 124, 149
  sipunculans 106
  tardigrades 176
  teloblasticans 99
  vertebrates 434

sensillae 150, 291
  arthropods 165, 166
  nematodes 278, 284
  tardigrades 178, 179
sensory cells 20, 49
  cnidarians 55
  sponges 36
sensory organs
  arthropods 165, 166
  chaetognaths 255
  entoprocts 186
  gastrotrichs 265
  nematodes 277–8
  onychophorans 153
  panarthropods 150
  rotifers 241
  tardigrades 178
  urochordate tadpoles 412
  vertebrates 434
  see also apical organs; eyes; sensillae
septate junctions 19–20, 21, 22
  cnidarians 54
  eumetazoans 45
  placozoans 21, 42, 43
  sponges 21, 36, 39
serosa 83–4, 104, 117
serosa larva, *see* pericalymma larva
serotonin 22, 338, 340
setae, *see* chaetae
shells
  brachiopods 346, 348
  ectoproct larvae 197, 199
  molluscs 113, 120
sister groups 444
skeleton
  brachiopods 347–8, 352
  echinoderms 366, 367
  gill 382, 383, 388, 423, 426
  hydrostatic 65
  proboscis 388
  sponges 36
  see also exoskeleton
slime glands 152, 156
solenocytes 340–1, 342
somites 426, 433
spermatophores 154, 300, 339
spermatozoa 20, 24, 25
  cilia 22, 23, 70
  cnidarians 55–6
  entry point 89, 115, 130
  onychophorans 154
  platyhelminths 218
  primitive type 24, 25
  sponges 37, 38, 39
spicules, calcareous 113, 118, 347–8
spinal cord 413–14, 423, 432
spines 254
spiracles 152
spiral cleavage 62, 71–2, 89–95, 222
  acanthocephalans 250, 252

annelids 90–1, 92, 93, 94, 130–1, 136–8, 141–2
arthropods 92, 93, 94, 125, 167–70, 173
ectoprocts 92, 93, 94, 197–9
entoprocts 92, 93, 94, 186–7, 189
molluscs 90–1, 92, 93, 94, 116
nemertines 92, 93, 95, 213, 227
platyhelminths 92, 93, 94–5, 213, 218–19
protostomes 75, 76
sipunculans 92, 93–4, 103–4
teloblasticans 99
spot desmosomes 19, 22
stalk
    brachiopods 346
    pterobranchs 361, 362
statocytes 412
sterroblastula 45, 48, 56
stigmata (urochordate gill slits) 382–4, 408–9,
    413
stolon 36   5, 416
stomo   50–1, 362, 363–4, 388, 404
stomo   281, 444
    an   31, 132
    arthropods 170
    nematodes 281
    nemertines 228, 229
stomo-proctodaeum 154–5
stone canal 367, 368
stretch-sensitive channels 22, 37
stylets
    nemertines 226
    platyhelminths 216
    tardigrades 177–8
sulphur bacteria, symbiotic 139
swarmers 43
symplesiomorphy 444
synapomorphy 444
synapses 20, 22, 49
    cnidarians 49, 55
synapticles 382, 388
syncytia
    acanthocephalans 248, 251
    nematodes 275
    platyhelminths 217
    rotifers 239, 241

tadpole larva, urochordates 407, 411, 412–15,
    417–18
    ascidian 402, 412–14, 418
tail, urochordate tadpoles 412, 413, 414, 415,
    417
tanycytes 292, 295, 300, 305
teeth
    chaetognaths 254
    gastrotrichs 264–5
    gizzard 142, 143, 194–5, 205
    molluscs 113
    nematodes 276
    priapulans 295

teloblasts 99, 124, 136–8, 170
telotroch 12, 80, 81, 444
    annelids 131, 132
    molluscs 117
    vs. actinotrocha perianal ring 324, 329
tentacle coelom 102–3
tentacles
    brachiopods 347, 348
    choanoflagellates 32
    ctenophores 310, 311, 315
    ectoprocts 192, 193, 194, 205
    entoprocts 185–6
    phoronids 337, 340, 343
    pterobranchs 360
tentillae 310, 313
test-cell larva, see pericalymma larva
thyroid 405, 432, 434
thyroxine 399
tight junctions 19, 21
toes, rotifers 240, 244
tongue bars 382, 388, 426
tornaea 10, 13, 14
tornaria larva 325, 402, 444
    ancestral deuterostome 14, 329–30, 331
    apical organs 47, 95–6
    ciliary bands 70, 324–6, 328
    enteropneusts 390–2
trabecular tissue 36
tracheae
    arthropods 162, 165–6
    onychophorans 152, 156
trimery 205
tritocerebrum 161, 178, 179
trochaea theory 9–16, 438–9
    ciliary bands 11–12, 85
    origins of deuterostomes 13–14, 323, 324,
        330–2, 334, 438–9
    Protornaeozoa and 307, 308
trochal discs 238, 240
trochoblasts 130
trochophora larva 14, 444
    annelids 132–3, 134, 135, 142
    apical organ 47
    entoprocts 188, 189
    molluscs 117
    nemertines 211, 214
    platyhelminths 211, 214
    protostomes 75, 76, 80–4
    and rotifers 240, 244–5
    sipunculans 104, 105
    vs. actinotrocha larva 83, 324, 329
trochozoon 244
trochus 238
trophamnion 24
trophi 240–1
trophosome 138, 139
tube feet 366, 371
tubes
    pogonophore 139–40

polychaete 129, 139–40
pterobranchs 359, 361
tunic 407, 408, 416

Ulrich, W. 61–2
undulipodia 22
urochord 396, 398, 403–4, 407
    adult larvaceans 408
    development 412, 413, 414, 415

valves, brachiopods 346, 348, 352
vane 25, 36
vegetal (blastoporal) pole 89, 442
    see also apical–blastoporal axis
veliger larva 117–18, 119, 444
velum 117, 239
ventral nerve cords
    annelids 129, 133, 142

nematodes 277, 282, 283
chophorans 153, 156
otostomes 76–80
sipunculan 106
viviparous development 154, 197, 202

water-vascular system 66, 67, 366, 368, 369
wheel organ
    cephalochordates 422
    rotifers (corona) 238, 239, 240, 244–5

Y-cells 265
yolk 76, 91, 154
Y-organs 165

zoanthina 54
zooids 185, 192, 359